Advances in Intelligent Systems and Computing

Volume 228

T0135268

Series Editor

J. Kacprzyk, Warsaw, Poland

For further volumes:
http://www.springer.com/series/11156

Advances in Intelligent Systems and Computing

Volume 218

Humberto Bustince · Javier Fernandez
Radko Mesiar · Tomasa Calvo
Editors

Aggregation Functions in Theory and in Practise

Proceedings of the 7th International Summer
School on Aggregation Operators at the Public
University of Navarra, Pamplona, Spain,
July 16–20, 2013

 Springer

Editors
Humberto Bustince
Departamento de Automática y
  Computación
Universidad Pública de Navarra
Pamplona
Spain

Javier Fernandez
Departamento de Automática y
  Computación
Universidad Pública de Navarra
Pamplona
Spain

Radko Mesiar
Department of Mathematics and Descriptive
  Geometry
Slovak University of Technology
Bratislava
Slovakia

and

Institute of Information Theory and
  Automation
Academy of Sciences of the Czech Republic
Prague
Czech Republic

Tomasa Calvo
Department of Computer Science
Universidad de Alcalá de Henares
Madrid
Spain

ISSN 2194-5357   ISSN 2194-5365   (electronic)
ISBN 978-3-642-39164-4   ISBN 978-3-642-39165-1   (eBook)
DOI 10.1007/978-3-642-39165-1
Springer Heidelberg New York Dordrecht London

Library of Congress Control Number: 2013941373

Printed on acid-free paper

Springer is part of Springer Science+Business Media (www.springer.com)

Preface

Aggregation functions are nowadays one of the most powerful theoretical tools in any field of research where merging or fusing information either from homogeneous or from heterogeneous sources is required. Their value has been established in many different fields, such as image processing, decision making, classification, robotics, control, and a very long etc.

This volume collects the extended abstracts of 45 contributions of participants to the Seventh International Summer School on Aggregation Operators (AGOP 2013), held at Pamplona in July, 16-20, 2013. These contributions cover a very broad range, from the purely theoretical ones to those with a more applied focus. Moreover, the summaries of the plenary talks and tutorials given at the same workshop are included. Together they provide a good overview of recent trends in research in aggregation functions which can be of interest to both researchers in Physics or Mathematics working on the theoretical basis of aggregation functions, and to engineers who require them for applications.

Humberto Bustince
Javier Fernandez
Radko Mesiar
Tomasa Calvo

The 7th International Summer School on Aggregation Operators has been partially supported by *Universidad Publica de Navarra: Vicerrectorado de Proyección Universitaria, Vicerrectorado de Investigación, Escuela Técnica Superior de Ingenieros Industriales y de Telecomunicación* and *Departamento de Auomtática y Computación*; European Society for Fuzzy Logic and Technology (EUSFLAT); *Ayuntamiento de Pamplona* and Apezteguia Architects.

Organization

General chairs
H. Bustince
J. Fernandez
R. Mesiar
T. Calvo

Organizing Chairs
E. Barrenechea
A. Jurio
D. Paternain

Local committee
A. Burusco
M. Galar
C. Lopez-Molina
M. Pagola
J. Sanz

Scientific Programme Committee

M. Baczyński (Poland)
B. Bedregal (Brazil)
G. Beliakov (Australia)
F. Chiclana (United Kingdom)
B. De Baets (Belgium)
G. Deschrijver (Belgium)
J. Dombi (Humgary)
D. Dubois (France)
J. Dujmović (USA)
J. Fodor (Hungary)
J.L. Garcia-Lapresta (Spain)
M. Grabisch (France)
S. Greco (Italy)
E. Herrera-Viedma (Spain)
E. Indurain (Spain)
B. Jayaram (India)
E.P. Klement (Austria)
A. Kolesárová (Slovakia)

J.L. Marichal (Luxembourg)
G. Mayor (Spain)
J. Montero (Spain)
S. Montes (Spain)
E. Pap (Serbia)
I. Perfilieva (Czech Republic)
H. Prade (France)
A. Pradera (Spain)
J. Quesada (Spain)
S. Saminger-Platz (Austria)
P. Sarkoci (Slovakia)
C. Sempi (Italy)
V. Torra (Spain)
J. Torrens (Spain)
E. Trillas (Spain)
L. Troiano (Italy)
R.R. Yager (USA)

Invited Speakers

Bernard De Baets
Ghent University (Belgium)

Michał Baczyński
University of Silesia (Poland)

Jozo Dujmović
San Francisco State University (USA)

Fabrizio Durante
Free University of Bozen/Bolzano (Italy)

Salvatore Greco
University of Catania (Italy)

Anna Kolesárová
Slovak University of Technology (Slovakia)

Gabriella Pigozzi
University Paris 9 Dauphine (France)

Ana Pradera
University Rey Juan Carlos (Spain)

Contents

Part I
Invited Talks

Functional Equations Involving Fuzzy Implications and Their Applications in Approximate Reasoning

Michał Baczyński

Abstract. Research on fuzzy implications, where the truth values belong to the unit interval $[0, 1]$, are carried out from the beginning of fuzzy set theory and fuzzy logic. In recent years, investigations has been deepened, which resulted in publishing some surveys [6] and two research monographs [1, 3] entirely devoted to this class of fuzzy connectives.

In our talk we concentrate on different functional equations (or inequalities) involving fuzzy implications and their role in approximate reasoning. Firstly, based on [3] and [5], we discuss the role of distributivity equations of fuzzy implications over other fuzzy connectives in equivalent transformation of the compositional rule of inference (CRI) or similarity based reasoning (SBR) to mitigate the computational cost. Secondary, based on [4] and [7], we discuss the role of the law of importation in equivalency of the Hirarchical CRI with the classical CRI proposed by Zadeh. Next, based on [8], we show the importance of T-conditionality inequalities in generalized modus ponens.

For each of the above functional equation with fuzzy implications, we describe the current state of theoretical research and we show what are the open problems. Finally, we also mention other functional equations (or inequalities) involving fuzzy implications and considered in the scientific literature.

References

[1] Baczyński, M., Beliakov, G., Bustince, H., Pradera, A.: Adv. in Fuzzy Implication Functions. STUDFUZZ, vol. 300. Springer, Heidelberg (2013)

[2] Baczyński, M., Jayaram, B.: Fuzzy Implications. STUDFUZZ, vol. 231. Springer, Heidelberg (2008)

Michał Baczyński
Institute of Mathematics, University of Silesia, 40-007 Katowice, ul. Bankowa 14, Poland
e-mail: michal.baczynski@us.edu.pl

H. Bustince et al. (eds.), *Aggregation Functions in Theory and in Practise*,
Advances in Intelligent Systems and Computing 228,
DOI: 10.1007/978-3-642-39165-1_1, © Springer-Verlag Berlin Heidelberg 2013

[3] Combs, W.E., Andrews, J.E.: Combinatorial rule explosion eliminated by a fuzzy rule configuration. IEEE Trans. Fuzzy Syst. 6(1), 1–11 (1998)

[4] Jayaram, B.: On the law of importation $(x \wedge y) \to z \equiv (x \to (y \to z))$ in fuzzy logic. IEEE Trans. Fuzzy Syst. 16(1), 130–144 (2008)

[5] Jayaram, B.: Rule reduction for efficient inferencing in similarity based reasoning. Internat. J. Approx. Reason. 48(1), 156–173 (2008)

[6] Mas, M., Monserrat, M., Torrens, J., Trillas, E.: A survey on fuzzy implication functions. IEEE Trans. Fuzzy Syst. 15(6), 1107–1121 (2007)

[7] Štěpnička, M., Jayaram, B.: On the suitability of the Bandler-Kohout subproduct as an inference mechanism. IEEE Trans. Fuzzy Syst. 18(2), 285–298 (2010)

[8] Trillas, E., Alsina, C., Pradera, A.: On MPT-implication functions for fuzzy logic. Rev. R. Acad. Cien. Serie A Mat. 98(1), 259–271 (2004)

Aggregation Operators and Observable Properties of Human Reasoning

Jozo Dujmović

Abstract. In this paper we investigate properties of aggregation operators that are necessary to create mathematical models that are consistent with observable properties of human reasoning. Such models are required in many applications where mathematical modeling is used to suggest a justifiable decision and a subsequent course of action. No decision can be acceptable unless it is fully compatible with intuitive reasoning, common sense, and expert knowledge. In this paper we focus on necessary properties of aggregators and aggregation structures that are used in the weighted compensative logic and applied in evaluation decision models, and investigate interactions between formal logic properties and semantic properties of aggregation models.

Keywords: Partial conjunction, partial disjunction, partial absorption, threshold andness, threshold orness, sensitivity analysis, missingness-tolerant aggregation.

1 Introduction

The theory of aggregation operators and the practice of aggregation operators are two different things. Theoretical developments are not restricted by conditions of applicability, and/or consistency with observable properties of human decision logic. Consequently, the theory of aggregation operators is a rather wide and expanding area. On the other hand, aggregation operators that are used in practical applications must satisfy restrictive conditions of the selected application area. In particular, a mandatory prerequisite for all applications is to show that aggregation operators are models of observable properties of human reasoning. It is not acceptable to build applied decision models using aggregators that behave in a way that is inconsistent

Jozo Dujmović
San Francisco State University, Computer Science Department, 1600 Holloway Ave., San Francisco, CA 94132, USA
e-mail: jozo@sfsu.edu

H. Bustince et al. (eds.), *Aggregation Functions in Theory and in Practise*,
Advances in Intelligent Systems and Computing 228,
DOI: 10.1007/978-3-642-39165-1_2, © Springer-Verlag Berlin Heidelberg 2013

with intuitive reasoning of decision makers. In this paper we focus on necessary properties of aggregation operators and aggregation processes that are used in decision engineering, primarily in the design of criteria for evaluation, comparison, and selection of alternatives. In this area we assume that each alternative is described by a set of n suitability degrees that are used as inputs for a stepwise aggregation process that generates an overall suitability degree.

Empirical research on aggregators is an area that attracts negligible attention of the research community -everything else belongs primarily to theory. Zimmermann's book [24] seems to be the only book in this field that has a chapter entitled "Empirical Research on Aggregators"' and should be commended for explicitly asking the fundamental question *"How do human beings aggregate subjective categories, and which mathematical models describe this procedure adequately?"* One would expect this crucial question in the first chapter of the book. It is indicative, however, that this question appears only on p. 390 of [24], in the very last chapter of the book.

Human reasoning is an observable process that includes formal logic and semantic components. Empirical research on aggregators assumes design of experiments with human subjects and a quantitative evaluation of mathematical models that are expected to adequately describe the analyzed aggregation process. Such models must include both *formal logic aspects* and *semantic aspects* of reasoning. In the context of evaluation logic the formal logic aspects reflect modeling of compound logic relationships by combining basic models of simultaneity (partial conjunction), replaceability (partial disjunction) and negation. As an addition to formal logic aspects, most realistic models of reasoning must also include semantic components that reflect the meaning of variables and their relationships with the environment in which the decisions are made. Such components depend on the goals of decision maker and an assessment of possible consequences of the decision. Thus, the semantic components include the overall importance of decision, the relative importance of individual attributes, classification of inputs as mandatory, sufficient, desired, and optional, etc. Even without sophisticated experiments, all significant properties of aggregation operators can be investigated, discussed, and validated from the standpoint of their compatibility with observed properties of human reasoning. This is the primary goal of this paper.

A general mathematical approach to aggregation functions and their properties can be found in [15], where aggregation functions are classified using a very fine mathematical granularity. Since applicability has little significance in mathematical investigations of various classes of aggregation functions, it is not surprising that the distribution of the aggregation function applicability is very nonuniform. In this paper we will focus on compensative aggregators that are most frequent in applications that use a continuous logic or a fuzzy logic for modeling decisions [13, 23]. The corresponding aggregators can be defined as mappings $A : I^n \to I$, $I = [0,1]$, $n > 1$, that are nondecreasing in each variable and satisfy the boundary conditions $A(0,\ldots,0) = 0, A(1,\ldots,1) = 1$. Of course, such mappings can have a spectrum of other mathematical properties that are carefully studied in [15, 1, 13, 3, 20].

A rather coarse and application oriented classification of fundamental logic aggregators that are models of simultaneity (conjunction) and replaceability (disjunction) can be based on three characteristic intensities of simultaneity and replaceability: *weak, medium, and strong* [7, 8]. This classification is suitable in the case of n-tuple $\mathbf{x} = (x_1, \ldots, x_n)$ whose all components belong to interval $D = [x_{\min}, x_{\max}] \subseteq I$, $0 \leq x_{\min} = x_1 \vee \cdots \vee x_n$, $x_1 \wedge \ldots, \wedge x_n = x_{\max} \leq 1$. *Medium aggregators* are compensative aggregators based on mappings $A : D^n \to D$. These are averaging aggregators used to model simultaneity and replaceability and implemented as various means, characterized by the global andness $\alpha \in I$ and the global orness $\omega = 1 - \alpha \in I$. In addition, we have $x_{\min} \leq A(x_1, \ldots, x_n) \leq x_{\max}$ and idempotency $A(x, \ldots, x) = x$. *Strong aggregators* are characterized by mapping $A : D^n \to [0, x_{\min}] \cup [x_{\max}, 1]$; these aggregators show very strong conjunctive and disjunctive properties and are implemented using t-norms and t-conorms [19]. *Weak aggregators* realize the mapping $A : D^n \to C$, $C \subseteq D$ where C is the reduced D interval limited by weighted conjunction and weighted disjunction [8]. Weak simultaneity/replaceability models are compensative functions based on means that use implicative weights [17, 18, 7].

While weak and strong aggregators have modest applicability [4, 18, 24], the vast majority of applications in the area of evaluation are based on medium aggregators, primarily because the internality, idempotency and compensativeness are very visible in human reasoning and desirable in mathematical models of evaluation decisions. In addition, these aggregators are means, and the theory of means is well developed and widely used in science. In particular, the literature on means [12, 21, 1] always includes a detailed presentation of properties of various classes of means, and that is useful when selecting aggregation functions.

The paper is organized as follows. In the next section we shortly discuss traditional mathematical properties of means and aggregation functions. The remaining sections are devoted to characteristic properties of human reasoning that affect the selection and use of aggregation structures: sensitivity analysis of aggregation operators, the problem of adjustability of threshold andness/orness, semantic properties (the concept of overall importance and its decomposition to derive andness, orness and relative importance), penalty controlled missingness tolerant aggregation (aggregation of incomplete inputs), and canonical aggregation structures.

2 Basic Mathematical Properties of Aggregators and Means

Basic mathematical properties of means and aggregators can be assessed from the standpoint of their relevance in evaluation decision models and their adequacy to describe human reasoning. Following are the most important properties:

- **Monotonicity**

$$\mathbf{x} = (x_1, \ldots, x_i, \ldots, x_n), \ \mathbf{x}' = (x_1, \ldots, x_i', \ldots, x_n), \ x_i < x_i' \ i \in \{1, \ldots, n\}$$

$$\Rightarrow \begin{cases} A(\mathbf{x}) \leq A(\mathbf{x}') & \text{(nondecreasing)} \\ A(\mathbf{x}) < A(\mathbf{x}') & \text{(strict)} \end{cases}$$

The monotonicity is a property that looks necessary, simple and self- evident: the overall suitability of a complex (multi-component) system must not decrease (or must strictly increase) when the suitability of any of its components increases. Monotonicity also affects the sensitivity to improvements. However, there are delicate questions related to the character of monotonicity because some forms of monotonicity do not occur in human reasoning and consequently cannot be acceptable. For example:

o For some aggregators the monotonicity must be strict (e.g. the soft partial conjunction/disjunction [7]), and for others it must be nondecreasing (e.g. the hard partial conjunction/disjunction).
o There are important aggregators where the nature of monotonicity is not the same for all inputs: e.g., if x_m is a mandatory input of a partial absorption [7] and all other inputs are optional, then $x_m = 0$ implies $\forall i \neq m \ \partial A/\partial x_i = 0$. However, if $x_m > 0$ then $\partial A/\partial x_i > 0$, $i = 1, \ldots, n$.
o The compatibility with human reasoning imposes limitations on the properties of derivatives $\partial A/\partial x_i \ i = 1, \ldots, n$. In human reasoning it is not observable that insignificant increments of input suitability can cause significant increments of output suitability. Consequently, the values of first derivatives must be limited: $|\partial A/\partial x_i| < L \ i = 1, \ldots, n$. Monotonicity beyond this limit is regularly not acceptable.

Therefore, the monotonicity is not a simple issue, because it quickly expands into the sensitivity analysis that investigates the acceptability of variations of $A(x_1, \ldots, x_n)$ caused by variations of x_i, $i = 1, \ldots, n$. In the next section we will investigate this issue in more detail.

- **Internality:** $x_{\min} \leq A(x_1, \ldots, x_n) \leq x_{\max}$. If the suitability is understood as a degree of satisfaction of justifiable requirements, then the overall suitability of a system in almost all cases cannot be greater than the maximum component suitability or less than the minimum component suitability. For example, it would be counterintuitive that the GPA of a student is greater than the highest course grade, or less than the lowest course grade of the student.
- **Idempotency (reflexivity).** $\forall x \in I, \ A(x, \ldots, x) = x$. This property is easily derived from internality, both formally and intuitively. Indeed, if we insert $x_1 = \cdots = x_n = x$ in the internality relation, then idempotency directly follows from $x = x_{\min} \leq A(x_1, \ldots, x_n) \leq x_{\max} = x$. If the degree of satisfaction of all component requirements is the same, then the degree of overall satisfaction with the system as a whole should not be different from that value (a student whose all course grades are x should also have a GPA equal to x). In some rare cases, however, both internality and idempotency can be questionable properties. For example, if suitability x_i is interpreted as a probability that a candidate can satisfy a component requirement, then for independent component requirements the overall suitability might be a t-norm Πx_i. Of course, proving independence is almost impossible (good students are good in most courses, and bad students are bad in most courses, making course grades, or component suitability, significantly correlated). However, this example shows a semantic dimension of internality and

idempotency: they depend on the nature or interpretation of the values $x_1, \ldots, x_n$. Aggregation operators should not be defined without understanding the identity of values that are being aggregated. The values that we call suitability or preference can be interpreted as degrees of truth of statements, degrees of membership in a fuzzy set, probabilities, percents of satisfied requirements, scores, etc. Before defining an aggregator it is necessary to precisely specify what is going to be aggregated, regardless the fact that $x_1, \ldots, x_n$ are just real numbers from $[0, 1]$ that have no physical unit. In evaluation logic, we assume the aggregation of degrees of truth. mean

- **Commutativity (Symmetry).** $\forall \mathbf{x} \in I^n, A(\mathbf{x}) = A(\mathbf{x}_{perm})$, where $\mathbf{x}_{perm}$ denotes any of $n!$ permutations of the n-tuple $\mathbf{x}$. Commutativity is a property primarily used in the theory of means. In the area of aggregation that property implies that all aggregated inputs have equal status (equal importance) and can be aggregated in any order of the inputs. Of course, the commutativity is generally not acceptable in any practical aggregation problem because it contradicts the basic semantic component visible in human reasoning: the perception of importance. Indeed, in all aggregations of degrees of truth we assume that each degree of truth corresponds to a specific value statement, and each value statement may have a different importance for a given decision maker. Adjustable importance is necessary in all multiple criteria decision problems [2].

- **Associativity and distributivity.** These properties enable aggregation of suitability in any grouping order and they are observable in human reasoning. Strictly increasing averaging aggregators cannot be associative, but the errors of associativity and distributivity relations are small [7] and do not reduce the applicability of averaging aggregators.

3 The Problem of Hypersensitivity

Variations of input suitability and/or selected parameters cause the variations of output suitability of aggregators. In human reasoning the observed sensitivity of output suitability with respect to input suitability (or model parameters) is always a limited value. It is highly unlikely that decision makers can perceive significant variations of output suitability caused by minute (indiscernible) variations of inputs or parameters. Consequently, if z denotes the output of an aggregator and x is one of inputs then the sensitivity coefficient $\partial z / \partial x$ should have a limited value, and aggregators that violate such limitations are inconsistent with observable properties of human reasoning. In addition, many types of aggregators that appear in literature (particularly in the families of t-norms, t-conorms, and mixed functions [1]) have points and lines where z or $\partial z / \partial x$ have discontinuities. This property needs verification (i.e. a proof that it reflects human reasoning).

Let δ be the indiscernibility zone in the sense that if degrees of suitability a and b are sufficiently close ($|a - b| < \delta$), then decision-makers cannot feel the difference and perceive $a \approx b$. The same holds for selecting andness and orness. For example, if $a = 0.66$ and $b = 0.67$, then decision-makers cannot discern the difference and only feel $a \approx b \approx 2/3$. The value of δ depends on evaluator's training and according to experiments reported in [6] it can grow up to 0.05.

Aggregation operators should not have regions of hypersensitivity where indiscernible variations of inputs cause significant variations of output. The hypersensitivity is visible even in the simplest case of geometric mean $z = x^w y^{1-w}$, $w \in]0, 1[$, because the sensitivity coefficient $\partial z / \partial x = w(y/x)^{1-w}$ has excessive values in the vicinity of $x = 0$. If $y = 1$ and w has a small value, then insignificant changes of x can cause explosive changes of z. For example, if $w = 0.02$ and x changes from 0 to 0.02 (which is an increase within a human indiscernibility zone) than z changes from 0 to 0.925 (the truth amplification of 46 times). Even for larger weights, e.g. $w = 0.05$ and discernible change of x from 0 to 0.05, z changes from 0 to 0.86 (the truth amplification of 17 times). The dramatic increase of output caused by a negligible increase of input (excessive truth amplification) is certainly not an observable property in human reasoning. Of course, this property of geometric mean is not the reason to disqualify the geometric mean as an aggregator, but it shows that even the most useful aggregators may have regions where their behavior is unacceptable and such regions should be identified and avoided in applications. That is attainable in two ways: (1) by avoiding inputs that have values in the domain of excessive truth amplification (e.g. binary inputs 0/1 cannot cause such problems), and (2) by ensuring the consistency of logic conditions (because the hypersensitivity is primarily a consequence of inconsistent logic conditions).

4   Inconsistent Logic Conditions

Formal logic and semantic aspects of aggregators are not independent and must be consistent. There are combinations of properties of aggregation operators that can be considered extreme, infrequent in human reasoning, or even counter-intuitive. Some extreme combinations of andness/orness and weights, and some combinations of andness/orness and mandatory/sufficient requirements may be questionable and interpreted as inconsistent logic conditions. They include the following:

(a) High andness (in the region of a hard partial conjunction) or a high orness (in the region of a hard partial disjunction) and a very low weight of an input.
(b) Very high andness and nonmandatory inputs.
(c) Very high orness and nonsufficient inputs.
(d) Low andness (slightly above the neutrality level $\alpha = \omega = 1/2$) and mandatory requirements.
(e) Low orness (slightly above $\omega = 1/2$) and sufficient requirements.

To illustrate these problems let us investigate a combination of high simultaneity (high andness, in the region of hard partial conjunction, $\alpha \geq 2/3$) and low relative importance. If we request a high simultaneity, then all inputs are expected to be positive and uniformly high. This indirectly implies that every input is rather important, because its low presence is not acceptable. If an evaluation model combines a very low weight and a high andness, then such combination is most likely a contradiction. The low weight asserts that the input is not important and the high andness claims that the input is very important. Obviously, such contradiction is inconsistent with observable properties of human reasoning.

A similar situation occurs if the desired andness is very high but the requirement is not mandatory. It does not seem logical to first firmly insist that all inputs must be strictly simultaneously satisfied, and then to fail to sufficiently punish alternatives that do not satisfy such requirements. Equally inconsistent might be the requirement for a low level of andness combined with the ultimate level of punishment in the case of failing to satisfy even the least important input. These combinations of andness and mandatory requirements are theoretically possible, but they are regularly not a part of human reasoning.

5 Semantic Properties: Decomposition of Overall Importance

If aggregation functions are considered abstract mappings $A : I^n \to I$ then we deal with aggregation of real numbers that have no additional meaning. On the other hand, if these real numbers represent the area of home, the number of bedrooms, the number of bathrooms and other parameters of a home, then each homebuyer can identify the overall importance of such a group of inputs, and there is a specific meaning associated with the aggregation process. The meaning of individual and aggregated arguments and the associated perception of importance of aggregated values represent semantic properties of aggregators and links the design of aggregators with the area of perceptual computing [22]. A careful observation of intuitive reasoning shows that in the case of compensative averaging aggregators both semantic and formal logic components originate in the evaluator's perception of the overall importance of aggregated inputs. The evidence that supports that claim is clearly visible in extreme and irregular cases of aggregators where the perception of high overall importance simultaneously necessitates the use of high andness/orness and prevents the use of low weights. That makes weights and andness/orness look dependent, and the combinations of high andness/orness and very low weighs incompatible and unacceptable. If semantic and formal logic aspects of aggregation originate in the perception of overall importance, then the perception of overall importance can be decomposed, in order to derive an appropriate degree of andness and an appropriate set of weights. A decomposition method proposed in [10] is illustrated in Fig. 1. In the case of Generalized Conjunction/Disjunction (GCD) based on weighted power means: $y = W_1 x_1 \diamond \cdots \diamond W_n x_n = \lim_{p \to r}(W_1 x_1^p + \cdots + W_n x_n^p)^{\frac{1}{p}}$, $-\infty \leq r \leq \infty$ the specification of the overall importance can be based on a verbalized table with levels of importance going from the lowest (0), to the highest (L). For example, if we want a model of simultaneity, and the inputs have the overall importance levels $S_i \in \{0,\ldots,L\}$, $i = 1,\ldots,n$, then the simplest of methods proposed in [10] generates the global andness $\alpha = (S_1 + \cdots + S_n)/nL$ and weights $W_i = S_i/(S_1 + \cdots + S_n)$.

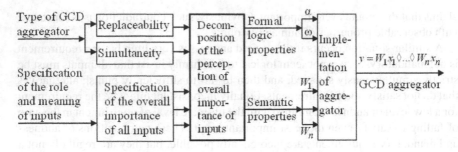

Fig. 1 Generating a GCD aggregator from the overall importance of inputs

6 Adjustability of Threshold Andness and Threshold Orness

The fundamental GCD aggregator has the following properties that depend on the threshold andness α_θ and the threshold orness ω_θ.

$$W_1x_1 \diamond \cdots \diamond W_n x_n \begin{cases} = 0, x_i = 0, x_j \geq 0, j \neq i, i \in \{1,\ldots,n\}, j \in \{1,\ldots,n\}, \alpha \geq \alpha_\theta \\ > 0, x_i > 0, x_j \geq 0, j \neq i, i \in \{1,\ldots,n\}, j \in \{1,\ldots,n\}, \alpha < \alpha_\theta \\ < 1, x_i < 1, x_j \leq 1, j \neq i, i \in \{1,\ldots,n\}, j \in \{1,\ldots,n\}, \omega < \omega_\theta \\ = 1, x_i = 1, x_j \leq 1, j \neq i, i \in \{1,\ldots,n\}, j \in \{1,\ldots,n\}, \omega < \omega_\theta \end{cases}$$

If $\alpha \geq \alpha_\theta$ then GCD is called the hard partial conjunction (HPC) [7], and if $1/2 < \alpha < \alpha_\theta$ then GCD is called the soft partial conjunction (SPC). Similarly, if $\omega \geq \omega_\theta$ then GCD is called the hard partial disjunction (HPD) and if $1/2 < \omega < \omega_\theta$ then GCD is called the soft partial disjunction (SPD). Generally, both the threshold andness α_θ and the threshold orness ω_θ can be adjustable.

Experiments that include both experts and non-experts show a distribution of desired threshold andness that has a mean value $\overline{\alpha_\theta} = 0.81$ and a standard deviation $\sigma = 0.1$. In this population, 80% of all participants propose the threshold andness/orness in the range $[0.71, 0.91]$. Obviously, aggregators based on weighted power means are not consistent with these experiments because their global threshold andness is insufficient: $\alpha_\theta = 2/3$ (for $n = 2$ and the geometric mean). A methodology for creating aggregators with adjustable threshold andness/orness can be found in [9]. For example, for $n = 2$ let us select $\alpha_\theta = \ln 16 - 2 \approx 0.77$ that corresponds to the harmonic mean. In the range $0.5 \leq \alpha \leq 1$ we can use the partial conjunction Δ_α, and in the range $0.5 \leq \omega \leq 1$ its symmetric De Morgan dual, the partial disjunction ∇_ω, as follows:

$$W_1x_1 \Delta_\alpha W_2x_2 = \begin{cases} \frac{\alpha_\theta - \alpha}{\alpha_\theta - 0.5}(W_1x_1 + W_2x_2) + \frac{\alpha - 0.5}{\alpha_\theta - 0.5}\left(\frac{W_1}{x_1} + \frac{W_2}{x_2}\right)^{-1} & 0.5 \leq \alpha \leq \alpha_\theta \\ \frac{1-\alpha}{1-\alpha_\theta}\left(\frac{W_1}{x_1} + \frac{W_2}{x_2}\right)^{-1} + \frac{\alpha - \alpha_\theta}{1 - \alpha_\theta}(x_1 \wedge x_2), & \alpha_\theta \leq \alpha \leq 1, \alpha_\theta = \ln 16 - 2 \end{cases}$$

$$W_1x_1 \nabla_\omega W_2x_2 = 1 - W_1(1-x_1)\nabla_\omega W_2(1-x_2), \ 0.5 \leq \omega \leq 1, \ \omega_\theta = \alpha_\theta$$

In this case the threshold orness and the threshold andness have the same value.

7 Missingness-Tolerant Aggregation

The availability of arguments is an implicit assumption for all aggregation operators. Unfortunately, in many real life decision problems we regularly face missing data. For example, Internet real estate web sites typically provide up to 40 input attributes characterizing homes for sale. Our experiments with Internet real estate data in San Francisco show the average input data availability of 70% (i.e. on the average 30% of attribute values are simply not available). Similarly, the majority of car buyers know car attributes that are undoubtedly relevant but their values are missing. Everybody would like to have a car that has good safety and reliability records, but majority of car buyers have difficulties to precisely define and get such data; they usually must select a car with incomplete or missing safety, reliability, and other data.

Human reasoning is normally performed with incomplete and imprecise data, and we cannot avoid aggregation only because of missing data. Thus, we have to be ready for missingness-tolerant aggregation. A method for solving this problem in the context of the LSP evaluation criteria was proposed in [11], based on the following concepts:

- Generally, any subset of input arguments can be missing. Consequently, a way to deal with this problem is to replace missing data with data that are expected to be the least damaging for the evaluation process.
- Whenever a missing input is replaced by an estimated value, we unavoidably make an error: we either reward an undeserving alternative, or we punish an innocent alternative. Thus, it is necessary to minimize such errors.
- The missing data replacement process also depends on the estimation of the reason for missingness. If the missingness is caused by deliberate attempt to hide inconvenient properties, such practice must be discouraged using an appropriate penalty. In cases where the missingness has other origins, the missing data should be replaced with neutral values that are selected so that the output values only depend on available inputs.

These concepts yield a *penalty controlled missingness tolerant aggregation*, where we aggregate known and unknown suitability degrees. Known suitability degrees belong to $[0, 1]$, and unknown suitability degrees belong to $[-1, 0]$, so that missingness tolerant aggregators are mappings $A : [-1, 1]^n \rightarrow [-1, 1]$. Let v_i be the value of input suitability degree x_i and let $P_i \in I$ be a penalty if v_i is unknown. If $P_i = 0$ there is no penalty for missing data, and $P_i = 1$ denotes the maximum missingness penalty. If v_i is known we use $x_i = v_i$ and if it is unknown then $x_i = P_i - 1$. So, $x_i = 0$ denotes either the zero suitability or the maximum penalty.

According to Kolmogoroff/Nagumo theorem [12] for all aggregators based on quasi-arithmetic means a subset of values can be replaced by their mean with no effect on the total mean: $A_n(x_1, \ldots, x_k, x_{k+1}, \ldots, x_n) = A_n(x, \ldots, x, x_{k+1}, \ldots, x_n)$, where $x = A_k(x_1, \ldots, x_k)$ (associativity property). A similar property [21] is the following: $A_n(x_1, \ldots, x_{n-1}, A_{n-1}(x_1, \ldots, x_{n-1})) = A_{n-1}(x_1, \ldots, x_{n-1})$. So, if x_n is missing/unknown then it can be replaced by the mean of remaining inputs and the result will be the same, $A_{n-1}(x_1, \ldots, x_{n-1})$. That is the result of aggregation in the case of zero penalty; if $P_n > 0$, then we return $A_n(x_1, \ldots, x_{n-1}, (1 - P_n)A_{n-1}(x_1, \ldots, x_{n-1}))$. The same approach can be extended for cases of multiple unknown variables. If all inputs of an aggregator are missing, then we return $-A_n(1 - P_1, \ldots, 1 - P_n)$ and that negative value propagates to the next layer of the aggregation structure. However, if at least one of n inputs is positive, the result of aggregation will be nonnegative.

8 Canonical Forms of Aggregation Structures

Some aggregation structures encountered in human reasoning occur with higher frequency than others. That is not surprising because it is natural that any successful form of reasoning becomes a repetitive pattern. A detailed analysis of canonical aggregation operators and structures can be found in [11]. The most important canonical structures are the following:

(a) Layered aggregation tree with andness that increases from SPC to HPC along the paths from leaves to the root.
(b) Layered aggregation tree with orness that increases from SPD to HPD along the paths from leaves to the root.
(c) Asymmetric logic aggregators: conjunctive partial absorption [7] (CPA, aggregating a mandatory input and an optional input) and disjunctive partial absorption (DPA, aggregating sufficient and optional inputs)
(d) Nested partial absorption (mandatory, desired, and optional inputs)
(e) Multi-input partial absorption (multiple mandatory inputs are aggregated using a HPC, and used as a mandatory input of CPA and multiple optional inputs are aggregated using a SPC and used as the optional input of CPA. A similar canonical form can be made using HPD, SPD, and DPA.

9 Conclusions

Aggregation operators in weighted compensative logic are predominantly used for creating decision models. Such models cannot be incompatible with common sense and intuitive expert reasoning. Quite contrary, in decision making practice the aggregation operators are used to derive decisions that are refinement and enhancement of sound and prudent intuitive human reasoning. Therefore, the compatibility with observable properties of human reasoning is a prerequisite for applicability of aggregation operators in the area of decision models.

Requirements for the applicability of aggregators generate various specific conditions that aggregators must satisfy. If we view aggregators strictly from the standpoint of applicability and compatibility with expert reasoning, then even the most elementary mathematical properties such as monotonicity, idempotency, and commutativity may become questionable and need proper justification.

Semantic aspects of aggregation operators are present in all applications. These are aspects related to interpretation of variables and their role in decision models.

Typical examples of the semantic aspects of aggregation are all issues related to the overall and relative importance of inputs in a given aggregation process. In addition, all aggregators should be verified from the standpoint of sensitivity analysis. In each intuitive aggregation process, human reasoning is based on a mix of formal logic and semantic components. This fact must affect the design of aggregation operators and the selection of their properties.

Properties of aggregation operators are also conditioned by the environment in which the aggregators are expected to operate. Consequently, it is useful to differentiate internal and environmental properties or aggregators. The missingness- tolerant aggregation is a typical example of the environmental conditions that affect aggregators. In addition, most aggregation operators do not function in isolation, but must operate as components in canonical aggregation structures.

The main message of this paper is that the properties of aggregation operators must be established taking simultaneously into account formal logic, semantic, and environmental aspects of the aggregation process. More empirical research on aggregators is necessary if we want to move further in that direction.

References

[1] Beliakov, G., Pradera, A., Calvo, T.: Aggregation Functions: A Guide for Practitioners. Springer (2007)

[2] Belton, V., Stewart, T.J.: Multiple Criteria Decision Analysis: an Integrated Approach. Kluwer Academic Publishers (2002)

[3] Bullen, P.S.: Handbook of means and their inequalities. Kluwer (2003)

[4] Carlsson, C., Fullér, R.: Fuzzy Reasoning in Decision Making and Optimization. Physica-Verlag (2002)

[5] Detyniecki, M.: Mathematical Aggregation Operators and their Application to Video Querying. Doctoral Thesis, Universite Pierre & Marie Curie, Paris (2000)

[6] Dujmović, J.J., Fang, W.Y.: An Empirical Analysis of Assessment Errors for Weights and Andness in LSP Criteria. In: Torra, V., Narukawa, Y. (eds.) MDAI 2004. LNCS (LNAI), vol. 3131, pp. 139–150. Springer, Heidelberg (2004)

[7] Dujmović, J.J.: Continuous Preference Logic for System Evaluation. IEEE Transactions on Fuzzy Systems 15(6), 1082–1099 (2007)

[8] Dujmović, J.J., Larsen, H.L.: Generalized conjunction/disjunction. International J. Approx. Reas. 46, 423–446 (2007)

[9] Dujmović, J.J.: Characteristic Forms of Generalized Conjunction/Disjunction. In: Proceedings of the IEEE World Congress on Computational Intelligence, pp. 1075b–1080b. IEEE Press, Hong Kong (2008) ISBN 978-1-4244-1819-0

[10] Dujmović, J.J.: Andness and Orness as a Mean of Overall Importance. In: Proceedings of the IEEE World Congress on Computational Intelligence, Brisbane, Australia, June 10-15, pp. 83–88 (2012)

[11] Dujmović, J.J.: The Problem of Missing Data in LSP Aggregation. In: Greco, S., Bouchon-Meunier, B., Coletti, G., Fedrizzi, M., Matarazzo, B., Yager, R.R. (eds.) IPMU 2012, Part III. CCIS, vol. 299, pp. 336–346. Springer, Heidelberg (2012)

[12] Dujmović, J.J., De Tré, G.: Multicriteria Methods and Logic Aggregation in Suitability Maps. International Journal of Intelligent Systems 26(10), 971–1001 (2011)

[13] Fodor, J., Roubens, M.: Fuzzy Preference Modeling and Multicriteria Decision Support. Kluwer Academic Publishers (1994)

[14] Gini, C. in collaboration with Barbensi, G., Galvani, L., Gatti, S., Pizzetti, E.: Le Medie. Unione Tipografico-Editrice Torinese, Torino (1958)

[15] Grabisch, M., Marichal, J.-L., Mesiar, R., Pap, E.: Aggegation Functions. Cambridge Univ. Press (2009)

[16] Klement, E.P., Mesiar, R., Pap, E.: Triangular Norms. Kluwer Academic Publishers, Dordrecht (2000)

[17] Larsen, H.L.: Efficient Andness-directed Importance Weighted Averaging Operators. International Journal of Uncertainty, Fuzziness and Knowledge-Based Systems 12(suppl.), 67–82 (2003)

[18] Larsen, H.L.: Importance weighting and andness control in De Morgan dual power means and OWA operators. Fuzzy Sets and Systems 196(1), 17–32 (2012)

[19] Marichal, J.-L.: Aggregation Operators for Multicriteria Decision Aid. Doctoral Thesis, Universite de Liege (1999)

[20] Mitrinović, D.S., Bullen, P.S., Vasić, P.M.: Means and their Inequalities. Publications of the University of Belgrade EE Department, Series: Mathematics and Physics, No. 600 (1977) (in Serbo-Croatian)

[21] Mendel, J.M., Wu, D.: Perceptual Computing. IEEE and J. Wiley (2010)

[22] Torra, V., Narukawa, Y.: Modeling Decisions. Springer (2007)

[23] Zimmermann, H.-J.: Fuzzy Set Theory and its Applications. Kluwer Academic Publishers (1996)

Copulas, Tail Dependence and Applications to the Analysis of Financial Time Series

Fabrizio Durante

Abstract. Tail dependence is an important property of a joint distribution function that has a huge impact on the determination of risky quantities associated to a stochastic model (Value-at-Risk, for instance). Here we aim at presenting some investigations about tail dependence including the following aspects: the determination of suitable stochastic models to be used in extreme scenarios; the notion of threshold copula, that helps in describing the tail of a joint distribution. Possible applications of the introduced concepts to the analysis of financial time series are presented with particular emphasis on cluster methods and determination of possible contagion effects among markets.

1 Introduction

Copulas are mathematical objects that fully capture the dependence structure among random variables and hence, offer a great flexibility in building multivariate stochastic models. As such, since their introduction in the early 50s, they have gained a lot of popularity in several fields like finance, insurance, reliability theory and environmental sciences (see, e.g., [3, 10, 11, 35] and the references therein).

Generally, the problem of constructing and/or selecting a suitable joint distribution function (and, hence, a suitable copula) describing a multivariate random vector (shortly, r.v.) $\mathbf{X}$ is a preliminary step in the determination of derived quantities associated to the model, like Value-at-Risk of a financial portfolio [31] or joint return periods [18]. In such cases, special care should be devoted to the description of the dependence in the tails of the distribution function (shortly, d.f.) $F_{\mathbf{X}}$ of $\mathbf{X}$. Especially for financial risk management, in fact:

> Extreme, synchronized crises and falls in financial markets occur infrequently but they do occur. The problem with the models is that they did not assign a high enough chance

Fabrizio Durante
School of Economics and Management, Free University of Bozen–Bolzano, Italy
e-mail: `fabrizio.durante@unibz.it`

H. Bustince et al. (eds.), *Aggregation Functions in Theory and in Practise*,
Advances in Intelligent Systems and Computing 228,
DOI: 10.1007/978-3-642-39165-1_3, © Springer-Verlag Berlin Heidelberg 2013

of occurrence to the scenario in which many things go wrong at the same time – the perfect storm scenario [31].

For such problems, it is necessary to consider models with increasing dependencies in the tails, since a dramatic underestimation of the risk can be obtained when one tries to use copulas that do not exhibit any peculiar behaviour in the tails. This was exactly one of the main pitfalls of the criticized Li's model (see [29]) for credit risk (see, for example, [26, 36]). In fact, as clarified, for instance, in [4], using Gaussian copulas (as Li did) "will always underestimate joint extremal events".

For a better specification of the tail of a distribution, Joe [25] introduced the so-called *tail dependence coefficients*, in order to quantify the amount of dependence on the tails of a joint bivariate distribution, and, hence, distinguish, among several possibilities, the joint d.f.'s that are able to describe such extreme situations. The formal definitions of these coefficients are given here.

Definition 1. Let X and Y be continuous r.v.'s with d.f.'s F_X and F_Y, respectively. The *upper tail dependence coefficient* λ_U of (X,Y) is defined by

$$\lambda_U = \lim_{t \to 1^-} P\left(Y > F_Y^{[-1]}(t) \mid X > F_X^{[-1]}(t)\right);$$ (1)

and the *lower tail dependence coefficient* λ_L of (X,Y) is defined by

$$\lambda_L = \lim_{t \to 0^+} P\left(Y \le F_Y^{[-1]}(t) \mid X \le F_X^{[-1]}(t)\right);$$ (2)

provided that the above limits exist. Note that $F_X^{[-1]}(t) = \inf\{s \in [0,1] \mid F_X(s) \ge t\}$ and $F_Y^{[-1]}(t) = \inf\{s \in [0,1] \mid F_Y(s) \ge t\}$ are the *quantile inverses* associated with F_X and F_Y, respectively.

Thus, λ_L and λ_U are defined in terms of limiting conditional probabilities of quantile exceedances. Now, it is important to note that the tail dependence coefficients are rank-invariant, and hence they can be calculated just from the copula C of (X,Y), by means of the following formulas:

$$\lambda_L = \lim_{u \to 0^+} \frac{C(u,u)}{u} \quad \text{and} \quad \lambda_U = \lim_{u \to 1^-} \frac{1 - 2u + C(u,u)}{1 - u}.$$ (3)

Here we concentrate our attention to some selected investigations about tail dependence (as described by means of copulas) and to possible applications in the description of extreme behaviour of financial time series, as we are going to illustrate.

2 Copulas with Given Diagonal Sections

As it can be seen from (3), the tail dependence coefficients are actually connected with the *diagonal section* of the bivariate copula C, defined as the function

$\delta_C \colon [0,1] \to [0,1]$, $\delta_C(t) = C(t,t)$. Thus, tail dependence lies in the behaviour of the diagonal of the copula.

Therefore, over the years, a number of investigations has been focused on the construction of copulas with different diagonals (and, hence, eventually admitting non-zero tail dependence coefficients). The related is quite vast; just to make few references, one may consider, [6, 7, 10, 18, 22, 28, 32] and the references therein.

Here we restrict to present a family of multivariate copulas, whose bivariate marginals are the so-called *semilinear copulas*, that are linear on specific segments of the unit square. Semilinear copulas have been considered in [9, 16], even if the general idea originated in a seminal paper by A. Marshall [30]. For an extension of this idea, see [27].

Specifically, for any natural number $d \geq 2$, we consider d–dimensional copulas C_d that can be written in the following form:

$$C_d(u_1, \ldots, u_d) = u_{[1]} \prod_{i=2}^{d} f(u_{[d]}), \tag{4}$$

where f is a function from $[0,1]$ to $[0,1]$, and $u_{[1]}, \ldots, u_{[d]}$ denote the components of $(u_1, \ldots, u_d)$ rearranged in increasing order.

The family (4) is a starting point of related ideas concerning the construction of copulas. For instance, in [19] it has been used in order to derive methods for generating multivariate extreme value laws that have a suitable number of parameters, and that can be efficiently simulated and easily fitted to empirical data. Finally, the probabilistic interpretation that can be derived for copulas of type (4) has originated other similar methods for obtaining copulas that can be interpreted in terms of suitable shock models [14].

3   Threshold Copulas and Financial Contagion

Apart from its advantages, the tail dependence coefficient is a single number, so it cannot contain the whole information concerning the tails of a bivariate copula. A more informative way for describing the dependence in the tails is represented by the concept of *threshold copula*, which has been recently considered in risk management and reliability theory (see, for instance, [3, 11]). Intuitively, a threshold copula is the copula that describes the dependence structure of two r.v.'s X and Y given that they belong to a suitable Borel set.

The use of threshold copulas has been recently applied to the determination of financial contagion among markets (see, e.g., [15]). In this case, contagion between two markets arises when significant increases in comovements across markets appear, conditional on a crisis occurring in one market or group of markets (see [2, 15, 21]), namely when the positive association among the markets increases in crisis period with respect to tranquil period.

Here, we introduce and discuss a non-parametric test and a related index to detect and measure the contagion effects. As an empirical application, the proposed test is exploited in order to detect contagion in the Euro area (for a complete overview about these methods, see [12, 13]).

4 Cluster Methods Based on Tail Dependence

In portfolio risk analysis a current practice for minimizing the whole risk consists of adopting some diversification techniques that are based, loosely speaking, on the selection of different assets from markets and/or regions that one believes to be weakly (or negatively) correlated. Such an approach tries to reduce the impact of joint losses that might occur simultaneously in different markets. To this end, cluster techniques for multivariate time series have been proposed in the literature in order to give a guideline to practitioners for the selection of a suitable portfolio. Such techniques span from the use of correlation coefficient (see, for instance, [1]) to the use of techniques based on the comparisons among the coefficients of the underlying processes (see, for instance, [20, 33]).

However, it has been stressed several times that diversification principle may fail when there is some tail dependence (or contagion) among the markets under consideration. To this end, it could be useful to introduce some clustering methods that focus their attention to the behaviour of financial markets in presence of extreme dependence scenarios. An innovative work in this direction has been recently done in [8], where it is proposed a clustering procedure that aims at grouping time series with an association between extremely low values, measured by a tail dependence coefficient.

Starting with these, we present some new clustering procedures for extreme scenario. Such a methodology is grounded on the conditional (Spearman's) correlation coefficient between time series. It aims at creating cluster of time series that are homogeneous, in the sense that they tend to be comonotone in their extreme low values (where the degree of extremeness is specified by a given threshold). The results have been discussed in details in [17] and are expected to be useful for portfolio management in crisis periods.

Acknowledgements. The author acknowledges the support of Free University of Bozen-Bolzano, School of Economics and Management, via the projects "Stochastic Models for Lifetimes" and "Risk and Dependence".

References

[1] Bonanno, G., Caldarelli, G., Lillo, F., Micciché, S., Vandewalle, N., Mantegna, R.: Networks of equities in financial markets. Eur. Phys. J. B 38(2), 363–371 (2004)

[2] Bradley, B., Taqqu, M.: Framework for analyzing spatial contagion between financial markets. Financ. Lett. 2(6), 8–16 (2004)

[3] Charpentier, A., Juri, A.: Limiting dependence structures for tail events, with applications to credit derivatives. J. Appl. Probab. 43(2), 563–586 (2006)

[4] Chavez-Demoulin, V., Embrechts, P.: An EVT primer for credit risk. In: Lipton, A., Rennie, A. (eds.) Handbook of Credit Derivatives. Oxford University Press (2010)

[5] Cherubini, U., Mulinacci, S., Gobbi, F., Romagnoli, S.: Dynamic Copula methods in finance. Wiley Finance Series. John Wiley & Sons Ltd., Chichester (2012)

[6] de Amo, E., Díaz-Carrillo, M., Fernández-Sánchez, J.: Absolutely continuous copulas and sub–diagonal sections. Fuzzy Sets and Systems (in press, 2013)

[7] De Baets, B., De Meyer, H., Mesiar, R.: Asymmetric semilinear copulas. Kybernetika (Prague) 43(2), 221–233 (2007)

[8] De Luca, G., Zuccolotto, P.: A tail dependence-based dissimilarity measure for financial time series clustering. Adv. Data Anal. Classif. 5(4), 323–340 (2011)

[9] Durante, F.: A new class of symmetric bivariate copulas. J. Nonparametr. Stat. 18(7-8), 499–510 (2006, 2007)

[10] Durante, F., Fernández-Sánchez, J.: On the classes of copulas and quasi-copulas with a given diagonal section. Internat. J. Uncertain. Fuzziness Knowledge-Based Systems 19(1), 1–10 (2011)

[11] Durante, F., Foschi, R., Spizzichino, F.: Threshold copulas and positive dependence. Statist. Probab. Lett. 78(17), 2902–2909 (2008)

[12] Durante, F., Foscolo, E.: An analysis of the dependence among financial markets by spatial contagion. Int. J. Intell. Syst. 28(4), 319–331 (2013)

[13] Durante, F., Foscolo, E., Sabo, M.: A spatial contagion test for financial markets. In: Kruse, R., Berthold, M., Moewes, C., Gil, M.A., Grzegorzewski, P., Hryniewicz, O. (eds.) Synergies of Soft Computing and Statistics. AISC, vol. 190, pp. 313–320. Springer, Heidelberg (2013)

[14] Durante, F., Hofert, M., Scherer, M.: Multivariate hierarchical copulas with shocks. Methodol. Comput. Appl. Probab. 12(4), 681–694 (2010)

[15] Durante, F., Jaworski, P.: Spatial contagion between financial markets: a copula-based approach. Appl. Stoch. Models Bus. Ind. 26(5), 551–564 (2010)

[16] Durante, F., Kolesárová, A., Mesiar, R., Sempi, C.: Semilinear copulas. Fuzzy Sets and Systems 159(1), 63–76 (2008)

[17] Durante, F., Pappadà, R., Torelli, N.: Clustering of financial time series in risky scenarios (2012) (submitted)

[18] Durante, F., Rodríguez-Lallena, J., Úbeda-Flores, M.: New constructions of diagonal patchwork copulas. Inform. Sci. 179(19), 3383–3391 (2009)

[19] Durante, F., Salvadori, G.: On the construction of multivariate extreme value models via copulas. Environmetrics 21(2), 143–161 (2010)

[20] D'Urso, P., Maharaj, E.: Autocorrelation-based fuzzy clustering of time series. Fuzzy Sets and Systems 160(24), 3565–3589 (2009)

[21] Forbes, K.J., Rigobon, R.: No contagion, only interdependence: measuring stock market comovements. J. Financ. 57(5), 2223–2261 (2002)

[22] Fredricks, G.A., Nelsen, R.B.: The Bertino family of copulas. In: Cuadras, C.M., Fortiana, J., Rodríguez-Lallena, J. (eds.) Distributions with given Marginals and Statistical Modelling, pp. 81–91. Kluwer, Dordrecht (2003)

[23] Jaworski, P., Durante, F., Härdle, W. (eds.): Copulae in Mathematical and Quantitative Finance. Lecture Notes in Statistics - Proceedings. Springer, Heidelberg (2013)

[24] Jaworski, P., Durante, F., Härdle, W., Rychlik, T. (eds.): Copula Theory and its Applications. Lecture Notes in Statistics - Proceedings, vol. 198. Springer, Heidelberg (2010)

[25] Joe, H.: Parametric families of multivariate distributions with given margins. J. Multivariate Anal. 46(2), 262–282 (1993)

[26] Jones, S.: Of couples and copulas. Financial Times (2009) (Published on April 24, 2009)

[27] Jwaid, T., De Baets, B., De Meyer, H.: Orbital semilinear copulas. Kybernetika (Prague) 45(6), 1012–1029 (2009)

[28] Klement, E.P., Kolesárová, A.: Intervals of 1-Lipschitz aggregation operators, quasi-copulas, and copulas with given affine section. Monatsh. Math. 152(2), 151–167 (2007)

[29] Li, D.: On default correlation: a copula function approach. J. Fixed Income 9, 43–54 (2001)

[30] Marshall, A.W.: Copulas, marginals, and joint distributions. In: Distributions with Fixed Marginals and Related Topics (Seattle, WA, 1993). IMS Lecture Notes Monogr. Ser., vol. 28, pp. 213–222. Inst. Math. Statist., Hayward (1996)

[31] McNeil, A.J., Frey, R., Embrechts, P.: Quantitative risk management. Concepts, Techniques and Tools. Princeton Series in Finance. Princeton University Press, Princeton (2005)

[32] Nelsen, R.B., Quesada-Molina, J.J., Rodríguez-Lallena, J.A., Úbeda-Flores, M.: On the construction of copulas and quasi-copulas with given diagonal sections. Insurance Math. Econom. 42(2), 473–483 (2008)

[33] Otranto, E.: Clustering heteroskedastic time series by model–based procedures. Comput. Statist. Data Anal. 52(10), 4685–4698 (2008)

[34] Salvadori, G., De Michele, C., Durante, F.: On the return period and design in a multivariate framework. Hydrol. Earth Syst. Sci. 15, 3293–3305 (2011)

[35] Salvadori, G., De Michele, C., Kottegoda, N.T., Rosso, R.: Extremes in Nature. An Approach Using Copulas. Water Science and Technology Library, vol. 56. Springer, Dordrecht (2007)

[36] Whitehouse, M.: How a formula ignited market that burned some big investors. The Wall Street Journal (2005) (Published on September 12, 2005)

On Quadratic Constructions of Copulas

Anna Kolesárová

1 Introduction

In the lecture we will present quadratic constructions of two-dimensional copulas (copulas for short). Basic properties of copulas can be found, e.g., in the monographs [1, 4]. We say that a *function K_C is obtained by a quadratic construction from a copula C*, if it is defined on $[0,1]^2$ by composition of a quadratic polynomial P of three variables with real coefficients,

$$P(x,y,z) = ax^2 + by^2 + cz^2 + dxy + exz + fyz + gx + hy + iz + j, \qquad (1)$$

and a copula C in the following way

$$K_C(x,y) = P(x,y,C(x,y)), \quad (x,y) \in [0,1]^2. \qquad (2)$$

Note that this construction requires for determining the values $K_C(x,y)$ of a new function K_C the knowledge of x, y and $C(x,y)$ only. In general, functions K_C defined in this way need not be copulas. There are polynomials whose composition with an arbitrary copula C is again a copula, but there are polynomials whose compositions only with some copulas lead to new copulas, and finally, there are also completely negative examples. As a completely positive example we can mention the polynomial $P(x,y,z) = xz + yz - z^2$ whose composition with a copula C, i.e., the function

$$K_C(x,y) = xC(x,y) + yC(x,y) - C^2(x,y) \qquad (3)$$

is a copula for any copula C. Note that this function is equal to the function D_C defined by

Anna Kolesárová

Institute of Information Engineering, Automation and Mathematics,

Faculty of Chemical and Food Technology,

Slovak University of Technology in Bratislava,

Radlinského 9, 812 37 Bratislava 1, Slovakia

e-mail: anna.kolesarova@stuba.sk

H. Bustince et al. (eds.), *Aggregation Functions in Theory and in Practise*,
Advances in Intelligent Systems and Computing 228,
DOI: 10.1007/978-3-642-39165-1_4, © Springer-Verlag Berlin Heidelberg 2013

$$D_C(x,y) = C(x,y)(x+y-C(x,y)), \quad (x,y) \in [0,1]^2, \tag{4}$$

studied in detail in [2]. As another completely positive example we recall the composition of the polynomial $P(x,y,z) = z^2 + xy - xz - yz + z$ and a copula C, giving the function

$$K_C(x,y) = C^2(x,y) + xy - xC(x,y) - yC(x,y) + C(x,y), \tag{5}$$

which is also a copula for each copula C, see [3]. It can be checked that the function K_C in (5) is equal to the function $\left(D_{C_{0,1}}\right)_{0,1}$, where $C_{0,1}$ is the flipped copula [4], $C_{0,1}(x,y) = x - C(x, 1-y)$, and $D_{C_{0,1}}$ is defined by (4). It will be shown that both functions given in (4) and (5) play a key role in a complete characterization of general quadratic constructions.

On the other hand, using, e.g., the polynomial $P(x,y,z) = z^2$, we never obtain by composition (2) a copula. As a partially positive example we can mention the polynomial $P(x,y,z) = z^2 - xz - yz + 2z$, whose composition with copula W, $W(x,y) = \max\{x+y-1,0\}$, is equal to W, i.e., $K_W = W$, while K_M, where $M(x,y) = \min\{x,y\}$ is the greatest copula, is not a copula. Indeed, as $K_M(x,y) = \min\{x,y\}(2 - \max\{x,y\})$, it holds $K_M > M$, thus K_M cannot be a copula [4]. Finally, let us note that composition of the considered polynomial P with the product copula Π, $\Pi(x,y) = xy$, gives the function K_Π, $K_\Pi(x,y) = xy + xy(1-x)(1-y)$, i.e., a member of the Farlie-Gumbel-Morgenstern family of copulas.

2 Results

In our contribution we will characterize all quadratic polynomials of the form (1) having the property that the composite function $P(x,y,C(x,y))$ is a copula for each copula C.

The complete characterization of quadratic constructions is given in the following theorem.

Theorem 1. *The function K_C defined on $[0,1]^2$ by $K_C(x,y) = P(x,y,C(x,y))$, where P is a quadratic polynomial given in (1) and C a copula, is a copula for any copula C if and only if it is of the form*

$$K_C(x,y) = cC^2(x,y) + dxy - cxC(x,y) - cyC(x,y) + (1+c-d)C(x,y), \tag{6}$$

with coefficients c, d satisfying conditions

$$0 \le d-c \le 1, \quad 0 \le d \le 1, \quad -1 \le c \le 1.$$

The set $\Omega = \{(c,d) \in \mathbb{R}^2 \mid 0 \le d-c \le 1, \ 0 \le d \le 1, \ -1 \le c \le 1\}$ of all possible ordered pairs (c,d) of coefficients described in the previous theorem is illustrated in Fig. 1. The restriction of coefficients of a polynomial P in (1) to the pairs $(c,d) \in \Omega$ and other coefficients equal to zero, has been obtained by requiring

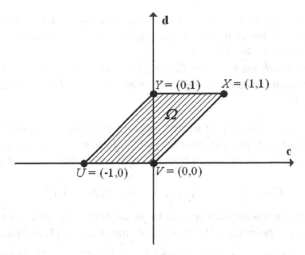

Fig. 1 The domain Ω of possible pairs of coefficients (c,d)

$K_C(x,y) = P(x,y,C(x,y))$ to be a copula for any copula C, i.e., Theorem 1 characterizes "universal" quadratic polynomials for the proposed construction. For some special copulas C the restrictions can be relaxed.

In the lecture we will also present invariant copulas with respect to quadratic constructions.

For a given pair $(c,d) \in \Omega$ and a copula C, let $K_C^{c,d}$ denote the copula given by (6). We will say that a *copula C is invariant with respect to the quadratic construction with coefficients (c,d)*, if the equation $K_C^{c,d}(x,y) = C(x,y)$ holds for all $(x,y) \in [0,1]^2$. The previous equation can be written as

$$cC^2(x,y) + dxy - cxC(x,y) - cyC(x,y) + (1+c-d)C(x,y) = C(x,y), \quad (x,y) \in [0,1]^2. \tag{7}$$

Clearly, if $c = 0$ and $d = 0$, the equation is trivially satisfied by all C. If $c = 0$ and $d \neq 0$, the only copula satisfying the equation, is the product copula $C = \Pi$. If we suppose that $c \neq 0$ and put $\theta = \frac{d}{c}$, then the previous equation is equivalent to the quadratic equation

$$C^2(x,y) - (x+y-1+\theta)C(x,y) + \theta xy = 0. \tag{8}$$

As $c \in [-1,0[\cup]0,1]$, $d \in [0,1]$ and $d \geq c$, we obtain $\theta \in]-\infty,0] \cup [1,\infty[$.

Solutions of this equation are functions $C = C_\theta$ given by

$$C_\theta(x,y) = \frac{1}{2}\left(x+y-1+\theta+\sqrt{(x+y-1+\theta)^2-4\theta xy}\right) \text{ if } \theta \leq 0, \tag{9}$$

$$C_\theta(x,y) = \frac{1}{2}\left(x+y-1+\theta-\sqrt{(x+y-1+\theta)^2-4\theta xy}\right) \text{ if } \theta \geq 1. \tag{10}$$

It can be proved that they are copulas and the following theorem holds.

Theorem 2. *Let (c,d) be in Ω, $c \neq 0$. Then the only copula invariant with respect to the quadratic construction with coefficients (c,d) is the copula C_θ with $\theta = \frac{d}{c}$, given by (9) if $\theta \leq 0$, or by (10) if $\theta \geq 1$.*

On the other hand, each copula C_θ, $\theta \in]-\infty, 0] \cup [1, \infty[$, is invariant with respect to the quadratic construction with an arbitrary pair of coefficients $(c,d) \in \Omega$ satisfying $\frac{d}{c} = \theta$.

Note that solving the problem of invariantness of copulas wrt. quadratic constructions has led us to two new parametric classes of copulas, $(C_\theta)_{\theta \leq 0}$ and $(C_\theta)_{\theta \geq 1}$. The relationship between their members is for all admissible values θ given by the formula

$$C_\theta(x,y) = x - C_{1-\theta}(x, 1-y), \quad (x,y) \in [0,1]^2.$$

Both classes are decreasing with respect to parameter θ. Marginal members of the class $(C_\theta)_{\theta \leq 0}$ are copulas $C_0 = W$ and $C_{-\infty} = \lim\limits_{\theta \to -\infty} C_{\theta \leq 0} = \Pi$, and marginal members of the class $(C_\theta)_{\theta \geq 1}$ are $C_1 = M$ and $C_\infty = \lim\limits_{\theta \to \infty} C_{\theta \geq 1} = \Pi$. These properties are illustrated in Fig. 2.

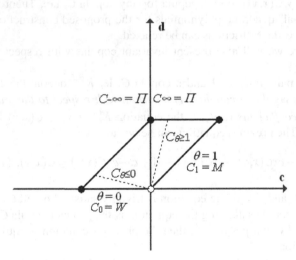

Fig. 2 All pairs $(c,d) \in \Omega$ with $\frac{d}{c} = \theta$ have the same invariant copula C_θ

Besides the complete characterization of quadratic constructions of copulas and determining invariant copulas with respect to these constructions, we also show several properties of newly constructed copulas a their statistical consequences. For more details we refer to [2, 3].

Acknowledgements. The author kindly acknowledges the support of the grant VEGA 1/0419/13. Moreover, she also thanks to the coauthors of papers [2, 3] for cooperation in investigation of quadratic constructions of copulas.

References

[1] Joe, H.: Multivariable Models and Dependence Concepts. Chapman & Hall, London (1997)

[2] Kolesárová, A., Mesiar, R., Kalická, J.: On a new construction of 1-Lipschitz aggregation functions, quasi-copulas and copulas. Fuzzy Sets and Systems (2013), doi:10.1016/j.fss.2013.01.005

[3] Kolesárová, A., Mayor, G., Mesiar, R.: Quadratic constructions of copulas. Inform. Sci. (submitted)

[4] Nelsen, R.B.: An Introduction to Copulas. Lecture Notes in Statistics, vol. 139. Springer, New York (1999)

Compatible Group Decisions

Gabriella Pigozzi

Abstract. Individuals may draw different conclusions from the same information. For example, members of a jury may disagree on the verdict even though each member possesses the same information regarding the case under discussion. This happens because individuals can hold different reasonable positions based on the information they share. The field of judgment aggregation studies how individual positions on the same information can be aggregated into a collective one. After a gentle introduction to judgment aggregation, I will offer an analysis of judgment aggregation problems using an argumentation approach. One of the principles of argumentation theory is that an argumentation framework can have several labellings. If the information the group shares is represented by an argumentation framework, and each agent's reasonable position is a labelling of that argumentation framework, the question becomes how to aggregate the individual positions into a collective one. Whereas judgment aggregation focuses on the observation that the aggregation of individual logically consistent judgments may lead to an inconsistent group outcome, I will present an approach that not only ensures collective rationality but also social outcomes that are 'compatible' with the individuals' evaluations. This ensures that no individual member has to become committed to a group position that is in conflict with his own individual position. (Part of my presentation will be based on a joint work with Martin Caminada and Mikolaj Podlaszewski.)

Gabriella Pigozzi
LAMSADE
Université Paris-Dauphine
e-mail: gabriella.pigozzi@dauphine.fr

H. Bustince et al. (eds.), *Aggregation Functions in Theory and in Practise*,
Advances in Intelligent Systems and Computing 228,
DOI: 10.1007/978-3-642-39165-1_5, © Springer-Verlag Berlin Heidelberg 2013

A Review of the Relationships between Aggregation, Implication and Negation Functions

Ana Pradera

Abstract. Aggregation functions, on one hand, and implication functions, on the other, play different although important roles in the field of fuzzy logic. Both have been intensively investigated in the last years, revealing the tide relationship that exists between them. The present work reviews the most relevant aspects of this relation, which most of the times also involves negation functions. In addition to the well-known use of aggregation and negation functions to build and to characterize implication functions, we analyze how new families of aggregation functions and negation functions can be obtained from implications, and we recall the main equations and inequations involving these three classes of functions.

Extended Abstract

Aggregation functions, which perform the combination of several inputs into a single output, are successfully used in many practical applications, and the interest on them is unceasingly growing (see e.g. the recent monographs on the topic [24, 5, 13]). Although they are defined for inputs of any cardinality, in this work we will only deal with *bivariate* aggregation functions, i.e., non-decreasing functions $A : [0,1]^2 \rightarrow [0,1]$ verifying the boundary conditions $A(0,0) = 0$ and $A(1,1) = 1$.

In turn, *fuzzy implication functions* (or implications, for short), which generalize to the fuzzy world the classical Boolean implication, have proved to be essential in many different fields, ranging from approximate reasoning or fuzzy control to fuzzy mathematical morphology and image processing (see the survey [17] and the monographs [4, 3] for details and appropriate references). Different definitions of fuzzy implication have been proposed in the literature, but nowadays the most established one is the following ([12, 17, 4]): a *(fuzzy) implication* is a function $I : [0,1]^2 \rightarrow [0,1]$

Ana Pradera

Departamento de Ciencias de la Computación. Universidad Rey Juan Carlos, Madrid, Spain

e-mail: ana.pradera@urjc.es

H. Bustince et al. (eds.), *Aggregation Functions in Theory and in Practise*,   31
Advances in Intelligent Systems and Computing 228,
DOI: 10.1007/978-3-642-39165-1_6, © Springer-Verlag Berlin Heidelberg 2013

which is non-increasing in the first variable, non-decreasing in the second variable and that satisfies the boundary conditions $I(0,0) = I(1,1) = 1$ and $I(1,0) = 0$.

Aggregation and implication functions appear to have a close relation which, in most of the cases, is set up via *negation functions* (non-increasing functions N : $[0,1] \rightarrow [0,1]$ verifying the boundary conditions $N(0) = 1$ and $N(1) = 0$; the most usual negation is $N_C(x) = 1 - x$). In this work we will review the main aspects of this relation, including the following ones:

- **Implications built from aggregation and negation functions.**

  Different classes of implications exist. The probably better-established ones are those built from logical formulae that, although being equivalent in the classical setting, turn out to be no longer equivalent in the fuzzy context: $\neg a \vee b$ (that provides the so-called (S,N)-implications), $\bigvee \{t : a \wedge t \leq b\}$ (R-implications), $\neg a \vee (a \wedge b)$ (QL-implications) and $(\neg a \wedge \neg b) \vee b$ (D-implications). Initially, the logical connectives $\wedge$ and $\vee$ were replaced with just two specific classes of aggregation functions, the well-known triangular norms and triangular conorms ([14, 2]), providing interesting families of implications that have been widely studied ([4]). But triangular norms and conorms are not the only aggregation functions that are able to model conjunctions and disjunctions, and hence many authors have considered the use of alternative ones: see e.g. [7, 15, 21, 22, 23, 16] for works dealing with fuzzy implications derived from uninorms, [9, 25] for implications built from (dual) copulas, quasi-copulas and semi-copulas, [1, 6] for implications generated from (dual) representable aggregation functions and [19] for implications generated from TS-functions. A general approach regarding the relations between fuzzy implications and other fuzzy connectives was first proposed by Fodor in [10, 11], while [8] concentrated on the case of R-implications generated from a large class of binary operators. Recently, the papers [18] and [20] pick up the subject again with general views on the construction of implications by means of aggregation functions, dealing with the cases of R-implications and (S,N)-implications. Note for example that the latter become (A,N)-implications, where A stands for an appropriate aggregation function:

  Proposition 1. *Let A be an aggregation function, let N be a negation and let* $I_{A,N} : [0,1]^2 \rightarrow [0,1]$ *be defined as* $I_{A,N}(x,y) = A(N(x),y)$ *for any* $x,y \in [0,1]$. *Then* $I_{A,N}$ *is a fuzzy implication if and only if* $A(1,0) = A(0,1) = 1$ *(i.e., if and only if A has absorbing element* 1*).*

  In this section we will summarize all these results, stressing the properties of the aggregation functions and/or the negations that lead to specific properties of the corresponding implications, and illustrating the issue with some new families of implications functions. In particular, the above Proposition shows that many other aggregation functions, in addition to triangular conorms or disjunctive uninorms, may be used in order to build (A,N)-implications. Some of them, generated from TS-functions, weighted quasi-arithmetic means or generalized OWAs, may be found in [19, 20].

- **Aggregation functions built from implication and negation functions.**

  The logical formulae mentioned in the previous item have been used to obtain implications from aggregation and negation functions, but they can also be used in the opposite direction, thus allowing to build aggregation functions by means of implications and negations. For example, the expressions $\neg a \vee b$ and $\bigvee \{t : a \wedge t \leq b\}$, that allow to build, respectively, (A,N)-implications and R-implications, can be shifted as follows in order to generate families of aggregation functions:

  Proposition 2. *Let I be a fuzzy implication and let N be a negation function. Then:*

  - *The function $A_{I,N} : [0,1]^2 \to [0,1]$, defined as $A_{I,N}(x,y) = I(N(x),y)$ for any $x,y \in [0,1]$, is an aggregation function. Moreover, it satisfies $A_{I,N}(x,1) = A_{I,N}(1,x) = 1$ for any $x \in [0,1]$.*
  - *If I is such that $I(1,y) \neq 1$ for all $y \in [0,1[$, then the function $A_I : [0,1]^2 \to [0,1]$, defined as $A_I(x,y) = \inf\{t \in [0,1] : I(x,t) \geq y\}$, is an aggregation function. Moreover, it satisfies $A_I(x,0) = A_I(0,x) = 0$ and $A_I(1,y) \neq 0$ for any $x,y \in [0,1]$.*

  In this section we will explore the use of results of this kind in order to obtain new families of aggregation functions, comparing their properties with the ones of the originating implications.

- **Negation functions built from implication/aggregation functions.**

  It is well known that any fuzzy implication I provides a fuzzy negation N_I, the so-called *natural negation of I*, defined as $N_I(x) = I(x,0)$ for any $x \in [0,1]$, and that the former can be generalized as follows (see [4]):

  Proposition 3. *Let I be a fuzzy implication and let $\alpha \in [0,1[$ verify $I(1,\alpha) = 0$. Then the function $N_I^\alpha : [0,1] \to [0,1]$, given by $N_I^\alpha(x) = I(x,\alpha)$ for any $x \in [0,1]$, is a fuzzy negation called the natural negation of I with respect to α.*

  Therefore, any implication has at least one natural negation associated to it (N_I, which is equal to N_I^0), but can have a family of them as long as there exists $\alpha > 0$ such that $I(1,\alpha) = 0$: indeed, in such cases, thanks to the non-decreasing monotonicity of I in its second variable, any N_I^β with $\beta \leq \alpha$ is also a natural negation of I, which, in addition, verifies $N_I^\beta \leq N_I^\alpha$. This result is not interesting in the case, for example, of (S,N)-implications, i.e., (A,N)-implications built from triangular conorms (they have only one natural negation, which in addition coincides with the original one), but can provide new families of negation functions in other situations ([4, 20]).

  On the other hand, it is also possible to obtain negations from aggregation functions. Indeed, similarly to what is usually done for triangular norms and conorms (see e.g. [14]), it is not difficult to find the conditions that aggregation functions A need to fulfill in order to ensure that the univariate functions $N_A^{\sup}$ and $N_A^{\inf}$, defined, respectively, as $N_A^{\sup}(x) = \sup\{t \in [0,1] : A(x,t) = 0\}$ and $N_A^{\inf}(x) = \inf\{t \in [0,1] : A(x,t) = 1\}$, are fuzzy negations.

- **Characterizations of implications by means of aggregation and negation functions.**

Another important issue when dealing with fuzzy implications is to characterize them, in the sense of obtaining the minimal set of properties required to a fuzzy implication in such a way that it belongs to a given family of implications. Such characterizations usually involve properties of the underlying aggregation and/or negation functions. For example (see [4]), the class of (A,N)-implications generated from triangular conorms and continuous negations has been characterized, as well as the family of R-implications built from left-continuous triangular norms. Recently ([20]) it has been pointed out that the class of (A,N)-implications coincides with the whole class of fuzzy implications, which means that actually any fuzzy implication has a logical representation based on the classical material implication $\neg a \vee b$:

Proposition 4. *For a function $I : [0,1]^2 \to [0,1]$, the following statements are equivalent:*

1. *I is a fuzzy implication.*
2. *I is an (A,N_C)-implication, i.e., there exists an aggregation function A verifying $A(0,1) = A(1,0) = 1$ such that $I(x,y) = I_{A,N_C}(x,y) = A(1-x,y)$ for any $x,y \in [0,1]$.*

Note that the above result implies that all the implications belonging to families such as R-implications, QL-implications, D-implications, etc, could also have been obtained as (A,N)-implications. Consider, for example, the popular Gödel implication ([4]), defined as

$$I_{GD}(x,y) = \begin{cases} 1 \text{ if } x \leq y, \\ y \text{ otherwise.} \end{cases}$$

As is well-known, I_{GD} is the R-implication obtained from the minimum triangular norm, and it does not belong to the class of (S,N)-implication, i.e., it may not be obtained from the scheme $\neg a \vee b$ if the connective $\vee$ is replaced with a triangular conorm (see e.g. [4]). Notwithstanding, it is a (A,N_C)-implication, i.e., there exists an aggregation function A such that $I_{GD}(x,y) = A(N_C(x),y) = A(1-x,y)$. Indeed, it suffices to take A as

$$A(x,y) = \begin{cases} 1 \text{ if } 1-x \leq y, \\ y \text{ otherwise.} \end{cases}$$

Note that in the above result the standard negation N_C can be replaced with any strong negation (i.e., any involutive negation).

References

[1] Aguiló, I., Carbonell, M., Suñer, J., Torrens, J.: Dual representable aggregation functions and their derived S-implications. In: Hüllermeier, E., Kruse, R., Hoffmann, F. (eds.) IPMU 2010. LNCS, vol. 6178, pp. 408–417. Springer, Heidelberg (2010)

[2] Alsina, C., Frank, M., Schweizer, B.: Associative functions. World Scientific, Singapore (2006)

[3] Baczyński, M., Beliakov, G., Bustince, H., Pradera, A. (eds.): Advances in Fuzzy Implication Functions. Springer, Heidelberg (2013)

[4] Baczyński, M., Jayaram, B.: Fuzzy Implications. Springer, Heidelberg (2008)

[5] Beliakov, G., Pradera, A., Calvo, T.: Aggregation Functions: A Guide for Practitioners. Springer, Heidelberg (2007)

[6] Carbonell, M., Torrens, J.: Continuous R-implications generated from representable aggregation functions. Fuzzy Sets and Systems 161(17), 2276–2289 (2010)

[7] De Baets, B., Fodor, J.: Residual operators of uninorms. Soft Comput. 3(2), 89–100 (1999)

[8] Demirli, K., De Baets, B.: Basic properties of implicators in a residual framework. Tatra Mountains Mathematical Publications 16, 31–46 (1999)

[9] Durante, F., Klement, E., Mesiar, R., Sempi, C.: Conjunctors and their residual implicators: Characterizations and construction methods. Mediterranean Journal of Mathematics 4, 343–356 (2007)

[10] Fodor, J.: On fuzzy implication operators. Fuzzy Sets and Systems 42, 293–300 (1991)

[11] Fodor, J.: A new look at fuzzy connectives. Fuzzy Sets and Systems 57, 141–148 (1993)

[12] Fodor, J., Roubens, M.: Fuzzy preference modelling and multicriteria decision support. Kluwer Academic Publishers (1994)

[13] Grabisch, M., Marichal, J., Mesiar, R., Pap, E.: Aggregation Functions. Cambridge University Press (2009)

[14] Klement, E., Mesiar, R., Pap, E.: Triangular Norms. Kluwer, Dordrecht (2000)

[15] Mas, M., Monserrat, M., Torrens, J.: Two types of implications derived from uninorms. Fuzzy Sets and Systems 158(23), 2612–2626 (2007), http://dx.doi.org/10.1016/j.fss.2007.05.007

[16] Mas, M., Monserrat, M., Torrens, J.: A characterization of (U, N), RU, QL and D-implications derived from uninorms satisfying the law of importation. Fuzzy Sets and Systems 161(10), 1369–1387 (2010)

[17] Mas, M., Monserrat, M., Torrens, J., Trillas, E.: A survey on fuzzy implication functions. IEEE T. Fuzzy Systems 15(6), 1107–1121 (2007)

[18] Ouyang, Y.: On fuzzy implications determined by aggregation operators. Information Sciences 193, 153–162 (2012)

[19] Pradera, A., Beliakov, G., Bustince, H., Fernández, J.: On (TS,N)-fuzzy implications. In: Proceedings of 6th International Summer School on Aggregation Operators, pp. 93–98 (2011)

[20] Pradera, A., Beliakov, G., Bustince, H., Galar, M., Bedregal, B.: Fuzzy implications seen as (A,N)-implications (submitted)

[21] Ruiz-Aguilera, D., Torrens, J.: Residual implications and co-implications from idempotent uninorms. Kybernetika 40(1), 21–38 (2004)

[22] Ruiz-Aguilera, D., Torrens, J.: Distributivity of residual implications over conjunctive and disjunctive uninorms. Fuzzy Sets and Systems 158(1), 23–37 (2007)

[23] Ruiz-Aguilera, D., Torrens, J.: S- and R-implications from uninorms continuous in $[0, 1]^2$ and their distributivity over uninorms. Fuzzy Sets and Systems 160(6), 832–852 (2009)

[24] Torra, V., Narukawa, Y.: Modeling Decisions: Information Fusion and Aggregation Operators. Springer (2007)

[25] Yager, R.: Modeling holistic fuzzy implication using co-copulas. Fuzzy Optim. and Decis. Making 5, 207–226 (2006)

Part II
Copulas

On Some Construction Methods for Bivariate Copulas

Radko Mesiar, Jozef Komorník, and Magda Komorníková

Abstract. We propose a rather general construction method for bivariate copulas, generalizing some construction methods known from the literature. In some special cases, the constraints ensuring the output of the proposed method to be a copula are given. Our approach opens several new problems in copula theory.

Keywords: Farlie–Gumbel–Morgenstern copulas, Mayor–Torrens copulas, construction method for copula.

1 Introduction

We suppose readers to be familiar with the basics of copula theory. In the opposite case, we recommend the lecture notes [12]. Recently, several construction methods for bivariate copulas have been proposed. Recall, for example, conic copulas [10], univariate conditioning method proposed in [15], UCS (univariate conditioning stable) copulas [8], a method proposed by Rodríguez–Lallena and Úbeda–Flores [17] and its generalization in [11], another method introduced by Aguilló et al. in [1], quadratic construction introduced in [13], several construction methods based on diagonal or horizontal (vertical) sections discussed in [5, 3, 7], etc.

Radko Mesiar · Magda Komorníková
Faculty of Civil Engineering, Slovak University of Technology, Radlinského 11
813 68 Bratislava, Slovakia
e-mail: {mesiar,magda}@math.sk

Jozef Komorník
Faculty of Management Comenius University, Odbojárov 10, P.O. BOX 95
820 05 Bratislava, Slovakia
e-mail: Jozef.Komornik@fm.uniba.sk

H. Bustince et al. (eds.), *Aggregation Functions in Theory and in Practise*,
Advances in Intelligent Systems and Computing 228,
DOI: 10.1007/978-3-642-39165-1_7, © Springer-Verlag Berlin Heidelberg 2013

We recall two of the above mentioned methods. Recall that a function $N : [0,1] \rightarrow$ $[0,1]$ which is a decreasing involution is called a *strong negation*. We denote its unique fixed point as e, $N(e) = e$. Due to [1, Theorem 23], the next result holds.

Proposition 1. *Let $N : [0,1] \rightarrow [0,1]$ be a strong negation such that it is 1–Lipschitz on the interval $[e,1]$. Then the function $C_N : [0,1]^2 \rightarrow [0,1]$ given by*

$$C_N(x,y) = \max(0, x \wedge y - N(x \vee y)) \tag{1}$$

is a copula.

Inspired by the form of the Farlie–Gumbel–Morgenstern copulas (FGM–copulas)

$$C_\lambda^{FGM}(x,y) = x \cdot y + \lambda x \cdot y \cdot (1-x) \cdot (1-y), \tag{2}$$

where $\lambda \in [-1,1]$, Kim et al. have studied in [11] the constraints for λ so that the function $C : [0,1]^2 \rightarrow [0,1]$ given by

$$C(x,y) = C^*(x,y) + \lambda f(x) \cdot g(y) \tag{3}$$

is a copula, where C^* is an a priori given copula, $f, g : [0,1] \rightarrow [0,1]$ are Lipschitz continuous functions satisfying $f(0) = g(0) = f(1) = g(1) = 0$.

Note that a special case of (3) when $C^* = \Pi$ under consideration was the product copula, was studied in [17].

The aim of this paper is to find a formula generalizing all above introduced formulae (5), (2), (3) and to study some of its new instances.

The paper is organized as follows. In the next section, we introduce our general formula and discuss a special case of (5) yielding Mayor–Torrens copulas [14] and open a related problem based on our generalized formula. Section 3 is focused on the product copula based constructions exploiting our generalized formula. Finally, in concluding remarks we sketch some problems for further investigations.

2 A General Formula for Constructing Bivariate Copulas

Observe that denoting by M the strongest bivariate copula, $M = \min$, formula (5) can be rewritten as

$$C_N(x,y) = \max(0, M(x,y) - M(N(x), N(y))). \tag{4}$$

Similarly, formulae (2) and (3) can be written as

$$C(x,y) = \max(0, C^*(x,y) + \lambda \Pi(f(x), g(y))). \tag{5}$$

Denote by **p** the pentuple $(C_1, C_2, \lambda, f, g)$, where $C_1, C_2 : [0,1]^2 \rightarrow [0,1]$ are bivariate copulas, λ is a real constant and $f, g : [0,1] \rightarrow [0,1]$ are real functions. It is evident that the formula

$$C_{\mathbf{p}}(x,y) = \max\left(0, C_1(x,y) + \lambda C_2(f(x),g(y))\right) \tag{6}$$

is a well defined real function, $C_{\mathbf{p}} : [0,1]^2 \to \Re$. Clearly, formulae (5) and (4) correspond to $\mathbf{p} = (M,M,-1,N,N)$, while formula (2) is linked to $\mathbf{p} = (\Pi,\Pi,\lambda,f,f)$ with $\lambda \in [-1,1]$ and $f(x) = x(1-x)$. Similarly, formulae (3) and (5) are related to $\mathbf{p} = (C^*,\Pi,\lambda,f,g)$. Recall a trivial result $C_{\mathbf{p}} = C_1$ whenever $\lambda = 0$.

Example 1. Consider $\mathbf{p} = (C^H,\Pi,\lambda,f,f)$, where C^H is the Hamacher product, i.e., a copula given by

$$C^H(x,y) = \frac{x \cdot y}{x+y-x \cdot y}$$

whenever $x \cdot y \neq 0$, and $f(x) = x^2 \cdot (1-x)$. After some processing with software *MATHEMATICA* it can be shown that $C_{\mathbf{p}}$ is a copula if and only if $\lambda \in [-2,1]$. Moreover, then $C_{\mathbf{p}}$ is an absolutely continuous copula given by

$$C_{\mathbf{p}}(x,y) = \frac{x \cdot y}{x+y-x \cdot y} + \lambda x^2 \cdot y^2 \cdot (1-x) \cdot (1-y).$$

On the other hand, putting $\mathbf{r} = (C^H,\Pi,\lambda,g,g)$, where $g(x) = x \cdot (1-x)$, compare (2), $C_{\mathbf{r}}$ is a copula only if $\lambda = 0$, i.e., when $C_{\mathbf{r}} = C^H$.

Consider an arbitrary copula $C : [0,1]^2 \to [0,1]$ and its diagonal section $\delta : [0,1] \to [0,1]$ given by $\delta(x) = C(x,x)$. Recall that δ is non–decreasing, 2–Lipschitz, $\delta(0) = 0$, $\delta(1) = 1$ and $\delta(x) \leq x$ for all $x \in [0,1]$. Then the function $f : [0,1] \to [0,1]$ given by $f(x) = x - \delta(x)$ is 1–Lipschitz, $f(0) = f(1) = 0$. For $\mathbf{p} = (M,M,-1,f,f)$, it holds

$$C_{\mathbf{p}}(x,y) = \max\left(0, M(x,y) - M(x - \delta(x), y - \delta(y))\right). \tag{7}$$

Applying formula (5) considering $N = f$, one gets

$$C_f(x,y) = \max\left(0, x \wedge y - f(x \vee y)\right) = \max\left(0, \delta(x \vee y) - |x-y|\right) = C_{\delta}^{MT}(x,y),$$

where C_{δ}^{MT} is a Mayor–Torrens copula [14] derived from the diagonal section δ.

On the other hand, $C_{\mathbf{p}}$ given by formula (7) for diagonal sections of 3 basic copulas W,Π,M yields copulas $W,C_{\mathbf{q}},M$, where $\mathbf{q} = (M,M,-1,f_{\Pi},f_{\Pi})$, $f_{\Pi}(x) = x \cdot (1-x)$. Observe that the copula $C_{\mathbf{q}} : [0,1]^2 \to [0,1]$ is described in Fig. 1.

On the other side, consider the ordinal sum copula $C = (\langle 0,\frac{1}{2},W\rangle, \langle\frac{1}{2},1,W\rangle)$, i.e., $f : [0,1] \to [0,1]$ given by

$$f(x) = \begin{cases} \min\left(x,\frac{1}{2}-x\right) & \text{if } x \in \left[0,\frac{1}{2}\right] \\ \min\left(x-\frac{1}{2},1-x\right) & \text{else}. \end{cases}$$

Then for $\mathbf{r} = (M,M,-1,f,f)$ the resulting function $C_{\mathbf{r}} : [0,1]^2 \to [0,1]$ satisfies

$$C_{\mathbf{r}}\left(x,\frac{3}{4}\right) = 2x-1 \quad \text{if } x \in \left[\frac{1}{4},\frac{1}{2}\right],$$

violating the 1–Lipschitz property of $C_{\mathbf{r}}$. Thus $C_{\mathbf{r}}$ is not a copula.

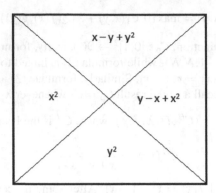

Fig. 1 Formulae for the copula C_q

We open a problem of characterizing all diagonal sections δ of bivariate semicopulas such that the formula (7) yields a copula. Note also that if a function C_p given by (7) is a copula, then C_q with $q = (M, M, \lambda, f, f)$ is a copula for any $\lambda \in [-1, 0]$.

3 Product–Based Construction of Copulas

Inspired by (5), consider a pentuple $p = (C, C, -1, N, N)$ where N is a strong negation, i.e., consider a function $C_p : [0, 1]^2 \to [0, 1]$ given by

$$C_p(x, y) = \max\left(0, C(x, y) - C(N(x), N(y))\right). \tag{8}$$

Evidently, C_p is non–decreasing in both coordinates and satisfies the boundary conditions for copulas, i.e., C_p is a semicopula [2, 4]. For arbitrary Frank copula [9] and the standard negation $N_s : [0, 1] \to [0, 1]$ given by $N_s(x) = 1 - x$, we see that $C_p = W$ is a copula.

For the 3 basic copulas, the case $C = M$ was discussed in [1], see Proposition 1. For the case $C = W$, observe that C_p is a copula if and only if $N \circ N_s \le N_s \circ N$ and then $C_p = W$ (this is, e.g., the case of a convex strong negation N). We focus now on the third basic copula $C = \Pi$, i.e., we will consider $p = (\Pi, \Pi, -1, N, N)$, i.e., $C_p : [0, 1]^2 \to [0, 1]$ given by

$$C_p(x, y) = \max\left(0, x \cdot y - N(x) \cdot N(y)\right). \tag{9}$$

Proposition 2. *Let $N : [0, 1] \to [0, 1]$ be a differentiable convex strong negation. Then the function C_p given by (9) is a negative quadrant dependent copula.*

Proof. Observe first that under requirements of this proposition, $C_p(x, y) = 0$ if and only if $y \le N(x)$. Moreover, if $x \cdot y = N(x) \cdot N(y)$, then $N'(x) \cdot N'(y) = 1$. As for as C_p is a semicopula, it is enough to show its 2–increasingness on its positive area. Consider $0 < x_1 < x_2 \le 1$, $0 < y_1 < y_2 \le 1$ such that $x_1 \cdot y_1 \ge N(x_1) \cdot N(y_1)$. Then the volume V_{C_p} of the rectangle $[x_1, x_2] \times [y_1, y_2]$ is non–negative if and only if

$$(x_2 - x_1) \cdot (y_2 - y_1) \geq (N(x_1) - N(x_2)) \cdot (N(y_1) - N(y_2)) =$$

$$= (x_1 - x_2) \cdot N'(x_0) \cdot (y_1 - y_2) \cdot N'(y_0),$$

where x_0 is some point from $]x_1, x_2[$ and y_0 is some point from $]y_1, y_2[$.

Equivalently, it should hold $N'(x_0) \cdot N'(y_0) \leq 1$. Due to the fact that $x_0 \cdot y_0 > x_1 \cdot y_1$ and $N(x_0) < N(x_1)$, $N(y_0) < N(y_1)$, it holds $x_0 \cdot y_0 > N(x_0) \cdot N(y_0)$.

Consequently, $x_0 \cdot y_0 > x_0 \cdot y = N(x_0) \cdot N(y)$, where $y = N(x_0) < y_0$. Due to the convexity and monotonicity of N, it holds

$$N'(y) < N'(y_0) < 0, \quad N'(x_0) < 0,$$

and hence

$$N'(x_0) \cdot N'(y_0) < N'(x_0) \cdot N'(y) = 1.$$

Thus $C_{\mathbf{p}}$ is a copula. Obviously, $C_{\mathbf{p}} \leq \Pi$, i.e., $C_{\mathbf{p}}$ is a NQD copula. $\qquad\square$

Remark 1. As a by–product of the proof of Proposition 2, we see that for a convex differentiable strong negation N, the copula $C_{\mathbf{p}}$ given by (9) has its zero–area bounded by the graph of the function N. The same zero area is also obtained by some other kinds of constructing copulas by means of N. For example, this is the case of conic copulas based on N [3], or UCS copulas introduced by Durante and Jaworski in [8], $C_N(x,y) = x \cdot N\left(\min\left(1, \frac{N(y)}{x}\right)\right)$.

We expect that Proposition 2 is also valid for convex strong negations N which are not differentiable.

Example 2. For $c \in]0, 1[$, define a function $N_c : [0, 1] \to [0, 1]$ by

$$N_c(x) = \begin{cases} 1 - \frac{1-c}{c} x & \text{if } x \in [0, c] \\ \frac{c(1-x)}{1-c} & \text{else} \end{cases}.$$

Then N_c is a strong negation which is convex if and only if $c \in]0, \frac{1}{2}]$.

Applying formula (9), we see that $C_{\mathbf{p}} : [0, 1]^2 \to [0, 1]$ is given by

$$C_{\mathbf{p}}(x, y) = \begin{cases} \frac{c^2 x + c^2 y - c^2 + (1 - 2c) x \cdot y}{(1-c)^2} & \text{if } (x, y) \in [c, 1]^2, \\ \frac{(1-c)x + (y-1)c}{1-c} & \text{if } x \in [0, c] \text{ and } y \geq \frac{c - (1-c)x}{c}, \\ \frac{(1-c)y + (x-1)c}{1-c} & \text{if } y \in [0, c] \text{ and } x \geq \frac{c - (1-c)y}{c}, \\ 0 & \text{else}. \end{cases}$$

Observe that for each $c \in]0, 1[$, $C_{\mathbf{p}}$ is a semicopula which is Lipschitz with constant $\max\left(1, \left(\frac{c}{1-c}\right)^2\right)$, i.e., for $c > \frac{1}{2}$, $C_{\mathbf{p}}$ is not a copula. On the other hand, for each $c \in]0, \frac{1}{2}]$, $C_{\mathbf{p}}$ is a copula.

Open problems:

i) For each convex strong negation N, putting $\mathbf{p} = (\Pi, \Pi, \lambda, N, N)$, the function $C_{\mathbf{p}}$
   is a copula for $\lambda \in \{-1, 0\}$. Is this claim valid for each $\lambda \in [-1, 0]$? Are there
   some other constant λ so that $C_{\mathbf{p}}$ is a copula?

ii) For two convex strong negations N_1, N_2, and some $\lambda \in \mathfrak{R}$, does $\mathbf{p} = (\Pi, \Pi, \lambda,$
    $N_1, N_2)$ generate a copula $C_{\mathbf{p}}$ applying (9)?

Example 3. Consider the standard negation N_s. Applying formula (9) to $\mathbf{p} = (\Pi, \Pi,$
$\lambda, N_s, N_s)$, it holds

$$C_{\mathbf{p}}(x, y) = \max\left(0, x \cdot y + \lambda(1 - x) \cdot (1 - y)\right) = \max\left(0, (1 + \lambda)x \cdot y - \lambda(x + y - 1)\right),$$

which is a copula (Sugeno–Weber t-norm, see [12]) for each $\lambda \in [-1, 0]$. For $\lambda >$
0, $C_{\mathbf{p}}$ is not monotone and thus not a copula (even it is not an aggregation function).
For $\lambda < -1$, $C_{\mathbf{p}}$ is Lipschitz with constant $-\lambda$, and thus not a copula.

As another interesting fact consider the pentuple $\mathbf{p} = (\Pi, M, \lambda, f, f)$ with $f(x) =$
$x(1 - x)$. After a short processing it is not difficult to check that then $C_{\mathbf{p}}$ given by (6)
is a copula if and only if $\lambda = 0$ and then $C_{\mathbf{p}} = \Pi$.

This observation opens another problem, namely whether it can be shown that
for any differentiable functions f, g such that $f(0) = f(1) = g(0) = g(1) = 0$ and
$f'(0), f'(1), g'(0), g'(1)$ are different from 0, compare [11], $C_{\mathbf{p}}$ for $\mathbf{p} = (\Pi, M, \lambda, f, g)$
is a copula only if $\lambda = 0$ (and then $C_{\mathbf{p}} = \Pi$).

4 Concluding Remarks

We have proposed a rather general formula (6) transforming a given copula into a
real function, which in several special cases leads to new parametric families of copulas. We have discussed some of such families, but also some negative cases leading
to trivial solutions only. Our proposal opens several problems for a deeper study. For
example, problems of fitting copulas with special properties, such as symmetric copulas which are NQD but with Spearman's rho close to 0 (then copulas discussed in
Example 1 can be of use). For several special types of $\mathbf{p}$ with fixed C_1, C_2, f, g, the
problem of characterizing all constants λ such that $C_{\mathbf{p}}$ is a copula generalizes the
problem opened by Kim et al. in [11]. For example, consider $\mathbf{p} = (M, M, \lambda, f, f)$
with $f : [0, 1] \to [0, 1]$ non–increasing and $f(0) = 0$. Obviously, $C_{\mathbf{p}}$ is then a semicopula if and only if $\lambda \leq 0$, independently of non–zero function f. As another particular problem, we can consider pentuples $\mathbf{p}_1, \mathbf{p}_2$ applied consecutively. Indeed, for
$\mathbf{p}_1 = (C_1, C_2, \lambda, f, g)$ such that $C_{\mathbf{p}_1}$ is a copula, and $\mathbf{p}_2 = (C_{\mathbf{p}_1}, C_3, \tau, h, q)$ one can
define $C_{\mathbf{p}_1, \mathbf{p}_2} = (C_{\mathbf{p}_1})_{\mathbf{p}_2}$, which in the case $\lambda, \tau \leq 0$ can be written as

$$C_{\mathbf{p}_1, \mathbf{p}_2}(x, y) = \max\left(0, C_1(x, y) + \lambda C_2(f(x), g(y)) + \tau C_3(h(x), q(y))\right).$$

Acknowledgements. The research summarized in this paper was supported by the Grants
VEGA 1/0143/11 and APVV–0496–10.

References

[1] Aguiló, I., Suñer, J., Torrens, J.: A construction method of semicopulas from fuzzy negations. Fuzzy Sets and Systems (to appear)

[2] Bassano, B., Spizzichino, F.: Relations among univariate aging, bivariate aging and dependence for exchangeable lifetimes. J. Multivariate Anal. 93(2), 313–339 (2005)

[3] De Baets, B., De Meyer, H., Kalická, J., Mesiar, R.: Flipping and cyclic shifting of binary aggregation functions. Fuzzy Sets and Systems 160(6), 752–765 (2009)

[4] Durante, F., Sempi, C.: Semicopulae. Kybernetika 41(3), 315–328 (2005)

[5] Durante, F., Kolesárová, A., Mesiar, R., Sempi, C.: Copulas with given diagonal sections: novel constructions and applications. International Journal of Uncertainty, Fuzziness and Knowlege-Based Systems 15(4), 397–410 (2007)

[6] Durante, F., Kolesárová, A., Mesiar, R., Sempi, C.: Copulas with given values on a horizontal and a vertical section. Kybernetika 43(2), 209–220 (2007)

[7] Durante, F., Kolesárová, A., Mesiar, R., Sempi, C.: Semilinear copulas. Fuzzy Sets and Systems 159(1), 63–76 (2008)

[8] Durante, F., Jaworski, P.: Invariant dependence structure under univariate truncation. Statistics 46, 263–267 (2012)

[9] Frank, M.J.: On the simultaneous associativity of F(x, y) and x+y-F(x, y). Aequationes Math. 19(2-3), 194–226 (1979)

[10] Jwaid, T., De Baets, B., Kalická, J., Mesiar, R.: Conic aggregation functions. Fuzzy Sets and Systems 167(1), 3–20 (2011)

[11] Kim, J.-M., Sungur, E.A., Choi, T., Heo, T.-Y.: Generalized Bivariate Copulas and Their Properties. Model Assisted Statistics and Applications-International Journal 6, 127–136 (2011)

[12] Klement, E.P., Mesiar, R., Pap, E.: Triangular norms. Trends in Logic-Studia Logica Library, vol. 8. Kluwer Academic Publishers, Dordrecht (2000)

[13] Kolesárová, A., Mesiar, R., Kalická, J.: On a new construction of 1-Lipschitz aggregation functions, quasi–copulas and copulas. Fuzzy Sets and Systems (2013), doi:10.1016/j.fss.2013.01.005

[14] Mayor, G., Torrens, J.: On a class of binary operations: Non–strict Archimedean aggregation functions. In: Proceedings of ISMVL 1988, Palma de Mallorca, Spain, pp. 54–59 (1988)

[15] Mesiar, R., Jágr, V., Juráňová, M., Komorníková, M.: Univariate conditioning of copulas. Kybernetika 44(6), 807–816 (2008)

[16] Nelsen, R.B.: An introduction to copulas, 2nd edn. Springer Series in Statistics. Springer, New York (2006)

[17] Rodríguez–Lallena, J.A., Úbeda–Flores, M.: A new class of bivariate copulas. Statistics & Probability Letters 66, 315–325 (2004)

On the Construction of Semiquadratic Copulas

Tarad Jwaid, Bernard De Baets, and Hans De Meyer

Abstract. We introduce several classes of semiquadratic copulas (i.e. copulas that are quadratic in at least one coordinate of any point of the unit square) of which the diagonal section or the opposite diagonal section are given functions. These copulas are constructed by quadratic interpolation on segments connecting the diagonal (resp. opposite diagonal) of the unit square to the boundaries of the unit square. We provide for each class the necessary and sufficient conditions on a diagonal (resp. opposite diagonal) function and two auxiliary real functions f and g to obtain a copula which has this diagonal (resp. opposite diagonal) function as diagonal (resp. opposite diagonal) section.

1 Introduction

Bivariate copulas (briefly copulas) [10] are binary operations on the unit interval having 0 as absorbing element and 1 as neutral element and satisfying the condition of 2-increasingness, i.e. a copula is a function $C : [0,1]^2 \to [0,1]$ satisfying the following conditions:

1. for all $x \in [0,1]$, it holds that

$$C(x,0) = C(0,x) = 0, \quad C(x,1) = C(1,x) = x;$$

2. for all $x, x', y, y' \in [0,1]$ such that $x \leq x'$ and $y \leq y'$, it holds that

Tarad Jwaid · Bernard De Baets
KERMIT, Department of Mathematical Modelling, Statistics and Bioinformatics,
Ghent University, Coupure links 653, B-9000 Gent, Belgium
e-mail: {tarad.jwaid,bernard.debaets}@ugent.be

Hans De Meyer
Department of Applied Mathematics and Computer Science, Ghent University,
Krijgslaan 281 S9, B-9000 Gent, Belgium
e-mail: hans.demeyer@ugent.be

H. Bustince et al. (eds.), *Aggregation Functions in Theory and in Practise*,
Advances in Intelligent Systems and Computing 228,
DOI: 10.1007/978-3-642-39165-1_8, © Springer-Verlag Berlin Heidelberg 2013

$$V_C([x,x'] \times [y,y']) := C(x,y) + C(x',y') - C(x,y') - C(x',y) \geq 0.$$

$V_C([x,x'] \times [y,y'])$ is called the C-volume of the rectangle $[x,x'] \times [y,y']$. The copulas M and W, defined by $M(x,y) = \min(x,y)$ and $W(x,y) = \max(x+y-1,0)$, are called the Fréchet–Hoeffding upper and lower bounds: for any copula C it holds that $W \leq C \leq M$. A third important copula is the non-singular product copula Π defined by $\Pi(x,y) = xy$.

The diagonal section of a copula C is the function $\delta_C : [0,1] \to [0,1]$ defined by $\delta_C(x) = C(x,x)$. A diagonal function δ is a $[0,1] \to [0,1]$ function that satisfies the following conditions:

(D1) $\delta(0) = 0$, $\delta(1) = 1$;
(D2)   for all $x \in [0,1]$, it holds that $\delta(x) \leq x$;
(D3) δ is increasing;
(D4) δ is 2-Lipschitz continuous, i.e. for all $x,x' \in [0,1]$, it holds that

$$|\delta(x') - \delta(x)| \leq 2|x' - x|.$$

Note that (D4) implies that δ is absolutely continuous, and hence differentiable almost everywhere. The diagonal section δ_C of a copula C is a diagonal function. Conversely, for any diagonal function δ there exists at least one copula C with diagonal section $\delta_C = \delta$. For example, the copula K_δ, defined by $K_\delta(x,y) = \min(x,y,(\delta(x) + \delta(y))/2)$, is the greatest symmetric copula with a given diagonal section δ [11] (see also [5, 7]). Moreover, the Bertino copula B_δ defined by

$$B_\delta(x,y) = \min(x,y) - \min\{t - \delta(t) \mid t \in [\min(x,y), \max(x,y)]\},$$

is the smallest copula with a given diagonal section δ. Note that B_δ is symmetric.

Similarly, the opposite diagonal section of a copula C is the function $\omega_C : [0,1] \to [0,1]$ defined by $\omega_C(x) = C(x,1-x)$. An opposite diagonal function [2] is a function $\omega : [0,1] \to [0,1]$ that satisfies the following conditions:

(OD1)   for all $x \in [0,1]$, it holds that $\omega(x) \leq \min(x,1-x)$;
(OD2) ω is 1-Lipschitz continuous, i.e. for all $x,x' \in [0,1]$, it holds that

$$|\omega(x') - \omega(x)| \leq |x' - x|.$$

Note that (OD2) implies that ω is absolutely continuous, and hence differentiable almost everywhere. The opposite diagonal section ω_C of a copula C is an opposite diagonal function. Conversely, for any opposite diagonal function ω, there exists at least one copula C with opposite diagonal section $\omega_C = \omega$. For instance, the copula F_ω defined by

$$F_\omega(x,y) = \max(x+y-1,0) + \min\{\omega(t) \mid t \in [\min(x,1-y), \max(x,1-y)]\},$$

is the greatest copula with opposite diagonal section ω [2, 9].

Diagonal functions and opposite diagonal functions have been used recently to construct several classes of copulas [1, 2, 3, 4, 5, 6, 7, 12].

Copulas with a given diagonal section are important tools for modelling upper (λ_U) and lower (λ_L) tail dependence [1, 3], which can be expressed as

$$\lambda_U = 2 - \delta_C'(1^-) \quad \text{and} \quad \lambda_L = \delta_C'(0^+).$$

On the other hand, copulas with a given opposite diagonal section are important tools for modelling upper-lower (λ_{UL}) and lower-upper (λ_{LU}) tail dependence [2], which can be expressed as

$$\lambda_{UL} = 1 + \omega_C'(1^-) \quad \text{and} \quad \lambda_{LU} = 1 - \omega_C'(0^+).$$

The above tail dependences are used in the literature to model the dependence between extreme events [13].

This paper is organized as follows. In the next section we introduce lower, upper, horizontal and vertical semiquadratic functions with a given diagonal section and characterize the corresponding classes of copulas. In Section 3 we introduce in a similar way lower-upper, upper-lower, horizontal and vertical semiquadratic functions with a given opposite diagonal section and characterize the corresponding classes of copulas. Finally, some conclusions are given.

2 Semiquadratic Copulas with a Given Diagonal Section

2.1 Lower and Upper Semiquadratic Copulas with a Given Diagonal Section

Lower (resp. upper) semiquadratic copulas with a given diagonal section are constructed by quadratic interpolation on segments connecting the diagonal of the unit square to the left (resp. right) and lower (resp. upper) boundary of the unit square. The quadratic interpolation scheme for theses classes is depicted in Fig. 1.

For any two functions $f, g :]0, 1] \rightarrow \mathbb{R}$ that are absolutely continuous and satisfy

$$\lim_{\substack{x \to 0 \\ 0 \leq y \leq x}} y(x - y)f(x) = 0 \quad \text{and} \quad \lim_{\substack{y \to 0 \\ 0 \leq x \leq y}} x(y - x)g(y) = 0, \tag{1}$$

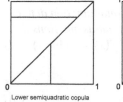 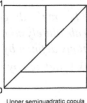 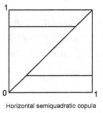 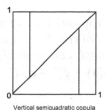

Lower semiquadratic copula · Upper semiquadratic copula · Horizontal semiquadratic copula · Vertical semiquadratic copula

Fig. 1 Semiquadratic copulas with a given diagonal section

and any diagonal function δ, the function $C_{l,\delta}^{f,g} : [0,1]^2 \to \mathbb{R}$ defined by

$$C_{l,\delta}^{f,g}(x,y) = \begin{cases} \dfrac{x}{y}\delta(y) - x(y-x)g(y) & ,\text{if } 0 < x \leq y, \\[3mm] \dfrac{y}{x}\delta(x) - y(x-y)f(x) & ,\text{if } 0 < y \leq x, \end{cases} \tag{2}$$

with $C_{l,\delta}^{f,g}(t,0) = C_{l,\delta}^{f,g}(0,t) = 0$ for all $t \in [0,1]$, is well defined. Note that the limit conditions on f and g ensure that $C_{l,\delta}^{f,g}$ is continuous. The function $C_{l,\delta}^{f,g}$ is called *a lower semiquadratic function with diagonal section* δ since it satisfies $C_{l,\delta}^{f,g}(t,t) = \delta(t)$ for all $t \in [0,1]$, and since it is quadratic in x on $0 \leq x \leq y \leq 1$ and quadratic in y on $0 \leq y \leq x \leq 1$. Obviously, symmetric functions are obtained when $f = g$. Note that for $f = g = 0$, the definition of a lower semilinear function [5] is retrieved.

We now investigate the conditions to be fulfilled by the functions f, g and δ such that the lower semiquadratic function $C_{l,\delta}^{f,g}$ is a copula. Note that f and g, being absolutely continuous, are differentiable almost everywhere.

Proposition 1. *Let δ be a diagonal function and let f and g be two absolutely continuous functions that satisfy conditions (1). Then the lower semiquadratic function $C_{l,\delta}^{f,g}$ defined by (2) is a copula with diagonal section δ if and only if*

(i) $f(1) = g(1) = 0$,

(ii) $\max\left(f(t) + t|f'(t)|, g(t) + t|g'(t)|\right) \leq \left(\dfrac{\delta(t)}{t}\right)'$,

(iii) $f(t) + g(t) \geq t\left(\dfrac{\delta(t)}{t^2}\right)'$,

for all $t \in\,]0,1]$ where the derivatives exist.

Example 1. Let δ_Π be the diagonal section of the product copula Π, i.e. $\delta_\Pi(t) = t^2$ for all $t \in [0,1]$. Let f and g be defined by $f(t) = g(t) = 1 - t$ for all $t \in\,]0,1]$. One easily verifies that the conditions of Proposition 1 are satisfied and hence, $C_{l,\delta_\Pi}^{f,g}$ is a lower semiquadratic copula with diagonal section δ_Π.

Upper semiquadratic copulas with a given diagonal section can be obtained easily from lower semiquadratic copulas with an appropriate given diagonal section.

Proposition 2. *Let δ be a diagonal function and $\hat{\delta}$ be the diagonal function defined by $\hat{\delta}(x) = 2x - 1 + \delta(1-x)$. Let $\hat{f}$ and $\hat{g}$ be two absolutely continuous functions that satisfy conditions (1) and f and g be two functions defined by $f(x) = \hat{f}(1-x)$ and $g(x) = \hat{g}(1-x)$. The function $C_{u,\delta}^{f,g} : [0,1]^2 \to [0,1]$ defined by*

$$C_{u,\delta}^{f,g}(x,y) = x + y - 1 + C_{l,\hat{\delta}}^{\hat{f},\hat{g}}(1-x,1-y), \tag{3}$$

is a copula with diagonal section δ if and only if

(i) $f(0) = g(0) = 0$,

(ii) $\max\left(f(t) + (1-t)|f'(t)|, g(t) + (1-t)|g'(t)|\right) \leq \left(\frac{t - \delta(t)}{1-t}\right)'$,

(iii) $f(t) + g(t) \geq (1-t)\left(\frac{2t - 1 - \delta(t)}{(1-t)^2}\right)'$,

for all $t \in [0,1[$ where the derivatives exist.

The function $C^{f,g}_{u,\delta}$ defined by (3) is called *an upper semiquadratic function with diagonal section δ.*

2.2 Horizontal and Vertical Semiquadratic Copulas with a Given Diagonal Section

Horizontal (resp. vertical) semiquadratic copulas with a given diagonal section are constructed by quadratic interpolation on segments connecting the diagonal of the unit square to the left (resp. lower) and right (resp. upper) boundary of the unit square. The quadratic interpolation scheme for theses classes is depicted in Fig. 1.

For any two functions $f : [0,1[\to \mathbb{R}$ and $g :]0,1] \to \mathbb{R}$ that are absolutely continuous and satisfy

$$\lim_{\substack{y \to 0 \\ 0 \leq x \leq y}} x(y-x)g(y) = 0 \quad \text{and} \quad \lim_{\substack{y \to 1 \\ 0 \leq y \leq x}} (1-x)(y-x)f(y) = 0, \tag{4}$$

and any diagonal function δ, the function $C^{f,g}_{h,\delta} : [0,1]^2 \to \mathbb{R}$ defined by

$$C^{f,g}_{h,\delta}(x,y) = \begin{cases} \dfrac{x}{y}\delta(y) - x(y-x)g(y) & \text{,if } 0 < x \leq y, \\[3mm] y - \dfrac{1-x}{1-y}(y - \delta(y)) - (1-x)(x-y)f(y) & \text{,if } y \leq x < 1, \end{cases} \tag{5}$$

with $C^{f,g}_{h,\delta}(0,t) = 0$ and $C^{f,g}_{h,\delta}(1,t) = t$ for all $t \in [0,1]$, is well defined. Note that the limit conditions on f and g ensure that $C^{f,g}_{h,\delta}$ is continuous. The function $C^{f,g}_{h,\delta}$ is called *a horizontal semiquadratic function with diagonal section δ* since it satisfies $C^{f,g}_{h,\delta}(t,t) = \delta(t)$ for all $t \in [0,1]$, and since it is quadratic in x on $[0,1]^2$. Note that for $f = g = 0$, the definition of a horizontal semilinear function [1] is retrieved.

We now investigate the conditions to be fulfilled by the functions f, g and δ such that the horizontal semiquadratic function $C^{f,g}_{h,\delta}$ is a copula.

Proposition 3. *Let δ be a diagonal function and let f and g be two absolutely continuous functions that satisfy conditions (4). Then the horizontal semiquadratic function $C^{f,g}_{h,\delta}$ defined by (5) is a copula with diagonal section δ if and only if*

(i) $f(0) = g(1) = 0$,

(ii) $f(t) + (1-t)|f'(t)| \leq \left(\frac{t-\delta(t)}{1-t}\right)'$,

(iii) $g(t) + t|g'(t)| \leq \left(\frac{\delta(t)}{t}\right)'$,

(iv) $tf(t) + (1-t)g(t) \geq \frac{t^2-\delta(t)}{t(1-t)}$,

for all $t \in]0, 1[$ where the derivatives exist.

Example 2. Let δ_Π be the diagonal section of the product copula. Let f be defined by $f(t) = t$ for all $t \in [0, 1[$ and g be defined by $g(t) = 1 - t$ for all $t \in]0, 1]$. One easily verifies that the conditions of Proposition 3 are satisfied and hence, $C_{h,\delta_\Pi}^{f,g}$ is a horizontal semiquadratic copula with diagonal section δ_Π.

Vertical semiquadratic copulas with a given diagonal section are trivially obtained as the converses of horizontal semiquadratic copulas with a given diagonal section.

Proposition 4. *Let δ be a diagonal function and let f and g be two absolutely continuous functions that satisfy conditions (4). The function $C_{v,\delta}^{f,g} : [0,1]^2 \to [0,1]$, defined by*

$$C_{v,\delta}^{f,g}(x,y) = C_{h,\delta}^{f,g}(y,x), \tag{6}$$

is a copula with diagonal section δ if and only if $C_{h,\delta}^{f,g}$ is a copula, i.e. under the conditions of Proposition 3.

The function $C_{v,\delta}^{f,g}$ defined by (6) is called *a vertical semiquadratic function with diagonal section δ.*

3 Semiquadratic Copulas with a Given Opposite Diagonal Section

3.1 Lower-Upper and Upper-Lower Semiquadratic Copulas with a Given Opposite Diagonal Section

Lower-upper (resp. upper-lower) semiquadratic copulas with a given opposite diagonal section are constructed by quadratic interpolation on segments connecting the opposite diagonal of the unit square to the left (resp. right) and upper (resp. lower) boundary of the unit square. The quadratic interpolation scheme for these classes is depicted in Fig. 2.

For any two functions $f :]0, 1] \to \mathbb{R}$ and $g : [0, 1[\to \mathbb{R}$ that are absolutely continuous and satisfy

$$\lim_{\substack{y\to 0\\0\leq 1-x\leq y}} (1-y)(x+y-1)f(x) = 0 \quad \text{and} \quad \lim_{\substack{y\to 1\\y\leq 1-x\leq 1}} x(1-x-y)g(y) = 0, \tag{7}$$

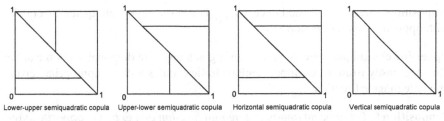

Lower-upper semiquadratic copula   Upper-lower semiquadratic copula   Horizontal semiquadratic copula   Vertical semiquadratic copula

Fig. 2 Semiquadratic copulas with a given opposite diagonal section

and any opposite diagonal function ω, the function $C_{lu,\omega}^{f,g} : [0,1]^2 \to \mathbb{R}$ defined by
$C_{lu,\omega}^{f,g}(x,y) =$

$$
\begin{cases}
\dfrac{x}{1-y}\omega(1-y) - x(1-x-y)g(y) & \text{,if } y \leq 1-x < 1, \\[4mm]
x+y-1+\dfrac{1-y}{x}\omega(x) - (1-y)(x+y-1)f(x) & \text{,if } 0 < 1-y \leq x,
\end{cases}
\tag{8}
$$

with $C_{lu,\omega}^{f,g}(0,t) = 0$ and $C_{lu,\omega}^{f,g}(1,t) = t$ for all $t \in [0,1]$, is well defined. Note that
the limit conditions on f and g ensure that $C_{lu,\omega}^{f,g}$ is continuous. The function $C_{lu,\omega}^{f,g}$
is called *a lower-upper semiquadratic function with opposite diagonal section ω*
since it satisfies $C_{lu,\omega}^{f,g}(t,1-t) = \omega(t)$ for all $t \in [0,1]$, and since it is quadratic in
x on $0 \leq x+y \leq 1$ and quadratic in y on $1 \leq x+y \leq 2$. Note that for $f = g = 0$,
the definition of a lower-upper semilinear function with a given opposite diagonal
section [8] is retrieved.

We now investigate the conditions to be fulfilled by the functions f, g and ω such
that the lower-upper semiquadratic function $C_{lu,\omega}^{f,g}$ is a copula.

Proposition 5. *Let ω be an opposite diagonal function and let f and g be two absolutely continuous functions that satisfy conditions (7). Then the lower-upper semiquadratic function $C_{lu,\omega}^{f,g}$ defined by (8) is a copula with opposite diagonal section ω
if and only if*

(i) $f(1) = g(0) = 0$,

(ii) $f(t) - t|f'(t)| \geq \left(\frac{\omega(t)}{t}\right)'$,

(iii) $g(t) - (1-t)|g'(t)| \geq -\left(\frac{\omega(1-t)}{1-t}\right)'$,

(iv) $f(t) + g(1-t) \leq t\left(\frac{\omega(t)-t}{t^2}\right)'$,

for all $t \in\,]0,1[$ where the derivatives exist.

Example 3. Let ω_Π be the opposite diagonal section of the product copula Π, i.e.
$\omega_\Pi(t) = t(1-t)$ for all $t \in [0,1]$. Let f be defined by $f(t) = 1-t$ for all $t \in\,]0,1]$ and
g be defined by $g(t) = -t$ for all $t \in [0,1[$. One easily verifies that the conditions of

Proposition 5 are satisfied and hence, $C_{lu,\omega_\Pi}^{f,g}$ is a lower-upper semiquadratic copula with opposite diagonal section ω_Π.

Upper-lower semiquadratic copulas with a given opposite diagonal section can be obtained easily from lower-upper semiquadratic copulas with an appropriate given opposite diagonal section.

Proposition 6. *Let ω be an opposite diagonal function and $\hat{\omega}$ be the opposite diagonal function defined by $\hat{\omega}(x) = \omega(1-x)$. Let f and g be two absolutely continuous functions that satisfy conditions (7). The function $C_{ul,\omega}^{f,g} : [0,1]^2 \to [0,1]$, defined by*

$$C_{ul,\omega}^{f,g}(x,y) = C_{lu,\hat{\omega}}^{f,g}(y,x), \tag{9}$$

is a copula with opposite diagonal section ω if and only if

(i) $f(1) = g(0) = 0$,

(ii) $f(t) - t|f'(t)| \geq \left(\frac{\omega(1-t)}{t}\right)'$,

(iii) $g(t) - (1-t)|g'(t)| \geq -\left(\frac{\omega(t)}{1-t}\right)'$,

(iv) $f(1-t) + g(t) \leq (1-t)\left(\frac{1-t-\omega(t)}{(1-t)^2}\right)'$,

for all $t \in]0,1[$ where the derivatives exist.

The function $C_{ul,\omega}^{f,g}$ defined by (9) is called *an upper-lower semiquadratic function with opposite diagonal section ω.*

3.2 Horizontal and Vertical Semiquadratic Copulas with a Given Opposite Diagonal Section

Horizontal (resp. vertical) semiquadratic copulas with a given opposite diagonal section are constructed by quadratic interpolation on segments connecting the opposite diagonal of the unit square to the left (resp. lower) and right (resp. upper) boundary of the unit square. The quadratic interpolation scheme for these classes is depicted in Fig. 2.

For any two functions $f :]0,1] \to \mathbb{R}$ and $g : [0,1[\to \mathbb{R}$ that are absolutely continuous and satisfy

$$\lim_{\substack{y \to 0 \\ 0 \leq 1-x \leq y}} (1-x)(x+y-1)f(y) = 0 \quad \text{and} \quad \lim_{\substack{y \to 1 \\ y \leq 1-x \leq 1}} x(1-x-y)g(y) = 0, \tag{10}$$

and any opposite diagonal function ω, the function $C_{h,\omega}^{f,g} : [0,1]^2 \to \mathbb{R}$ defined by
$C_{h,\omega}^{f,g}(x,y) =$

$$\begin{cases} \dfrac{x}{1-y}\omega(1-y) - x(1-x-y)g(y) & \text{,if } y \leq 1-x < 1, \\[4mm] x+y-1+\dfrac{1-x}{y}\omega(1-y) - (1-x)(x+y-1)f(y) & \text{,if } 0 < y \leq 1-x, \end{cases} \quad (11)$$

with $C_{h,\omega}^{f,g}(0,t) = 0$ and $C_{h,\omega}^{f,g}(1,t) = t$ for all $t \in [0,1]$, is well defined. Note that the limit conditions on f and g ensure that $C_{h,\omega}^{f,g}$ is continuous. The function $C_{h,\omega}^{f,g}$ is called *a horizontal semiquadratic function with opposite diagonal section ω* since it satisfies $C_{h,\omega}^{f,g}(t,1-t) = \omega(t)$ for all $t \in [0,1]$, and since it is quadratic in x on $[0,1]^2$. Note that for $f = g = 0$, the definition of a horizontal semilinear function with a given opposite diagonal section [8] is retrieved.

We now investigate the conditions to be fulfilled by the functions f, g and ω such that the horizontal semiquadratic function $C_{h,\omega}^{f,g}$ is a copula.

Proposition 7. *Let ω be an opposite diagonal function and let f and g be two absolutely continuous functions that satisfy conditions (10). Then the horizontal semiquadratic function $C_{h,\omega}^{f,g}$ defined by (11) is a copula with opposite diagonal section ω if and only if*

(i) $f(1) = g(0) = 0$,

(ii) $f(t) - t|f'(t)| \geq \left(\dfrac{\omega(1-t)}{t} \right)'$,

(iii) $g(t) - (1-t)|g'(t)| \geq -\left(\dfrac{\omega(1-t)}{1-t} \right)'$,

(iv) $(1-t)f(1-t) + tg(1-t) \leq 1 - \dfrac{\omega(t)}{t(1-t)}$,

for all $t \in]0,1[$ where the derivatives exist.

Example 4. Let ω_Π be the opposite diagonal section of the product copula Π. Let f be defined by $f(t) = 1-t$ for all $t \in]0,1]$ and g be defined by $g(t) = -t$ for all $t \in [0,1[$. One easily verifies that the conditions of Proposition 7 are satisfied and hence, $C_{h,\omega_\Pi}^{f,g}$ is a horizontal semiquadratic copula with opposite diagonal section ω_Π.

Vertical semiquadratic copulas with a given opposite diagonal section are trivially obtained as the converses of horizontal semiquadratic copulas with an appropriate given opposite diagonal section.

Proposition 8. *Let ω be an opposite diagonal function and $\hat{\omega}$ be the opposite diagonal function defined by $\hat{\omega}(x) = \omega(1-x)$. Let f and g be two absolutely continuous functions that satisfy conditions (10). The function $C_{v,\hat{\omega}}^{f,g} : [0,1]^2 \to [0,1]$, defined by*

$$C_{v,\hat{\omega}}^{f,g}(x,y) = C_{h,\hat{\omega}}^{f,g}(y,x), \quad (12)$$

is a copula with opposite diagonal section ω if and only if

(i) $f(1) = g(0) = 0$,

(ii) $f(t) + t|f'(t)| \geq \left(\frac{\omega(t)}{t}\right)'$,

(iii) $g(t) + (1-t)|g'(t)| \geq -\left(\frac{\omega(t)}{1-t}\right)'$,

(iv) $tf(1-t) + (1-t)g(1-t) \leq 1 - \frac{\omega(t)}{t(1-t)}$,

for all $t \in \,]0,1[$ *where the derivatives exist.*

The function $C_{v,\tilde{\omega}}^{f,g}$ defined by (12) is called *a vertical semiquadratic function with opposite diagonal section* ω.

4 Conclusions

We have introduced the classes of lower, upper, horizontal and vertical semi-quadratic functions with a given diagonal section as well as the classes of lower-upper, upper-lower, horizontal and vertical semiquadratic functions with a given opposite diagonal section. Moreover, we have identified the necessary and sufficient conditions on a diagonal (resp. opposite diagonal) function and two auxiliary real functions f and g to obtain a copula that has this diagonal (resp. opposite diagonal) function as diagonal (resp. opposite diagonal) section.

References

[1] De Baets, B., De Meyer, H., Mesiar, R.: Asymmetric semilinear copulas. Kybernetika 43, 221–233 (2007)

[2] De Baets, B., De Meyer, H., Úbeda-Flores, M.: Opposite diagonal sections of quasi-copulas and copulas. Internat. J. Uncertainty, Fuzziness and Knowledge-Based Systems 17, 481–490 (2009)

[3] De Baets, B., De Meyer, H., Úbeda-Flores, M.: Constructing copulas with given diagonal and opposite diagonal sections. Communications in Statistics: Theory and Methods 40, 828–843 (2011)

[4] Durante, F., Jaworski, P.: Absolutely continuous copulas with given diagonal sections. Communications in Statistics: Theory and Methods 37, 2924–2942 (2008)

[5] Durante, F., Kolesárová, A., Mesiar, R., Sempi, C.: Copulas with given diagonal sections, novel constructions and applications. Internat. J. Uncertainty, Fuzziness and Knowledge-Based Systems 15, 397–410 (2007)

[6] Durante, F., Kolesárová, A., Mesiar, R., Sempi, C.: Semilinear copulas. Fuzzy Sets and Systems 159, 63–76 (2008)

[7] Durante, F., Mesiar, R., Sempi, C.: On a family of copulas constructed from the diagonal section. Soft Computing 10, 490–494 (2006)

[8] Jwaid, T., De Baets, B., De Meyer, H.: Orbital semilinear copulas. Kybernetika 45, 1012–1029 (2009)

[9] Klement, E., Kolesárová, A.: Extension to copulas and quasi-copulas as special 1-Lipschitz aggregation operators. Kybernetika 43, 329–348 (2005)

[10] Nelsen, R.: An Introduction to Copulas. Springer, New York (2006)

[11] Nelsen, R., Fredricks, G.: Diagonal copulas. In: Beneš, V., Štěpán, J. (eds.) Distributions with given Marginals and Moment Problems, pp. 121–127. Kluwer Academic Publishers, Dordrecht (1997)

[12] Nelsen, R., Quesada-Molina, J., Rodríguez-Lallena, J., Úbeda-Flores, M.: On the construction of copulas and quasi-copulas with given diagonal sections. Insurance: Math. Econ. 42, 473–483 (2008)

[13] Schmidt, R., Stadtmüller, U.: Non-parametric estimation of tail dependence. Scand. J. Statist. 33(2), 307–335 (2006)

Copulas and Self-affine Functions

Enrique de Amo, Manuel Díaz Carrillo, and Juan Fernández Sánchez

Abstract. We characterize self-affine functions whose graphs are the support of a copula using the fact that the functions defined on the unit interval whose graphs support a copula are those that are Lebesgue-measure-preserving. This result allows the computation of the Hausdorff, packing, and box-counting dimensions. The discussion is applied to a classic example such as the Peano curve.

1 Introduction and Preliminaries

The notion of copula was introduced by Sklar [20] when he proved his celebrated theorem in 1959. His aim was to express the relationship between multivariate distribution functions and their univariate margins. For an introduction to copulas, see [16].

Many authors in various fields have drawn attention to methods to generate fractal sets and to describe the concept of "size" for sets in the plane, computing different types of fractal dimensions (in particular, Hausdorff, packing, and box-counting dimensions). Fractal features are often exhibited by measures. This allows the investigation of the connection between fractals and measure-preserving transformations, and the use of methods from Probability Theory and Ergodic Theory.

In addition, some authors describe several ways in which fractal geometry interacts with the notion of copula. Specifically, recent studies have been carried out on

Enrique de Amo
Universidad de Almería, Departamento de Matemáticas
e-mail: edeamo@ual.es

Manuel Díaz Carrillo
Universidad de Granada, Departamento de Análisis Matemático
e-mail: madiaz@ugr.es

Juan Fernández Sánchez
Universidad de Almería, Grupo de Investigación de Análisis Matemático
e-mail: juanfernandez@ual.es

H. Bustince et al. (eds.), *Aggregation Functions in Theory and in Practise*,
Advances in Intelligent Systems and Computing 228,
DOI: 10.1007/978-3-642-39165-1_9, © Springer-Verlag Berlin Heidelberg 2013

examples where the copula has a fractal support, and on the relationship between copulas and measure-preserving transformations on the Borel sets of the unit interval in [3, 6, 7, 12]. Moreover, sufficient conditions for the graph of a function supports a copula are given in [12], and a necessary and sufficient condition is given in [3].

Finally, fractals that are invariant under simple families of transformations include self-similar and self-affine sets. In particular, Kamae [11], using a definition of self-affine function that generalizes that given by Kôno in [13], gives a characterization of them as functions generated by finite automata. Urbański [23] has given conditions to determine dimensions of the graphs of continuous self-affine functions.

In this paper we establish closer relations between the notions of copulas and measure-preserving transformations, self-affine functions whose graphs are supported by a copula, and their applications to computing several fractal dimensions.

We recall some notions and definitions used bellow.

Let $\mathbb{I} := [0, 1]$ be the closed unit interval and let $\mathbb{I}^2$ be the unit square. We can say that a *two-dimensional copula* (or a *copula*, for brevity) is a bidimensional distribution whose restriction to $\mathbb{I}^2$ has its marginal distribution functions uniformly distributed (see [16]). Therefore, each copula C induces a probability measure μ_C on $\mathbb{I}^2$ via the formula

$$\mu_C([a,b] \times [c,d]) = C(b,d) - C(b,c) - C(a,d) + C(a,c)$$

in a similar fashion to joint distribution functions. Through standard measure-theoretical techniques, μ_C can be extended from the semi-ring of rectangles in $\mathbb{I}^2$ to the σ-algebra $\mathscr{B}(\mathbb{I}^2)$ of Borel sets in the unit square. We denote by λ the standard Lebesgue measure on the σ-algebra $\mathscr{B}(\mathbb{I})$ of Borel sets in the unit interval. The *support of a copula C* is the complement of the union of all open subsets of $\mathbb{I}^2$ with μ_C-measure equal to zero.

We use Mandelbrot's original definition of *fractal set* (i.e. a set whose topologial dimension is less than its Hausdorff dimension $\dim_{\mathscr{H}}$). Dimensions of different types are particularly useful in describing the concept of "size" of sets in the plane, in particular, sets of zero Lebesgue measure. For basic properties concerning dimensions (Hausdorff, box-counting, and packing), and other useful notions for expressing the fractal properties of sets, the reader is referred to [8, 9].

Fredricks et al. [10], using an iterated function system, construct the first example of a family of copulas whose supports are fractals. In particular, they give sufficient conditions for the support of a self-similar copula to be a fractal whose Hausdorff dimension is between 1 and 2. The main result of these authors states that for every $s \in [1, 2]$ there exists a copula whose support has dimension equals to s. New results concerning to copulas of fractal support can be found in [3, 4, 5, 21, 22].

Given a measurable space (X, Ω, μ), a measurable function $F : X \to X$ is said to be *measure-preserving* (or F preserves μ) iff $\mu(F^{-1}(A)) = \mu(A)$, for all $A \in \Omega$.

If the σ-algebra Ω is generated by a family Ω_0 that is closed for finite intersections (i.e. a π-system), a sufficient condition for F to be measurable and measure-preserving (see [2, Sec.24]) is that $F^{-1}(A) \in \Omega$ and $\mu\left(F^{-1}(A)\right) = \mu(A)$, for all $A \in \Omega_0$. We are interested in the case $(X, \Omega, \mu) = (\mathbb{I}, \mathscr{B}(\mathbb{I}), \lambda)$.

Many authors have established a correspondence between copulas and measure-preserving transformations f, g on the unit interval via the formula

$$C_{f,g}(u,v) = \lambda\left(f^{-1}[0,u] \cap g^{-1}[0,v]\right)$$

(as we can see in [6, 7, 12, 17, 24]). In [3], the authors investigate the hardest implication in this correspondence; that is, for a given copula C, the goal is to find a pair of measure-preserving transformations (f, g) such that $C = C_{f,g}$.

On the other hand, for the general problem of determining just what functions satisfy that their graph concentrates the mass of a copula, in [7] it is proven that, for every copula obtained as a shuffle of Min, there is a piece-wise linear function whose graph supports the probability mass.

In 1986 Kôno [13] introduced the notion of a self-affine function f of order $\alpha > 0$, whose paradigm is the component functions of the Peano curve.

"Self-affinity" properties have been studied by different authors, with definitions that generalize the Kôno notion using different methods (e.g. Kamae [11] or Peitgen et al. [18]). The main fact is that the graphs of self-affine functions are expected to show strong fractal features. In [14], the author obtains Hausdorff, box, and packing dimensions for graphs of self-affine functions under some conditions. We note that [1, 6, 19] have related results. In particular, with [14, 23], for a given continuous and self-affine function $f : \mathbb{I} \to \mathbb{I}$, a necessary and sufficient condition for the probability distribution $\lambda \circ f^{-1}$ to be absolutely continuous with respect to Lebesgue measure is that the Hausdorff and box dimensions of the graph of f be equal to $2 - \alpha$.

2 Self-affine Functions Whose Graphs Support a Copula

We start with the general problem of determining just what functions in $\mathbb{I}$ have graphs that can concentrate the associated mass with a copula. We recall that [7] gives an answer to this problem using the notion of shuffle of an arbitrary copula. In [3], we have the following general result:

Proposition 1. *Let $f : \mathbb{I} \to \mathbb{I}$ be a Borel measurable function. Then, there exists a copula C whose associated measure μ_C has its mass concentrated in the graph of f (denoted by Γ, $\mu_C(\Gamma(f)) = 1$) if and only if the function f preserves the Lebesgue measure λ.*

For the sake of brevity, we say that f supports C.

Now, we introduce a family of self-affine functions on $\mathbb{I}$. It is adapted from those of Kamae [11] that generalizes the previous concept given by Kôno [13]. See also Peitgen et al. [18].

We use the following notations: For $k \in \mathbb{Z}^+$, let us denote by $[k]$ the set $\{0, 1, 2, \ldots, k-1\}$, and by $[k]^* = \left\{a_1 \ldots a_k : k \in \mathbb{Z}^+, 1 \leq j \leq k, a_j \in [k]\right\}$.

Definition 1. A family of functions $x_0, x_1, \ldots, x_{N-1} : \mathbb{I} \longrightarrow \mathbb{I}$ is called self-affine of order $\alpha \in \,]0,1[$ and with base $m \in \mathbb{Z}^+\backslash\{1\}$ (or simply, (m, α)-self-affine) iff the following conditions are satisfied:

a) $x_j(0), x_j(1) \in \{0,1\}$ for all $j \in [N]$.

b) There is an application $\theta : [N] \to [N]^*$ of constant length m (i.e. $\theta(j)$ has the same number m of terms for all $j \in [N]$) such that, for all $(j,h) \in [N] \times [m]$ and for $t \in \mathbb{I}$, we have

$$x_j\left(\frac{h+t}{m}\right) - x_j\left(\frac{h}{m}\right) = \frac{x_{\theta_h(j)}(t) - x_{\theta_h(j)}(0)}{m^\alpha},$$

where $\theta_h(j)$ is the element in $[k]$ in the h-th position in $\theta(j)$.

We say that each one of the functions x_j is self-affine.

Observe that any self-affine function is continuous, because for all $z, z' \in \mathbb{I}$, $|z - z'| < 2m^{-n}$ implies that $\left|x_j(z) - x_j(z')\right| < m^{-n\alpha}$.

A typical example of a self-affine function is each coordinate function in the Peano curve (see for instance [13, 18]). Let us see the details below.

Example 1. (**Coordinate functions for the Peano curve**). Let us define the operator $k(\beta) = 2 - \beta$, with $\beta \in \{0,1,2\}$. If $t = \sum_{n=1}^{\infty} \frac{t_n}{3^n}$, then

$$\begin{cases} x(t) = \frac{t_1}{3} + \frac{k^{t_2}(t_1)}{3^2} + \frac{k^{t_2+t_4}(t_5)}{3^3} + \cdots \\ y(t) = \frac{k^{t_1}(t_2)}{3} + \frac{k^{t_1+t_3}(t_4)}{3^2} + \frac{k^{t_1+t_3+t_5}(t_6)}{3^3} + \cdots \end{cases}$$

The x coordinate for the Peano curve is self-affine with values:

a) $N = 2;\quad m = 9;\quad \alpha = 1/2$

b) $x_0(t) = x(t),\quad x_1(t) = 1 - x(t)$

c) $\theta(0) = 010010010;\quad \theta(1) = 101101101$.

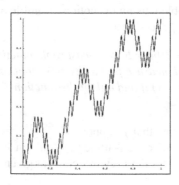

Fig. 1 x coordinate function

In order to characterize the self-affine functions whose graphs can support a copula, we use the set $\mathbb{I}^2 \times [N]$. We can define the next metric on it:

$$d\left((x,y,j),(x',y',j')\right) = \begin{cases} \rho, & \text{if } j \neq j' \\ \sqrt{(x-x')^2 + (y-y')^2}, & \text{if } j = j' \end{cases}$$

with $\rho > \sqrt{2}$. Let $\kappa(\mathbb{I}^2 \times [N])$ be the space of compact sets in $\mathbb{I}^2 \times [N]$, endowed with the Hausdorff metric given by d. Let us consider the attractor F given by the Contraction Mapping Theorem (see for example [9, Chap.9]) for the contraction τ given in the following form. Let us introduce the functions:

$$\tau_{jh} : \mathbb{I}^2 \times \{\theta_h(j)\} \longrightarrow$$
$$\left[\frac{h}{m}, \frac{h+1}{m}\right] \times \left[x_j\left(\frac{h}{m}\right) - \frac{x_{\theta_h(j)}(0)}{m^\alpha}, x_j\left(\frac{h}{m}\right) + \frac{-x_{\theta_h(j)}(0)+1}{m^\alpha}\right] \times \{j\}$$

given by

$$(x,y) \times \{\theta_h(j)\} \rightarrow \left(\frac{h+x}{m}, x_j\left(\frac{h}{m}\right) + \frac{y - x_{\theta_h(j)}(0)}{m^\alpha}\right) \times \{j\};$$

and let us define:

$$\tau : \kappa(\mathbb{I}^2 \times [N]) \longrightarrow \kappa(\mathbb{I}^2 \times [N])$$

$$D \rightarrow \bigcup_{jh} \tau_{jh}(D \cap \mathbb{I}^2 \times \{\theta_h(j)\}) \tag{1}$$

Now, we can characterize the self-affine functions in Definition 1:

Proposition 2. *The function x_j is self-affine if and only if its graph is the intersection of the square $\mathbb{I}^2 \times \{j\}$ and the attractor F, in the space $\kappa(\mathbb{I}^2 \times [N])$, given by (2.1.1).*

Now, we establish the main result

Theorem 1. *A family of self-affine functions can support a copula if and only if $m^{1-\alpha} \in \mathbb{Z}^+$ and, for each $j \in [N]$ and $r \in [m^\alpha]$:*

$$\text{Card}\left\{h : 0 \leq h \leq m-1, x_j\left(\frac{h}{m}\right) - \frac{x_j(0)}{m^\alpha} = \frac{r}{m^\alpha}\right\} = m^{1-\alpha}.$$

Theorem 2. *Let us consider a family of self-affine functions with order of self-affinity α. If their graphs can support a copula, then their Hausdorff dimension is not less than $2 - \alpha$.*

We summarize with this result

Corollary 1. *Let us consider a self-affine function with affinity order α. If its graph Γ supports a copula, then the packing, Hausdorff, and box-counting dimensions for Γ are exactly $2 - \alpha$.*

Acknowledgements. The authors thank the support of the Ministerio de Ciencia e Innovación (Spain), under Research Project No. MTM2011-22394

References

[1] Bedford, T.: The box dimension of self-affine graphs and repellers. Nonlinearity 2, 53–71 (1989)

[2] Billingsley, P.: Probability and measure, 3rd edn. Wiley Series in Probability and Mathematical Statistics. A Wiley-Interscience Publication, John Wiley & Sons, Inc., New York (1995)

[3] de Amo, E., Díaz Carrillo, M., Fernández-Sánchez, J.: Measure-preserving functions and the independence copula. Mediterr. J. Math. 8(4), 431–450 (2011)

[4] de Amo, E., Díaz Carrillo, M., Fernández Sánchez, J.: Copulas and associated fractal sets. J. Math. Anal. Appl. 386, 528–541 (2012)

[5] de Amo, E., Díaz Carrillo, M., Fernández-Sánchez, J., Salmerón, A.: Moments and associated measures of copulas with fractal support. Appl. Math. Comput. 218, 8634–8644 (2012)

[6] Durante, F., Klement, E.P., Quesada-Molina, J.J., Sarkoci, P.: Remarks on two product-like constructions for copulas. Kybernetika 43, 235–244 (2007)

[7] Durante, F., Sarkoci, P., Sempi, C.: Shuffles of copulas. J. Math. Anal. Appl. 352, 914–921 (2009)

[8] Edgar, G.A.: Measure, Topology, and Fractal Geometry. Undergraduate Texts in Mathematics. Springer, Heidelberg(1990)

[9] Falconer, K.J.: Fractal Geometry, 2nd edn. Mathematical Foundations and Applications. John Wiley & Sons (2003)

[10] Fredricks, G.A., Nelsen, R.B., Rodríguez-Lallena, J.A.: Copulas with fractal supports. Insurance Math. Econom. 37(1), 42–48 (2005)

[11] Kamae, T.: A characterization of self-affine functions. Japan J. Appl. Math. 3, 271–280 (1986)

[12] Kolesárová, A., Mesiar, R., Sempi, C.: Measure-preserving transformations, copulas and compatibility. Mediterr. J. Math. 5, 325–339 (2008)

[13] Kôno, N.: On self-affine functions. Japan J. Appl. Math. 3, 259–269 (1986)

[14] Kôno, N.: On self-affine functions II. Japan J. Appl. Math. 5, 441–454 (1988)

[15] McClure, M.: The Hausdorff dimension of Hilbert's coordinate functions. Real Anal. Exchange 24(2), 875–883 (1998/1999)

[16] Nelsen, R.B.: An introduction to copulas, 2nd edn. Springer Series in Statistics. Springer, New York (2006)

[17] Olsen, E.T., Darsow, W.F., Nguyen, B.: Copulas and Markov operators. In: Distributions with Fixed Marginals and Related Topics, Seattle, WA, 1993. IMS Lecture Notes Monogr. Ser., vol. 28, pp. 244–259. Inst. Math. Statist., Hayward (1996)

[18] Peitgen, H.O., Rodenhausen, A., Skordev, G.: Self-affine curves and sequential machines. Real Anal. Exchange 22(2), 446–491 (1996-1997)

[19] Przytycki, F., Urbański, M.: On Hausdorff dimension of some fractal sets. Studia Math. 93, 155–186 (1989)

[20] Sklar, A.: Fonctions de répatition à n dimensions et leurs marges. Publ. Inst. Statidt. Univ. Paris 8, 229–231 (1959)

[21] Trutschnig, W.: On a strong metric on the space of copulas and its induced dependence measure. J. Math. Anal. Appl. 384(2), 690–705 (2011)

[22] Trutschnig, W., Fernández Sánchez, J.: Idempotent and multivariate copulas with fractal support. J. Statist. Plann. Inference 142, 3086–3096 (2012)

[23] Urbański, M.: The Hausdorff dimension of the graphs of continuous self-affine function. Proc. Amer. Math. Soc. 108(4), 921–930 (1990)

[24] Vitale, R.A.: Parametrizing doubly stochastic measures. In: Distributions with Fixed Marginals and Related Topics, Seattle, WA. IMS Lecture Notes Monogr. Ser., vol. 28, pp. 358–364. Inst. Math. Statist., Hayward (1996)

Diagonal Copulas

Radko Mesiar and Jana Kalická

Abstract. Copulas constructed from the respective diagonal sections were deeply studied in 2-dimensional case. For general case of dimension n, only few results are known so far. We recall known results in this area and focus on some constructions for $n > 2$. Several examples considering $n = 3$ are included.

Keywords: copula, diagonal copula, diagonal section, Bertino copula.

1 Introduction

We suppose the readers are familiar with the basics of copula theory, see, e.g., lecture notes [12] or monograph [9]. Recall that for an n-dimensional copula $C : [0,1]^n \to [0,1], n \geq 2$, its diagonal section $\delta_C(x) = C(x, ..., x)$. From the statistical point of view if the considered copula C describes the stochastic dependence structure of random vector $(X_1, ..., X_n)$, where all marginals $X_1, ..., X_n$ are uniformly distributed over the unit interval $[0,1]$, the diagonal section δ_C is (a restriction on $[0,1]$ of) the distribution function of the random variable $Y = \bigvee_{i=1}^{n} X_i$, i.e., δ_C describes the randomness of the extremes of $X_1, ..., X_n$. The aim of this contribution is to discuss the reverse problem, i.e., how to find for an a priori given diagonal section $\delta : [0,1]^n \to [0,1]$ (of some unknown copula) an n-dimensional copula $C : [0,1]^n \to [0,1]$ so that $\delta = \delta_C$. The paper is organized as follows. In the next section, we characterize axiomatically diagonal sections of n-dimensional copulas and we recall some construction methods for 2-dimensional copulas with an a priori given diagonal section. In Section 3, we recall a method for constructing n-dimensional copula from an a priori given diagonal section, and we introduce a

Radko Mesiar · Jana Kalická
Faculty of Civil Engineering, Slovak University of Technology, Radlinského 11
813 68 Bratislava, Slovakia
e-mail: {radko.mesiar,jana.kalicka}@stuba.sk

H. Bustince et al. (eds.), *Aggregation Functions in Theory and in Practise*,
Advances in Intelligent Systems and Computing 228,
DOI: 10.1007/978-3-642-39165-1_10, © Springer-Verlag Berlin Heidelberg 2013

new method of this type. Based on these methods, a parametric class of construction methods is obtain. Moreover, some new problems are opened. In Section 4, some examples are included. Finally, some concluding remarks are added.

2 Diagonal Sections of Copulas and 2-Dimensional Copulas

For a fixed $n \in \{2,3,...\}$, we denote as $\mathscr{C}_n$ the class of all n-dimensional copulas, and as $\mathscr{D}_n$ the class of all diagonal sections of copulas from $\mathscr{C}_n$. It is easy to check that if the function $d : [0,1] \to [0,1]$ is an element of $\mathscr{D}_n$ then it satisfies the next conditions:

(D1) d is non-decreasing,
(D2) $d \leq id_{[0,1]}$,
(D3) $d(1) = 1$,
(D4) d is n-Lipschitz, i.e., $|d(x) - d(y)| \leq n|x - y|$ for all $x,y \in [0,1]$.

The next proposition follows from the construction method proposed in [8], see also Section 3.

Proposition 1. *Let $d : [0,1] \to [0,1]$ be a function and $n \in \{2,3,...\}$ be a fixed dimension. Then d is a diagonal section of some n-dimensional copula, i.e., $d \in \mathscr{D}_n$ if and only if d satisfies conditions (D1) - (D4).*

Observe that classes $\mathscr{C}_n$ and $\mathscr{D}_n$ are convex. $\mathscr{D}_n$ is closed under suprema (infima), and its smallest element is given by $d_n^-(x) = max(0, nx - n + 1)$, while its greatest element is given by $d_n^+(x) = x$. On the other hand, the classes $\mathscr{C}_n$ are not closed under suprema (infima). The greatest element of $\mathscr{C}_n$ is the comonotonicity copula M, $M(x_1,...,x_n) = min(x_1,...,x_n)$, while the smallest element in $\mathscr{C}_n$, $n > 2$ does not exist. In the case of $\mathscr{C}_2$, the smallest element is the countermonotonicity copula W, $W(x_1,x_2) = max(0, x_1 + x_2 - 1)$. We recall some construction methods for copulas from $\mathscr{C}_2$, when an a priori given diagonal section $d \in \mathscr{D}_2$ is known.

Bertino Copulas ([1], [6])

For any $d \in \mathscr{D}_2$, the function $B_d : [0,1]^2 \to [0,1]$ given by

$$B_d(x,y) = \bigvee_{t \in [x \wedge y, x \vee y]} \left(d(t) - (t-x)^+ - (t-y)^+ \right)^+, \tag{1}$$

where $u^+ = max(u,0)$ for $u \in R$ is a copula. Observe that B_d is the smallest copula with diagonal section, and it is simultaneously the smallest quasi-copula ([7], [14]) possessing diagonal section d.

Diagonal Copulas ([5], [13])
For any $d \in \mathscr{D}_2$, the function $K_d : [0,1]^2 \to [0,1]$ given by

$$K_d(x,y) = min\left(x, y, \frac{d(x) + d(y)}{2} \right) \tag{2}$$

is a copula. K_d is the greatest symmetric copula with diagonal section d, but not necessarily the greatest one, see [13]. There are several other constructions of a diagonal copula with an a priori given diagonal section d, however, these methods are not universal, they can be applied to diagonal sections from some special subdomains of $\mathscr{D}_2$. This is, for example, the case of semilinear copulas discussed in [2], biconic copulas studied in [10], several methods introduced in [3], [14] etc. Observe also an important fact based on patchwork techniques introduced in [3], Theorem 1, see also [14].

Proposition 2. *Let A and B be symmetric copulas from $\mathscr{C}_2$ with the same diagonal section $d \in \mathscr{D}_2$. Then the function $C_{A,B} : [0,1]^2 \rightarrow [0,1]$ given by*

$$C_{A,B}(x,y) = \begin{cases} A(x,y) & \text{if } x \leq y, \\ B(x,y) & \text{else,} \end{cases} \tag{3}$$

is a copula form $\mathscr{C}_2$, and $d_{C_{A,B}} = d_A = d_B = d$.

The above proposition allow to introduce for any $d \in \mathscr{D}_2$ two copulas C_{B_d,K_d} and C_{K_d,B_d} with diagonal section d. Note that for any $d \in \mathscr{D}_2$ different from d^+, $\text{card}\left\{K_d, B_d, C_{B_d,K_d}, C_{K_d,B_d}\right\} = 4$.

Remark 1. Proposition 2 can be modified, replacing the symmetry of copulas A, B by their comparability.

3 N-Dimensional Diagonal Copulas

Based on results of Rychlik [15], Jaworski [8] has introduced a method of constructing n-dimensional copulas with an a priori given diagonal section $d \in \mathscr{D}_n, n \geq 2$.

Proposition 3. *For a fixed $n \in \{2,3,...\}$, let $d \in \mathscr{D}_n$. Then the function $J_d : [0,1]^n \rightarrow [0,1]$ given by*

$$J_d(x_1,...,x_n) = \frac{1}{n} \sum_{i=1}^{n} \min\left(f(x_{i+1}),...,f(x_{i+n-1}), d(x_{i+n})\right) \tag{4}$$

where $f : [0,1] \rightarrow [0,1]$ is given by $f(x) = \frac{nx - d(x)}{n-1}$ and $x_j = x_{j-n}$ for $j \in \{n+1,...2n\}$, is a copula, $J_d \in \mathscr{C}_n$.

Observe that for $n = 2, f(x) = 2x - d(x)$, and

$$J_d(x_1,x_2) = \frac{1}{2}\left(\min(2x_2 - d(x_2), d(x_1)) + \min(2x_1 - d(x_1), d(x_2))\right) =$$

$$= \min\left(x_1, x_2, \frac{d(x_1) + d(x_2)}{2}\right) = K_d(x_1,x_2),$$

i.e., copula introduced by Jaworski coincide with diagonal copula K_d.

Considering the generalization of Bertino copula for $n > 2$, for $d \in \mathscr{D}_n$, the smallest n-dimensional quasi-copula $B_d : [0,1]^n \rightarrow [0,1]$ is given by

$$B_d(x_1,...,x_n) = \bigvee_{[\wedge x_i, \vee x_i]} \left(d(t) - (t-x_1)^+ - ... - (t-x_n)^+\right)^+. \tag{5}$$

However then $B_{d^-}(x_1,...x_n) = W(x_1,...x_n) = max\left(0, \sum_{i=1}^{n} x_i - (n-1)\right)$ is not a copula, i.e., (5) is not a universal construction method for n-dimensional copulas.

Open Problem 1. Characterize all diagonal section $d \in \mathscr{D}_n$ such that $B_d \subset \mathscr{C}_n$.

Similarly, one can consider the generalization of diagonal copulas K_d for higher dimensions, i.e., for a fixed $d \in \mathscr{D}_n, n > 2$, to consider a function $K_d : [0,1]^n \to [0,1]$ given by

$$K_d(x_1,...x_n) = min\left(x_1,...,x_n, \frac{d(x_1),...,d(x_n)}{n}\right). \tag{6}$$

It is not difficult to check that K_d is a symmetric quasi-copula for any $d \in \mathscr{D}_n$. However, as shown in [4], K_d is an n-dimensional copula only if $d = d^+$, and then $K_{d^+} = M$ is the greatest copula from $\mathscr{C}_n$. To illustrate this fact, consider the diagonal section $d \in \mathscr{D}_3$ given by $d(x) = x^3$. Then $K_d\left(\frac{1}{2}, \frac{1}{2}, 1\right) = min\left(\frac{1}{2}, \frac{1}{2}, \frac{\frac{1}{8}+\frac{1}{8}+1}{3}\right) = \frac{5}{12}$ and the volume $V_{K_d}\left([0,\frac{1}{2}] \times [\frac{1}{2},1] \times [\frac{1}{2},1]\right) = \frac{1}{2} - 2\frac{5}{12} + \frac{1}{8} = -\frac{5}{24} < 0$, violating the 3-increasingness of K_d. Observe that due to the ordinal sum representation of copulas discussed in [11], we can introduce a corresponding notion of the ordinal sums of diagonal sections, $d = (\langle a_k, b_k, d_k \rangle | k \in \mathscr{K})$, where $\mathscr{K}$ is an index system, $(]a_k, b_k[)_{k \in \mathscr{K}}$ is a disjoint system of open subintervals of $[0,1]$, and $d_k \in \mathscr{D}_n$ for each $k \in \mathscr{K}$. Then $d : [0,1] \to [0,1]$ is given by

$$d(x) = \begin{cases} a_k + (b_k - a_k) d_k\left(\frac{x-a_k}{b_k-a_k}\right) & if \quad x \in]a_k, b_k[\quad for\ some \quad k \in \mathscr{K}, \\ x & else. \end{cases}$$

Note that the corresponding function $f : [0,1] \to [0,1]$ given by $f(x) = \frac{nx-d(x)}{n-1}$ can be written in the form

$$f(x) = \begin{cases} a_k + (b_k - a_k) f_k\left(\frac{x-a_k}{b_k-a_k}\right) & if \quad x \in]a_k, b_k[\quad for\ some \quad k \in \mathscr{K}, \\ x & else. \end{cases}$$

Proposition 4. *For a fixed* $n \in \{2,3,...\}$, *let* $d \in \mathscr{D}_n$ *be an ordinal sum,* $d = (\langle a_k, b_k, d_k \rangle | k \in \mathscr{K})$. *Then* J_d *is an ordinal sum copula* $J_d = (\langle a_k, b_k, J_{d_k} \rangle | k \in \mathscr{K})$.

As we see in Proposition 4, construction (4) and ordinal sum constructions commute.

As we have already mentioned, $\mathscr{D}_n$ and $\mathscr{C}_n$ are convex classes. It is not difficult to check, that construction (4) does not commute with convex sums construction. Consider, for $n = 2, d = \frac{d^- + d^+}{2} \in \mathscr{D}_2$,

$$d(x) = \begin{cases} \frac{x}{2} & if \quad x \leq \frac{1}{2}, \\ \frac{3x-1}{2} & else. \end{cases}$$

Then $J_{d^-}\left(\frac{1}{6},\frac{1}{2}\right)=0, J_{d^+}\left(\frac{1}{6},\frac{1}{2}\right)=\frac{1}{6}=J_d\left(\frac{1}{6},\frac{1}{2}\right)\neq\frac{1}{2}\left(J_{d^-}\left(\frac{1}{6},\frac{1}{2}\right)+J_{d^+}\left(\frac{1}{6},\frac{1}{2}\right)\right)=\frac{1}{12}$, though both copulas J_d and $\frac{1}{2}\left(J_{d^-}+J_{d^+}\right)$ have the same diagonal section. It can be shown, that the only elements of $\mathscr{D}_n$ which do not admit a non-trivial convex sum decomposition are the ordinal sums of type $(\langle a_k,b_k,d^-\rangle\,|k\in\mathscr{K})$. We denote their class by $\mathscr{E}_n$.

Proposition 5. *For a fixed $n\in\{2,3,...\}$, let $d\in\mathscr{D}_n\setminus\mathscr{E}_n$, i.e., $d=\lambda d_1+(1-\lambda)d_2$ for some $d_1,d_2\in\mathscr{D}_n$, $d_1\neq d_2$, $\lambda\in\,]0,1[$. Then $J_{\lambda,d_1,d_2}=\lambda J_{d_1}+(1-\lambda)J_{d_2}$ is a copula from $\mathscr{C}_n$ with diagonal section d.*

Note that for each $d\in\mathscr{D}_n\setminus\mathscr{E}_n$ we can introduce a parametric class of copulas having d as its diagonal section. Indeed, it is enough to consider the class $\left(J_{\lambda,d_1,d_2}\right)_{\lambda\in[0,1]}$, using the same notations as in the last proposition.

Remark 2. For $n=2$, any construction of a binary copula from an a priori given diagonal section $d\in\mathscr{D}_2$ can be "dualized", using the notion of a survival diagonal section.

Indeed, for any copula $C\in\mathscr{C}_2$ with a diagonal section $d\in\mathscr{D}_2$, the corresponding survival copula $\hat{C}$ has a diagonal section $\hat{d}\in\mathscr{D}_2$ given by $\hat{d}(x)=2x-1+d(1-x)$. Then the "dualized" construction is given as follows:

1. consider $d\in\mathscr{D}_2$
2. for $\hat{d}$, construct a copula D with diagonal section $\hat{d}$
3. the survival copula $\hat{D}$ has d as its diagonal section.

This approach cannot be applied when $n>2$, as the the diagonal section $d\in\mathscr{D}_n$ of a copula $C\in\mathscr{C}_n$ is not determining the diagonal section of the survival copula $\hat{C}\in\mathscr{C}_n$ (knowledge of values of C on some subdomains of $[0,1]^n$ is necessary). However considering the construction (4), for $d\in\mathscr{D}_n$ we can define $\hat{d}\in\mathscr{D}_n$ as the diagonal section of the survival copula $\hat{J}_d$. Observe that, for $n>2$, the equality $\widehat{(\hat{d})}=d$ holds only if $d=d^+$.

4   Examples

Example 1. Consider the weakest diagonal section $d^-\in\mathscr{D}_3$. Then J_{d^-} and K_{d^-} are described in Table 1.

Evidently, $J_{d^-}\leq K_{d^-}$. J_{d^-} is singular copula from $\mathscr{C}_3$. Its support consists of 3 segments connecting the point $\left(\frac{2}{3},\frac{2}{3},\frac{2}{3}\right)$ with vertices $(1,0,0)$, $(0,1,0)$ and $(0,0,1)$ and the mass 1 is uniformly distributed over the support of J_{d^-}. On the other hand, the proper quasi-copula K_{d^-} has a negative mass $-\frac{1}{3}$ on each of rectangle $\left[0,\frac{1}{3}\right]\times\left[\frac{2}{3},1\right]\times\left[\frac{2}{3},1\right],\left[\frac{2}{3},1\right]\times\left[0,\frac{1}{3}\right]\times\left[\frac{2}{3},1\right]$ and $\left[\frac{2}{3},1\right]\times\left[\frac{2}{3},1\right]\times\left[0,\frac{1}{3}\right]$.

Example 2. For the product copula $\Pi\in\mathscr{C}_n,n\geq2$, the corresponding diagonal section $d\in\mathscr{D}_n$ is given by $d_\Pi(x)=x^n$. For $0\leq x_1\leq x_2\leq...\leq x_n\leq1$, it holds

Table 1 Formulae for copula J_{d^-} and quasi-copula K_{d^-}, $n = 3$

domain	J_{d^-}	K_{d^-}
$\left[0, \frac{2}{3}\right]^3$	0	0
$\left[\frac{2}{3}, 1\right]^3$	$x_1 + x_2 + x_3 - 2$	$x_1 + x_2 + x_3 - 2$
$\left[0, \frac{2}{3}\right] \times \left[0, \frac{2}{3}\right] \times \left[\frac{2}{3}, 1\right]$	$min\left(\frac{x_1}{2}, \frac{x_2}{2}, x_3 - \frac{2}{3}\right)$	$min\left(x_1, x_2, x_3 - \frac{2}{3}\right)$
$\left[0, \frac{2}{3}\right] \times \left[\frac{2}{3}, 1\right] \times \left[0, \frac{2}{3}\right]$	$min\left(\frac{x_1}{2}, x_2 - \frac{2}{3}, \frac{x_3}{2}\right)$	$min\left(x_1, x_2 - \frac{2}{3}, x_3\right)$
$\left[\frac{2}{3}, 1\right] \times \left[0, \frac{2}{3}\right] \times \left[0, \frac{2}{3}\right]$	$min\left(x_1 - \frac{2}{3}, \frac{x_2}{2}, \frac{x_3}{2}\right)$	$min\left(x_1 - \frac{2}{3}, x_2, x_3\right)$
$\left[0, \frac{2}{3}\right] \times \left[\frac{2}{3}, 1\right] \times \left[\frac{2}{3}, 1\right]$	$min\left(\frac{x_1}{2}, \frac{x_2}{2} - \frac{2}{3}\right) + min\left(\frac{x_1}{2}, x_3 - \frac{2}{3}\right)$	$min\left(x_1, x_2 + x_3 - \frac{4}{3}\right)$
$\left[\frac{2}{3}, 1\right] \times \left[0, \frac{2}{3}\right] \times \left[\frac{2}{3}, 1\right]$	$min\left(\frac{x_2}{2}, x_1 - \frac{2}{3}\right) + min\left(\frac{x_2}{2}, x_3 - \frac{2}{3}\right)$	$min\left(x_2, x_1 + x_3 - \frac{4}{3}\right)$
$\left[\frac{2}{3}, 1\right] \times \left[\frac{2}{3}, 1\right] \times \left[0, \frac{2}{3}\right]$	$min\left(\frac{x_3}{2}, x_1 - \frac{2}{3}\right) + min\left(\frac{x_3}{2}, x_2 - \frac{2}{3}\right)$	$min\left(x_3, x_1 + x_2 - \frac{4}{3}\right)$

$J_{d_\Pi}(x_1, ..., x_n) = \frac{1}{n}\left(x_1^n + \sum_{i=2}^{n} min\left(\frac{nx_1 - x_1^n}{n-1}, x_i^n\right)\right)$. Consider diagonal sections $d_1, d_2 \in \mathscr{D}_3$ given by

$$d_1(x) = \begin{cases} 0 & if \quad x \leq \frac{1}{4}, \\ \frac{x}{2} - \frac{1}{8} & if \quad \frac{1}{4} \leq x \leq \frac{3}{4}, \\ 3x - 2 & else. \end{cases}$$

and

$$d_2(x) = \begin{cases} 2x^3 & if \quad x \leq \frac{1}{4}, \\ 2x^3 - \frac{x}{2} + \frac{1}{8} & if \quad \frac{1}{4} \leq x \leq \frac{3}{4}, \\ 2x^3 - 3x + 2 & else. \end{cases}$$

Then $\frac{d_1 + d_2}{2} = d_\Pi$ and thus the copula $\frac{1}{2}\left(J_{d_1} + J_{d_2}\right)$ has d_Π as its diagonal section.

Example 3. Define a mapping $s: \mathscr{D}_3 \to \mathscr{D}_3$ by $s(d) = \hat{d}$, where for a diagonal section $d \in \mathscr{D}_3$, $\hat{d}$ is the diagonal section of the survival copula $\hat{J}_d$. Then $s(d) = \hat{J}_d(x, x, x) = V_{J_d}\left([1 - x, 1]^3\right) = 1 - 3(1 - x) + 3J_d(1 - x, 1 - x, 1 - x) - J_d(1 - x, 1 - x, 1 - x) = 3x - 2 + 3\frac{1 - x + d(1 - x)}{2} - d(1 - x) = \frac{3x - 1 + d(1 - x)}{2}$. Then $s(d) = d$ if and only if $d = d^+$, i.e., $d(x) = x, x \in [0, 1]$. For the next iterations we have:

$$s^2(d)(x) = \frac{3x + d(x)}{4},$$

$$s^3(d)(x) = \frac{9x - 1 + d(1 - x)}{8},$$

$$s^4(d)(x) = \frac{15x + d(x)}{16},$$

In general, for $k = 1, 2, ...,$

$$s^{2k-1}(d)(x) = \frac{(1 + 2^{2k-1})x - 1 + d(1 - x)}{2^{2k-1}},$$

$$s^{2k}(d)(x) = \frac{\left(2^k - 1\right)x + d(x)}{2^k},$$

and for each $d \in \mathscr{D}_3$ it holds $\lim_{n \to \infty} s^n(d) = d^+$. Then for the corresponding copulas it holds $\lim_{n \to \infty} J_{s^n(d)} = M$.

5 Concluding Remarks

We have opened the problem of constructing n-dimensional copulas with a pre-described diagonal section, with the stress on higher dimensions, i.e., $n \in \{3, 4, \dots\}$. Though there are some similarities with well developed case $n = 2$, several techniques cannot be used for higher dimensions. Especially, there is no universal construction leading to a smallest copula having a given diagonal section. Obviously, this problem is related to the fact that, for $n > 2$, there is no smallest copula in $\mathscr{C}_n$. For the future investigation in this domain, we aim to focus on extension of particular methods known for the case $n = 2$, starting from a diagonal section $d \in \mathscr{D}_n$ with some specific properties, such as semilinear copulas [2] or biconic copulas [10] in the 2-dimensional case.

Acknowledgements. The research summarized in this paper was supported by the Grants VEGA 1/0143/11 and APVV–0073–10.

References

[1] Bertino, S.: Sulla dissomiglianza tra mutabili cicliche. Metron 35, 53–88 (1977)

[2] De Baets, B., De Meyer, H., Mesiar, R.: Asymmetric semilinear copulas. Kybernetika 43(2), 221–233 (2007)

[3] Durante, F., Kolesárová, A., Mesiar, R., Sempi, C.: Copulas with given values on a horizontal and a vertical section. Kybernetika 43(2), 209–220 (2007)

[4] Erdely, A., González Barrios, J., Hernández Cedillo, M.M.: Frank's condition for multivariate Archimedean copulas (submitted)

[5] Fredricks, G.A., Nelsen, R.B.: Copula constructed from diagonal section. In: Distributions with Given Marginals and Moment Problems, pp. 129–136. Kluwer, Dordrecht (1997)

[6] Fredricks, G.A., Nelsen, R.B.: The Bertino family of copulas. In: Distributions with Given Marginals and Statistical Modelling, pp. 81–91. Kluwer, Dordrecht (2002)

[7] Genest, C., Quesada Molina, J.J., Rodríguez Lallena, J.A., Sempi, C.: A characterization of quasi-copulas. Journal of Multivariate Analysis 69, 193–205 (1999)

[8] Jaworski, P.: On copulas and their diagonals. Information Sciences 179, 2863–2871 (2009)

[9] Joe, H.: Multivariate Models and Dependence Concepts. Chapman & Hall, New York (1997)

[10] Jwaid, T., De Baets, B., De Meyer, H.: Biconic aggregation functions. Information Sciences 187, 129–150 (2011)

[11] Mesiar, R., Sempi, C.: Ordinal sums and idempotents of copulas. Aequationes Mathematics 79(1-2), 39–52 (2010)

[12] Nelsen, R.B.: An introduction to copulas, 2nd edn. Springer Series in Statistics. Springer, New York (2006)
[13] Nelsen, R.B., Fredricks, G.A.: Diagonal copulas. In: Distributions with Given Marginals and Moment Problems, pp. 121–127. Kluwer, Dordrecht (1997)
[14] Nelsen, R.B., Quesada Molina, J.J., Rodríguez Lallena, J.A., Úbeda Flores, M.: On the construction copulas and quasi-copulas with given diagonal sections. Insur. Math. Econ. 42, 473–483 (2008)
[15] Rychlik, T.: Distribution and expectations of order statistics for possibly depend random variables. Journal of Multivariate Analysis 48, 31–42 (1994)

R Package to Handle Archimax or Any User-Defined Continuous Copula Construction: *acopula*

Tomáš Bacigál

Abstract. We introduce *acopula* package (run under R) that aims at researchers as well as practitioners in the field of modelling stochastic dependence. Description of tools with examples are given, namely several probability related functions, estimation and testing procedures, and two utility functions.

1 Introduction

Copula is a function that can combine any univariate cumulative distribution functions to form a joint distribution function of a random vector. Copula itself is a joint distribution function with uniform marginals. Since the turn of century when copulas began to attract attention of masses, several software tools arose. The first public yet commercial to mention was EVANESCE library [7] included in FinMetrics extension to S programming environment (predecessor of R), that provided a rich battery of copula classes, though only bivariate. With emergence of R (free software environment for statistical computing and graphics, [11]) there came open-source packages like *copula* [8] (recently incorporating *nacopula*) and *CDVine* [4] with successor *VineCopula*, that are still under vivid development. For further reading about recent copula software see, e.g., [1].

Here we introduce an R package that extends current offerings on the one hand by class of *Archimax* copulas [5] and on the other by several handy tools to test, modify, manipulate and inference from them and *arbitrary* user-defined continuous copulas, thus making copulas ready for *application*. That explains the initial letter of the package name. In the next section particular functions are detailed with the help of examples.

Tomáš Bacigál
Slovak University of Technology in Bratislava, Slovak Republic
e-mail: tomas.bacigal@stuba.sk

H. Bustince et al. (eds.), *Aggregation Functions in Theory and in Practise*,
Advances in Intelligent Systems and Computing 228,
DOI: 10.1007/978-3-642-39165-1_11, © Springer-Verlag Berlin Heidelberg 2013

2   Manual to the Package

To the date of this paper submission the package is available in single text file[1] that can be sourced to R workspace via command line (console) by typing

```
> source('acopula.r')
```

assumed the working directory has been set. Notice here the initial character > indicating the console awaits new command and as long as it is not finished, every new line starts with +. Any other lines will signify console output. However the user need not operate from console directly, instead there exist many front-ends (editors, integrated development environments) such as *RStudio* that simplify overall programming and code manipulation.

Coming back to the topic, the package is expected to be published in R Comprehesive Archive Network[2] (CRAN) during this year, so it can be installed

```
> install.package("acopula")
```

and loaded

```
> library(acopula)
```

with documentation at hand.

2.1   Definition Lists

Structure of the program is relatively simple, does NOT use object-oriented S4 classes and is comprehensible from the source code accompanied by explanation notes, so that even inexperienced user can, e.g., track erroneous behaviour if any occurs. Also it does not depend on any additional packages.

Every parametric class/family of copulas is defined within a list, either by its generator (in case of Archimedean copulas), Pickand's dependence function (Extreme-Value copulas) or directly by cumulative distribution function (CDF) with/or its density. Example of one such definition list follows[3] for generator of Gumbel-Hougaard family of Archimedean copulas

```
> genGumbel()
$parameters
[1] 4

$pcopula
function (t, pars) exp(-sum((-log(t))^pars[1])^(1/pars[1]))

$gen
function (t, pars) (-log(t))^pars[1]

$gen.der
```

---

[1] Available at www.math.sk/wiki/bacigal

[2] cran.r-project.org/

[3] Output printing is simplified whenever contains irrelevant parts.

```
function (t, pars) -pars[1]*(-log(t))^(pars[1]-1)/t

$gen.der2
function (t, pars) pars[1]*(-log(t))^(pars[1]-2)*(pars[1]-1-log(t))/t^2

$gen.inv
function (t, pars) exp(-t^(1/pars[1]))

$gen.inv.der
function (t, pars) -exp(-t^(1/pars[1]))*t^(1/pars[1]-1)/pars[1]

$gen.inv.der2
function (t, pars)
exp(-t^(1/pars[1]))*t^(1/pars[1]-2)*(pars[1]+t^(1/pars[1])-1)/pars[1]^2

$lower
[1] 1

$upper
[1] Inf

$id
[1] "Gumbel"
```

where, though some items may be fully optional (here \$pcopula and \$id), they can contribute to better performance or transparency. The user is encouraged to define new parametric families of Archimedean copula generator (likewise dependence function or copula in general) according to his/her needs, bounded only by this convention and allowed to add pcopula (stands for probability distribution function or CDF), dcopula (density) and rcopula (random sample generator) items, however compatibility with desired dimension has to be kept in mind. Currently implemented generators can be listed.

```
> ls("package:acopula",pattern="gen")
[1] "genAMH" "genClayton" "generator" "genFrank" "genGumbel" "genJoe" "genLog"
```

Notice the generic function generator which links to specified definition lists.

Similarly, Pickand's dependence functions are defined, namely Gumbel-Hougaard, Tawn, Galambos, Hüsler-Reiss (last three form only bivariate EV), extremal dep. functions and generalized convex combination of arbitrary valid dep. functions (see [10]). So are definition lists available for generic (i.e., not necessarily Archimax) copula, e.g. normal, Farlie-Gumbel-Morgenstern, Plackett and Gumbel-Hougaard parametric family. Their corresponding function names starts with dep and cop, respectively.

As the class of Archimax copulas contains Archimedean and EV class as its special cases, the setting depfu = dep1() and genrator = genLog() can distinguish them, respectively.

Because there are not many dependence function parametric families capable of producing more-than-2-dimensional EV or even Archimax copula, the (generalized) convex combination may come useful, e.g., in partition-based approach introduced by [2]. Thus we get special parametric class depGCC(ldepPartition3D(),dim=3) with 3×5 parameters leeding to 3-dimensional copula.

Any definition list item can be replaced already during the function call as shown in the next subsections. Thus one can set starting value of parameter(s) and their range in estimation routine, for instance.

2.2 Probability Functions

First thing one would expect from a copula package is to obtain a value of desired copula in some specific point. To show variability in typing commands, consider again Gumbel-Hougaard copula with parameter equal to 3.5 in point (0.2,0.3). Then the following commands give the same result.

```
> pCopula(data=c(0.2,0.3),generator=genGumbel(),gpars=3.5)
> pCopula(data=c(0.2,0.3),generator=genGumbel(parameters=3.5))
> pCopula(data=c(0.2,0.3),generator=generator("Gumbel"),gpars=3.5)
> pCopula(data=c(0.2,0.3),generator=generator("Gumbel",parameters=3.5))
> pCopula(data=c(0.2,0.3),copula=copGumbel(),pars=3.5)
> pCopula(data=c(0.2,0.3),copula=copGumbel(parameters=3.5))
> pCopula(data=c(0.2,0.3),generator=genLog(),depfun=depGumbel(),dpars=3.5)
> pCopula(data=c(0.2,0.3),generator=genLog(),depfun=depGumbel(parameters=3.5))
[1] 0.1723903
```

If we need probabilities that a random vector would not exceed several points, those can be supplied to data in rows of matrix or data frame.

Conversely, given an incomplete point and a probability, the corresponding quantile emerge.

```
> pCopula(c(0.1723903,0.3),gen=genGumbel(),gpar=3.5,quantile=1)
> pCopula(c(NA,0.3),gen=genGumbel(),gpar=3.5,quan=1,prob=0.1723903)
> qCopula(c(0.3),quan=1,prob=0.1723903,gen=genGumbel(),gpar=3.5)
[1] 0.1999985
```

Conditional probability $P(X < x|Y = y)$ of a random vector (X,Y) has similar syntax.

```
> cCopula(c(0.2,0.3),conditional.on=2,gen=genGumbel(),gpar=3.5)
[1] 0.2230437
> qCopula(c(0.3),quan=1,prob=0.2230437,cond=c(2),gen=genGumbel(),gpar=3.5)
[1] 0.200005
```

Sometimes the density of a copula is of interest, perhaps for visualisation purposes, such as in the following example

```
x <- seq(0,1,length.out=30)
y <- seq(0,1,length.out=30)
z <- dCopula(expand.grid(x,y),generator=genGumbel(),gpars=3.5)
dim(z) <- c(30,30)
persp(x,y,z)
```

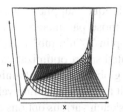

where instead of `persp` from package *graphics* a more impressive output is given by package *rgl* with function `persp3D`.

If definition lists do not contain explicit formulas for (constructing) density, the partial derivatives are approximated linearly. This is mostly the case with 3- and more-dimensional copulas.

Sampling from the copula is, unsurprisingly, also provided.

```
sample <- rCopula(n=1000,dim=2, generator=genGumbel(), gpars=3.5)
plot(sample)
```

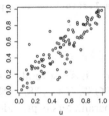

Sometimes no assumption about parametric family of copula is made, instead an empirical distribution is of more interest. Then for a given data, say, the previous random sample, one may ask for value of empirical copula in specific point(s) and more easily in the points of its discontinuity.

```
> pCopulaEmpirical(c(0.2,0.3),base=sample)
[1] 0.14
> empcop <- pCopulaEmpirical(sample)
> scatterplot3d::scatterplot3d(cbind(sample,empcop),type="h",angle=70)
```

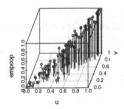

2.3 Estimation

Currently, there are two universal methods for parameters estimation implemented in the package (named `technique`): "ML", maximum (pseudo)likelihood method employing copula density, and "LS", least squares method minimizing distance to empirical copula. Each 'technique' supplies function to perform optimization procedure over, thus finding those parameters that correspond to an optimum. The 'procedures' are three: "optim", "nlminb" and "grid". First two are system native, based on well-documented smart optimization methods, the third one uses brute force to get approximate global maximum/minimum and can be useful with multi-parameter copulas, at least to provide starting values for the other two 'procedures'. The next few examples sketch various options one has got for copula fitting.

```
> eCopula(sample,gen=genClayton(),dep=depGumbel(),
+ technique="ML",procedure="optim",method="L-BFGS-B")
generator parameters:  0.09357958
   depfun parameters:  3.52958
   ML function value:  82.63223
   convergence code:  0
> eCopula(sample,gen=genClayton(),dep=depGumbel(),tech="ML",proc="nlminb")
generator parameters:  0.09183014
   depfun parameters:  3.533706
   ML function value:  82.63228
   convergence code:  0
> eCopula(sample,gen=genClayton(),dep=depGumbel(), tech="ML",proc="grid",
+ glimits=list(c(0),c(5)),dlimits=list(c(1),c(10)),pgrid=10)
generator parameters:  0.5555556
   depfun parameters:  3
   ML function value:  80.63322
   convergence code:
```

In addition, "optim" procedure has several methods to choose from: "L-BFGS-B", "Nelder-Mead", "BFGS", "CG", "SANN", "Brent".

So far, no precision for copula parameters is provided.

2.4 Testing

Having set of observations, it is often of great interest to test whether the estimated copula suffices to describe dependence structure in the data. For this purpose many goodness-of-fit tests were proposed, yet the principle remains to use different criterion than was employed with estimation of the copula parameters. Here we implement one of the 'blanket' tests described in [6] that is based on Kendall's transform. In the example below normal copula is tested on the Gumbel copula sample data.

```
> gCopula(sample,cop=copNormal(),
+ etechnique="ML",eprocedure="optim",ncores=1,N=100)
Loading required package: mvtnorm
   |============================================================| 100%

        Blanket GOF test based on Kendall's transform

statistic        q95    p.value
0.1195500 0.1658125 0.1800000
-----------------------------
data:  sample
copula:  normal
estimates:
      pars      fvalue
 0.9155766 80.3420886
```

Although the p-value does not lead to rejection of the copula adequacy, its low value and small data length arouse suspicion. As for the other arguments, N sets number of bootstrap cycles whereas their parallel execution can be enabled by setting number of processor cores in ncores. Package *mvtnorm* has been loaded to assist with simulation from normal copula, and when missing, internal but slower routine would be performed instead.

The traditional parametric bootstrap-based procedure to approximate p-value, when theoretical probability distribution of the test statistic is unknown, is reliable yet computationally very exhaustive, therefore recently a method based on multiplier central limit theorem and proposed by [9] becomes popular with large-sample testing. Its implementation to testing goodness of parametric copula fit is scheduled for next package update. Nevertheless, the multiplier method takes part here in another test comparing two empirical copulas, i.e. dependence structure of two data sets, see [12]. In the following example, random sample of the above Gumbel-Hougaard copula is tested for sharing common dependence structure with sample simulated from Clayton copula, parameter of which corresponds to the same Kendall's rank correlation ($\tau = 0.714$)

```
> sampleCl <- rCopula(n=100,dim=2,generator=genClayton(),gpars=5)
> gCopula(list(sample,sampleCl),ncores=1,N=100)
  |===========================================================| 100%

Test of equality between 2 empirical copulas

 statistic        q95    p.value
0.09791672 0.52893392 0.66000000
----------------------------
data:  sample sampleCl
copula:
estimates:
NULL
```

Obviously, the test fails to distinguish copulas with differing tail dependence, at least having small and moderate number of observations, however it is sensitive enough to a difference in rank correlation.

The last procedure to mention checks the properties of a d-dimensional copula ($d \geq 2$), that is, being d-increasing as well as having 1 as neutral element and 0 as annihilator. The purpose is to assist approval of new copula constructs when theoretical proof is too complicated. The procedure examines every combination of discrete sets of copula parameters, in the very same fashion as within "grid" procedure of eCopula, by computing a) first differences recursively over all dimensions of an even grid of data points,i.e., C-volumes of subcopulas, b) values on the margin where one argument equals zero and c) where all arguments but one equals unity. Then whenever the result is a) negative, b) non-zero or c) other than the one particular argument, respectively, a record is made and first 5 are printed as shown below. In the example we examine validity of an assumed Archimedean copula generated by Gumbel-Hougaard generator family, only with a parameter being out of bounds.

```
> isCopula(generator=genGumbel(lower=0),dim=3,glimits=list(0.5,2),
+ dagrid=10,pgrid=4,tolerance=1e-15)

Does the object appears to be a copula(?):   FALSE

Showing 2 of 2 issues:

  dim property      value gpar
1   2    monot -0.1534827  0.5
2   3    monot -0.1402209  0.5
```

Three parameter values $(0.5, 1, 1.5, 2)$ were used, each supposed copula were evaluated in 10^3 grid nodes, and every violation of copula properties (the most extremal value per dimension and exceeding `tolerance`) were reported. Thus it is seen, that parameter value 0.5 does not result in copula because 3-monotonicity is not fulfilled (negative difference already in the second-dimension run). Note that without redefinition of lower bound the parameter value 0.5 would be excluded from the set of Gumbel-Hougaard copula parameters.

2.5 Utilities

For the *acopula* package to work many utility functions were created during development that were neither available in the basic R libraries nor they were found in contributed package under CRAN. Most of them are hidden within the procedures described above, however the two following are accessible on demand. The first to mention is a linear approximation of partial derivative of any-dimensional function and of any order with specification of increment (theoretically fading to zero) and area (to allow semi-differentiability)

```
> fun <- function(x,y,z) x^2*y*exp(z)
> nderive(fun,point=c(0.2,1.3,0),order=c(2,0,1),difference=1e-04,area=0)
[1] 2.600004
```

whereas the second utility function numerically approximates integration (by trapezoidal rule) such as demonstrated on example of joint standard normal density with zero correlation parameter

```
> nintegrate(function(x,y) mvtnorm::dmvnorm(c(x,y)),
+            lower=c(-5.,-5.),upper=c(0.5,1),subdivisions=30)
[1] 0.5807843
> pnorm(0.5)*pnorm(1)
[1] 0.5817583
```

fine-tuned by number of subdivisions.

3 Conclusion

All the introduced and exemplified procedures are (a) extendible to arbitrary dimension, which is one of the significant contributions of the package. If explicit formulas are unavailable (through definition lists) then numerical approximation does the job. Another significant benefit is brought by (b) conditional probability and quantile function of the copula, as well as estimation methods based on least squares and grid complementing the usual maximum-likelihood method. Together with implementing (c) generalization of Archimedean and Extreme-Value by Archimax class with a (d) construction method of Pickand's dependence function, (e) test of equality between two empirical copulas, (f) numerical check of copula properties useful in new parametric families development, and (g) parallelized goodness-of-fit test based on Kendall's transform, these all (and under one roof) make the package competitive among both proprietary and open-source software tools for copula based analysis, to the date.

Yet because the routines are written solely in R language and rely on no non-standard packages (optionally), some tasks may take longer to perform. Nevertheless the source code is easy to access, understand and modify if necessary.

Future improvement is seen mainly in providing additional methods for parameters estimation (based on various dependence measures) and GoF tests, as well as connecting with other copula packages to simplify practical analysis.

Author appreciates any comments, bug reports or suggestions.

Acknowledgements. The work on this contribution was supported by grants APVV-0073-10, APVV-0496-10, and 1/0143/11.

References

[1] Bacigál, T.: Recent tools for modelling dependence with copulas and R. Forum Statisticum Slovacum 8(1), 62–67 (2012)

[2] Bacigál, T., Mesiar, R.: 3-dimensional Archimax copulas and their fitting to real data. In: COMPSTAT 2012, 20th International Conference on Computational Statistics, Limassol, Cyprus, August 27-31. The International Statistical Institute, pp. 81–88 (2012)

[3] Broy, M.: Software engineering — from auxiliary to key technologies. In: Broy, M., Dener, E. (eds.) Software Pioneers, pp. 10–13. Springer, Heidelberg (2002)

[4] Brechmann, E.C., Schepsmeier, U.: Modeling Dependence with C- and D-Vine Copulas: The R Package CDVine. Journal of Statistical Software 52(3), 1–27 (2013)

[5] Capéraà, P., Fougères, A.-L., Genest, C.: Bivariate distributions with given extreme value attractor. J. Multivariate Anal. 72(1), 30–49 (2000)

[6] Genest, C., Rémillard, B., Beaudoin, D.: Goodness-of-fit tests for copulas: A review and a power study. Insurance: Mathematics and Economics 44, 199–213 (2009)

[7] Insightful Corp.: EVANESCE Implementation in S-PLUS FinMetrics Module (2002), http://faculty.washington.edu/ezivot/book/QuanCopula.pdf (cited February 15, 2013)

[8] Kojadinovic, I., Yan, J.: Modeling Multivariate Distributions with Continuous Margins Using the copula R Package. Journal of Statistical Software 34(9), 1–20 (2010)

[9] Kojadinovic, I., Yan, J., Holmes, M.: Fast large-sample goodness-of-fit for copulas. Statistica Sinica 21(2), 841–871 (2011)

[10] Mesiar, R., Jágr, V.: *d*-Dimensional dependence functions and Archimax copulas (2012) (in press)

[11] R Development Core Team. R: A language and environment for statistical computing. R Foundation for Statistical Computing, Vienna, Austria (2012), http://www.R-project.org/ (cited February 15, 2013)

[12] Rémillard, B., Scaillet, O.: Testing for equality between two copulas. Journal of Multivariate Analysis 100(3), 377–386 (2009)

How to Prove Sklar's Theorem

Fabrizio Durante, Juan Fernández-Sánchez, and Carlo Sempi

Abstract. In this contribution we stress the importance of Sklar's theorem and present a proof of this result that is based on the compactness of the class of copulas (proved via elementary arguments) and the use of mollifiers. More details about the procedure can be read in a recent paper by the authors.

1 Introduction

The concept of copula was introduced by A. Sklar [20] in order to describe in a convenient way the class of distribution functions with given marginals. After Sklar's paper, copulas have been used in the study of the aggregation of information at different levels and under diverse aspects. Specifically, just to refer to a few examples, the following directions have been pursued in the literature.

- Firstly, copulas are useful in order to aggregate (e.g., join) univariate marginals distribution functions into multivariate models of distribution functions that have more flexibility (tail dependency, asymmetry) than standard (e.g., Gaussian) models (see, for instance, [10, 11, 12]).
- The copula structure of a random vector is a key ingredient in order to estimate and make inference about the risk connected with the aggregation of different random variables, especially, when they represent losses and/or assets of financial markets (see, for instance, [3, 7, 18]). In particular, upper and lower bounds

Fabrizio Durante
School of Economics and Management, Free University of Bozen–Bolzano, Italy
e-mail: fabrizio.durante@unibz.it

Juan Fernández-Sánchez
Grupo de Investigación de Análisis Matemático, Universidad de Almería, La Cañada de San Urbano, Almería, Spain
e-mail: juanfernandez@ual.es

Carlo Sempi
Dipartimento di Matematica e Fisica "Ennio De Giorgi", Università del Salento, Lecce, Italy
e-mail: carlo.sempi@unisalento.it

H. Bustince et al. (eds.), *Aggregation Functions in Theory and in Practise*,
Advances in Intelligent Systems and Computing 228,
DOI: 10.1007/978-3-642-39165-1_12, © Springer-Verlag Berlin Heidelberg 2013

for risk measures of random vectors can be derived by using Hoeffding-Fréchet bounds for copulas or similar inequalities (see, for instance, [21]).

- Aggregation of different inputs into a single numerical output in order to handle decisions, possibly in presence of general uncertainty models (see, for instance, [8, 9]) has benefited from the use of copulas, which has allowed robust (i.e., Lipschitzian) aggregation procedures. Moreover, recent trends in non–additive integrals have used copulas in order to aggregate preferences in a convenient way [13, 14].

These new directions in copula theory underline the need for underpinning the basic ideas about copulas and developing theoretical research about the foundations of the concept and its implications in different areas of mathematics.

Here we aim at discussing the first issue about copula theory, namely the representation of random vectors in terms of a copula and the univariate margins. Such result goes under the name of "Sklar's Theorem" since it was discovered by Abe Sklar [20] in its seminal paper of 1959. In particular, we present here a new proof of Sklar's Theorem that underlines some non–standard perspectives on the problem (for more details, see [6]).

2   Sklar's Theorem

Before proceeding we briefly recall the definition of a copula; a d–copula C is the restriction to the unit hypercube $[0,1]^d$ of a distribution function (d.f., for short) on $\overline{\mathbb{R}}^d$ that has uniform univariate margins on $[0,1]^d$. Therefore (a) $C(\mathbf{u}) = 0$ whenever at least one of the components of $\mathbf{u} = (u_1,\dots,u_d)$ equals zero, (b) $C(1,\dots,1,t,1,\dots,1) = t$, when all the components of $\mathbf{u}$ equal 1, with the possible exception of the j–th one and (c) the C volume V_C of every d–box $[\mathbf{a},\mathbf{b}] := [a_1,b_1] \times \cdots \times [a_d,b_d]$ contained in $[0,1]^d$ is non-negative, namely

$$V_C([\mathbf{a},\mathbf{b}]) := \sum_{\mathbf{v}} \mathrm{sign}(\mathbf{v}) C(\mathbf{v}) \geq 0,$$

where the sum is taken over the 2^d vertices $\mathbf{v}$ of the rectangle $[\mathbf{a},\mathbf{b}]$ and

$$\mathrm{sign}(\mathbf{v}) = \begin{cases} 1, & \text{if } v_j = a_j \text{ for an even number of indices,} \\ -1, & \text{if } v_j = a_j \text{ for an odd number of indices.} \end{cases}$$

As a consequence of the definition of copula, one has that every d–copula C satisfies the Lipschitz condition

$$|C(\mathbf{x}) - C(\mathbf{y})| \leq \|\mathbf{x} - \mathbf{y}\|_1,$$

where $\|\cdot\|_1$ is the ℓ^1–norm on $[0,1]^d$. For more details, see the monographs [11, 12, 16].

Nowadays, Sklar's theorem represents the building block of the modern theory of multivariate d.f.'s. It can be formulated as follows.

Theorem 1 (Sklar's Theorem). *Let* $(X_1, \ldots, X_d)$ *be a random vector with joint d.f.* H *and univariate marginals* $F_1, \ldots, F_d$. *Then there exists a copula* $C \colon [0,1]^d \to [0,1]$ *such that, for all* $\mathbf{x} = (x_1, \ldots, x_d) \in \mathbb{R}^d$,

$$H(x_1, \ldots, x_d) = C(F_1(x_1), \ldots, F_d(x_d)).$$

C *is uniquely determined on* $\mathrm{Range}(F_1) \times \cdots \times \mathrm{Range}(F_d)$ *and, hence, it is unique when* $F_1, \ldots, F_d$ *are continuous.*

Because of its importance in applied probability and statistics, Sklar's theorem has received a great deal of attention and has been proved several times (and with different techniques). It was announced, but not proved, by [20], who provided (together with Schweizer) a complete proof in [19].

Other proofs of Sklar's Theorem have been given in the literature. These are based either on analytical arguments, trying to extend a so-called sub-copula to a copula in some ways (see [2, 4]), or on probabilistic techniques, based on the modifications of the probability integral transform (see [5, 15, 17]).

Remark 1. It should be stressed that most of the proofs of Sklar's Theorem have slightly different settings. In the original proof by Schweizer and Sklar [19], the authors considered d.f.'s defined on $\overline{\mathbb{R}}^d$. In most of the probabilistic approaches to Sklar's Theorem [15, 17] (included the above described method), d.f.'s are defined on $\mathbb{R}^d$ (with some suitable margins). Depending on which settings is involved, both definitions have their pros and cons.

Actually, in the case of continuous d.f.'s, Sklar's Theorem admits an easy proof, since the following result holds.

Lemma 1. *For every d–dimensional d.f. H with continuous marginals $F_1, \ldots, F_d$ there exists a unique d–copula C such that, for all $\mathbf{x} = (x_1, \ldots, x_d) \in \mathbb{R}^d$,*

$$H(\mathbf{x}) = C(F_1(x_1), \ldots, F_d(x_d)). \tag{1}$$

Such a C is determined, for all $\mathbf{u} \in {]0,1[}^d$, via the formula

$$C(\mathbf{u}) = H\left(F_1^{[-1]}(u_1), \ldots, F_d^{[-1]}(u_d)\right),$$

where, for $i \in \{1, \ldots, d\}$ $F_i^{[-1]}$ is the quasi–inverse of F_i defined by $F_i^{[-1]}(t) := \inf\{x : F_i(x) \geq t\}$.

The hard part of Sklar's Theorem consists in the extension to the case when at least one of the marginals has a discrete component.

Here we aim at presenting a different proof of Sklar's Theorem following [6]. This proof is based on the compactness of the class of copulas that is usually proved by Ascoli–Arzelà Theorem, which we circumvent by the following result.

Theorem 2. *The set of all copulas $\mathscr{C}_d$ is a compact subset in the class of all continuous functions from $[0,1]^d$ to $[0,1]$.*

Proof. Since $[0,1]^d$ is a metric space under the metric of the norm of uniform convergence, it is enough to show that $\mathscr{C}_d$ is sequentially compact.

Let $(C_n)_n$ be a sequence of copulas in $\mathscr{C}_d$ and let $(\mathbf{x}_j) = (x_j^{(1)}, \ldots, x_j^{(d)})$ be a dense sequence of points in $[0,1]^2$; for instance take the sequence

$$\left(\frac{i_1}{n}, \ldots, \frac{i_d}{n} \right) \qquad (n \in \mathbb{N}; i_1, \ldots, i_d = 0, 1, \ldots, n),$$

and order it according to lexicographical order.

Consider the bounded sequence of real numbers $(C_n(\mathbf{x}_1))$; then there exists a convergent subsequence $C_{1,n}(\mathbf{x}_1)$. For the same reason, the sequence $(C_{1,n}(\mathbf{x}_2))$ contains a convergent subsequence $(C_{2,n}(\mathbf{x}_2))$. Notice that both sequences $(C_{2,n}(\mathbf{x}_1))$ and $(C_{2,n}(\mathbf{x}_2))$ converge. Proceeding in this way one constructs, for every $k \geq 1$, a subsequence $(C_{k,n})$ of $(C_{k-1,n})$ that converges at the points $\mathbf{x}_j$ with $j \leq k$. Consider now the diagonal sequence $(C_{k,k})_{k \in \mathbb{N}}$; this converges at every point of the sequence $(\mathbf{x}_j)$. Define the function $C : (\mathbf{x}_j)_{j \in \mathbb{N}} \to [0,1]$ via

$$C(\mathbf{x}_j) := \lim_{k \to +\infty} C_{k,k}(\mathbf{x}_j).$$

Since $(C_{k,k})_{k \in \mathbb{N}}$ satisfies the Lipschitz condition $|C_{k,k}(\mathbf{x}_i) - C_{k,k}(\mathbf{x}_j)| \leq \|\mathbf{x}_i - \mathbf{x}_j\|_1$, one has, on taking the limit as k goes to $+\infty$,

$$|C(\mathbf{x}_i) - C(\mathbf{x}_j)| \leq \|\mathbf{x}_i - \mathbf{x}_j\|_1,$$

which proves that C is uniformly continuous on the sequence $(\mathbf{x}_n)$; and since this latter is dense in $[0,1]^d$, the definition of C can be extended by continuity to the whole of $[0,1]^d$. It immediately follows from its very definition that C satisfies the boundary conditions of a copula. Consider now the d–box $[\mathbf{a}, \mathbf{b}]$, where the points $\mathbf{a}$ and $\mathbf{b}$ belong to the sequence $(\mathbf{x}_n)$. Then the C–Volume of $[\mathbf{a}, \mathbf{b}]$ is given by

$$\begin{aligned} V_C([\mathbf{a}, \mathbf{b}]) &:= \sum_{\mathbf{v}} \text{sign}(\mathbf{v}) C(\mathbf{v}) \\ &= \lim_{k \to +\infty} \sum_{\mathbf{v}} \text{sign}(\mathbf{v}) C_{k,k}(\mathbf{v}) = \lim_{k \to +\infty} V_{C_{k,k}}([\mathbf{a}, \mathbf{b}]) \geq 0. \end{aligned}$$

Finally, consider a d–box $[\mathbf{a}, \mathbf{b}]$ contained in $[0,1]^d$ and 2^d subsequences $(\mathbf{v}_n^{(s)})$ of $(\mathbf{x}_j)$ $(s = 1, 2, \ldots, 2^d)$ that converge to the vertices $\mathbf{v}^{(s)}$ of $[\mathbf{a}, \mathbf{b}]$. The continuity of C yields

$$V_C([\mathbf{a}, \mathbf{b}]) = \sum_{j=1}^{2^d} \text{sign}(\mathbf{v}^{(s)}) C(\mathbf{v}^{(s)}) = \lim_{n \to +\infty} \sum_{j=1}^{2^d} \text{sign}(\mathbf{v}_n^{(s)}) C(\mathbf{v}_n^{(s)}) \geq 0,$$

which proves that C is indeed a d–copula.   □

3 Proof of Sklar's Theorem by Means of Mollifiers

Here we sketch the main arguments presented in [6].

Let $B_r(\mathbf{a})$ denote the open ball in $\mathbb{R}^d$ of centre $\mathbf{a}$ and radius r and consider the function $\varphi : \mathbb{R}^d \to \mathbb{R}$ defined by

$$\varphi(\mathbf{x}) := k \exp\left(\frac{1}{|\mathbf{x}|^2 - 1}\right) \mathbf{1}_{B_1(\mathbf{0})}(\mathbf{x}),$$

where the constant k is such that the L^1 norm $\|\varphi\|_1$ of φ is equal to 1. Further, for $\varepsilon > 0$, define $\varphi_\varepsilon : \mathbb{R}^d \to \mathbb{R}$ by

$$\varphi_\varepsilon(\mathbf{x}) := \frac{1}{\varepsilon^d} \varphi\left(\frac{\mathbf{x}}{\varepsilon}\right).$$

It is known (see, e.g., [1, Chapter 4]) that φ_ε belongs to $C^\infty(\mathbb{R}^d)$, and that its support is the closed ball $\overline{B}_\varepsilon(\mathbf{0})$. Functions like φ_ε are sometimes called *mollifiers*.

If a d–dimensional d.f. H is given, then the convolution

$$H_n(\mathbf{x}) := \int_{\mathbb{R}^d} H(\mathbf{x} - \mathbf{y})\, \varphi_{1/n}(\mathbf{y})\, d\mathbf{y} = \int_{\mathbb{R}^d} \varphi_{1/n}(\mathbf{x} - \mathbf{y}) H(\mathbf{y})\, d\mathbf{y} \qquad (2)$$

is well defined for every $\mathbf{x} \in \mathbb{R}^d$ and finite; in fact, H_n is bounded below by 0 and above by 1.

The following facts hold.

Lemma 2. *For every d–dimensional d.f. H and for every $n \in \mathbb{N}$, the function H_n defined by (2) is a d–dimensional continuous d.f..*

Lemma 3. *If H is continuous at $\mathbf{x} \in \mathbb{R}^d$, then $\lim_{n \to +\infty} H_n(\mathbf{x}) = H(\mathbf{x})$.*

We are now ready for the final step.

Proof of Sklar's Theorem. We only sketch the main idea, the details being presented in [6]. For any given d.f. H, construct, for every $n \in \mathbb{N}$, the continuous d.f. H_n defined by eq. (2); its marginals $F_{n,1}, \ldots, F_{n,d}$ are also continuous; therefore, Lemma 1 ensures that there exists a d–copula C_n such that

$$H_n(\mathbf{x}) = C_n\left(F_{n,1}(x_1), \ldots, F_{n,d}(x_d)\right).$$

Because of the compactness of $\mathscr{C}_d$, there exists a subsequence $(C_{n(k)})_{k \in \mathbb{N}} \subset (C_n)_{n \in \mathbb{N}}$ that converges to a copula C. Now, the thesis follows by showing that such a C is actually a possible copula of H. $\qquad \square$

Acknowledgements. The first author acknowledges the support of Free University of Bozen-Bolzano, School of Economics and Management, via the project "Risk and Dependence". The first and second author have been supported by the Ministerio de Ciencia e Innovación (Spain) under research project MTM2011-22394.

References

[1] Brezis, H.: Functional analysis, Sobolev spaces and partial differential equations. Universitext. Springer, New York (2011)

[2] Carley, H., Taylor, M.D.: A new proof of Sklar's theorem. In: Cuadras, C.M., Fortiana, J., Rodriguez-Lallena, J.A. (eds.) Distributions with given Marginals and Statistical Modelling, pp. 29–34. Kluwer Acad. Publ., Dordrecht (2002)

[3] Cherubini, U., Mulinacci, S., Gobbi, F., Romagnoli, S.: Dynamic Copula methods in finance. Wiley Finance Series. John Wiley & Sons Ltd., Chichester (2012)

[4] de Amo, E., Díaz-Carrillo, M., Fernández-Sánchez, J.: Characterization of all copulas associated with non-continuous random variables. Fuzzy Sets and Systems 191, 103–112 (2012)

[5] Deheuvels, P.: Caractérisation complète des lois extrêmes multivariées et de la convergence des types extrêmes. Publ. Inst. Stat. Univ. Paris 23(3-4), 1–36 (1978)

[6] Durante, F., Fernández-Sánchez, J., Sempi, C.: Sklar's theorem obtained via regularization techniques. Nonlinear Anal. 75(2), 769–774 (2012)

[7] Embrechts, P., Puccetti, G.: Risk aggregation. In: Jaworski, P., Durante, F., Härdle, W., Rychlik, T. (eds.) Copula Theory and its Applications. Lecture Notes in Statistics - Proceedings, vol. 198, pp. 111–126. Springer, Heidelberg (2010)

[8] Figueira, J., Greco, S., Ehrgott, M.: Multiple Criteria Decision Analysis: State of the Art Surveys. Springer, Boston (2005)

[9] Grabisch, M., Marichal, J.L., Mesiar, R., Pap, E.: Aggregation functions. In: Encyclopedia of Mathematics and its Applications (No. 127). Cambridge University Press, New York (2009)

[10] Jaworski, P., Durante, F., Härdle, W. (eds.): Copulae in Mathematical and Quantitative Finance. Lecture Notes in Statistics - Proceedings. Springer, Heidelberg (2013)

[11] Jaworski, P., Durante, F., Härdle, W., Rychlik, T. (eds.): Copula Theory and its Applications. Lecture Notes in Statistics - Proceedings, vol. 198. Springer, Heidelberg (2010)

[12] Joe, H.: Multivariate models and dependence concepts. In: Monographs on Statistics and Applied Probability, vol. 73. Chapman & Hall, London (1997)

[13] Klement, E.P., Kolesárová, A., Mesiar, R., Stupnanová, A.: A generalization of universal integrals by means of level dependent capacities. Knowledge-Based Systems 38, 14–18 (2013)

[14] Klement, E.P., Mesiar, R., Pap, E.: A universal integral as common frame for Choquet and Sugeno integral. IEEE Trans. Fuzzy Systems 18(1), 178–187 (2010)

[15] Moore, D.S., Spruill, M.C.: Unified large-sample theory of general chi-squared statistics for tests of fit. Ann. Statist. 3, 599–616 (1975)

[16] Nelsen, R.B.: An introduction to copulas, 2nd edn. Springer Series in Statistics. Springer, New York (2006)

[17] Rüschendorf, L.: On the distributional transform, Sklar's Theorem, and the empirical copula process. J. Statist. Plan. Infer. 139(11), 3921–3927 (2009)

[18] Salvadori, G., De Michele, C., Durante, F.: On the return period and design in a multivariate framework. Hydrol. Earth Syst. Sci. 15, 3293–3305 (2011)

[19] Schweizer, B., Sklar, A.: Operations on distribution functions not derivable from operations on random variables. Studia Math. 52, 43–52 (1974)

[20] Sklar, A.: Fonctions de répartition à n dimensions et leurs marges. Publ. Inst. Statist. Univ. Paris 8, 229–231 (1959)

[21] Tankov, P.: Improved fréchet bounds and model-free pricing of multi-asset options. J. Appl. Probab. 48(2), 389–403 (2011)

Part III
Ordered Aggregation

OM3: Ordered Maxitive, Minitive, and Modular Aggregation Operators: Axiomatic Analysis under Arity-Dependence (I)

Anna Cena and Marek Gagolewski

Abstract. Recently, a very interesting relation between symmetric minitive, maxitive, and modular aggregation operators has been shown. It turns out that the intersection between any pair of the mentioned classes is the same. This result introduces what we here propose to call the OM3 operators. In the first part of our contribution on the analysis of the OM3 operators we study some properties that may be useful when aggregating input vectors of varying lengths. In Part II we will perform a thorough simulation study of the impact of input vectors' calibration on the aggregation results.

1 Introduction

The process called aggregation of quantitative (numeric) data is of great importance in many practical domains. e.g. in mathematical statistics, engineering, operational research, and quality control. For instance, in scientometrics we are often interested in assessing scholars via aggregation of the citations number of each of their articles, or by using some other measures of their quality, see e.g. [11]. On the other hand, in marketing we sometimes need to synthesize the sales of a product with multiple model-ranges, i.e. to aggregate the number of versions (assessment of the product

Anna Cena · Marek Gagolewski
Systems Research Institute, Polish Academy of Sciences, ul. Newelska 6,
01-447 Warsaw, Poland
e-mail: {Anna.Cena,gagolews}@ibspan.waw.pl

Marek Gagolewski
Faculty of Mathematics and Information Science, Warsaw University of Technology,
ul. Koszykowa 75, 00-662 Warsaw, Poland

H. Bustince et al. (eds.), *Aggregation Functions in Theory and in Practise*,
Advances in Intelligent Systems and Computing 228,
DOI: 10.1007/978-3-642-39165-1_13, © Springer-Verlag Berlin Heidelberg 2013

diversification) and the number of units sold for each model-range (assessment of market penetration) [6, 10].

The two above-presented examples are similar: they both concern summarization of quantitative data sets of nonuniform sizes. In classical approach to aggregation, however, the number of input values is fixed, see [3, 16, 15, 14]. To take into account such domains of applications, we shall rather consider arity-dependent [9] aggregation operators, cf. also [4, 3, 13, 19, 21].

In a recent article [7] some desirable properties of aggregation operators were considered: maxitivity, minitivity [14], and modularity [20, 18]. This result introduces a very appealing class of functions which we call here the OM3 operators. The OM3 operators include i.a. the well-known h-index [15], order statistics, and OWMax/OWMin operators [5]. In this paper we explore these functions under arity-dependence.

The paper is organized as follows. In Sec. 2 we introduce the OM3 operators and recall their most fundamental properties. Then, in Sec. 3 we study some desirable arity-dependent properties which are of interest in many practical situations, including their insensitivity to addition of elements equal to 0 or $F(\mathbf{x})$ to the input vector $\mathbf{x}$, and sensitivity to addition of elements strictly greater than $F(\mathbf{x})$. Finally, Sec. 4 concludes the paper. Moreover, in the second part of our contribution we perform a simulation study of the effect of inputs' calibration on the ranking of vectors by means of OM3 operators.

2   The OM3 Aggregation Operators

From now on let $\mathbb{I} = [0, b]$ denote a closed interval of the extended real line, possibly with $b = \infty$. The set of all vectors of arbitrary length with elements in $\mathbb{I}$, i.e. $\bigcup_{n=1}^{\infty} \mathbb{I}^n$, is denoted by $\mathbb{I}^{1,2,\cdots}$. Moreover, let $\mathscr{E}(\mathbb{I})$ denote the set of all *aggregation operators* (also called extended aggregation functions) in $\mathbb{I}^{1,2,\cdots}$, i.e. $\mathscr{E}(\mathbb{I}) = \{F : \mathbb{I}^{1,2,\cdots} \to \mathbb{I}\}$.

We see that the notion of an aggregation operator is very general: the only restriction is that it is a function into $\mathbb{I}$. Let us then focus our attention on operators that are nondecreasing (in each variable) and, additionally, symmetric (i.e. which do not depend on the order of elements' presentation) [10, 9].

Definition 1. We say that $F \in \mathscr{E}(\mathbb{I})$ is *symmetric*, denoted $F \in \mathscr{P}_{(\text{sym})}$, if

$$(\forall n \in \mathbb{N}) \, (\forall \mathbf{x}, \mathbf{y} \in \mathbb{I}^n) \, \mathbf{x} \cong \mathbf{y} \Longrightarrow F(\mathbf{x}) = F(\mathbf{y}),$$

where $\mathbf{x} \cong \mathbf{y}$ if and only if there exists a permutation σ of $[n] := \{1, 2, \ldots, n\}$ such that $\mathbf{x} = (y_{\sigma(1)}, \ldots, y_{\sigma(n)})$

Definition 2. We say that $F \in \mathscr{E}(\mathbb{I})$ is *nondecreasing*, denoted $F \in \mathscr{P}_{(\text{nd})}$, if

$$(\forall n \in \mathbb{N}) \, (\forall \mathbf{x}, \mathbf{y} \in \mathbb{I}^n) \, \mathbf{x} \le \mathbf{y} \Longrightarrow F(\mathbf{x}) \le F(\mathbf{y}),$$

where $\mathbf{x} \le \mathbf{y}$ if and only if $(\forall i \in [n]) \, x_i \le y_i$.

We see that for each $F \in \mathscr{P}_{(\mathrm{nd})}$ it holds $0 \leq F(n * 0) \leq F(\mathbf{x}) \leq F(n * b) \leq b$ for all $\mathbf{x} \in \mathbb{I}^n$, where $(n * y)$, $y \in \mathbb{I}$, denotes a vector $(y, y, \ldots, y) \in \mathbb{I}^n$.

Let us recall the notion of symmetrized maxitivity, minitivity, and modularity used in [7], cf. also [14, 20, 18]. For $\mathbf{x}, \mathbf{y} \in \mathbb{I}^n$ let $\mathbf{x} \overset{S}{\vee} \mathbf{y} = (x_{(1)} \vee y_{(1)}, \ldots, x_{(n)} \vee y_{(n)})$ and $\mathbf{x} \overset{S}{\wedge} \mathbf{y} = (x_{(1)} \wedge y_{(1)}, \ldots, x_{(n)} \wedge y_{(n)})$, where $x_{(i)}$ denotes the ith order statistic of $\mathbf{x} \in \mathbb{I}^n$, i.e. the i-th smallest value in $\mathbf{x}$.

Definition 3. Let $F \in \mathscr{E}(\mathbb{I})$. Then we call F a *symmetric maxitive* aggregation operator (denoted $F \in \mathscr{P}_{(\mathrm{smax})}$), whenever $(\forall n)$ $(\forall \mathbf{x}, \mathbf{y} \in \mathbb{I}^n)$ $F(\mathbf{x} \overset{S}{\vee} \mathbf{y}) = F(\mathbf{x}) \vee F(\mathbf{y})$.

Definition 4. Let $F \in \mathscr{E}(\mathbb{I})$. Then F is *symmetric minitive* (denoted $F \in \mathscr{P}_{(\mathrm{smin})}$), if $(\forall n)$ $(\forall \mathbf{x}, \mathbf{y} \in \mathbb{I}^n)$ $F(\mathbf{x} \overset{S}{\wedge} \mathbf{y}) = F(\mathbf{x}) \wedge F(\mathbf{y})$.

Definition 5. Let $F \in \mathscr{E}(\mathbb{I})$. Then F is *symmetric modular* (denoted $F \in \mathscr{P}_{(\mathrm{smod})}$) whenever $(\forall n)$ $(\forall \mathbf{x}, \mathbf{y} \in \mathbb{I}^n)$ $F(\mathbf{x} \overset{S}{\vee} \mathbf{y}) + F(\mathbf{x} \overset{S}{\wedge} \mathbf{y}) = F(\mathbf{x}) + F(\mathbf{y})$.

It may be easily shown that $\mathscr{P}_{(\mathrm{smax})}, \mathscr{P}_{(\mathrm{smin})}, \mathscr{P}_{(\mathrm{smod})} \subseteq \mathscr{P}_{(\mathrm{sym})} \cap \mathscr{P}_{(\mathrm{nd})}$. Moreover, each symmetric modular aggregation operator is also symmetric additive (i.e. $F(\mathbf{x} \overset{S}{+} \mathbf{y}) = F(\mathbf{x}) + F(\mathbf{y})$, where $\mathbf{x} \overset{S}{+} \mathbf{y} = (x_{(1)} + y_{(1)}, \ldots, x_{(n)} + y_{(n)})$), cf. [10, 14].

Let us introduce the following class of aggregation operators.

Definition 6. Given $\mathbf{w} = (w_1, w_2, \ldots)$, $w_i : \mathbb{I} \to \mathbb{I}$, and a triangle of coefficients $\triangle = (c_{i,n})_{i \in [n], n \in \mathbb{N}}$, $c_{i,n} \in \mathbb{I}$, for any $\mathbf{x} \in \mathbb{I}^n$, let

$$M_{\triangle,\mathbf{w}}(\mathbf{x}) = \bigvee_{i=1}^{n} w_n(x_{(n-i+1)}) \wedge c_{i,n}.$$

We see that the above contains i.a. all order statistics (whenever $w_n(x) = x$, and $c_{i,n} = 0$, $c_{j,n} = b$ for $i < k$, $j \geq k$, and some k), OWMax operators (for $w_n(x) = x$), and the famous Hirsch h-index ($w_n(x) = \lfloor x \rfloor$, $c_{i,n} = i$).

It turns out that in case of nondecreasingness, with no loss in generality, we may assume that such aggregation operators are defined by $w_1, w_2, \ldots$ and $\triangle$ of a specific form.

Lemma 1 (Reduction). $M_{\triangle,\mathbf{w}} \in \mathscr{P}_{(\mathrm{nd})}$ *if and only if there exist* $\mathbf{w}' = (w_1', w_2', \ldots)$, $w_i' : \mathbb{I} \to \mathbb{I}$, *and a triangle of coefficients* $\triangledown = (c_{i,n}')_{i \in [n], n \in \mathbb{N}}$ *satisfying the following conditions:*

(i) $(\forall n)$ w_n' *is nondecreasing,*
(ii) $(\forall n)$ $c_{1,n}' \leq c_{2,n}' \leq \cdots \leq c_{n,n}'$,
(iii) $(\forall n)$ $0 \leq w_n'(0) \leq c_{1,n}'$,
(iv) $(\forall n)$ $w_n'(b) = c_{n,n}'$,

such that $M_{\triangle,\mathbf{w}} = M_{\triangledown,\mathbf{w}'}$.

Proof. ($\Longrightarrow$) Let us fix n. Let $x,y \in \mathbb{I}$ be such that $x \le y$. By $\mathscr{P}_{(\mathrm{nd})}$, $M_{\triangle,\mathbf{w}}(n *$ $x) = \bigvee_{i=1}^{n} \mathsf{w}_n(x) \wedge c_{i,n} = \mathsf{w}_n(x) \wedge \bigvee_{i=1}^{n} c_{i,n} \le \mathsf{w}_n(y) \wedge \bigvee_{i=1}^{n} c_{i,n} = M_{\triangle,\mathbf{w}}(n * y)$, where $\bigvee_{i=1}^{n} c_{i,n}$ is constant. Thus, we may set $\mathsf{w}_n'(z) := \mathsf{w}_n(z) \wedge \bigvee_{i=1}^{n} c_{i,n}$ for all $z \in \mathbb{I}$. Necessarily, w_n' is nondecreasing. Moreover, we may set $c_{i,n}' := c_{i,n} \wedge \mathsf{w}_n(b)$ and hence $\mathsf{w}_n'(b) = \bigvee_{i=1}^{n} c_{i,n}'$. Please note that $M_{\triangle,\mathbf{w}} = M_{\triangledown,\mathbf{w}'}$, where $\triangledown = (c_{i,n}')_{i \in [n]}$.

Let $d_n := M_{\triangle,\mathbf{w}'}(n * 0) = \bigvee_{i=1}^{n} \mathsf{w}_n'(0) \wedge c_{i,n}' \ge 0$. Therefore, as $M_{\triangle,\mathbf{w}'} \in \mathscr{P}_{(\mathrm{nd})}$, for all $\vec{x} \in \mathbb{I}^n$ it holds $M_{\triangle,\mathbf{w}'}(\vec{x}) \ge d_n \ge 0$. As a consequence,

$$M_{\triangle,\mathbf{w}'}(\vec{x}) = \bigvee_{i=1}^{n} \mathsf{w}_n'(x_{(n-i+1)}) \wedge c_{i,n}' = \left(\bigvee_{i=1}^{n} \mathsf{w}_n'(x_{(n-i+1)}) \wedge c_{i,n}' \right) \vee d_n =$$
$$= \bigvee_{i=1}^{n} (\mathsf{w}_n'(x_{(n-i+1)}) \vee d_n) \wedge (c_{i,n}' \vee d_n).$$

Therefore, we may set $\mathsf{w}_n'(y) := \mathsf{w}_n'(y) \vee d_n$ for all $y \in \mathbb{I}$, $c_{i,n}' := c_{i,n}' \vee d_n$, still with $M_{\triangle,\mathbf{w}} = M_{\triangledown,\mathbf{w}'}$, where $\triangledown = (c_{i,n}')_{i \in [n]}$. Since $c_{i,n}' \ge d_n$ for all i, then $M_{\triangledown,\mathbf{w}'}(n * 0) = \mathsf{w}_n'(0)$, hence, $\mathsf{w}_n'(0) \le \bigwedge_{i=1}^{n} c_{i,n}'$.

Fix any $\mathbf{x} \in \mathbb{I}^n$. We have:

$$M_{\triangle',\mathbf{w}'}(\vec{x}) = M_{\triangle',\mathbf{w}'}\big(x_{(n)} \vee x_{(n-1)} \vee \cdots \vee x_{(2)} \vee x_{(1)},$$
$$x_{(n-1)} \vee \cdots \vee x_{(2)} \vee x_{(1)},$$
$$\cdots,$$
$$x_{(2)} \vee x_{(1)},$$
$$x_{(1)}\big).$$

As w_n' is nondecreasing, we get $\mathsf{w}_n'(x_{(n)} \vee \cdots \vee x_{(1)}) = \mathsf{w}_n'(x_{(n)}) \vee \cdots \vee \mathsf{w}_n'(x_{(1)})$. This implies

$$M_{\triangledown,\mathbf{w}'}(\vec{x}) = \bigvee_{i=1}^{n} \left[(\mathsf{w}_n'(x_{(n-i+1)})) \wedge \big(\bigvee_{j=1}^{i} c_{j,n}' \big) \right].$$

Now we may set $c_{i,n}' := \bigvee_{j=1}^{i} c_{j,n}'$, and still $M_{\triangle,\mathbf{w}} = M_{\triangledown,\mathbf{w}'}$. It is clear to see that $\mathsf{w}_n'(0) \le c_{1,n}' \le \cdots \le c_{n,n}' = \mathsf{w}_n'(b)$.

($\Longleftarrow$) Let us fix n. It suffices to show that if w_n' and $\triangledown = (c_{i,n}')_{i \in [n], n \in \mathbb{N}}$ fulfill conditions (i)–(iv) then $M_{\triangledown,\mathbf{w}'}$ is nondecreasing. Let $\mathbf{x}, \mathbf{y} \in \mathbb{I}^n$ be such that $\mathbf{x} \le \mathbf{y}$. It is clear to see that $x_{(n-i+1)} \le y_{(n-i+1)}$ for all i. Since w_n' is nondecreasing, we have $\mathsf{w}_n'(x_{(n-i+1)}) \wedge c_{i,n}' \le \mathsf{w}_n'(y_{(n-i+1)}) \wedge c_{i,n}'$. Thus, $\bigvee_{i=1}^{n} \mathsf{w}_n'(x_{(n-i+1)}) \wedge c_{i,n}' \le \bigvee_{i=1}^{n} \mathsf{w}_n'(y_{(n-i+1)}) \wedge c_{i,n}'$, which completes the proof. $\square$

Most importantly, we have the following result, see [7, Theorem 20].

Theorem 1. *Let* $\mathbf{w}$ *and* $\triangle$ *be of the form given in Lemma 1. Then for all* $\mathbf{x} \in \mathbb{I}^{1,2,\cdots}$

$$M_{\triangle,\mathbf{w}}(\mathbf{x}) = \bigvee_{i=1}^{n} w_n(x_{(n-i+1)}) \wedge c_{i,n}$$

$$= \bigwedge_{i=1}^{n} (w_n(x_{(n-i+1)}) \vee c_{i-1,n}) \wedge c_{n,n}$$

$$= \sum_{i=1}^{n} \left((w_n(x_{(n-i+1)}) \vee c_{i-1,n}) \wedge c_{i,n} - c_{i-1,n} \right).$$

with convention $c_{0,n} = 0$.

We see that $M_{\triangle,\mathbf{w}}$ are symmetric maxitive, minitive and modular. What is more, by [7, Theorem 19], these are the only aggregation operators that belong to $\mathscr{P}_{(\mathrm{smax})} \cap \mathscr{P}_{(\mathrm{smin})} = \mathscr{P}_{(\mathrm{smax})} \cap \mathscr{P}_{(\mathrm{smod})} = \mathscr{P}_{(\mathrm{smin})} \cap \mathscr{P}_{(\mathrm{smod})} = \mathscr{P}_{(\mathrm{smax})} \cap \mathscr{P}_{(\mathrm{smod})} \cap \mathscr{P}_{(\mathrm{smin})}$. This is the reason why from now on we propose to call all $M_{\triangle,\mathbf{w}}$ the *OM3* operators, i.e. *ordered maxitive, minitive, and modular* aggregation operators.

3   Some Arity-Dependent Properties

Note that up to now our discussion concerned a fixed sample size n. Here we consider some properties that take into account the behavior of the aggregation operator when a new element is added to the input vector. This situation often occurs in practice: a "producer" whose quality has to be assessed "outputs" yet another "product" and we have to reevaluate his/her rating.

3.1   Zero-Insensitivity

For each $\mathbf{x} \in \mathbb{I}^n$ and $\mathbf{y} \in \mathbb{I}^m$, let $(\mathbf{x},\mathbf{y})$ denote the concatenation of the two vectors, i.e. $(x_1,\dots,x_n,y_1,\dots,y_m) \in \mathbb{I}^{n+m}$. In some applications, it is desirable to guarantee that if we add an element with rating 0, then the valuation of the vector does not change. It is because 0 may denote a minimal quality measure needed for an item to be taken into account in the aggregation process (note, however, a very different approach e.g. in [3] where *averaging* is considered). Such a property is called zero-insensitivity, see [9] and also [22]. More formally:

Definition 7. We call $F \in \mathscr{E}(\mathbb{I})$ a *zero-insensitive* aggregation operator, denoted $F \in \mathscr{P}_{(\mathrm{a}0)}$, if for each $\mathbf{x} \in \mathbb{I}^{1,2,\dots}$ it holds $F(\mathbf{x},0) = F(\mathbf{x})$.

In other words, if F is zero-insensitive, then 0 is its so-called extended neutral element, see [14, Def. 2.108]. What is more, 0 is an idempotent element of every zero-insensitive function F such that $F(0) = 0$.

It is easily seen that $\mathscr{P}_{(\mathrm{a}0)} \cap \mathscr{P}_{(\mathrm{nd})} \subseteq \mathscr{P}_{(\mathrm{am})} \cap \mathscr{P}_{(\mathrm{nd})}$, where $\mathscr{P}_{(\mathrm{am})}$ denotes arity-monotonic aggregation operators, see [9], such that $(\forall \mathbf{x},\mathbf{y} \in \mathbb{I}^{1,2,\dots}) \ F(\mathbf{x}) \leq F(\mathbf{x},\mathbf{y})$.

Concerning OM3 aggregation operators, we have what follows.

Theorem 2. *Let* **w** *and* $\triangle$ *be of the form given in Lemma 1. Then* $\mathsf{M}_{\triangle,\mathbf{w}} \in \mathscr{P}_{(a0)}$ *if and only if* $(\forall n)\ (\forall i \in [n])\ c_{i,n} = c_{i,n+1}$, *and*

(i) if x *s.t.* $\mathsf{w}_n(x) < c_{n,n}$, *then* $\mathsf{w}_n(x) = \mathsf{w}_{n+1}(x)$,
(ii) if x *s.t.* $\mathsf{w}_n(x) = c_{n,n}$, *then* $\mathsf{w}_{n+1}(x) \geq c_{n,n}$.

In other words, in such case $\mathsf{M}_{\triangle,\mathbf{w}} \in \mathscr{P}_{(a0)}$ if and only if there exists a nondecreasing function w and a sequence $(c_1, c_2, \ldots)$ such that $(\forall n)\ \mathsf{w}_n = \mathsf{w} \wedge c_n$ and

$$\triangle = \begin{pmatrix} c_{1,1} & & & \\ c_{1,2} & c_{2,2} & & \\ c_{1,3} & c_{2,3} & c_{3,3} & \\ \vdots & \vdots & \vdots & \ddots \\ \| & \| & \| & \cdots \cdots \\ c_1 & c_2 & c_3 & \cdots \cdots \end{pmatrix}$$

Proof. ($\Longrightarrow$) Take any $\mathsf{M}_{\triangle,\mathbf{w}} \in \mathscr{P}_{(a0)}$ from Lemma 1. Let $n = 1$. Then $\mathsf{M}_{\triangle,\mathbf{w}}(x) = \mathsf{w}_1(x) \wedge c_{1,1}$ and $\mathsf{M}_{\triangle,\mathbf{w}}(x,0) = (\mathsf{w}_2(x) \wedge c_{1,2}) \vee (\mathsf{w}_2(0) \wedge c_{2,2})$ for any $x \in \mathbb{I}$. Please note that as $\mathsf{M}_{\triangle,\mathbf{w}}$ is nondecreasing, we have $\mathsf{w}_2(0) \leq c_{1,2} \leq c_{2,2}$ and w_2 is nondecreasing. Thus, $\mathsf{M}_{\triangle,\mathbf{w}}(x,0) = \mathsf{w}_2(x) \wedge c_{1,2}$. As $\mathsf{M}_{\triangle,\mathbf{w}} \in \mathscr{P}_{(a0)}$, it holds $\mathsf{M}_{\triangle,\mathbf{w}}(x) = \mathsf{M}_{\triangle,\mathbf{w}}(x,0)$, hence $\mathsf{w}_1(x) \wedge c_{1,1} = \mathsf{w}_2(x) \wedge c_{1,2}$. Let $x_1 = \inf\{x : \mathsf{w}_1(x) \geq c_{1,1}\}$. We shall consider two cases.

1. Let $x \leq x_1^-$.

    (a) If $\mathsf{w}_1(x) < c_{1,2}$, we have $\mathsf{w}_1(x) = \mathsf{w}_2(x)$.
    (b) If $\mathsf{w}_1(x) \geq c_{1,2}$, then $c_{1,1} > c_{1,2}$. Please note that $\mathsf{w}_1(x_1) \geq c_{1,1}$. Thus, $\mathsf{M}_{\triangle,\mathbf{w}}(x_1) = c_{1,1} = \mathsf{w}_2(x_1) \wedge c_{1,2} = \mathsf{M}_{\triangle,\mathbf{w}}(x_1,0)$. Monotonicity of w_2 and case (a) implies $\mathsf{w}_2(x_1) \geq c_{1,2}$. Thus, $c_{1,1} = c_{1,2}$, a contradiction.

    Hence, for all x such that $\mathsf{w}_1(x) < c_{1,1}$ we have $\mathsf{w}_1(x) = \mathsf{w}_2(x)$.
2. Now let us consider $x \geq x_1^+$. Please note that, as w_1 is nondecreasing and $\mathsf{w}_1(b) = c_{1,1}$, we have $\mathsf{w}_1(x) = c_{1,1}$. Thus, $c_{1,1} = \mathsf{w}_2(x) \wedge c_{1,2}$. From previous case and the fact that w_2 is nondecreasing, we have $\mathsf{w}_2(x) \geq c_{1,2}$. Hence, $c_{1,2} = c_{1,1}$.

Let $n = 2$. By $\mathscr{P}_{(a0)}$, we have $\mathsf{M}_{\triangle,\mathbf{w}}(x,0) = \mathsf{M}_{\triangle,\mathbf{w}}(x,2*0)$. Thus, $\mathsf{w}_2(x) \wedge c_{1,2} = \mathsf{w}_3(x) \wedge c_{1,3}$. By similar steps as the above-performed, we get $c_{1,3} = c_{1,2} = c_{1,1}$ and $\mathsf{w}_3 = \mathsf{w}_2 = \mathsf{w}_1$ for $x < x_1$. For $\mathbf{x} = (x,x)$ we have $\mathsf{M}_{\triangle,\mathbf{w}}(x,x) = \mathsf{M}_{\triangle,\mathbf{w}}(2*x,0) \Leftrightarrow \mathsf{w}_2(x) \wedge c_{2,2} = \mathsf{w}_3(x) \wedge c_{2,3}$. Likewise, we get $c_{2,2} = c_{2,3}$, $\mathsf{w}_2(x) = \mathsf{w}_3(x)$ for x such that $\mathsf{w}_2(x) < c_{2,2}$, and $\mathsf{w}_3(x) \geq c_{2,2}$ for x for which $\mathsf{w}_2(x) = c_{2,2}$.

The above reasoning may easily be extended for all other n.

($\Longleftarrow$) Please note that when conditions given in the right side of Theorem 2 hold, we may set $\mathsf{w}_n := \mathsf{w} \wedge c_{n,n}$ and $c_{i,n} := c_i$ for all n and some nondecreasing w and c_i. This notion generates **w** and $\triangle$ fulfill conditions given in Lemma 1 such that $\mathsf{M}_{\triangle,\mathbf{w}} \in \mathscr{P}_{(nd)}$.

We will now show that $M_{\triangle,\mathbf{w}} \in \mathscr{P}_{(a0)}$. Assume contrary. There exists $\mathbf{x} \in \mathbb{I}^{1,2,\cdots}$ such that $M_{\triangle,\mathbf{w}}(\mathbf{x}) \neq M_{\triangle,\mathbf{w}}(\mathbf{x},0)$. As $\mathbf{w}$ is nondecreasing and $w(0) \leq c_1$, we have $M_{\triangle,\mathbf{w}}(\mathbf{x},0) = \left(\bigvee_{i=1}^{n}(w(x_{(n-i+1)}) \wedge c_i) \right) \vee (w(0) \wedge c_{n+1}) = \bigvee_{i=1}^{n} w(x_{(n-i+1)}) \wedge c_i = M_{\triangle,\mathbf{w}}(\mathbf{x})$, a contradiction, and the proof is complete. □

3.2 F-insensitivity

Zero-sensitivity may be strengthened as follows, cf. [9] and [23, Axiom A1].

Definition 8. $F \in \mathscr{E}(\mathbb{I})$ is *F-insensitive*, denoted $F \in \mathscr{P}_{(F0)}$, if

$$(\forall \mathbf{x} \in \mathbb{I}^{1,2,\cdots}) \, (\forall y \in \mathbb{I}) \, y \leq F(\mathbf{x}) \Longrightarrow F(\mathbf{x},y) = F(\mathbf{x}).$$

Thus, we see that in this property we do not want to distinct a "producer" in any special way if he/she outputs a "product" with valuation not greater than his/her current overall rating.

Please note that $\mathscr{P}_{(F0)} \cap \mathscr{P}_{(nd)} \subseteq \mathscr{P}_{(a0)} \cap \mathscr{P}_{(nd)}$. Moreover, if $F \in \mathscr{P}_{(a0)} \cap \mathscr{P}_{(nd)}$, then $F \in \mathscr{P}_{(F0)}$ iff $(\forall \mathbf{x} \in \mathbb{I}^{1,2,\cdots}) \, F(\mathbf{x},F(\mathbf{x})) = F(\mathbf{x})$. Note also that the property $F(\mathbf{x},F(\mathbf{x})) = F(\mathbf{x})$, introduced in [24], is known as self-identity. A similar property, called *stability*, was also considered in [1].

Theorem 3. *Let $\mathbf{w}$ and $\triangle$ be of the form given in Lemma 1. Then $M_{\triangle,\mathbf{w}} \in \mathscr{P}_{(F0)}$ if and only if there exists:*

(i) a nondecreasing function w, for which if there exists x such that $w(x) > x$, then
$(\forall y \in [x, w(x)]) \, w(y) = w(x)$,
(ii) a nondecreasing sequence $(c_1, c_2, \ldots)$, such that $(\forall i) \, c_i \notin \{x \in \mathbb{I} : x < w(x)\}$,

such that $w_n = w \wedge c_n$ and $c_{i,n} = c_i$.

Proof. ($\Longrightarrow$) Let c_n be such that $c_n < w_n(c_n)$ and $c_{n+1} > c_n$ for some n. Take $\mathbf{x} = (n * c_n)$, then $M_{\triangle,\mathbf{w}}(\mathbf{x}) = w_n(c_n) \wedge c_n = c_n$. Since $M_{\triangle,\mathbf{w}} \in \mathscr{P}_{(F0)}$, we have $M_{\triangle,\mathbf{w}}(\mathbf{x}) = M_{\triangle,\mathbf{w}}(\mathbf{x},c_n) = w_n(c_n) \wedge c_{n+1}$, a contradiction, because $c_{n+1} > c_n$ and $w_n(c_n) > c_n$. Thus, it is easily seen that for all i we have $w(c_i) \leq c_i$.

Let us now consider $y \in \mathbb{I}$ such that $w(y) > y$. There is no loss in generality in assuming that $w(y) \in (c_{n-1}, c_n]$. $M_{\triangle,\mathbf{w}}((n-1) * b, y) = (w(b) \wedge c_{n-1}) \vee (w(y) \wedge c_n) = c_{n-1} \vee w(y) = w(y)$. Moreover, $M_{\triangle,\mathbf{w}}((n-1) * b, w(y), y) = (w(b) \wedge c_{n-1}) \vee (w(w(y)) \wedge c_n) \vee (w(y) \wedge c_{n+1}) = (w(w(y)) \wedge c_n) \vee w(y) = w(w(y)) \wedge c_n$. Since $M_{\triangle,\mathbf{w}} \in \mathscr{P}_{(F0)}$, $M_{\triangle,\mathbf{w}}((n-1) * b, y) = M_{\triangle,\mathbf{w}}((n-1) * b, w(y), y)$. Thus, $w(y) = w(w(y)) \wedge c_n$. This implies that either $w(y) = w(w(y))$ or $w(y) = c_n$. Hence, $w(y) \leq c_n$. We shall consider the $w(y) = w(w(y))$ case.

Let us take the largest interval $\mathbb{L}$, $y \in \mathbb{L}$, such that $(\forall x \in \mathbb{L}) \, w(x) > x$. Denote the bounds of this interval by x_1, x_2, respectively. $\mathbb{L}$ may be either left-open or left-closed depending on the kind of potential discontinuity of w at x_1, but this will not affect our reasoning here. Surely, however, it is right-open (by definition of $\mathbb{L}$ and nondecreasingness of w), we have $w(x_2) = x_2$. Take any $y_1 \in \mathbb{L}$. From the previous paragraph, $y_1 < w(y_1)$. Let $y_2 = w(y_1)$. If $y_2 \notin \mathbb{L}$, then $w(y_2) \leq y_2$ and from nondecreasingness of w we have that $y_2 = x_2$. On the other hand, let $y_2 \in \mathbb{L}$. By the definition of $\mathbb{L}$, we get $y_2 < w(y_2)$ and, from the above-performed reasoning, as $M_{\triangle,\mathbf{w}} \in \mathscr{P}_{(F0)} \cap \mathscr{P}_{(nd)}$, it follows $w(y_2) = w(w(y_2))$. Let $y_3 = w(y_2)$. Then $y_2 < y_3 = w(y_3)$ and therefore $y_3 = w(y_2) \notin \mathbb{L}$, a contradiction. Thus, $y_2 = x_2$ and $(\forall x \in \mathbb{L}) \, w(x) = w(x_2)$.

($\Longleftarrow$) Assume otherwise. Thus, there exists $\mathbf{x} \in \mathbb{I}^{n-1}$ for some n such that $y := M_{\triangle,\mathbf{w}}(\mathbf{x}) \neq M_{\triangle,\mathbf{w}}(\mathbf{x}, y)$. Moreover, let $\mathbf{x}' = (\mathbf{x}, 0)$ and $\mathbf{x}'' = (\mathbf{x}, y)$. By $\mathscr{P}_{(a0)}$, we have $M_{\triangle,\mathbf{w}}(\mathbf{x}) = M_{\triangle,\mathbf{w}}(\mathbf{x}')$. From definition, there exists k such that $y = w(x'_{(n-k+1)}) \wedge c_k$. We shall now consider two cases.

1. If $y > x'_{(n-k+1)}$, then for some $l < k$ we have $M_{\triangle,\mathbf{w}}(\mathbf{x}'') = M_{\triangle,\mathbf{w}}(x'_{(n)}, \ldots, x'_{(n-l+2)},$
    $y, x'_{(n-l+1)}, \ldots, x'_{(n-k+1)}, \ldots, x'_{(2)}) = \left(\bigvee_{i=1}^{l-1} w_n(x'_{(n-i+1)}) \wedge c_{i,n} \right) \vee (w(y) \wedge c_{l,n}) \vee$
    $\left(\bigvee_{i=l+1}^{k} w_n(x'_{(n-i+2)}) \wedge c_{i,n} \right) \vee \left(\bigvee_{i=k+1}^{n} w_n(x'_{(n-i+2)}) \wedge c_{i,n} \right)$.

    (a) If $w_n(x'_{(n-k+1)}) \leq c_{k,n}$, then $y = w_n(x'_{(n-k+1)}) > x'_{(n-k+1)}$. Moreover, $w_n(y) = w_n(w_n(x'_{(n-k+1)})) = w_n(x'_{(n-k+1)})$. This implies $M_{\triangle,\mathbf{w}}(\mathbf{x}'') = w_n(x'_{(n-k+1)}) \vee$
    $\left(\bigvee_{i=l+1}^{k} w_n(x'_{(n-i+2)}) \wedge c_{i,n} \right) \neq w_n(x'_{(n-k+1)}) = y$. But $w_n(x'_{(n-i+2)})$ for $i = l+1, \ldots, k$ is equal to $w_n(x'_{(n-k+1)})$, a contradiction.

    (b) If $w_n(x'_{(n-k+1)}) > c_{k,n}$, then by (ii), $y = c_{k,n} \not> x'_{(n-k+1)}$.

2. Now assume that $y \leq x'_{(n-k+1)}$. Then for some $l \geq k$ we obviously have
    $M_{\triangle,\mathbf{w}}(\mathbf{x}'') = M_{\triangle,\mathbf{w}}(x'_{(n)}, \ldots, x'_{(n-k+1)}, \ldots, x'_{(n-l+2)}, y, x'_{(n-l+1)}, \ldots, x'_{(2)})$
    $= (w_n(y) \wedge c_{l,n}) \vee \left(\bigvee_{i=1}^{l-1} w_n(x'_{(n-i+1)}) \wedge c_{i,n} \right) \vee \left(\bigvee_{i=l+1}^{n} w_n(x'_{(n-i+2)}) \wedge c_{i,n} \right) =$
    $(w_n(x_{(n-k+1)}) \wedge c_{k,n}) \vee (w_n(y) \wedge c_{l,n}) \vee \left(\bigvee_{i=l+1}^{n} w_n(x'_{(n-i+2)}) \wedge c_{i,n} \right)$.

    (a) If $w_n(x'_{(n-k+1)}) \leq c_{k,n}$, then for all $i > k$ we have $w_n(x''_{(n-i+1)}) \leq w_n(x'_{(n-k+1)})$ $= y \leq c_{k,n}$. Therefore, $M_{\triangle,\mathbf{w}}(\mathbf{x}'') = y$, a contradiction.

    (b) If $w_n(x'_{(n-k+1)}) > c_{k,n}$, then by (ii), we have $w_n(y) \leq c_{k,n}$. It implies that for all $i > l$ we have $w_n(x_{(n-i+2)}) \leq w_n(y) \leq c_{k,n}$. Thus, $M_{\triangle,\mathbf{w}}(\mathbf{x}'') = c_{k,n} = y$, a contradiction.

Hence, $M_{\triangle,\mathbf{w}}(\mathbf{x}) = M_{\triangle,\mathbf{w}}(\mathbf{x}, M_{\triangle,\mathbf{w}}(\mathbf{x}))$ for any $\mathbf{x}$, and the proof is complete. □

3.3   F+Sensitivity

Clearly, F-insensitivity does not guarantee that if a "producer" outputs an element with valuation greater than $F(\mathbf{x})$, then his/her overall valuation is raised. As such situation may sometimes be desirable, let us then consider the property discussed in [9] and [23, Axiom A2].

Definition 9. $F \in \mathscr{E}(\mathbb{I})$ is F+sensitive, denoted $F \in \mathscr{P}_{(F+)}$, if

$$(\forall \mathbf{x} \in \mathbb{I}^{1,2,\ldots}) \, (\forall y \in \mathbb{I}) \, y > F(\mathbf{x}) \Longrightarrow F(\mathbf{x}, y) > F(\mathbf{x}).$$

Let us study when this property holds in case of OM3 operators. We will consider it under $\mathscr{P}_{(a0)}$, as otherwise the form of $\triangle$ and $\mathbf{w}$ gets very complicated and too inconvenient to be used in practical applications.

Theorem 4. *Let $\mathbf{w}$ and $\triangle$ be of the form given in Lemma 1. Then $M_{\triangle,\mathbf{w}} \in \mathscr{P}_{(a0)} \cap \mathscr{P}_{(F+)}$ if and only if there exist:*

(i) a function $\mathbf{w}$ such that $\mathbf{w}(x) \geq x$ for all x, and strictly increasing for $x : \mathbf{w}(x) < \mathbf{w}(b)$,

(ii) a sequence $(c_1, c_2, \ldots)$ such that for $c_i < \mathbf{w}(b)$ we have $\mathbf{w}(x) < c_i$ for all $x : \mathbf{w}(x) < \mathbf{w}(b)$ and $c_i < c_{i+1}$

such that $\mathbf{w}_n = \mathbf{w} \wedge c_n$ and $c_{i,n} = c_i$.

Before proceeding to the proof, note that if $\mathbf{w}$ is continuous at $x_d = \sup\{x \in \mathbb{I} : \mathbf{w}(x) < \mathbf{w}(b)\}$, then $c_1 = \mathbf{w}(b)$ and $M_{\triangle,\mathbf{w}}(\mathbf{x}) = \mathbf{w}(x_{(n)})$.

Proof. ($\Longrightarrow$) First we will show that $\mathbf{w}(x) \geq x$ for all $x \in \mathbb{I}$. Assume otherwise. Take any x, $\mathbf{w}(x) < \mathbf{w}(b)$, such that $\mathbf{w}(x) < x$, and the smallest $c_k > \mathbf{w}(x)$. Therefore, $M_{\triangle,\mathbf{w}}(k * x) = \mathbf{w}(x)$. Let us take $\varepsilon > 0$ such that $\mathbf{w}(x) \leq \mathbf{w}(x) + \varepsilon < x$. This implies $M_{\triangle,\mathbf{w}}(k * x, \mathbf{w}(x) + \varepsilon) = (\mathbf{w}(x) \wedge c_k) \vee (\mathbf{w}(\mathbf{w}(x) + \varepsilon) \wedge c_{k+1}) = \mathbf{w}(x)$, a contradiction, since $\mathbf{w}$ is nondecreasing and $M_{\triangle,\mathbf{w}}(\mathbf{x}) \in \mathscr{P}_{(F+)}$.

Take any $x \in \mathbb{I}$ such that $\mathbf{w}(x) < \mathbf{w}(b)$ and the smallest $c_k > \mathbf{w}(x)$. Then $M_{\triangle,\mathbf{w}}(k * x) = \mathbf{w}(x)$. Let $\varepsilon > 0$. Then $\mathbf{w}(x) + \varepsilon > x$ and $M_{\triangle,\mathbf{w}}(\mathbf{w}(x) + \varepsilon, k * x) = (\mathbf{w}(\mathbf{w}(x) + \varepsilon) \wedge c_1) \vee (\mathbf{w}(x) \wedge c_{k+1}) = (\mathbf{w}(\mathbf{w}(x) + \varepsilon) \wedge c_1) \vee \mathbf{w}(x) > \mathbf{w}(x)$. This implies $\mathbf{w}(\mathbf{w}(x) + \varepsilon) > \mathbf{w}(x)$. Thus, $\mathbf{w}$ must be strictly increasing. Moreover, $\mathbf{w}(x) < c_1$. We shall now consider two cases.

If $\mathbf{w}$ is continuous at x_d, then it is easily seen that $c_1 = \mathbf{w}(b)$ and $M_{\triangle,\mathbf{w}} \in \mathscr{P}_{(F+)}$ for all $\mathbf{x}$ since $\mathbf{w}$ is nondecreasing and $\mathbf{w}(x) \geq x$.

If $\mathbf{w}$ is discontinuous at x_d, then for $x \in \mathbb{I}$ such that $\mathbf{w}(x) > c_1$ we have $\mathbf{w}(x) = \mathbf{w}(b)$. Therefore, $M_{\triangle,\mathbf{w}}(x) = \mathbf{w}(x) \vee c_1 = c_1$. Let us take $\varepsilon > 0$. If $c_1 + \varepsilon < x$, then $M_{\triangle,\mathbf{w}}(c_1 + \varepsilon, x) = (\mathbf{w}(c_1 + \varepsilon) \vee c_1) \wedge (\mathbf{w}(x) \wedge c_2) > c_1$. This implies $c_2 > c_1$. Otherwise $M_{\triangle,\mathbf{w}}(x, c_1 + \varepsilon) = (\mathbf{w}(x) \vee c_1) \wedge (\mathbf{w}(c_1 + \varepsilon) \wedge c_2)$ and from $\mathscr{P}_{(F+)}$ we get $c_2 > c_1$. We continue in this fashion by considering vectors $(i * x)$ and we obtain $c_i < c_{i+1}$ for all i.

($\Longleftarrow$) As noted above, if $\mathbf{w}$ is continuous at x_d, then $M_{\triangle,\mathbf{w}}(\mathbf{x}) = \mathbf{w}(x_{(n)})$. Thus, $M_{\triangle,\mathbf{w}} \in \mathscr{P}_{(F+)}$ for all $\mathbf{x}$ since $\mathbf{w}$ is strictly increasing and $\mathbf{w}(x) \geq x$.

Let us now consider discontinuity at x_d. Assume that $\mathscr{P}_{(F+)}$ does not hold. Take $\mathbf{x} \in \mathbb{I}^{n-1}$ such that $M_{\triangle,\mathbf{w}}(\mathbf{x}) < \mathbf{w}(b)$ and $M_{\triangle,\mathbf{w}}(\mathbf{x}') = M_{\triangle,\mathbf{w}}(\mathbf{x}'')$, where $\mathbf{x}' = (\mathbf{x}, 0)$ and $\mathbf{x}'' = (\mathbf{x}, M_{\triangle,\mathbf{w}}(\mathbf{x}') + \varepsilon)$ for some $\varepsilon > 0$. Please note that as $M_{\triangle,\mathbf{w}} \in \mathscr{P}_{(a0)}$, we have $M_{\triangle,\mathbf{w}}(\mathbf{x}) = M_{\triangle,\mathbf{w}}(\mathbf{x}')$. If $\mathbf{w}(x'_{(n)}) < c_1$, then $M_{\triangle,\mathbf{w}}(\mathbf{x}') = \mathbf{w}(x'_{(n)}) < c_1$. Thus for $\varepsilon > 0$ we have $\mathbf{w}(x'_{(n)}) + \varepsilon > x'_{(n)}$ and either $M_{\triangle,\mathbf{w}}(\mathbf{x}'') = \mathbf{w}(\mathbf{w}(x'_{(n)}) + \varepsilon) > \mathbf{w}(x'_{(n)})$ if $\mathbf{w}(\mathbf{w}(x'_{(n)}) + \varepsilon) < c_1$ or $M_{\triangle,\mathbf{w}}(\mathbf{x}'') \geq c_1 > \mathbf{w}(x'_{(n)})$ if $\mathbf{w}(\mathbf{w}(x'_{(n)}) + \varepsilon) \geq c_1$. In both cases we have a contradiction. Therefore, if $(F+)$ does not hold, we surely have $\mathbf{w}(x'_{(n)}) \geq c_1$. This implies $\mathbf{w}(x'_{(n)}) = \mathbf{w}(b)$ and since $M_{\triangle,\mathbf{w}}(\mathbf{x}') < \mathbf{w}(b)$ and $M_{\triangle,\mathbf{w}}(\mathbf{x}') > c_1$, it must holds $M_{\triangle,\mathbf{w}}(\mathbf{x}') = c_k$ for some k. Therefore, $\mathbf{w}(x'_{(n-k+1)}) \geq c_1$

and $w(x'_{(n-k)}) < c_1$. Take $\varepsilon > 0$. As $w(c_k + \varepsilon) \geq c_k + \varepsilon > c_k > c_1$, we have $w(c_k + \varepsilon) = w(b)$. Hence, $M_{\triangle,\mathbf{w}}(\mathbf{x}'') = \bigvee_{i=1}^{k+1}(w(x'_{(n-k+1)}) \wedge c_i) = c_{k+1} > c_k$, a contradiction. Thus, $M_{\triangle,\mathbf{w}} \in \mathscr{P}_{(F+)}$, QED. $\qquad\square$

4 Conclusions

In this paper we have considered a very interesting class of symmetric maxitive, minitive and modular aggregation operators, that is the OM3 operators. Our investigation was focused here on some properties useful when it comes to aggregation of vectors of different lengths. We have developed conditions required for the OM3 operators to be zero-insensitive, F-insensitive, and F+sensitive.

It is worth mentioning that in many applications it is more natural to consider OM3 operators that are continuous. A sufficient condition for that is the continuity of $\mathbf{w}$. In such case, the theorems presented in this paper have much simpler form. An OM3 operator is F-insensitive if and only if $w(x) \leq x$ for all $x \in \mathbb{I}$. On the other hand, we get F+sensitivity together with zero-insensitivity if and only if $w(x) \geq x$ for all $x \in \mathbb{I}$ and $c_1 = w(b)$. From this we easily get, quite surprisingly, that the only continuous OM3 operator that fulfills all the properties discussed here is the Max operator.

Please note that the famous Hirsch index $H(\mathbf{x}) = \max\{i = 1, \ldots, |\mathbf{x}| : x_{(n-i+1)} \geq i\} = \bigvee_{i=1}^{n} \lfloor x_{(n-i+1)} \rfloor \wedge i$, which is the most widely used tool in scientometrics, is an OM3 operator ($w(x) = \lfloor x \rfloor$, $c_i = i$) fulfilling $\mathscr{P}_{(F0)}$ (and $\mathscr{P}_{(a0)}$).

Moreover, it is easily seen that an OM3 operator in $\mathscr{P}_{(a0)}$ is asymptotically idempotent [14], iff $w(x) = x$ and $c_i \to b$ as $i \to \infty$. Additionally, each such operator is effort-dominating [7].

In the second part of our contribution we are going to perform a simulation study to asses behavior of OM3 operators for samples following different distributions. It will turn out that in order to study the effects of input vectors' ranking by means of OM3 operators, it suffices to consider a fixed triangle of coefficients. This makes the construction of OM3 operators quite easy in practical applications. However, what is still left for further research, is the method of automated construction of these operators.

Acknowledgements. The contribution of Marek Gagolewski was partially supported by FNP START Scholarship from the Foundation for Polish Science.

References

[1] Beliakov, G., James, S.: Stability of weighted penalty-based aggregation functions. Fuzzy Sets and Systems (2013), doi:10.1016/j.fss.2013.01.007

[2] Beliakov, G., Pradera, A., Calvo, T.: Aggregation Functions: A Guide for Practitioners. STUDFUZZ, vol. 221. Springer, Heidelberg (2007)

[3] Calvo, T., Mayor, G., Torrens, J., Suner, J., Mas, M., Carbonell, M.: Generation of weighting triangles associated with aggregation functions. International Journal of Uncertainty, Fuzziness and Knowledge-based Systems 8(4), 417–451 (2000)

[4] Calvo, T., Kolesarova, A., Komornikova, M., Mesiar, R.: Aggregation operators: Properties, classes and construction methods. In: Calvo, T., Mayor, G., Mesiar, R. (eds.) Aggregation Operators. New Trends and Applications. STUDFUZZ, vol. 97, pp. 3–104. Physica-Verlag, New York (2002)

[5] Dubois, D., Prade, H., Testemale, C.: Weighted fuzzy pattern matching. Fuzzy Sets and Systems 28, 313–331 (1988)

[6] Franceschini, F., Maisano, D.A.: The Hirsch index in manufacturing and quality engineering. Quality and Reliability Engineering International 25, 987–995 (2009)

[7] Gagolewski, M.: On the relation between effort-dominating and symmetric minitive aggregation operators. In: Greco, S., Bouchon-Meunier, B., Coletti, G., Fedrizzi, M., Matarazzo, B., Yager, R.R. (eds.) IPMU 2012, Part III. CCIS, vol. 299, pp. 276–285. Springer, Heidelberg (2012)

[8] Gagolewski, M.: On the relationship between symmetric maxitive, minitive, and modular aggregation operators. Information Sciences 221, 170–180 (2013)

[9] Gągolewski, M., Grzegorzewski, P.: Arity-monotonic extended aggregation operators. In: Hüllermeier, E., Kruse, R., Hoffmann, F. (eds.) IPMU 2010. CCIS, vol. 80, pp. 693–702. Springer, Heidelberg (2010)

[10] Gagolewski, M., Grzegorzewski, P.: Axiomatic characterizations of (quasi-) L-statistics and S-statistics and the Producer Assessment Problem. In: Galichet, S., Montero, J., Mauris, G. (eds.) Proc. Eusflat/LFA 2011, pp. 53–58 (2011)

[11] Gagolewski, M., Grzegorzewski, P.: Possibilistic analysis of arity-monotonic aggregation operators and its relation to bibliometric impact assessment of individuals. International Journal of Approximate Reasoning 52(9), 1312–1324 (2011)

[12] Gagolewski, M., Mesiar, R.: Aggregating different paper quality measures with a generalized h-index. Journal of Informetrics 6(4), 566–579 (2012)

[13] Ghiselli Ricci, R., Mesiar, R.: Multi-attribute aggregation operators. Fuzzy Sets and Systems 181(1), 1–13 (2011)

[14] Grabisch, M., Marichal, J.L., Mesiar, R., Pap, E.: Aggregation functions, Cambridge (2009)

[15] Grabisch, M., Marichal, J.L., Mesiar, R., Pap, E.: Aggregation functions: Construction methods, conjunctive, disjunctive and mixed classes. Information Sciences 181, 23–43 (2011)

[16] Grabisch, M., Marichal, J.L., Mesiar, R., Pap, E.: Aggregation functions: Means. Information Sciences 181, 1–22 (2011)

[17] Hirsch, J.E.: An index to quantify individual's scientific research output. Proceedings of the National Academy of Sciences 102(46), 16,569–16,572 (2005)

[18] Klement, E., Manzi, M., Mesiar, R.: Ultramodular aggregation functions. Information Sciences 181, 4101–4111 (2011)

[19] Mayor, G., Calvo, T.: On extended aggregation functions. In: Proc. IFSA 1997, vol. 1, pp. 281–285. Academia, Prague (1997)

[20] Mesiar, R., Mesiarová-Zemánková, A.: The ordered modular averages. IEEE Transactions on Fuzzy Systems 19(1), 42–50 (2011)

[21] Mesiar, R., Pap, E.: Aggregation of infinite sequences. Information Sciences 178, 3557–3564 (2008)

[22] Woeginger, G.J.: An axiomatic analysis of Egghe's g-index. Journal of Informetrics 2(4), 364–368 (2008)

[23] Woeginger, G.J.: An axiomatic characterization of the Hirsch-index. Mathematical Social Sciences 56(2), 224–232 (2008)

[24] Yager, R., Rybalov, A.: Noncommutative self-identity aggregation. Fuzzy Sets and Systems 85, 73–82 (1997)

OM3: Ordered Maxitive, Minitive, and Modular Aggregation Operators. A Simulation Study (II)

Anna Cena and Marek Gagolewski

Abstract. This article is a second part of the contribution on the analysis of the recently-proposed class of symmetric maxitive, minitive and modular aggregation operators. Recent results (Gagolewski, Mesiar, 2012) indicated some unstable behavior of the generalized h-index, which is a particular instance of OM3, in case of input data transformation. The study was performed on a small, carefully selected real-world data set. Here we conduct some experiments to examine this phenomena more extensively.

1 Introduction

In the first part of our contribution on OM3 aggregation operators, see [4], we carried out their axiomatic analysis under arity-dependence. Our motivation was that in many applications the "classical" assumption about fixed length of input vectors being aggregated, cf. [3, 14], is too restrictive. For example, in the Producer Assessment Problem (PAP), cf. [10], we wish to evaluate a set of producers according to their productivity and – simultaneously – the quality of the items they create. In Table 1 we list some typical instances of such situation, see also e.g. [6, 11]. It is easily seen that the number of artifacts varies from producer to producer. Thus, our main aim was to determine conditions required for the OM3 operators to poses some desirable properties such as zero- and F-insensitivity, or F+sensitivity.

The mentioned class of aggregation operators was of our interest, because these are the only functions which are symmetric modular, minitive, and – at the same

Anna Cena · Marek Gagolewski
Systems Research Institute, Polish Academy of Sciences, ul. Newelska 6,
01-447 Warsaw, Poland
e-mail: {Anna.Cena,gagolews}@ibspan.waw.pl

Marek Gagolewski
Faculty of Mathematics and Information Science, Warsaw University of Technology,
ul. Koszykowa 75, 00-662 Warsaw, Poland

H. Bustince et al. (eds.), *Aggregation Functions in Theory and in Practise,*
Advances in Intelligent Systems and Computing 228,
DOI: 10.1007/978-3-642-39165-1_14, © Springer-Verlag Berlin Heidelberg 2013

Table 1 Typical instances of the Producer Assessment Problem (PAP)

Producer	Products	Rating method
Scientist	Scientific articles	Number of citations
Scientific institute	Scientists	The h-index
Web server	Web pages	Number of targeting web-links
R package author	R packages	Number of dependencies
Artist	Paintings	Auction price

time – maxitive, see [7]. To recall, given a closed interval of the extended real line $\mathbb{I} = [0, b]$ (possibly with $b = \infty$), the OM3 operators are defined as follows. Note that we assume that the reader is familiar with notation convention introduced in [4].

Definition 1. A sequence of nondecreasing functions $\mathbf{w} = (w_1, w_2, \dots)$, $w_i : \mathbb{I} \to \mathbb{I}$, and a triangle of coefficients $\triangle = (c_{i,n})_{i \in [n], n \in \mathbb{N}}$, $c_{i,n} \in \mathbb{I}$ such that $(\forall n)\, c_{1,n} \leq c_{2,n} \leq \cdots \leq c_{n,n}$, $0 \leq w_n(0) \leq c_{1,n}$, and $w_n(b) = c_{n,n}$, generates a nondecreasing OM3 operator $\mathsf{M}_{\triangle,\mathbf{w}} \in \mathscr{P}_{(\mathrm{nd})}$ such that for $\mathbf{x} \in \mathbb{I}^n$ we have:

$$\mathsf{M}_{\triangle,\mathbf{w}}(\mathbf{x}) = \bigvee_{i=1}^n w_n(x_{(n-i+1)}) \wedge c_{i,n} = \bigwedge_{i=1}^n \left(w_n(x_{(n-i+1)}) \vee c_{i-1,n}\right) \wedge c_{n,n}$$

$$= \sum_{i=1}^n \left(\left(w_n(x_{(n-i+1)}) \vee c_{i-1,n}\right) \wedge c_{i,n} - c_{i-1,n}\right).$$

Please note that this class includes i.a. the well-known h-index [15], all order statistics, and OWMax/OWMin operators [5].

In the second part of our contribution we perform a simulation study of OM3 operators. Recently, it was noted in [11] that the generalized h-index (which is also an OM3 operator) exhibits a very unstable behavior upon some simple input elements' tranformations. The study was performed on a small-sized, but carefully selected bibliometric data set. We therefore pose a question: does this undesirable behavior is also observed in a large-scale study?

The paper is organized as follows. In Sec. 2 we present some theoretical results connecting the issue of ranking of vectors using OM3 operators. The simulation results, concerning both fixed- and variable-length scenarios, are discussed in Sec. 3. Finally, Sec. 4 concludes the paper.

2 Theoretical Results

We are going to analyze the correlation/association between rankings naturally created by aggregation with OM3 operators to assess their "global" change caused by vector "calibration". This is because precise values of OM3 operators applied to variously transformed input vectors are rather meaningless. Such approach is often encountered in many domains in which aggregation operators are applied. For example, in scientometrics, we sometimes wish to order a set of authors according to the value of some citation-based quality measure, just to indicate a potential group of prominent scientists.

Keeping this in mind, let us present some theoretical results that may be useful when it comes to *comparing* OM3 operators' values. From now on we assume that $\mathbb{I} = [0, \infty]$.

First of all, it turns out that – as far as the ranking problem is concerned – we may assume with no loss in generality that $\triangle$ is of the following, very simple form.

Proposition 1. *Let* $\mathsf{M}_{\triangle,\mathbf{w}} \in \mathscr{P}_{\text{(nd)}} \cap \mathscr{P}_{\text{(a0)}}$ *(see [4]) such that* $\mathsf{M}_{\triangle,\mathbf{w}}(x_1,\ldots,x_n) = \bigvee_{i=1}^n \mathsf{w}(x_{(n-i+1)}) \wedge c_i$, *where* w *is strictly increasing and* $c_1 < c_2 < \ldots$. *Then there exist increasing functions* $\mathsf{f}, \mathsf{w}' : \mathbb{I} \to \mathbb{I}$ *for which for all* $\mathbf{x} \in \mathbb{I}^{1,2,\ldots}$ *it holds* $\mathsf{M}_{\triangle,\mathbf{w}}(\mathbf{x}) = \mathsf{f}\left(\mathsf{M}_{\triangledown,\mathbf{w}'}(\mathbf{x})\right) = \mathsf{f}\left(\bigvee_{i=1}^n \left(\mathsf{w}'(x_{(n-i+1)}) \wedge i\right)\right)$.

Proof. Let f be a piecewise linear continuous function such that for $i = 1, 2, \ldots$ we have $\mathsf{f}(i) = c_i$. It is obvious that f is a strictly increasing function, since the sequence $(c_i)_{i\in\mathbb{N}}$ is strictly increasing, and onto $\mathbb{I}$. Hence, there exists its (also strictly increasing) inverse, f^{-1}, for which we have $\mathsf{f}^{-1}(c_i) = i$. Thus, $\mathsf{f}^{-1}\left(\mathsf{M}_{\triangle,\mathbf{w}}(\mathbf{x})\right) = \bigvee_{i=1}^n \left(\mathsf{f}^{-1}(\mathsf{w}(x_{(n-i+1)})) \wedge \mathsf{f}^{-1}(c_i)\right) = \bigvee_{i=1}^n \left((\mathsf{f}^{-1} \circ \mathsf{w})(x_{(n-i+1)}) \wedge i\right)$ for any $\mathbf{x} \in \mathbb{I}^{1,2,\ldots}$. We may therefore set $\mathsf{w}' = \mathsf{f}^{-1} \circ \mathsf{w}$, which completes the proof. $\square$

Moreover, please note that for $\mathsf{M}_{c\mathsf{w}}(\mathbf{x}) = \bigvee_{i=1}^n c\mathsf{w}(x_{(n-i+1)}) \wedge i$, where $\mathsf{w} : \mathbb{I} \to \mathbb{I}$ is increasing, $\mathsf{w}(\infty) < \infty$, we may easily show that the following results hold.

Remark 1. For any $\mathbf{x} \in \mathbb{I}^n$, $x_{(n)} < \infty$, we have $\lim_{c \to 0^+} \mathsf{M}_{c\mathsf{w}}(\mathbf{x}) \sim \mathsf{MAX}(\mathsf{w}(\mathbf{x}))$.

Remark 2. For any $\mathbf{x} \in \mathbb{I}^n$, $x_{(1)} > 0$, it holds $\lim_{c \to \infty} \mathsf{M}_{c\mathsf{w}}(\mathbf{x}) = n$.

Therefore, we see that, intuitively, the rankings generated by some zero-insensitive OM3 operators "fall somewhere between" those generated by two very simple functions, one concerning only the producer's ability to output artifacts of high quality, and the other reflecting solely his/her productivity.

3 Simulation Study

We conducted simulation studies to assess the impact of input vector calibration on the output values of OM3 operators. We considered the following classes of functions:

- $\mathsf{M}_c(\mathbf{x}) = \bigvee_{i=1}^n cx_{(n-i+1)} \wedge i$,
- $\mathsf{M}_{c\log}(\mathbf{x}) = \bigvee_{i=1}^n c\log\left(1 + x_{(n-i+1)}\right) \wedge i$,
- $\mathsf{M}_{\log c}(\mathbf{x}) = \bigvee_{i=1}^n \log\left(1 + cx_{(n-i+1)}\right) \wedge i$,

where $c \in \mathbb{R}_+$ is a scaling parameter. Note that the scaling operation is often performed on real-world data. For example, in scientometrics one may be interested in "normalizing" citations so that they reflect various characteristics of different fields (e.g. a citation in mathematics may be "worth" more than in biology), cf. [1]. The use of the logarithm is motivated by the fact that in most instances of the Producers Assessment Problem we encounter heavily-tailed and skewed data distributions.

Additionally, we considered four reference aggregation operators:

- $MAX(\mathbf{x}) = x_{(n)}$,
- generalized h-index given by $HIRSCH(\mathbf{x}) = \bigvee_{i=1}^{n} x_{(n-i+1)} \wedge i$, cf. [9],
- $MED(\mathbf{x})$ (sample median),
- $\Sigma \log(\mathbf{x}) = \sum_{i=1}^{n} (\log(1 + x_{(n-i+1)}))$.

Note that the first and the second function belongs to OM3. Moreover, as it was mentioned in [4], MAX is the only OM3 operator satisfying properties $\mathscr{P}_{(F+)}$ and $\mathscr{P}_{(F0)}$ (and $\mathscr{P}_{(a0)}$). HIRSCH, on the other hand, fulfills $\mathscr{P}_{(F0)}$ (which implies $\mathscr{P}_{(a0)}$). For $c \in \mathbb{R}_+$ such that $c \leq 1$, M_c fulfills $\mathscr{P}_{(F0)}$, but when $c > 1$, then it belongs only to $\mathscr{P}_{(a0)}$. Operators $M_{c\log}$ and $M_{\log c}$ satisfy $\mathscr{P}_{(a0)}$.

Spearman's Rank Correlation Coefficient. The effect of input vector calibration was evaluated by measuring the correlation between rankings created by OM3 values calculated for different scaling parameters. To assess the strength of correlation, we used Spearman's correlation coefficient, which is a rank-based measure of association between two vectors. Technically, it is defined as the Pearson correlation coefficient between the ranks of elements. However, unlike Pearson's r, which gives good results only when there is linear dependency, Spearman's ρ gives sensible results when $\mathbf{y}$ is a monotonic transformation of $\mathbf{x}$. What is more, since it is a nonparametric measure, it releases us from assumptions about variables' distribution. In this paragraph we recall the definition of Spearman's ρ and its basic properties.

Definition 2. Let $((x_1, y_1), \ldots, (x_n, y_n))$ be a two-dimensional sample and let $R_i = r(x_i)$ and $S_i = r(y_i)$ denote the ranks of x_i and y_i, respectively, i.e. $x_i = x_{(R_i)}$ and $y_i = y_{(S_i)}$. Then Spearman's rank correlation coefficient is given by

$$\rho(\mathbf{x}, \mathbf{y}) = \frac{\sum_{i=1}^{n} (R_i - \frac{n+1}{2})(S_i - \frac{n+1}{2})}{\sqrt{\sum_{i=1}^{n} (R_i - \frac{n+1}{2})^2 \sum_{i=1}^{n} (S_i - \frac{n+1}{2})^2}}.$$

Spearman's ρ takes its values in $[-1, 1]$ and represents the degree of correlation between $\mathbf{x}$ and $\mathbf{y}$. In particular, the closer Spearman's ρ is either to 1 or -1, the stronger the correlation between $\mathbf{x}$ and $\mathbf{y}$ is. The sign of the Spearman correlation indicates the direction of association between $\mathbf{x}$ and $\mathbf{y}$. Moreover, in the context of probability, when variables are independent, the distribution of ρ not only does not depend on the joint probability distribution of $(\mathbf{x}, \mathbf{y})$, but also it holds $\mathbb{E}\rho(\mathbf{x}, \mathbf{y}) = 0$.

Experimental Data. Input vectors were generated from type II Pareto (Lomax) distribution family, $P2(k, s)$ (where $s > 0$ and $k > 0$), given by density function $f(x) = \frac{ks^k}{(s+x)^{k+1}}$, for $x \in \mathbb{I} = [0, \infty]$. This class of heavy-tailed, right-skewed distributions is often used in e.g. scientometrical modeling (where sometimes $k \in [1, 2]$ and $s = 1$ is assumed), see e.g. [2, 12, 13]. In this setting, a Pareto distribution describes a producer's ability to produce artifacts of various quality measures. Therefore, our knowledge of the producer's skills are given solely by k and s here.

For the sake of simplicity, we assumed that $s = 1$. The shape parameter k was randomly generated for every vector from the uniform distribution on interval $(1, 2)$, i.e. $k \sim U(1, 2)$, or the $P2(1, 1)$ distribution shifted by one, i.e. $k \sim P2(1, 1) + 1$.

These model a population of producers of different abilities. Additionally, we considered the cases of producers of equal skills, with k equal to 1, 1.5, or 2. The calibration parameters c were taken from $[0.001, 10000]$.

We considered three simulation scenarios. In the first one, input vectors' length n was the same for all vectors. In the second one, we will examine the correlation for vectors of equal lengths and their expanded versions (cf. the arity-monotonicity property from [4]). In the last scenario, for each vector we generated their lengths randomly. In each step, $MC = 100000$ Monte Carlo samples were generated. The computations were performed with the agop package [8] for R.

3.1 Vectors of Fixed Lengths

First we analyzed fixed input vectors' lengths which were set to $n = 25, 100, 250$, and 1000 elements. Please note that this may be interpreted as an evaluation of producers of the same productivity.

Let us examine the sensitivity of OM3-generated rankings to vectors calibration. We calculated Spearman's rank correlation for $(M_c(\mathbf{x}_1), M_{c+\delta}(\mathbf{x}_1)), \ldots, (M_c(\mathbf{x}_{MC}), M_{c+\delta}(\mathbf{x}_{MC}))$, and the same for the other operators. Two plots in Fig. 1 depict some exemplary, but representative results concerning, respectively, the functions $M_{\log c}$ and $M_{c \log}$ for $n = 25, k \sim U(1, 2)$.

We note that for small δ, the value of $\rho(M_c, M_{c+\delta})$ if relatively high (≥ 0.9 for $n = 25$). However, in most of the analyzed cases we observe a decrease in correlation strength for $c \simeq 0.4$, which may indicate some sort of ranking instability. Therefore, as far as applications are concerned, the scaling parameters should be chosen with care.

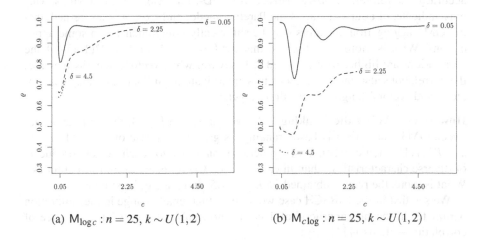

(a) $M_{\log c} : n = 25, k \sim U(1, 2)$ (b) $M_{c \log} : n = 25, k \sim U(1, 2)$

Fig. 1 The effect of adding small values to the calibration parameter

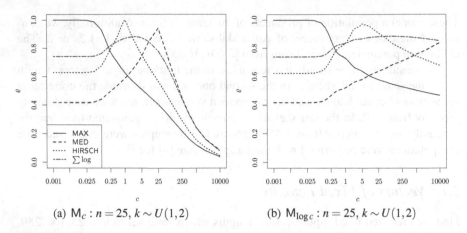

(a) $M_c : n = 25, k \sim U(1,2)$ (b) $M_{\log c} : n = 25, k \sim U(1,2)$

Fig. 2 Spearman's rank correlation coefficient between OM3 and the reference rankings

Let us now consider the correlation between M_c, $M_{\log c}$, and $M_{c\log}$, and the reference rankings, i.e. those generated by MAX, MED, HIRSCH, and Σlog. Two exemplary cases are depicted in Fig. 2. Please note the log scale for c on the x axis.

Obviously, in each case for small c we get the same ranking as for the MAX function (see Remark 1). On the other hand, as c approaches ∞, the OM3 rankings are uncorrelated with the reference ones (see Remark 2, e.g MAX and n are independent random variables). Also note that M_1 = HIRSH.

In all the analyzed cases we observed quite similar behavior of the four functions. Interestingly, with each of the three OM3 classes we can obtain, with a good accuracy, the reference, MAX-, HIRSCH-, MED-, and Σlog-based rankings. This may indicate, of course as far as the Paretian model and fixed n is concerned, that the OM3 aggregation operators may be sufficiently comprehensive in some applications. What is more, we observed that for $k \sim U(1,2)$ or $k \sim P2(1,1)+1$ the correlations are higher then for fixed k. Likewise, when vectors' lengths increase, the correlations also increase. Let us now investigate the influence of shape parameter's and vectors' lengths n selection deeper.

How Does k Affect the Rankings? As we can see in Fig. 3, the correlation between OM3- and HIRSCH-based rankings is greater in a case of $k \sim U(1,2)$ and $k \sim P2(1,1)+1$, i.e. when k was generated randomly for each vector (producers of diverse characteristics), than in case of fixed k (producers of uniform abilities). What is more, the results obtained for $k = 1$, 1.5, and 2 are quite similar.

We see that in the HIRSCH case we observe that small change in the calibration parameter in the neighborhood of 1 causes noticeable decrease of the degree of correlation – cf. also [11].

How Does n Affect the Rankings? In Fig. 4 we depict the case of producers of different productivity. For HIRSCH we observe that for randomly generated k, the

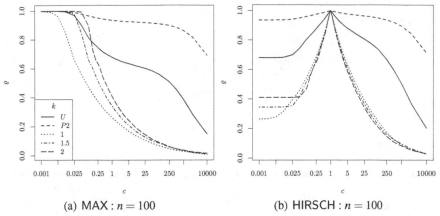

(a) MAX : $n = 100$ (b) HIRSCH : $n = 100$

Fig. 3 Spearman's rank correlation coefficient between OM3- and, respectively, MAX- and HIRSCH-based rankings for different k generation methods.

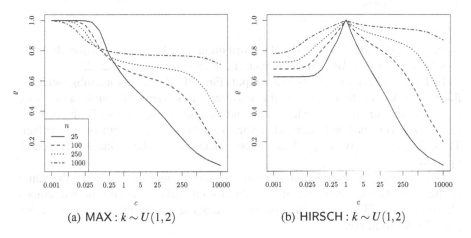

(a) MAX : $k \sim U(1,2)$ (b) HIRSCH : $k \sim U(1,2)$

Fig. 4 Spearman's ρ between OM3- and, respectively, MAX- and HIRSCH-based rankings for different n.

bigger n is, the larger correlations we get. However, for fixed k the behavior is more complicated. For small c we notice larger ρ for smaller n.

3.2 Vector Expansion

In the next scenario we represent the case in which we have a set of producers with $n_0 = 25$ artifacts. They are assessed with different OM3 operators. Then, to each vector describing the producer, we add new elements. Of course, according to the arity-monotonicity property, their valuation does not decrease (cf. [4]). The number of added elements, Δn, was independently generated for each producer

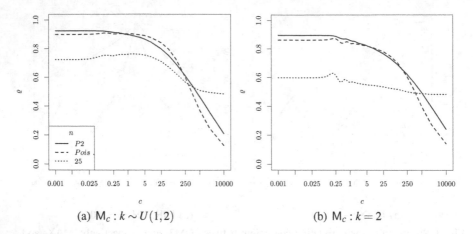

(a) $M_c : k \sim U(1,2)$ (b) $M_c : k = 2$

Fig. 5 Spearman's rank correlation coefficient between OM3-ranked original vectors and their expanded versions.

from the heavy-tailed $\lfloor P2(1,1)+1 \rfloor$ distribution and the shifted Poisson distribution $Pois(5)+1$ ($\mathrm{Var}\Delta n = 5$). Moreover, $\Delta n = 25$ was also considered.

In Fig. 5 we presented a typical output. First of all, there is no substantial influence of the Δn distribution in the analyzed cases. Here, of course, as $c \to \infty$, $\rho \to 0$ (n and Δn are independent). For small and moderate values of c the correlation between original and extended vectors' valuations are high, but yet not perfect. Thus, the productivity of a producer indeed affects also his/her valuation with OM3 operators.

For $\Delta n = 25$ the correlation is lower, but much more insensitive to the value of the calibration coefficient. Note that the default ranking method for tied (equal) observations in R's cor() function uses averaging, therefore for $c \to \infty$ we get $\rho \to 0.5$ for fixed Δn.

3.3 Vectors of Random Lengths

In the last scenario let us examine a set of producers of random productivity. We considered $n \sim \lfloor P2(1,1)+1 \rfloor$, $n \sim Pois(5)+1$, and $n \sim \lfloor U[1,500] \rfloor$.

Fig. 6ab depicts the correlation between OM3 and reference-based rankings in the first two cases. Note that the density functions of these distributions are decreasing. Therefore, there is a relatively large probability of obtaining small values of n: for $n \sim \lfloor P2(1,1)+1 \rfloor$ we have $\mathrm{Med}\, n = 1$ and for $n \sim Pois(5)+1$ we get $\mathrm{Med}\, n = 6$.

Fig. 6c depicts the $n \sim \lfloor U[1,500] \rfloor$ and the fixed k case. It may be shown that if $(X_1, \ldots, X_n)$ i.i.d. $P2(k,1)$ then $Y_n := \sum_{i=1}^{n} \log(X_i + 1) \sim \Gamma(n, 1/k)$ and $\mathbb{E}Y_n = n/k$. This explains a high degree of correlation between $\Sigma\log$ and M_c for $c \to \infty$.

While describing the results instantiated in Fig. 2 (fixed n) we noted that OM3 class is quite flexible in terms of approximating the two reference aggregation

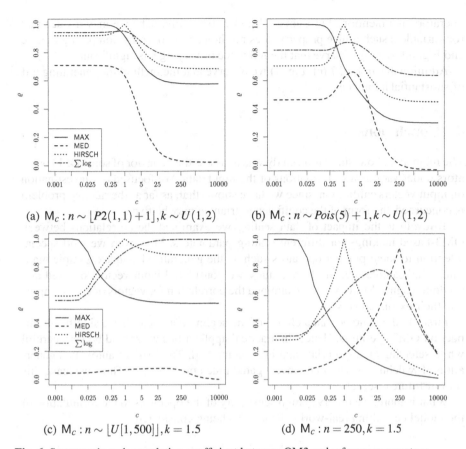

(a) $M_c : n \sim \lfloor P2(1,1)+1 \rfloor, k \sim U(1,2)$

(b) $M_c : n \sim Pois(5)+1, k \sim U(1,2)$

(c) $M_c : n \sim \lfloor U[1,500] \rfloor, k = 1.5$

(d) $M_c : n = 250, k = 1.5$

Fig. 6 Spearman's rank correlation coefficient between OM3 and references operators

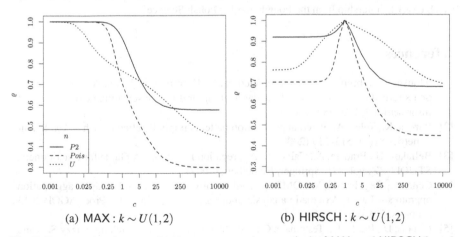

(a) $MAX : k \sim U(1,2)$

(b) $HIRSCH : k \sim U(1,2)$

Fig. 7 Spearman's ρ between OM3- operators and, respectively, MAX- and HIRSCH-based rankings for different n generation methods.

operators, not mentioned in Remark 1 and 2. From Fig. 6abc we may deduce that for variable n such a nice property does not hold. Moreover, by comparing Fig. 6c and Fig. 6d we may observe that the influence of varying n is significant.

Additionally, in Fig. 7 (cf. Fig. 4) we observe that the method for generating n if of substantial influence.

4 Conclusions

The main aim of our simulation study was to assess the behavior of some OM3 operators under various transformations of the input data. We focused our investigation on input vectors calibration, since we have shown that, as far as the ranking problem is concerned, the form of the coefficients' triangle may be fixed.

To evaluate the impact of data scaling we examined the correlations between OM3-based rankings for different scaling parameters. Moreover, we paid special attention to some popular operators such as the generalized h-index, sample maximum and sample median. In our study we considered input vectors of fixed and random lengths. Moreover, we examined the correlation for vectors of equal lengths and their expanded versions.

First of all, we noted that a choice of the scaling parameter has a significant impact on OM3 operators. Hence, in practical applications we should be very careful while selecting an appropriate aggregation method. The issue of automated generation of w definitely should be investigated much more deeply. Thus, we leave this for our future research.

What is more, we observed high sensitivity of the operators to the formulation of the model describing real-world phenomena being considered.

Acknowledgements. The contribution of Marek Gagolewski was partially supported by FNP START Scholarship from the Foundation for Polish Science.

References

[1] Alonso, S., Cabrerizo, F.J., Herrera-Viedma, E., Herrera, F.: h-index: A review focused on its variants, computation and standardization for different scientific fields. Journal of Informetrics 3, 273–289 (2009)

[2] Barcza, K., Telcs, A.: Paretian publication patterns imply Paretian Hirsch index. Scientometrics 81(2), 513–519 (2009)

[3] Beliakov, G., Pradera, A., Calvo, T.: Aggregation Functions: A Guide for Practitioners. STUDFUZZ, vol. 221. Springer, Heidelberg (2007)

[4] Cena, A., Gagolewski, M.: OM3: ordered maxitive, minitive, and modular aggregation operators – Part I: Axiomatic analysis under arity-dependence. In: Proc. AGOP 2013 (in press, 2013)

[5] Dubois, D., Prade, H., Testemale, C.: Weighted fuzzy pattern matching. Fuzzy Sets and Systems 28, 313–331 (1988)

[6] Franceschini, F., Maisano, D.A.: The Hirsch index in manufacturing and quality engineering. Quality and Reliability Engineering International 25, 987–995 (2009)

[7] Gagolewski, M.: On the relationship between symmetric maxitive, minitive, and modular aggregation operators. Information Sciences 221, 170–180 (2013)

[8] Gagolewski, M., Cena, A.: agop: Aggregation Operators Package for R (2013), http://www.ibspan.waw.pl/~gagolews/agop/

[9] Gągolewski, M., Grzegorzewski, P.: Arity-monotonic extended aggregation operators. In: Hüllermeier, E., Kruse, R., Hoffmann, F. (eds.) IPMU 2010. CCIS, vol. 80, pp. 693–702. Springer, Heidelberg (2010)

[10] Gagolewski, M., Grzegorzewski, P.: Possibilistic analysis of arity-monotonic aggregation operators and its relation to bibliometric impact assessment of individuals. International Journal of Approximate Reasoning 52(9), 1312–1324 (2011)

[11] Gagolewski, M., Mesiar, R.: Aggregating different paper quality measures with a generalized h-index. Journal of Informetrics 6(4), 566–579 (2012)

[12] Glänzel, W.: H-index concatenation. Scientometrics 77(2), 369–372 (2008a)

[13] Glänzel, W.: On some new bibliometric applications of statistics related to the h-index. Scientometrics 77(1), 187–196 (2008b)

[14] Grabisch, M., Marichal, J.L., Mesiar, R., Pap, E.: Aggregation functions, Cambridge (2009)

[15] Hirsch, J.E.: An index to quantify individual's scientific research output. Proceedings of the National Academy of Sciences 102(46), 16,569–16,572 (2005)

Duplication in OWA-Generated Positional Aggregation Rules

José Luis García-Lapresta and Miguel Martínez-Panero

Abstract. In this paper we deal with positional aggregation rules where the alternatives are socially ordered according to their aggregated positions. These positional values are generated by means of a predetermined aggregation function from the positions in the corresponding individual orderings. Specifically, our interest is focused on OWA-generated positional aggregation rules and, as a first step in our research, we characterize those ones satisfying duplication and propose an overall social order induced by them.

1 Introduction

According to Gärdenfors [13], "positionalist voting functions are those social choice functions where the positions of the alternatives in the agents' preference orders crucially influence the social ordering of the alternatives". This is a vague notion that can be understood in different ways[1]. The most popular case of positional aggregation rules are scoring rules[2], where a score is associated with each position and alternatives are socially ordered by the sum of scores obtained from the individual orderings.

However, scoring rules are not exclusive to capture positionalist features of voting. In fact, our proposal based on aggregation functions (mainly through OWAs) sheds light to some aspects not taken into account in the scoring approach. One of

José Luis García-Lapresta · Miguel Martínez-Panero
PRESAD Research Group, IMUVA, Dep. of Applied Economics,
University of Valladolid, Spain
e-mail: {lapresta,panero}@eco.uva.es

---

[1] See Pattanaik [21], specially Section 3.

[2] See Chebotarev and Shamis [7] for a referenced survey on scoring rules and their characterizations.

H. Bustince et al. (eds.), *Aggregation Functions in Theory and in Practise*,   117
Advances in Intelligent Systems and Computing 228,
DOI: 10.1007/978-3-642-39165-1_15, © Springer-Verlag Berlin Heidelberg 2013

these interesting properties, not satisfied by the scoring rules, is the duplication principle. This property, appearing naturally in several contexts, entails a sort of irrelevance of clone voters in the final result and might not seem suitable at all in voting scenarios. But it will be shown that it is related to some concrete OWA operators inducing positional voting rules and intended to be used under complete ignorance.

The paper is organized as follows. In Section 2 we introduce the basic notation for the preferences of the agents over the alternatives and their related positions. Section 3 is devoted to aggregation rules and aggregation functions; specifically, we focus our attention on OWAs and show their connections with some well-known voting systems appearing in the literature. The need of taking into account a variable electorate leads us to use extended OWAs (EOWAs) and, with this background, in Section 4 we define duplication and then we characterize those OWA-generated positional aggregation rules satisfying this property. An illustrative example is also presented and, finally, a proposal of an overall social order based on the characterized rules is obtained in a unifying way.

2 Preliminaries

Consider a set of agents $V = \{1, \ldots, m\}$, with $m \in \mathbb{N}$, who show their preferences on a set of alternatives $X = \{x_1, \ldots, x_n\}$, with $n \geq 2$. With $L(X)$ we denote the set of *linear orders* on X, and with $W(X)$ the set of *weak orders* (or *complete preorders*) on X. Given $R \in W(X)$, with $\succ$ and $\sim$ we denote the asymmetric and the symmetric parts of R, respectively. A *profile* is a vector $\boldsymbol{R} = (R_1, \ldots, R_m)$ of weak orders, where R_v contains the preferences of the agent v, with $v = 1, \ldots, m$. Vectors in $\mathbb{R}^n$ are denoted as $\boldsymbol{a} = (a_1, \ldots, a_n)$. Given $\boldsymbol{a}, \boldsymbol{b} \in \mathbb{R}^n$, with $\boldsymbol{a} \leq \boldsymbol{b}$ we mean $a_i \leq b_i$ for every $i \in \{1, \ldots, n\}$.

Definition 1. Given $R \in W(X)$, the *position* of alternative $x_i \in X$ is defined as

$$p(x_i) = n - \#\{x_j \in X \mid x_i \succ x_j\} - \frac{1}{2}\#\{x_j \in X \setminus \{x_i\} \mid x_j \sim x_i\}. \tag{1}$$

It is equivalent to linearize the weak order and to assign each alternative the average of the positions of the alternatives within the same equivalence class (see, for instance, Smith [23] for a similar procedure in the context of scoring rules).

Example 1. Consider $R \in W(\{x_1, \ldots, x_7\})$ given by

$$\begin{array}{c} \hline R \\ \hline x_2 \ x_3 \ x_5 \\ x_1 \\ x_4 \ x_7 \\ x_6 \end{array}$$

Then,

$$p(x_2) = p(x_3) = p(x_5) = \frac{1+2+3}{3} = 2 = 7 - 4 - \frac{1}{2}2,$$

$$p(x_1) = 4 = 7 - 3 - \frac{1}{2}0,$$

$$p(x_4) = p(x_7) = \frac{5+6}{2} = 5.5 = 7 - 1 - \frac{1}{2}1,$$

$$p(x_6) = 7 = 7 - 0 - \frac{1}{2}0.$$

Consequently, R is codified by the *positions vector*

$$(p(x_1), p(x_2), p(x_3), p(x_4), p(x_5), p(x_6), p(x_7)) = (4, 2, 2, 5.5, 2, 7, 5.5).$$

Taking into account the positions of the alternatives, every profile $\boldsymbol{R} \in W(X)^m$ has associated a *position matrix* containing the positions of the alternatives for all the agents

$$\begin{pmatrix} p_1(x_1) & p_1(x_2) & \cdots & p_1(x_n) \\ p_2(x_1) & p_2(x_2) & \cdots & p_2(x_n) \\ \cdots & \cdots & \cdots & \cdots \\ p_m(x_1) & p_m(x_2) & \cdots & p_m(x_n) \end{pmatrix},$$

where $p_v(x_i)$ is the position of x_i for agent v. Thus, row v contains the positions of the alternatives according to agent v, and column i contains the positions of the alternative x_i.

3 The Aggregation Process

Given a domain $D \subseteq W(X)^m$ with $m \in \mathbb{N}$, an *aggregation rule on D* is a mapping $F : D \longrightarrow W(X)$ that satisfies the following conditions:

1. *Anonymity*: For every permutation π on $\{1, \ldots, m\}$ and every profile $\boldsymbol{R} \in D$, it holds

$$F\left(R_{\pi(1)}, \ldots, R_{\pi(m)}\right) = F\left(R_1, \ldots, R_m\right).$$

2. *Neutrality*: For every permutation σ on $\{1, \ldots, n\}$ and every profile $\boldsymbol{R} \in D$, it holds

$$F\left(R_1^\sigma, \ldots, R_m^\sigma\right) = \left(F(R_1, \ldots, R_m)\right)^\sigma,$$

where R_v^σ and $(F(R_1, \ldots, R_m))^\sigma$ are the orders obtained from R_v and $F(R_1, \ldots, R_m)$, respectively, by relabeling the alternatives according to σ, i.e., $x_{\sigma(i)} R_v^\sigma x_{\sigma(j)} \Leftrightarrow x_i R_v x_j$ and $x_{\sigma(i)} (F(R_1, \ldots, R_m))^\sigma x_{\sigma(j)} \Leftrightarrow x_i F(R_1, \ldots, R_m) x_j$.

3. *Unanimity*: For every profile $\boldsymbol{R} \in D$ and all $x_i, x_j \in X$, it holds

$$(\forall v \in V \ x_i R_v x_j) \Rightarrow x_i F(\boldsymbol{R}) x_j.$$

Anonymity means a symmetric consideration for the agents; neutrality means a symmetric consideration for the alternatives; and unanimity means that if all the individuals consider an alternative as good as another one, then the social preference coincides with the individual preferences on this issue.

It is worth mentioning that the setting of aggregation rules, where the outcome is a social order (as in Smith [23]), is not the unique framework in Social Choice Theory. Other possibilities can be taken into account, such as social choice correspondences, where the result is the (nonempty) subset of the best alternatives (as in Young [29, 30]; see also Laslier [16] for further rank-based and pairwise-based approaches), or even social choice functions, where a single alternative is assigned to each profile[3].

3.1 Aggregation Functions

In our proposal, we have extended the notion of aggregation function to the unbounded interval $[1, \infty)$. On aggregation functions in the standard unit interval, see Calvo *et al.* [5], Beliakov *et al.* [4] and Grabisch *et al.* [14].

Definition 2. $A : [1, \infty)^m \longrightarrow [1, \infty)$ is an *aggregation function* if it satisfies the following conditions:

1. *Boundary condition*: $A(1, \ldots, 1) = 1$.
2. *Monotonicity*: $a \leq b \Rightarrow A(a) \leq A(b)$, for all $a, b \in [1, \infty)^m$.

If, additionally, A satisfies *idempotency*, i.e., $A(a, \ldots, a) = a$ for every $a \in [1, \infty)$, then A is called *averaging aggregation function*.

It is easy to see that averaging aggregation functions satisfy *compensativeness*:

$$\min\{a_1, \ldots, a_m\} \leq A(a_1, \ldots, a_m) \leq \max\{a_1, \ldots, a_m\},$$

for every $(a_1, \ldots, a_m) \in [1, \infty)^m$. Typical averaging aggregation functions are the arithmetic mean, trimmed means, the median, the maximum, the minimum, etc. In fact, we can gather all these aggregation functions as particular cases of OWA operators[4].

A *weighting vector* of dimension m is a vector $w = (w_1, \ldots, w_m) \in [0, 1]^m$ such that $\sum_{i=1}^{m} w_i = 1$.

Definition 3. Given a weighting vector w of dimension m, the *OWA operator associated with* w is the mapping $A_w : [1, \infty)^m \longrightarrow [1, \infty)$ defined by

---

[3] As pointed out by Courtin *et al.* [9], differences in the axiomatic treatment arise depending on the type of social mechanism considered.

[4] The initials in OWA stand for *ordered weighted averaging*, see Yager [25], Yager and Kacprzyk [27] and Yager *et al.* [28]. A characterization of the OWA operators has been given by Fodor *et al.* [10].

$$A_w(a_1, \ldots, a_m) = \sum_{i=1}^{m} w_i \cdot a_{[i]},$$

where $a_{[i]}$ is the i-th greatest number of $\{a_1, \ldots, a_m\}$.

As noted before, some well-known aggregation functions are specific cases of OWA operators.

With appropriate weighting vectors $w = (w_1, \ldots, w_m)$ we obtain

1. The *maximum*, for $w = (1, 0, \ldots, 0)$.
2. The *minimum*, for $w = (0, \ldots, 0, 1)$.
3. The *arithmetic mean*, for $w = \left(\frac{1}{m}, \ldots, \frac{1}{m}\right)$.
4. The *k-trimmed means*:

   - If $k = 1$, $w = \left(0, \frac{1}{m-2}, \ldots, \frac{1}{m-2}, 0\right)$.
   - If $k = 2$, $w = \left(0, 0, \frac{1}{m-4}, \ldots, \frac{1}{m-4}, 0, 0\right)$.
   - ...

5. The *median*:

   a. If m is odd, $w_i = \begin{cases} 1, & \text{if } i = \frac{m+1}{2}, \\ 0, & \text{otherwise.} \end{cases}$

   b. If m is even, $w_i = \begin{cases} \frac{1}{2}, & \text{if } i \in \{\frac{m}{2}, \frac{m}{2}+1\}, \\ 0, & \text{otherwise.} \end{cases}$

6. The *mid-range*, for $w = (0.5, 0, \ldots, 0, 0.5)$.

3.2 Positional Aggregation Rules

Definition 4. Given an aggregation function $A : [1, \infty)^m \longrightarrow [1, \infty)$ and a profile $R \in W(X)^m$, the *aggregated position* of the alternative $x_i \in X$ is defined as

$$p_A(x_i) = A(p_1(x_i), \ldots, p_m(x_i)),$$

where $p_v(x_i)$ is the position of x_i for agent $v \in V$.

Definition 5. Given an aggregation function $A : [1, \infty)^m \longrightarrow [1, \infty)$, the *positional aggregation rule associated with A* is the mapping $F_A : W(X)^m \longrightarrow W(X)$ defined by $F_A(R) = \succcurlyeq_A$, where

$$x_i \succcurlyeq_A x_j \Leftrightarrow p_A(x_i) \leq p_A(x_j).$$

For example, taking into account some of the OWA operators introduced above, we obtain positional aggregation rules which are connected to (or even replicate) well-known procedures appearing in the literature:

- The arithmetic mean as aggregation operator induces the Borda rule. And it is worth mentioning that the arithmetic mean is also the basis for the *Range Voting* method (Smith [24]), in a decisional context where the alternatives receive numerical assessments one by one.
- The median instead of the arithmetic mean, and linguistic terms instead of numerical values, are used in the *Majority Judgment* voting system supported by Balinski and Laraki [2]. An extension of this procedure using centered OWA operators (Yager [26]) appears in García-Lapresta and Martínez-Panero [12]. Again, in a different scenario, Basset and Persky [3] already proposed to select the alternative with best median evaluation (see also Laslier [18]).
- The maximum leads to a voting system in which each alternative is evaluated according to the worst reached position. Those with the best assigned value are then elected. Such a *maximin* voting system, which advocates the maximin principle of normative economics[5], is characterized by Congar and Merlin [8] and the same idea is also the key for the *leximin* voting system appearing in Laslier [17], although in a different decisional framework (see also Laslier [18]). This is also the case for the Simpson-Kramer method (see Levin and Nalebuff [19]) in a pair-wise comparison context. Furthermore, the procedure obtained through the maximum as aggregation operator is also related to the *Coombs* method (where the alternative with the largest number of last positions is sequentially withdrawn), as well as to the antiplurality rule (see Baharad and Nitzan [1]).
- The minimum entails a voting system called *maximax*[6] by Congar and Merlin [8], also characterized by them. Its conception is similar to that of the *Hare* system, also known as *Alternative Vote* (where the alternative with the fewest first positions is sequentially withdrawn). It is also related to the most used (and criticized) system: plurality rule (see Laslier [17]).
- The mid-range OWA operator is related to the basic *best-worst* voting system (see García-Lapresta *et al.* [11]).

It is easy to check the following result.

Proposition 1. F_A *is an aggregation rule for every aggregation function A.*

In order to take into account a variable electorate (for example, to deal with the clonation or appearance of new agents), we introduce some extended notions of those already defined throughout the paper.

Definition 6. An *extended aggregation rule* is a mapping

$$\widetilde{F} : \bigcup_{m \in \mathbb{N}} W(X)^m \longrightarrow W(X)$$

such that $F_m = \widetilde{F}|_{W(X)^m}$ is an aggregation rule for each $m \in \mathbb{N}$ and $F_1(R) = R$.

---

[5] Rawls [22, p. 328]: "the basic structure is perfectly just when the prospects of the least fortunate are as great as they can be".

[6] The apparent discordance leading the maximum to the maximin voting system, as well as the minimum to the maximax, relies on our positional approach where, contrary to the scoring context, the smallest value is associated with the best position.

Definition 7. An *extended OWA operator (EOWA)* is a family of OWA operators with associated weighting vectors $w^m = (w_1^m, \ldots, w_m^m)$, one for each dimension $m = 1, 2, 3, \ldots$

Following Calvo and Mayor [6] and Mayor and Calvo [20] (see also Beliakov *et al.* [4, pp. 54-56]), we can show graphically an EOWA operator as a weighting triangle

$$
\begin{array}{ccccc}
 & & w_1^1 & & \\
 & w_1^2 & & w_2^2 & \\
 & w_1^3 & w_2^3 & w_3^3 & \\
w_1^4 & w_2^4 & w_3^4 & w_4^4 & \\
w_1^5 & w_2^5 & w_3^5 & w_4^5 & w_5^5
\end{array}
$$

$$\cdots \ \cdots \ \cdots \ \cdots \ \cdots \ \cdots \ \cdots \ \cdots \ \cdots$$

For simplicity, from now on superindexes will be avoided when confusion is not possible.

4 Duplication

Here we introduce a property which, broadly speaking, states that new voters replicating the same preferences of already existing voters will not affect the outcome. This (at first sight) non-compelling property appears as *duplication* in Congar and Merlin [8], where they characterize the maximin procedure.

Definition 8. An extended aggregation rule $\widetilde{F}$ satisfies *duplication* if

$$F_{m+1}(\mathbf{R}, R_i) = F_m(\mathbf{R})$$

for every profile $\mathbf{R} = (R_1, \ldots, R_i, \ldots, R_m) \in W(X)^m$ and every $i \in \{1, \ldots, m\}$.

4.1 A Characterization Result

It is interesting to find those procedures satisfying duplication, and the following result shows the answer for aggregation rules associated with EOWAs.

Proposition 2. *Given an EOWA operator A, the extended aggregation rule $\widetilde{F}_A$ satisfies duplication if and only if A is a rational convex combination of the maximum and the minimum EOWA operators, i.e., $w = \alpha(1, \ldots, 0) + (1 - \alpha)(0, \ldots, 1)$ for some $\alpha \in [0, 1] \cap \mathbb{Q}$.*

Proof. $\Leftarrow$) It is straightforward that aggregation rules associated with $w = (1, 0, \ldots, 0)$ (i.e., maximin), $w = (0, 0, \ldots, 1)$ (i.e., maximax), and convex combinations of them, $w = (\alpha, 0, \ldots, 0, 1 - \alpha)$ with $0 \le \alpha \le 1$, satisfy duplication.

$\Rightarrow$) We first prove that if duplication holds, all intermediate weights $w_2, \ldots, w_{m-1}$ should be zero. Our reasoning will deal with a profile consisting in all circular permutations of three ordered alternatives, but it is extensible to any order[7]. Thus, consider the profile

R_1	R_2	R_3
x_1	x_2	x_3
x_2	x_3	x_1
x_3	x_1	x_2

where the associated position matrix is

$$\begin{pmatrix} 1 & 2 & 3 \\ 3 & 1 & 2 \\ 2 & 3 & 1 \end{pmatrix}.$$

As each alternative occupies each position exactly once, a global tie arises and the aggregated position for each is $p_A(x_i) = 3w_1^3 + 2w_2^3 + w_3^3$, $i = 1, 2, 3$, so that $x_1 \sim x_2 \sim x_3$, being A the aggregation rule corresponding to any EOWA with $w^3 = (w_1^3, w_2^3, w_3^3)$.

Now suppose that agent 1 is replicated, becoming the new situation

R_1	R_2	R_3	$R_4 = R_1$
x_1	x_2	x_3	x_1
x_2	x_3	x_1	x_2
x_3	x_1	x_2	x_3

where the new associated position matrix is

$$\begin{pmatrix} 1 & 2 & 3 \\ 3 & 1 & 2 \\ 2 & 3 & 1 \\ 1 & 2 & 3 \end{pmatrix}.$$

Then, the aggregated positions for each alternative are

$$p_A(x_1) = 3w_1^4 + 2w_2^4 + w_3^4 + w_4^4,$$
$$p_A(x_2) = 3w_1^4 + 2w_2^4 + 2w_3^4 + w_4^4,$$
$$p_A(x_3) = 3w_1^4 + 3w_2^4 + 2w_3^4 + w_4^4.$$

Taking into account duplication, the tie among all three alternatives holds; hence

$$x_1 \sim_A x_2 \Leftrightarrow w_3^4 = 0,$$
$$x_1 \sim_A x_3 \Leftrightarrow w_2^4 + w_3^4 = 0,$$
$$x_1 \sim_A x_3 \Leftrightarrow w_2^4 = 0.$$

---

[7] These circular permutations yield a *Condorcet cycle*.

Then, $w_2^4 = w_3^4 = 0$. Once proven that central weights are null (this fact will be taken into account in what follows), what remains is to show that lateral weights should the same at any level, i.e., $w_1^m = \alpha$ and $w_m^m = 1 - \alpha$, for all $m \geq 2$. To do this, consider $\alpha = \frac{p}{q}$ with $p, q \in \mathbb{N}$ and $p < q$, expressed as an irreducible fraction, and any profile with m agents and $q+1$ alternatives where the alternative x_1 is at least the best for one agent and the worst for another one, while x_2 occupies the position $p+1$ for all of them. A sketch of such *ad hoc* profile would be

position	R_1	...	R_i	...	R_j	...	R_m
1	...	...	x_1	...	...	...	...
...	...	...	...	...	...	...	...
$p+1$	x_2	x_2	x_2	x_2	x_2	x_2	x_2
...	...	...	...	...	...	...	...
$q+1$	...	...	...	...	x_1	...	...

The aggregated positions for the selected alternatives would be

$$p_A(x_1) = \frac{p}{q}(q+1) + \left(1 - \frac{p}{q}\right) = p+1,$$

$$p_A(x_2) = \frac{p}{q}(p+1) + \left(1 - \frac{p}{q}\right)(p+1) = p+1,$$

so that $x_1 \sim_A x_2$, being A the aggregation rule corresponding to any EOWA with such weights.

But now, if we replicate any subset of agents becoming the new weights $\beta \neq \alpha$ and hence $1 - \beta \neq 1 - \alpha$, then the new aggregated positions would be

$$p_A(x_1) = \beta(q+1) + (1-\beta) \neq p+1,$$

$$p_A(x_2) = \beta(p+1) + (1-\beta)(p+1) = p+1,$$

so that $x_1 \sim_A x_2$ does not hold. Hence, if lateral weights change from one dimension to another, duplication fails.   □

In conclusion, under duplication we obtain the class of weighting triangles

$$1$$
$$\alpha \qquad 1-\alpha$$
$$\alpha \qquad 0 \qquad 1-\alpha$$
$$\alpha \qquad 0 \qquad 0 \qquad 1-\alpha$$

$$\cdots \cdots \cdots \cdots \cdots \quad \cdots \quad \cdots \quad \cdots \cdots \cdots$$

As particular cases we have:

$\alpha = 1$: maximum (maximin procedure),

$\alpha = 0$: minimum (maximax procedure),

$\alpha = 0.5$: mid-range.

It is worth mentioning that duplication is related to the *Hurwicz criterion* [15] used in decision making under complete uncertainty, where the value of a decision is a convex combination of its lowest possible expected value (pessimistic assessment) and of its highest one (optimistic assessment).

4.2 An Illustrative Example

Consider the profile

R_1	R_2	R_3
x_2	$x_2\ x_3$	$x_1\ x_3$
x_3	x_1	x_2
x_1		

where the associated position matrix is

$$\begin{pmatrix} 3 & 1 & 2 \\ 3 & 1.5 & 1.5 \\ 1.5 & 3 & 1.5 \end{pmatrix}.$$

If we choose an OWA $A_{w(\alpha)}$ associated with weights $w(\alpha) = (\alpha, 0, 1 - \alpha)$, the corresponding social positions for the alternatives would be:

$$p_{A_{w(\alpha)}}(x_1) = 3\alpha + 1.5(1 - \alpha) = 1.5\alpha + 1.5,$$

$$p_{A_{w(\alpha)}}(x_2) = 3\alpha + 1(1 - \alpha) = 2\alpha + 1,$$

$$p_{A_{w(\alpha)}}(x_3) = 2\alpha + 1.5(1 - \alpha) = 0.5\alpha + 1.5.$$

According to the possible values of α, the corresponding social orders are shown in the following table:

$\alpha = 0$	$0 < \alpha < \frac{1}{3}$	$\alpha = \frac{1}{3}$	$\frac{1}{3} < \alpha < 1$	$\alpha = 1$
x_2	x_2	$x_2\ x_3$	x_3	x_3
$x_1\ x_3$	x_3	x_1	x_2	$x_1\ x_2$
	x_1		x_1	

As one could expect, different social orders appear depending on α. In the following section we propose a integrating method to obtain a unified result for each alternative taking into account the different outcomes when α ranges from 0 to 1.

4.3 Overall Positions and Social Order

For the general case with n alternatives and using in a first stage the positional voting rule associated with the OWA $w(\alpha) = (\alpha, 0, \ldots, 0, 1 - \alpha)$, it is possible to assign the corresponding social position $p_{A_{w(\alpha)}}(x_i)$ to the alternative x_i. Thus, we can introduce the function $\mu_i : [0, 1] \longrightarrow \mathbb{R}$ given by $\mu_i(\alpha) = p_{A_{w(\alpha)}}(x_i)$. Such function is always piecewise constant, and hence Riemann integrable. This fact allows us to define the *overall position of x_i* as

$$p(x_i) = \int_0^1 \mu_i(\alpha)d\alpha.$$

Easy computations lead to the following results in the previous example:

$$p(x_1) = \int_0^1 \mu_1(\alpha)d\alpha = 3,$$

$$p(x_2) = \int_0^1 \mu_2(\alpha)d\alpha = 5/3,$$

$$p(x_3) = \int_0^1 \mu_3(\alpha)d\alpha = 4/3.$$

Thus, the overall social order is $x_3 \succ x_2 \succ x_1$.

In conclusion, for each $\alpha \in [0, 1]$ the corresponding positional aggregation rule associated with $p_{A_{w(\alpha)}}$ only takes into account the best and worst positions for each alternative, yielding different social orders in each case. However, the possible criticism on the influence of the choice of α in the result can be mitigated under this overall approach, where a social order is obtained not corresponding with any specific α, but amalgamating all allowable values for this parameter.

Acknowledgements. The authors would like to thank Jorge Alcalde-Unzu, Jean-François Laslier, Vincent Merlin and Marina Núñez for their comments and suggestions, as well as the Spanish Ministerio de Ciencia e Innovación (Project ECO2012-32178) and ERDF for the funding support.

References

[1] Baharad, E., Nitzan, S.: The inverse plurality rule – an axiomatization. Social Choice and Welfare 25, 173–178 (2005)
[2] Balinski, M., Laraki, R.: Majority Judgment: Measuring Ranking and Electing. MIT Press, Cambridge (2011)
[3] Bassett, G.W., Persky, J.: Robust voting. Public Choice 99, 299–310 (1999)
[4] Beliakov, G., Pradera, A., Calvo, T.: Aggregation Functions: A Guide for Practitioners. STUDFUZZ, vol. 221. Springer, Heidelberg (2007)
[5] Calvo, T., Kolesárova, A., Komorníková, M., Mesiar, R.: Aggregation operators: Properties, classes and construction methods. In: Calvo, T., Mayor, G., Mesiar, R. (eds.) Aggregation Operators: New Trends and Applications, pp. 3–104. Physica-Verlag, Heidelberg (2002)
[6] Calvo, T., Mayor, G.: Remarks on two types of extended aggregation functions. Tatra Mountains Mathematical Publications 16, 235–253 (1999)
[7] Chebotarev, P.Y., Shamis, E.: Characterizations of scoring methods for preference aggregation. Annals of Operations Research 80, 299–332 (1998)

[8] Congar, R., Merlin, V.: A characterization of the maximin rule in the context of voting. Theory and Decision 72, 131–147 (2012)

[9] Courtin, S., Mbih, B., Moyouwou, I.: Are Condorcet procedures so bad according to the reinforcement axiom? Thema Working Paper 2012-37, Université de Cergy Pontoise (2012)

[10] Fodor, J., Marichal, J.L., Roubens, M.: Characterization of the Ordered Weighted Averaging Operators. IEEE Transtactions on Fuzzy Systems 3, 236–240 (1995)

[11] García-Lapresta, J.L., Marley, A.A.J., Martínez-Panero, M.: Characterizing best-worst voting systems in the scoring context. Social Choice and Welfare 34, 487–496 (2010)

[12] García-Lapresta, J.L., Martínez-Panero, M.: Linguistic-based voting through centered OWA operators. Fuzzy Optimization and Decision Making 8, 381–393 (2009)

[13] Gärdenfors, P.: Positionalist voting functions. Theory and Decision 4, 1–24 (1973)

[14] Grabisch, M., Marichal, J.L., Mesiar, R., Pap, E.: Aggregation Functions. Cambridge University Press, Cambridge (2009)

[15] Hurwicz, L.: A class of criteria for decision-making under ignorance. Cowles Commission Discussion Paper: Statistics 356 (1951)

[16] Laslier, J.F.: Rank-based choice correspondences. Economics Letters 52, 279–286 (1996)

[17] Laslier, J.F.: And the loser is … plurality voting. In: Felsenthal, D.S., Machover, M. (eds.) Electoral Systems: Paradoxes, Assumptions, and Procedures, pp. 327–351. Springer, Berlin (2012)

[18] Laslier, J.F.: On choosing the alternative with the best median evaluation. Public Choice 153, 269–277 (2012)

[19] Levin, J., Nalebuff, B.: An introduction to vote-counting schemes. Journal of Economic Perspectives 9, 3–26 (1995)

[20] Mayor, G., Calvo, T.: On extended aggregation functions. In: Proceedings of IFSA 1997, Prague, vol. I, pp. 281–285 (1997)

[21] Pattanaik, P.K.: Positional rules of collective decision-making. In: Arrow, K.J., Sen, A.K., Suzumura, K. (eds.) Handbook of Social Choice and Welfare, vol. 1, pp. 361–394. Elsevier, Amsterdam (2002)

[22] Rawls, J.: Distributive justice. In: Phelps, E. (ed.) Economic Justice: Selected Readings, pp. 319–362. Penguin Education, Harmondsworth (1973)

[23] Smith, J.H.: Aggregation of preferences with variable electorate. Econometrica 41, 1027–1041 (1973)

[24] Smith, W.D.: Range Voting website, http://rangevoting.org/

[25] Yager, R.R.: On ordered weighted averaging operators in multicriteria decision making. IEEE Transactions on Systems Man and Cybernetics 8, 183–190 (1988)

[26] Yager, R.R.: Centered OWA operators. Soft Computing 11, 631–639 (2007)

[27] Yager, R.R., Kacprzyk, J. (eds.): The Ordered Weighted Averaging Operators: Theory and Applications. Kluwer Academic Publishers, Boston (1997)

[28] Yager, R.R., Kacprzyk, J., Beliakov, G. (eds.): Recent Developments in the Ordered Weighted Averaging Operators: Theory and Practice. Springer, Berlin (2011)

[29] Young, H.P.: An axiomatization of Borda's rule. Journal of Economic Theory 9, 43–52 (1974)

[30] Young, H.P.: Social choice scoring functions. SIAM Journal on Applied Mathematics 28, 824–838 (1975)

Application of OWA Operators in the L-Fuzzy Concept Analysis

C. Alcalde, A. Burusco, and R. Fuentes-González

Abstract. In some cases, the relationship between an object set X and an attribute set Y is set up by means of a fuzzy context sequence. A particular case of this situation appears when we want to study the evolution of an L-fuzzy context in time.

In this work, we analyze these situations. First we introduce the fuzzy context sequence definition. With the aid of the OWA operators, we propose an exhaustive study of the different contexts values of the sequence using some new relations. In the second part, we also study the fuzzy context sequences establishing tendencies. Finally, we illustrate all the results by means of an example.

1 Introduction

The L-Fuzzy Concept Analysis studies the information from an L-fuzzy context by means of the L-fuzzy concepts. These L-fuzzy contexts are tuples (L,X,Y,R), with L a complete lattice, X and Y sets of objects and attributes, and $R \in L^{X \times Y}$ an L-fuzzy relation between the objects and the attributes.

In some situations, we have several relations between the object set X and the attribute set Y, forming a fuzzy context sequence. When this sequence represents an evolution in time we can be more ambitious and try to study future tendencies besides past behaviors. The analysis of this fuzzy context sequences will be the main target of this work.

C. Alcalde
Departamento de Matemática Aplicada, Universidad del País Vasco UPV/EHU,
Plaza de Europa 1, 20018 San Sebastián, Spain
e-mail: `c.alcalde@ehu.es`

A. Burusco · R. Fuentes-González
Departamento de Automática y Computación. Universidad Pública de Navarra,
Campus de Arrosadía,31006 Pamplona, Spain
e-mail: `burusco,rfuentes@unavarra.es`

H. Bustince et al. (eds.), *Aggregation Functions in Theory and in Practise*,
Advances in Intelligent Systems and Computing 228,
DOI: 10.1007/978-3-642-39165-1_16, © Springer-Verlag Berlin Heidelberg 2013

We take as starting point a sequence formed by the L-fuzzy contexts $(L,X,Y,R_i)_{i \in I}$, with $I \subseteq \mathbb{N}$ a finite set, where X and Y are the sets of objects and attributes respectively, and R_i represents the ith relation between the objects of X and the attributes of Y.

The final goal is the study of the fuzzy context sequence and the derived information by the L-fuzzy concepts. To do this, we analyze two different situations: in the first one, we study the values that emphasize in the L-fuzzy contexts regardless of the context in which they are, and in the second one, it is important to maintain the order of the contexts since it represents an evolution in time.

In this second case, it will be of special interest the study of the evolution of the attributes by means of the search of patterns. Works in this line to analyze the course of time in a Formal context can be found in [8, 13, 14].

In [13, 14] K.E. Wolff defines the Temporal Concept Analysis where a Conceptual Time System is introduced such that the state and phase spaces are defined as concept lattices which represent the meaning of the states with respect to the chosen time description. On the other hand, the authors define the hidden evolution patterns in [8, 11] using temporal matching in the case of Formal Concept Analysis.

In this paper, we show a new method for L-Fuzzy Contexts with quantitative data that allows the detection of some kind of regularity.

In Section 2, we see some important results in the L-Fuzzy Concept Analysis. In Section 3, with the aid of the OWA operators, we propose an exhaustive study of the different contexts values of the Fuzzy context sequence using some new relations and establishing tendencies. Finally, we illustrate all the results by means of an example.

2 L-Fuzzy Contexts

The Formal Concept Analysis of R. Wille [12] extracts information from a binary table that represents a formal context (X,Y,R) with X and Y finite sets of objects and attributes respectively and $R \subseteq X \times Y$. The hidden information consists of pairs (A,B) with $A \subseteq X$ and $B \subseteq Y$, called formal concepts, verifying $A^* = B$ and $B^* = A$, where $(\cdot)^*$ is a derivation operator that associates the attributes related to the elements of A with every object set A, and the objects related to the attributes of B with every attribute set B. These formal Concepts can be interpreted as a group of objects A that shares the attributes of B.

In previous works [2, 3] we have defined the L-fuzzy contexts (L,X,Y,R), with L a complete lattice, X and Y sets of objects and attributes respectively and $R \in L^{X \times Y}$ a fuzzy relation between the objects and the attributes. This is an extension of Wille's formal contexts to the fuzzy case when we want to study the relations between the objects and the attributes with values in a complete lattice L, instead of binary values.

In our case, to work with these L-fuzzy contexts, we have defined the derivation operators 1 and 2 given by means of these expressions:

$\forall A \in L^X, \forall B \in L^Y$

$$A_1(y) = \inf_{x \in X}\{\mathscr{I}(A(x), R(x,y))\}$$

$$B_2(x) = \inf_{y \in Y}\{\mathscr{I}(B(y), R(x,y))\}$$

with $\mathscr{I}$ a fuzzy implication operator defined in the lattice $(L, \leq)$.

Although this implication operator can be any extension of the classical implication to $[0,1]$, in this work we will use residuated implication operators.

The information stored in the context is visualized by means of the L-fuzzy concepts that are pairs $(M, M_1) \in (L^X, L^Y)$ with $M \in fix(\varphi)$, set of fixed points of the operator φ, being defined from the derivation operators 1 and 2 as $\varphi(M) = (M_1)_2 = M_{12}$. These pairs, whose first and second components are said to be the fuzzy extension and intension respectively, represent a group of objects that share a group of attributes in a fuzzy way.

Using the usual order relation between fuzzy sets, we define the set $\mathscr{L} = \{(M, M_1)/M \in fix(\varphi)\}$ with the order relation $\preceq$ defined as:
$\forall (M, M_1), (N, N_1) \in \mathscr{L}$,

$$(M, M_1) \preceq (N, N_1) \text{ if } M \leq N(\text{ or } N_1 \leq M_1)$$

$(\mathscr{L}, \preceq)$ is a complete lattice that is said to be [2, 3] the L-fuzzy concept lattice. On the other hand, given $A \in L^X$, (or $B \in L^Y$) we can obtain the associated L-fuzzy concept applying the derivation operators. In the case of using a residuated implication, as we do in this work, the associated L-fuzzy concept is (A_{12}, A_1) (or (B_2, B_{21})).

3 Fuzzy Context Sequences

In this section we are interested in the study of the fuzzy context sequences. We are going to see the formal definition:

Definition 1. A fuzzy context sequence is a tuple $(L, X, Y, R_i)_{i \in I}$ with $L = [0, 1]$, X and Y sets of objects and attributes respectively and $R_i \in L^{X \times Y}, \forall i \in I$, with $I \subseteq \mathbb{N}$ a finite set.

In the case that we want to define a new L-fuzzy context that summarizes the information from the different contexts of the sequence, we can aggregate the observations of the relations R_i. Thus, either we can use the average (with or without weight), or obtain the intervals whose lower bound is the minimum of the observations and the upper one the maximum of them, obtaining an interval-valued L-fuzzy context, or working with multivalued contexts. We have developed these ideas in previous works [4, 5].

The use of weighted averages [9, 10] to summarize the information stored in the different relations allows us to associate different weights with the L-fuzzy contexts highlighting some of them. Thus, the new relation R is defined as:

$$R(x,y) = \sum_{i \in I} w_i . R_i(x,y), \forall x \in X, y \in Y$$

verifying, as is required by the definition, that $\sum_{i\in I} w_i = 1$, $\forall (w_i)_{i\in I}$,

However, it is possible that some observations of an L-fuzzy context of the sequence are interesting whereas others not so much. For instance, as we studied in [1], the used methods for obtaining the L-fuzzy concepts do not give good results when we have very low values in some relations.

On the other hand, to study similar situations by means of multivalued contexts in [4] we used multisets and expertons[7]. In that case, all the observations were analyzed globally without the establishment of different studies based on different requirement levels. This is one of the new contributions of this work.

Let us see the following example.

Example 1. Let $(L, X, Y, R_i)_{i\in I}$ be a fuzzy context sequence that represents the sales of sports articles (X) in some establishments (Y) throughout a period of time (I), and we want to study the places where the main sales hold taking into account that there are seasonal sporting goods (for instance skies, bathing suits) and of a certain zone (it is more possible to sale skies in Colorado than in Florida).

In this case, the weighted average model is not valid since it is very difficult to associate a weight with an L-fuzzy context (in some months more bath suits are sold whereas, in others, skies are).

To analyze this situation, it could be interesting the use of the OWA[15, 5] operators with the most of the weights near the largest values. In this way, we give more relevance to the largest observations, independently of the moment when they have taken place and, on the other hand, we would avoid some small values in the resulting relations (that could give problems in the calculation of the L-fuzzy concepts as has been already studied in [1]).

These are the definitions of these operators given by Yager [15]:

Definition 2. A mapping $F : L^n \longrightarrow L$, where $L = [0, 1]$, is called an OWA operator of dimension n if associated with F is a weighting n-tuple $W = (w_1, w_2 \ldots w_n)$ such that $w_i \in [0, 1]$ and $\sum_{1 \le i \le n} w_i = 1$, where $F(a_1, a_2, \ldots a_n) = w_1.b_1 + w_2.b_2 + \cdots + w_n.b_n$, with b_i the ith largest element in the collection $a_1, a_2, \ldots a_n$.

There are two particular cases of special interest:

W_* defined by the weighting n-tuple with $w_n = 1$ and $w_j = 0, \forall j \ne n$, and W^* defined by the weighting n-tuple such that $w_1 = 1$ and $w_j = 0, \forall j \ne 1$.

It is proved that $F_*(a_1, a_2, \ldots a_n) = \min_j(a_j)$ and $F^*(a_1, a_2, \ldots a_n) = \max_j(a_j)$. These operators are said to be *and* and *or*, respectively.

In order to do a more general study of the fuzzy context sequence, we are interested in the use of operators close to *or*. To measure this proximity we can use the orness degree definition given by [15]:

Definition 3. Let F be an OWA aggregation operator with an n-tuple of weights $W = (w_1, w_2, \ldots w_n)$. The orness degree associated with this operator is defined as:

$$orness(W) = (1/n - 1) \sum_{i=1}^{n} ((n - i).w_i)$$

Returning to the initial situation and using these OWA operators, we can give the following definition that summarizes the information stored in the fuzzy context sequence:

Definition 4. Let $(L,X,Y,R_i)_{i \in I}$ be the fuzzy context sequence and F an OWA aggregation operator. We can define an L-fuzzy relation R_F that aggregates the information from the different L-fuzzy contexts, in the case that we want to study the largest values, by means of this expression:

$$R_F(x,y) = F(R_1(x,y),R_2(x,y)\dots R_{|I|}(x,y)) = w_1.b_1 + w_2.b_2 + \cdots + w_{|I|}.b_{|I|}$$

$\forall x \in X, y \in Y$, where $W = (w_1,w_2,\dots w_{|I|})$ is the weighting tuple associated with F.

There are two special interesting cases:

- W verifying that $orness(W)$ is larger than a fixed threshold.

- W such that $w_i = 1/k$, if $i \le k$ and $w_i = 0$, if $i > k$. That is, the average of the k largest values (with $k \in \mathbb{N}, k \le |I|$)

In the next section we apply these OWA operators to the L-fuzzy contexts to study the values that stand out in the L-fuzzy contexts and to analyze tendencies when the sequence represents the evolution in time.

3.1   The Fuzzy Context Sequence General Study

For a more exhaustive study of the fuzzy context sequence, we can define $|I|$ relations associated with the different requirement levels using OWA operators where the weighting tuple W has just one non-null value $w_k = 1$, for a certain $k \le |I|$.

Definition 5. Given a fuzzy context sequence $(L,X,Y,R_i)_{i \in I}$ with X and Y sets of objects and attributes respectively and $R_i \in L^{X \times Y}, \forall i \in I$, and given a certain $k \in \mathbb{N}, k \le |I|$, we define the relation R_{F_k} using an OWA operator F_k with the weighting tuple W such that $w_k = 1$ and $w_i = 0, \forall i \ne k$.

$$R_{F_k}(x,y) = F_k(R_1(x,y),R_2(x,y)\dots R_{|I|}(x,y)), \forall x \in X, y \in Y.$$

To simplify the following notations, we will denote by $R^{(k)}$ this relation R_{F_k}.
Another way to express this definition is:

$$R^{(k)}(x,y) = \min J^k_{xy}$$

where J^k_{xy} is the set formed by the k largest values associated with the pair (x,y) in the R_i relations.

In this way, we are saying that there are at least k observations larger than or equal to the values of the relation $R^{(k)}$. So, this relation measures the degree in which x is at least k times related to y.

We have chosen the minimum OWA operator in this definition, but it could be possible to use another one if we want to be less demanding.

Example 2. We come back to the fuzzy context sequence $(L,X,Y,R_i)_{i\in I}$ of Example 1 that represents the sales of some sports articles $X = \{x_1,x_2,x_3\}$ in some establishment $Y = \{y_1,y_2,y_3\}$ during a period of time. In the following relations $R_i(x,y), i \in I, x \in X, y \in Y$ takes values in $L = [0,1]$ and represents the percentage of sales (based on the stock) of product x, in the establishment y, in the month i.

$$R_1 = \begin{pmatrix} 0.7 & 1 & 0.8 \\ 0 & 0.1 & 0.1 \\ 0 & 0.1 & 0 \end{pmatrix} R_2 = \begin{pmatrix} 1 & 0.8 & 1 \\ 0.2 & 0.4 & 0.1 \\ 0 & 0 & 0.2 \end{pmatrix} R_3 = \begin{pmatrix} 1 & 1 & 1 \\ 0.6 & 0.5 & 0.7 \\ 0 & 0.1 & 0.2 \end{pmatrix}$$

$$R_4 = \begin{pmatrix} 0.5 & 0.4 & 0.6 \\ 0.1 & 0.5 & 0.3 \\ 0.6 & 0.8 & 0.8 \end{pmatrix} R_5 = \begin{pmatrix} 0.1 & 0 & 0 \\ 0 & 0.1 & 0 \\ 0.8 & 1 & 0.9 \end{pmatrix}$$

First, by means of the L-Fuzzy Concept Analysis, we want to study in what establishments there are greater sales of each product without mattering when the sale has been carried out.

As we have expressed before, there are seasonal sporting goods that are sold in certain periods of time and not in others (skies, bathing suits ...). Therefore, try to summarize the information from the family of L-fuzzy concepts by means of the average, for instance, would not give good results (if a product is only sold during a pair of months in the year, the average with the other months would give a value close to 0 and we would not obtain good results applying the L-Fuzzy Concept Analysis).

On the other hand, if we fix the demand level for instance to $k = 2$ and use Definition 5, then we have the following relation:

$$R^{(2)} = \begin{pmatrix} 1 & 1 & 1 \\ 0.2 & 0.5 & 0.3 \\ 0.6 & 0.8 & 0.8 \end{pmatrix}$$

Now, we take the L-fuzzy context $(L,X,Y,R^{(2)})$ and obtain the L-fuzzy concepts associated with the crisp singletons $\{x_1\}$ and $\{x_3\}$ using the Lukasiewicz implication operator ($\mathscr{I}(a,b) = \min(1,1-a+b)$):

$\{x_1\} \longrightarrow (\{x_1/1, x_2/0.2, x_3/0.6\}, \{y_1/1, y_2/1, y_3/1\})$
$\{x_3\} \longrightarrow (\{x_1/1, x_2/0.5, x_3/1\}, \{y_1/0.6, y_2/0.8, y_3/0.8\})$

In this case, we can say that article x_1 has been successfully sold in the three establishments, at least during two months, and that there are, at least in two months, high sales of articles x_1 and x_3, more in the establishments y_2 and y_3.

As the chosen implication operator is residuated, the membership degree of the fuzzy intension of the L-fuzzy concepts is coincident with the rows of the L-fuzzy relation. Moreover, in all the cases we obtain a more complete information by means of the fuzzy extension.

Analogously, we can take other different k levels.

In particular, the computation of the L-fuzzy concepts associated with $R^{(1)}$ allows to analyze in what stores the main sales of each article during a month (independent of the month) have taken place. If we take relation $R^{(k)}$ with $k > 1$ we are relaxing the requirements taking the k greater sales for our study.

These studies allow us to ignore the small values of the relations (the sales of a non-seasonal sporting goods are close to 0) since, in this case, if we take the average of the relations, the results will be biased.

The observation of these L-fuzzy concepts gives the idea for the following propositions:

Proposition 1. *Consider $k \in \mathbb{N}$, with $k \leq |I|$. If (A,B) is an L-fuzzy concept of the L-fuzzy context $(L,X,Y,R^{(k)})$, then $\forall h \in \mathbb{N}, h \leq k$, there exists an L-fuzzy concept (C,D) of the L-fuzzy context $(L,X,Y,R^{(h)})$ such that $A \leq C$ and $B \leq D$.*

Proof. If $k = h$, then it is obvious.

Otherwise, when $h < k$, $R^{(k)}(x,y) \leq R^{(h)}(x,y)$ $\quad \forall (x,y) \in X \times Y$.

That is, $R^{(k)} \leq R^{(h)}$.

Thus, the L-fuzzy set B derived from A in $(L,X,Y,R^{(k)})$ is a subset of the L-fuzzy set D derived from A in $(L,X,Y,R^{(h)})$. Therefore, $B \leq D$.

Now, we derive again D in $(L,X,Y,R^{(h)})$, obtaining the set C ($C=D_2$) and, applying the properties of this closure operator formed by the composition of the derivation operators: $A \leq A_{12} = D_2 = C$. Therefore, the other inequality also holds. Moreover, it is obvious that if we use a residuated implication operator the obtained pair (C,D) is an L-fuzzy concept.

The following result sets relations up between the L-fuzzy concepts associated with the same starting set (see section 2) in the different L-fuzzy contexts.

Proposition 2. *Consider $k,h \in \mathbb{N}$, with $k,h \leq |I|$ and consider $A \in L^X$. (A^k,B^k) and (A^h,B^h) are the L-fuzzy concepts associated with A in the L-fuzzy contexts $(L,X,Y,R^{(k)})$ and $(L,X,Y,R^{(h)})$ respectively. If $k \leq h$ then $B^k \geq B^h$.*

Moreover, if $\mathscr{I}$ is a residuated implication operator and the set A is the crisp singleton $\{x_i\}$

$$A(x) = \begin{cases} 1 & if\ x = x_i \\ 0 & otherwise \end{cases}$$

then, $A^k(x_i) = A^h(x_i) = 1$.

A similar result is obtained taking as a starting point an L-fuzzy set of attributes $B \in L^Y$.

Proof. Consider $A \in L^X$. Unfolding the fuzzy extensions of both L-fuzzy concepts, and taking into account that a fuzzy implication operator is increasing on its second argument:

$$B^k(y) = \inf_{x \in X}\{\mathscr{I}(A(x), R^{(k)}(x,y))\} \geq \inf_{x \in X}\{\mathscr{I}(A(x), R^{(h)}(x,y))\} = B^h(y)$$

This result holds for every A and for every implication operator.

On the other hand, if we take a crisp singleton:

$$A(x) = \begin{cases} 1 & \text{if } x = x_i \\ 0 & \text{otherwise} \end{cases}$$

and a residuated implication, then the membership degree of x_i in the fuzzy extension of the L-fuzzy concepts is equal to 1:

$$B^k(y) = \inf_{x \in X}\{\mathscr{I}(A(x), R^{(k)}(x,y))\} = R^{(k)}(x_i, y)$$

$$A^k(x) = \inf_{y \in Y}\{\mathscr{I}(B^k(y), R^{(k)}(x,y))\} = \inf_{y \in Y}\{\mathscr{I}(R^{(k)}(x_i,y), R^{(k)}(x,y))\}$$

Therefore, $A^k(x_i) = 1$.

Similarly, the result for the L-fuzzy set $B \in L^Y$ can be proved.

However, the inequality $A^k \leq A^h$ does not always hold, as can be seen if we come back to the previous example and we compare the fuzzy extension $A(x) = \{x_1/1, x_2/0, x_3/0\}$ of the derived L-fuzzy concept in the L-fuzzy contexts $(L, X, Y, R^{(2)})$ and $(L, X, Y, R^{(4)})$:

In $(L, X, Y, R^{(2)})$, the result is $A^2 = \{x_1/1, x_2/0.2, x_3/0.6\}$ whereas in $(L, X, Y, R^{(4)})$ we get $A^4 = \{x_1/1, x_2/0.5, x_3/0.5\}$.

In the following section, we introduce the variable time in our study.

3.2 Temporal Analysis of the Fuzzy Context Sequence

Fixed $k \in I$, and a pair (x, y), with $x \in X$ and $y \in Y$, Definition 5 uses the minimum of the k largest observations $R_i(x,y), i \in I$, of the fuzzy context sequence, but does not allow to make an analysis of their evolution in time.

In this section, we approach this subject by means of studies that analyze tendencies.

We begin with a definition that takes the minimum value of the relations between each object and each attribute from an instant h.

Definition 6. Let $(L, X, Y, R_i)_{i \in I}$ be a fuzzy context sequence with X and Y sets of objects and attributes respectively and $R_i \in L^{X \times Y}$. We define an L-fuzzy relation $\bar{R}^{(h)}$ (with the notation adopted in Definition 5), using an OWA operator F with a weighting tuple W of dimension $k = |I| - h + 1$ with $w_k = 1$ the only non-null value:

$$\bar{R}^{(h)}(x,y) = F(R_h(x,y), R_{h+1}(x,y) \ldots R_{|I|}(x,y)), \quad \forall x \in X, y \in Y.$$

In other words:

$$\bar{R}^{(h)}(x,y) = \min_{i \geq h}\{R_i(x,y)\}, \forall x \in X, y \in Y$$

As in the previous section, instead of the minimum (that is very demanding) we can take the maximum, the average or other aggregation operators changing the weighting tuple W of the OWA operator.

Example 3. If we come back to the previous example and we want to study tendencies of the sequence, we can take a value h and analyze the L-fuzzy concepts.

For instance, if $h = 4$, we have the L-fuzzy relation:

$$\bar{R}^{(4)} = \begin{pmatrix} 0.1 & 0 & 0 \\ 0 & 0.1 & 0 \\ 0.6 & 0.8 & 0.8 \end{pmatrix}$$

and, taking as L-fuzzy context $(L, X, Y, \bar{R}^{(4)})$ and using the Lukasiewicz implication to obtain the L-fuzzy concepts associated with the crisp singletons, we have the following results:

$$\{x_1\} \longrightarrow (\{x_1/1, x_2/0.9, x_3/1\}, \{y_1/0.1, y_2/0, y_3/0\})$$
$$\{x_2\} \longrightarrow (\{x_1/0.9, x_2/1, x_3/1\}, \{y_1/0, y_2/0.1, y_3/0\})$$
$$\{x_3\} \longrightarrow (\{x_1/0.2, x_2/0.2, x_3/1\}, \{y_1/0.6, y_2/0.8, y_3/0.8\})$$

We can say that the future tendency is that article x_3 will have good sales in all the establishments whereas x_1 and x_2 will not be sold much and always associated with x_3, the first one in the establishment y_1 essentially, and the second one in y_2.

Obviously, the smaller is the value of h, the safer will be the "prediction" that we do because we have more information of the behaviour of the sales.

Moreover, we can establish comparisons between the different L-fuzzy concepts obtained from the different relations $\bar{R}^{(i)}, \forall i \in I$.

Proposition 3. *Consider $A \in L^X$. Let $(\bar{A}^k, \bar{B}^k)$ and $(\bar{A}^h, \bar{B}^h)$ be the L-fuzzy concepts associated with A in the L-fuzzy contexts $(L, X, Y, \bar{R}^{(k)})$ and $(L, X, Y, \bar{R}^{(h)})$ respectively, with $k, h \leq |I|$. If $k \leq h$ then $\bar{B}^k \leq \bar{B}^h$.*

Moreover, if we use a residuated implication operator $\mathscr{I}$ and a crisp singleton A, then

$$\bar{A}^k(x_i) = \bar{A}^h(x_i) = 1$$

with x_i the element of X where the crisp singleton A takes value 1.

A similar result is obtained taking as a starting point an L-fuzzy set of attributes $B \in L^Y$.

Proof. Similar to Proposition 2 taking into account that, in this case, if $k \leq h$ then $\bar{R}^{(k)} \leq \bar{R}^{(h)}$.

The meaning of this result is that if we look at the fuzzy intensions obtained for the different L-fuzzy contexts of the sequence, then they form a non-decreasing chain $\forall y \in Y$.

Example 4. In our example, the L-fuzzy concepts obtained taking as a starting point the crisp singleton $\{x_3\}$ in the L-fuzzy contexts $(L, X, Y, \bar{R}^{(4)})$ and $(L, X, Y, \bar{R}^{(5)})$, using the Lukasiewicz implication operator, are:

$$\bar{R}^{(4)} : (\{x_1/0.2, x_2/0.2, x_3/1\}, \{y_1/0.6, y_2/0.8, y_3/0.9\})$$
$$\bar{R}^{(5)} : (\{x_1/0, x_2/0.1, x_3/1\}, \{y_1/0.8, y_2/1, y_3/0.9\})$$

verifying the previous proposition.

On the other hand, since if an object and an attribute are related from instant h, they are related at least $|I| - h + 1$ times, hence a similar result between the L-fuzzy concepts obtained using Definition 5 and 6 can be seen.

Proposition 4. *If we take as starting point $A \in L^X$, then for any $h \in I$, the fuzzy intension $\bar{B}^h$ of the L-fuzzy concept $(\bar{A}^h, \bar{B}^h)$ obtained in $(L, X, Y, \bar{R}^{(h)})$ is included in the fuzzy intension B^k of the L-fuzzy concept (A^k, B^k) obtained in $(L, X, Y, R^{(k)})$ with $k = |I| - h + 1$. That is,*
$$\bar{B}^h(y) \leq B^k(y), \quad \forall y \in Y$$
We have also a similar result from $B \in L^Y$.

Proof. Immediate using the previous proposition proof and the inequality $\bar{R}^{(h)} \leq R^{(k)}$ with $k = |I| - h + 1$.

Example 5. If we take $h = 4$ and $k = 2$, and the L-fuzzy contexts $(L, X, Y, \bar{R}^{(4)})$ and $(L, X, Y, R^{(2)})$, then taking as a starting point the object x_1 and using the Lukasiewicz implication operator, the following L-fuzzy concepts are obtained:

$$\bar{R}^{(4)} : (\{x_1/1, x_2/0.9, x_3/1\}, \{y_1/0.1, y_2/0, y_3/0\})$$
$$R^{(2)} : (\{x_1/1, x_2/0.2, x_3/0.6\}, \{y_1/1, y_2/1, y_3/1\})$$

And the previous proposition holds.

An important result is the one that allows the study of the attributes associated with some elements of X from an instant h in two different ways:

Theorem 1. *Given A a crisp subset of X, and $\mathscr{I}$ a residuated implication.*
The fuzzy intension $\bar{B}^h \in L^Y$ of the L-fuzzy concept derived from A in $(L, X, Y, \bar{R}^{(h)})$ is equal to the intersection of the fuzzy intensions B_i of the L-fuzzy concepts obtained in the L-fuzzy contexts (L, X, Y, R_i) with $i \geq h$. That is,

$$\bar{B}^h(y) = \min_{i \geq h} B_i(y), \quad \forall y \in Y$$

Proof. If we use a residuated implication operator $\mathscr{I}$, then we have that $\forall y \in Y$:

$$\bar{B}^h(y) = \inf_{x \in X}\{\mathscr{I}(A(x), \bar{R}^{(h)}(x,y))\} = \min_{x \in X/A(x)=1} \bar{R}^{(h)}(x,y)$$

By the definition of $\bar{R}^h(x,y)$ we can say that:

$$\bar{B}^h(y) = \min_{x \in X/A(x)=1}\{\min_{i \geq h}\{R_i(x,y)\}\} = \min_{i \geq h}\{\min_{x \in X/A(x)=1}\{R_i(x,y)\}\} = \min_{i \geq h} B_i(y).$$

This result can be generalized replacing the minimum by any OWA operator in the proposition and in the definition of the relations $\bar{R}^{(h)}$.

Remark 1. This proposal justifies the utility of the defined relations $\bar{R}^{(h)}$ since allows the study of the attributes associated with some objects from an instant h looking only at the L-fuzzy context $(L, X, Y, \bar{R}^{(h)})$ instead of all the L-fuzzy contexts of the sequence.

4 Conclusions and Future Work

In this work, we have used OWA operators to study the L-fuzzy context sequence and the derived information by means of the L-fuzzy contexts.

After that, we have studied tendencies that we find when the sequence represents the evolution in the course of time of an L-fuzzy context. In the future we want to study temporal patterns to identify the evolution of the contexts.

On the other hand, these L-fuzzy contexts that evolve with time can be generalized if we study L-fuzzy contexts where the observations are other L-fuzzy contexts. This is the task that we will study in the future.

Finally we will try to apply these results to the interval-valued L-fuzzy context sequences.

Acknowledgements. This paper is partially supported by the Group of research of the Public University of Navarra *Adquisición de conocimiento y minería de datos, funciones especiales y métodos numéricos avanzados* and by the Research Group "Intelligent Systems and Energy (SI+E)" of the Basque Government, under Grant IT519-10.

References

[1] Alcalde, C., Burusco, A., Fuentes-González, R., Zubia, I.: The use of linguistic variables and fuzzy propositions in the L-Fuzzy Concept Theory. Computers and Mathematics with Applications 62, 3111–3122 (2011)

[2] Burusco, A., Fuentes-González, R.: The Study of the L-fuzzy Concept Lattice. Mathware and Soft Computing 1(3), 209–218 (1994)

[3] Burusco, A., Fuentes-González, R.: Construction of the L-fuzzy Concept Lattice. Fuzzy Sets and Systems 97(1), 109–114 (1998)

[4] Burusco, A., Fuentes-González, R.: Contexts with multiple weighted values. The International Journal of Uncertainty, Fuzziness and Knowledge-based Systems 9(3), 355–368 (2001)

[5] Burusco, A., Fuentes-González, R.: The study of the interval-valued contexts. Fuzzy Sets and Systems 121, 69–82 (2001)

[6] Fodor, J., Marichal, J.L., Roubens, M.: Characterization of the Ordered Weighted Averaging Operators. IEEE Transactions on Fuzzy Systems 3(2), 236–240 (1995)

[7] Kaufmann, A.: Theory of Expertons and Fuzzy Logic. Fuzzy Sets and Systems 28, 295–304 (1988)

[8] Neouchi, R., Tawfik, A.Y., Frost, R.A.: Towards a Temporal Extension of Formal Concept Analysis. In: Proceedings of Canadian Conference on Artificial Intelligence, pp. 335–344 (2001)

[9] Calvo, T., Mesiar, R.: Weighted triangular norms-based aggregation operators. Fuzzy Sets and Systems 137, 3–10 (2003)

[10] Calvo, T., Mesiar, R.: Aggregation operators: ordering and bounds. Fuzzy Sets and Systems 139, 685–697 (2003)

[11] Tawfik, A.Y., Scott, G.: Temporal Matching under Uncertainty. In: Proceedings of the Eighth International Workshop on Artificial Intelligence and Statistics (2001)

[12] Wille, R.: Restructuring lattice theory: an approach based on hierarchies of concepts. In: Rival, I. (ed.) Ordered Sets, pp. 445–470. Reidel, Dordrecht-Boston (1982)

[13] Wolff, K.E.: Transitions, and Life Tracks in Temporal Concept Analysis. In: Ganter, B., Stumme, G., Wille, R. (eds.) Formal Concept Analysis. LNCS (LNAI), vol. 3626, pp. 127–148. Springer, Heidelberg (2005)

[14] Wolff, K.E.: Temporal Relational Semantic Systems. In: Croitoru, M., Ferré, S., Lukose, D. (eds.) ICCS 2010. LNCS, vol. 6208, pp. 165–180. Springer, Heidelberg (2010)

[15] Yager, R.R.: On ordered weighted averaging aggregation operators in multi-criteria decision making. IEEE Transactions on Systems, Man and Cibernetics 18, 183–190 (1988)

Norm Aggregations and OWA Operators

José M. Merigó and Ronald R. Yager

Abstract. The ordered weighted average (OWA) is an aggregation operator that provides a parameterized family of aggregation operators between the minimum and the maximum. This paper studies the use of the OWA operator with norms. Several extensions and generalizations are suggested including the use of the induced OWA operator and the OWA weighted average. This approach represents a general frameworkof the aggregation operators when dealing with distance and similarity measures. Some key particular cases are studied including the addition OWA and the subtraction OWA operator

1 Introduction

The ordered weighted average (OWA) [17] is an aggregation operator that provides a parameterized family of aggregation operators between the minimum and the maximum. It has been used in a wide range of applications [1, 28] and has been extended and generalized in a wide range of directions. For example, Fodor et al. [2] presented a generalization by using quasi-arithmetic means. Yager and Filev [27] introduced the induced OWA (IOWA) operator providing a more general reordering process. Merigó and Gil-Lafuente [9] extended this approach by using generalized and quasi-arithmetic means. Other authors have studied the use of distance measures with the OWA operator [3, 10, 15, 29]. In this direction, it is worth noting a recent development by Yager [28] regarding the use of OWA operators with norms that generalizes distance and similarity measures under the same framework.

José M. Merigó
Manchester Business School, University of Manchester, Booth Street West, M15 6PB,
Manchester, UK
e-mail: `jose.merigolindahl@mbs.ac.uk`

Ronald R. Yager
Machine Intelligence Institute, Iona College 10801, New Rochelle, NY, USA
e-mail: `yager@panix.com`

H. Bustince et al. (eds.), *Aggregation Functions in Theory and in Practise*,
Advances in Intelligent Systems and Computing 228,
DOI: 10.1007/978-3-642-39165-1_17, © Springer-Verlag Berlin Heidelberg 2013

A further interesting approach is those aggregation operators that integrate the OWA operator with the weighted average. Several approaches have been proposed in this direction by many authors including the weighted OWA (WOWA) [12], the hybrid average [19], the importance OWA [21] and the immediate weights [11, 6]. Recently, Merigó [6] has suggested the OWA weighted average (OWAWA) as a generalization that unifies both concepts in the same formulation and considering the degree of importance they have in the specific aggregation taken into account.

The aim of this paper is to develop further extensions regarding the use of OWA operators with norms. It is presented the use of the IOWA operator with norms forming the IOWA norm (IOWAN) aggregation. Thus, it is possible to represent a wide range of norm aggregations from the minimum to the maximum and under complex reordering processes. Next, it is introduced the use of the OWAWA operator obtaining the OWAWA norm (OWAWAN) that provides a unified framework between the usual weighted average norm and the OWAN operator. Several families and particular cases are studied including the addition OWAWA (A-OWAWA), the subtraction OWAWA (S-OWAWA) and many other cases. These operators seem to be of great importance because they may provide a new methodology for dealing with arithmetic operations. The use of the variance [20, 25] as a particular type of norm is also considered.

This paper is organized as follows. Section 2 reviews some basic preliminaries regarding the OWA and the OWAWA operator and norm aggregations. Section 3 introduces the use of the OWAWA operator with norms and Section 4 studies a wide range of particular cases. Section 5 ends the paper summarizing the main conclusions of the paper.

2   Preliminaries

This section briefly reviews the OWA operator, the OWAWA operator and norm aggregations.

2.1   The OWA Operator

The OWA operator is an aggregation operator that considers a wide range of averaging operators that move between the minimum and the maximum. It permits to aggregate the information considering the degree of optimism or pessimism that a decision maker wants to use in the aggregation. It has been used in a wide range of applications including soft computing, decision making and statistics [1, 28]. It can be defined as follows.

Definition 1. An OWA operator of dimension n is a mapping $OWA : R^n \rightarrow R$ that has an associated weighting W and $\sum_{j=1}^{n} w_j = 1$, such that:

$$OWA(a_1,...,a_n) = \sum_{j=1}^{n} w_j b_j \tag{1}$$

where b_j is the jth largest of the a_i.

Several properties could be studied including different families of OWA operators and measures for characterizing the weighting vector [17, 18]. Note that in most of the OWA literature, the arguments are reordered according to a weighting vector. However, it is also possible to reorder the weighting vector according to the initial positions of the arguments a_i [22] and this is of great importance in order to integrate the weighted average and the OWA in the same formulation.

The OWA operator can be extended by using induced aggregation operators [9, 24] forming the induced OWA (IOWA) operator [27]. Therefore, it is possible to consider a more general reordering process that deals with complex situations.

2.2 The OWAWA Operator

The ordered weighted averaging - weighted average (OWAWA) [6] is a model that unifies the OWA operator and the weighted average in the same formulation considering the degree of importance that each concept has in the analysis. Therefore, both concepts can be seen as a particular case of a more general one. One of its key advantages is that it can be reduced to the usual weighted average or to the OWA. Therefore, any study that uses the OWA or the weighted average can be revised and extended with the OWAWA operator provides a more complete analysis of the information considered. It can be defined as follows.

Definition 2. An OWAWA operator of dimension n is a mapping $OWAWA : R^n \rightarrow R$ that has an associated weighting W of dimension n such that $w_j \in [0,1]$ and $\sum_{j=1}^{n} w_j = 1$, according to the following formula:

$$OWAWA(a_1,...,a_n) = \sum_{j=1}^{n} \hat{v}_j b_j \tag{2}$$

where b_j is the jth largest of the a_i, each argument a_i has an associated weight (WA) v_i with $\sum_{i=1}^{n} v_i = 1$ and $v_i \in [0,1]$, $\hat{v}_j = \beta w_j + (1-\beta)w_j$ with $\beta \in [0,1]$ and v_j is the weight (WA) v_i ordered according to b_j, that is, according to the jth largest of the a_i.

It provides a parameterized family of aggregation operators from the minimum to the maximum. The difference against the OWA is that it also considers subjective information. Thus, it is possible to consider a partial boundary condition that considers the minimum and maximum adjusted with the weighted average [6]. Note that if $\beta = 1$, we get the OWA operator and if $\beta = 0$, the WA. The OWAWA operator accomplishes similar properties than the usual OWA aggregation operators including the symmetry, the use of mixture operators and so on [6].

2.3   Norm Aggregations

Norm aggregations provide a more general representation of the aggregation when dealing with distance measures because they allow us to include more complex operations in the analysis. A norm associates with some vector or tuple $X = (x_1, x_2, ..., x_n)$ a unique non-negative scalar. A norm is a function $f : R^n \to [0, \infty)$ that has the following properties [1, 28]:

1. $f(x_1, x_2, ..., x_n) = 0$ if and only if all $x_i = 0$.
2. $f(aX) = |a| f(X)$.
3. $f(X) + f(Y) \geq f(X + Y)$, that is, the triangle inequality.

When dealing with averaging functions, norms can be used following a similar methodology as it is used with distance measures [7]. Thus, with the weighted average it can be formulated the following expression:

$$f(a_1, a_2, ..., a_n) = G(|a_1|, |a_2|, ..., |a_n|) = \sum_{i=1}^{n} w_i |a_i|, \tag{3}$$

Recently, Yager [25] has suggested the use of norms in the OWA operator by using:

$$f(a_1, a_2, ..., a_n) = G(|a_1|, |a_2|, ..., |a_n|) = \sum_{j=1}^{n} w_j N_j, \tag{4}$$

where N_j is the jth largest of the $|a_i|$.

Note that norms can be used in order to get a distance or a metric function assuming that if f is a norm then $d(X, Y) = f(a = |X - Y|)$.

3   Norms with OWAWA Operators

Norms are useful in a wide range of situations because they include many aggregation operators and distance measures as particular cases. Among others, it is worth noting the usual average, the Hamming distance [4] and the variance. This paper presents several extensions and generalizations by using a wide range of averaging aggregation operators with norms. First, let us consider the use of the induced OWA (IOWA) operator with norms forming the IOWA norm (IOWAN) operator. Its main advantage is that it considers complex reordering processes in the aggregation of norms providing a parameterized family of norms from the minimum to the maximum one. Note that this is of great interest because when dealing with norms, not always the highest or the lowest one is the preferred one. It can be defined as follows.

Definition 3. An IOWAN operator of dimension n is a mapping $IOWAN : R^n \times R^n \to R$ that has an associated weighting W with $w_j \in [0, 1]$ and $\sum_{j=1}^{n} w_j = 1$, such that:

$$IOWAN\left(\langle u_1, |a_1| \rangle, \langle u_2, |a_2| \rangle, ..., \langle u_n, |a_n| \rangle\right) = \sum_{j=1}^{n} w_j N_j, \qquad (5)$$

where N_j is the a_i value of the IOWAN pair $\langle u_i, a_i \rangle$ having the jth largest u_i, u_i is the order-inducing variable and $|a_i|$ is the argument variable represented in the form of individual norms.

Note that the IOWAN operator is reduced to the usual IOWA operator when $|a_i| = a_i$. It can also be seen as a distance measure when $d(X,Y) = f(a = |X - Y|)$ becoming the induced OWA distance (IOWAD) operator [7]. Furthermore, it is also possible to formulate it by using generalized means forming the induced generalized OWA (IGOWA) operator ($|a| = a^\lambda$) [9] and the induced Minkowski OWA distance (IMOWAD) ($d(X,Y) = f(a = |X - Y|^\lambda)$) [8].

Next, let us look into a more general framework by using the OWAWA operator with norms forming the OWAWA norm (OWAWAN) operator. Thus, it is possible to integrate norms with weighted averages (Eq. 3) and OWA operators in the same formulation and considering the degree of importance that each concept has in the aggregation. Thus, all the particular types of norms used previously can also be included in this framework including the use of distance and similarity measures. It can be defined as follows.

Definition 4. An OWAWAN operator is a mapping $OWAWAN : R^n \to R$ of dimension n, if it has an associated weighting vector W, with $\sum_{j=1}^{n} w_j = 1$ and $w_j \in [0,1]$ and a weighting vector V that affects the WA, with $\sum_{i=1}^{n} v_i = 1$ and $v_i \in [0,1]$, such that:

$$OWAWAN\left(|a_1|, |a_2|, ..., |a_n|\right) = \beta \sum_{j=1}^{n} w_j N_j + (1 - \beta) \sum_{i=1}^{n} v_i |a_i| \qquad (6)$$

where N_j is the jth smallest of the $|a_i|$, each argument $|a_i|$ is the argument variable represented in the form of individual norms and $\beta \in [0,1]$.

Note that the OWAWAN operator can be formulated integrating both equations into a single one as it is done with the OWAWA operator [6] as follows:

$$OWAWAN\left(|a_1|, |a_2|, ..., |a_n|\right) = \sum_{j=1}^{n} \hat{v}_j N_j, \qquad (7)$$

where N_j is the jth smallest of the $|a_i|$, each argument $|a_i|$ is the argument variable represented in the form of individual norms and has an associated weight v_i with $\sum_{i=1}^{n} v_i = 1$ and $v_i \in [0,1]$, $\hat{v}_j = \beta w_j + (1 - \beta) v_j$ with $\beta \in [0,1]$ and v_j is the weight (WA) v_i ordered according to N_j, that is, according to the jth smallest of the $|a_i|$.

Observe that this is possible only when dealing with arithmetic averaging functions. If it is used a generalized or quasi-arithmetic mean with $\lambda \neq 1$ or $g(a) \neq a$, it is not possible to integrate it in this way.

As we can see, if $\beta = 1$, the OWAWAN operator becomes the OWAN operator and if $\beta = 0$, the weighted averaging norm (WAN). The OWAWAN operator accomplishes similar properties than the norm aggregation operators [28].

Note that Eq. (6) has been presented adapting the ordering of the weighted average to the OWA operator. However, it is also possible to formulate the OWAWA operator adapting the ordering of the OWA operator to the weighted average as:

$$OWAWAN\left(|a_1|,|a_2|,...,|a_n|\right) = \beta \sum_{i=1}^{n} w_i |a_i| + (1 - \beta) \sum_{i=1}^{n} v_i |a_i|, \qquad (8)$$

where each argument a_i has an associated weight w_i that represents the weight w_j ordered according to the ordering of the arguments a_i and $\beta \in [0,1]$.

A further interesting issue appears when the weighting vector is not normalized, i.e., $W = \sum_{j=1}^{n} w_j \neq 1$ or $V = \sum_{i=1}^{n} v_i \neq 1$. In these situations and without considering the concept of heavy aggregations [23], the OWAWAN operator can be formulated in the following way:

$$OWAWAN\left(|a_1|,|a_2|,...,|a_n|\right) = \frac{\beta}{W} \sum_{j=1}^{n} w_j N_j + \frac{(1 - \beta)}{V} \sum_{i=1}^{n} v_i |a_i|, \qquad (9)$$

Similarly to the IOWAN, the OWAWAN operator can also be seen as a distance metric by using $d(X,Y) = f(a = |X - Y|)$. Thus, it becomes the OWAWA distance (OWAWAD) operator [11] that can be formulated as follows:

$$OWAWAD\left(|x_1,y_1|,|x_2,y_2|,...,|x_n,y_n|\right) = \beta \sum_{j=1}^{n} w_j D_j + (1 - \beta) \sum_{i=1}^{n} v_i |x_i - y_i|, \quad (10)$$

where D_j is the jth smallest of the $|x_i - y_i|$, each argument $|x_i - y_i|$ is the argument variable represented in the form of individual distances and $\beta \in [0,1]$.

Note that the main advantage of the OWAWAD is that it integrates the weighted Hamming distance (WHD) and the OWA distance (OWAD) [10, 15] in the same formulation considering the degree of importance that each concept has in the formulation. As we can see, if $\beta = 1$, the OWAWAD operator becomes the OWAD operator and if $\beta = 0$, the WHD.

Furthermore, it is also possible to use generalized and quasi-arithmetic means in the analysis. Thus, the OWAWAN operator becomes the GOWAWA and the Quasi-OWAWA operator. With generalized means this is obtained when $|a| = a^{\lambda}$ and with quasi-arithmetic means if $|a| = g(a)$.

An interesting issue when analysing these aggregation operators is to characterize the weighting vector. This can be done following the OWA literature where it is considered the degree of orness (attitudinal character) [6, 17], the entropy of dispersion [6, 17] and the divergence of the weights [6, 23]. When dealing with OWAWA operators, the degree of orness can be formulated from two different perspectives. A first perspective assumes that the weighted average can also be studied with this measure being the objective to determine the tendency of the aggregation to the minimum or to the maximum:

$$\alpha(\hat{V}) = \beta \sum_{j=1}^{n} w_j \left(\frac{n-j}{n-1} \right) + (1-\beta) \sum_{j=1}^{n} v_j \left(\frac{n-j}{n-1} \right). \tag{11}$$

It is straightforward to calculate the andness measure by using the dual. That is, $Andness(\hat{V}) = 1 - \alpha(\hat{V})$. The other perspective is focussed on the attitudinal character. In this case, the weighted average is seen as a neutral aggregation because it only considers the subjective opinion. Thus, it is reasonable to assume that the orness measure for this part of the equation should be 0.5 obtaining the following expression:

$$\alpha(\hat{V}) = \beta \sum_{j=1}^{n} w_j \left(\frac{n-j}{n-1} \right) + (1-\beta) \times 0.5. \tag{12}$$

In this case it is also trivial to form the andness measure or the degree of pessimism.

The entropy of dispersion measures the amount of information being used in the aggregation. If we extend this approach to the OWAWAN operator, it is obtained the following formulation:

$$H(\hat{V}) = - \left(\beta \sum_{j=1}^{n} w_j \ln(w_j) + (1-\beta) \sum_{i=1}^{n} v_i \ln(v_i) \right). \tag{13}$$

As we can see, if $\beta = 1$, we obtain the Yager entropy of dispersion for the OWAN operator and if $\beta = 0$, we get the classical Shannon entropy [13].

The divergence [6, 23] measures the divergence of the weights against the attitudinal character. It is useful in various situations, especially when the attitudinal character and the entropy of dispersion are not enough to correctly analyse the weighting vector of an aggregation. If we extend the divergence to the OWAWAN operator, we get the following divergence:

$$Div(\hat{V}) = \beta \left(\sum_{j=1}^{n} w_j \left(\frac{n-j}{n-1} - \alpha(W) \right)^2 \right) + (1-\beta) \left(\sum_{j=1}^{n} v_j \left(\frac{n-j}{n-1} - \alpha(V) \right)^2 \right). \tag{14}$$

Note that if $\beta = 1$, we get the OWAN divergence and if $\beta = 0$, the WAN divergence. Moreover, it is also possible to consider a variation of Eq. (14) by using Eq. (12). In this case, the divergence of the weighted average is 0 and it is only considered the divergence of the OWAN operator.

Finally, let us consider the use of norms under other frameworks that unifies the OWA operator and the weighted average in the same formulation. Among others, let us consider the use of hybrid averages [19], WOWA operators [15] and immediate weights [11, 6]. By using the hybrid average it is formed the hybrid averaging norm (HAN) that can be formulated as follows.

$$HAN(|a_1|, |a_2|, ..., |a_n|) = \sum_{j=1}^{n} w_j N_j, \tag{15}$$

where N_j is the jth smallest of the $|\hat{a}_i|$ ($\hat{a}_i = n\omega_i|a_i|, i = 1, 2, ..., n$), $\omega = (\omega_1, \omega_2, ..., \omega_n)$ is the weighting vector of the $|a_i|$, with $\omega_i \in [0, 1]$ and $\sum\limits_{i=1}^{n} \omega_i = 1$. With the WOWA operator it is obtained the WOWA norm (WOWAN) operator. It is formulated in the following way.

Definition 5. Let P and W be two weighting vectors of dimension n, with $P = (p_1, p_2, ..., p_n)$ and $W = (w_1, w_2, ..., w_n)$, such that $p_i \in [0, 1]$ and $\sum\limits_{i=1}^{n} p_i = 1$, and $w_j \in [0, 1]$ and $\sum\limits_{j=1}^{n} w_j = 1$. A mapping $WOWAN : R^n \times R^n \to R$ is a WOWAN operator of dimension n if:

$$WOWAN(|a_1|, |a_2|, ..., |a_n|) = \sum_{i=1}^{n} \omega_i |a_{\sigma(i)}|, \tag{16}$$

where $\{\sigma(1), ..., \sigma(n)\}$ is a permutation of $\{1, ..., n\}$ such that $a_{\sigma(i-1)} \geq a_{\sigma(i)}$ for all $i = 2, ..., n$, and the weight ω_i is defined as:

$$\omega_i = w * \left(\sum_{j \leq i} p_{\sigma(j)} \right) - w * \left(\sum_{j < i} p_{\sigma(j)} \right), \tag{17}$$

with $w*$ a monotone increasing function that interpolates the points $(i/n, \sum_{j \leq i} w_j)$ together with the point $(0, 0)$. $w*$ is required to be a straight line when the points can be interpolated in this way.

The use of immediate weights forms the immediate weighted averaging norm (IWAN) and it is constructed by using the following expression:

$$IWAN(|a_1|, |a_2|, ..., |a_n|) = \sum_{j=1}^{n} \hat{v}_j |N_j|, \tag{18}$$

where N_j is the jth smallest of the $|a_i|$, each $|a_i|$ has associated a weight v_i, v_j is the associated weight of N_j, and $\hat{v}_j = (w_j v_j / \sum_{j=1}^{n} w_j v_j)$.

Note that this expression can only be used when dealing with arithmetic averaging operators as in the OWAWAN operator because with generalized aggregation operators, the formulation may have some incorrect deviations.

Furthermore, observe that similar extensions and generalizations could also be studied with induced and generalized aggregation operators [9, 30] and by using distance measures as it has been explained before.

4   Families of OWAWAN Operators

The OWAWAN operator includes a wide range of particular cases. First, it is possible to study several families by analyzing the weighting vector [6, 18]. Thus, it is possible to form the following cases:

- Simple averaging norm: $w_j = 1/n$ and $v_i = 1/n$, for all i, j.
- Arithmetic WAN (AWAN): $w_j = 1/n$, for all j.
- Arithmetic OWAN (AOWAN): $v_i = 1/n$, for all i.
- Min-WAN: $w_1 = 1$ and $w_j = 0$, for all $j \neq 1$.
- Max-WAN: $w_n = 1$ and $w_j = 0$, for all $j \neq n$.
- Step OWAWAN: $w_k = 1$ and $w_j = 0$, for all $j \neq k$.
- Hurwicz WAN: $w_1 = 1 - \alpha$, $w_n = \alpha$ and $w_j = 0$ for all $j \neq 1, n$.
- Window OWAWAN: $w_j = 1/m$ for $k \leq j \leq k + m - 1$ and $w_j = 0$ for all $j > k + m, j < k$.
- Median odd OWAWAN: If n is odd we assign $w_{(n+1)/2} = 1$ and $w_j = 0$ for all others.
- Median even OWAWAN: If n is even we assign $w_{n/2} = w_{(n/2)+1} = 0.5$ and $w_j = 0$ for all others.
- Olympic OWAWAN: $w_1 = w_n = 0$, and for others $w_j = 1/(n-2)$.

Some other interesting cases are found by analyzing a different expression in the norm. Among others, it is worth noting the following ones:

- If $OWAWAN(X, Y) = f(X + Y)$, we obtain the addition OWAWA (A-OWAWA) operator that can be formulated as:

$$A - OWAWA\left(|x_1 + y_1|, \ldots, |x_n + y_n|\right) = \beta \sum_{j=1}^{n} w_j A_j + (1 - \beta) \sum_{i=1}^{n} v_i |x_i + y_i|, \quad (19)$$

where A_j is the jth smallest of the $|x_i + y_i|$ and $\beta \in [0, 1]$.

- If $OWAWAN(X, Y) = f(X - Y)$, the subtraction OWAWA (S-OWAWA) operator and it is expressed as follows:

$$S - OWAWA\left(|x_1 - y_1|, \ldots, |x_n - y_n|\right) = \beta \sum_{j=1}^{n} w_j S_j + (1 - \beta) \sum_{i=1}^{n} v_i (x_i - y_i), \quad (20)$$

where S_j is the jth smallest of the $(x_i - y_i)$ and $\beta \in [0, 1]$.

- If $OWAWAN(X, Y) = f(X \times Y)$, we get the multiplication OWAWA (M-OWAWA) operator that is defined in the following way:

$$M - OWAWA\left(|x_1 \times y_1|, \ldots, |x_n \times y_n|\right) = \beta \sum_{j=1}^{n} w_j M_j + (1 - \beta) \sum_{i=1}^{n} v_i |x_i \times y_i|, \quad (21)$$

where M_j is the jth smallest of the $|x_i \times y_i|$ and $\beta \in [0, 1]$.

- If $OWAWAN(X, Y) = f(X \div Y)$, we obtain the division OWAWA (D-OWAWA) operator:

$$D - OWAWA\left(|x_1 \div y_1|, \ldots, |x_n \div y_n|\right) = \beta \sum_{j=1}^{n} w_j D_j + (1 - \beta) \sum_{i=1}^{n} v_i |x_i \div y_i|, \quad (22)$$

where D_j is the jth smallest of the $|x_i \div y_i|$ and $\beta \in [0, 1]$.

- If $OWAWAN(X,Y) = f((X - Y)^2)$, it is formed the variance OWAWA (Var-OWAWA) operator:

$$Var - OWAWA\left((x_1 - y_1)^2, ..., (x_n - y_n)^2\right) = \beta \sum_{j=1}^{n} w_j A_j + (1 - \beta) \sum_{i=1}^{n} v_i (x_i - y_i)^2,$$

$$(23)$$

where A_j is the jth smallest of the $(x_i - y_i)^2$ and $\beta \in [0,1]$.

Note that all these cases can be reduced to the OWA and the weighted average version forming the addition OWA, subtraction OWA, multiplication OWA, addition WA, and so on. Moreover, observe that many other cases could be studied including the OWAWAD operator explained in Section 3 and OWAWA operators with other similarity measures or operations.

5 Conclusions

This paper has suggested the use of OWAWA operators with norms. The main advantage of this approach is that it considers a unified framework between the OWA and the weighted average when aggregating information with norms. Thus, it is possible to consider subjective opinions and the attitudinal character of the decision maker in the same formulation. Several fundamental properties have been studied. It has been shown that the OWAWAD operator is also a particular case of this approach.

Some other extensions have also been considered including the use of induced aggregation operators (IOWAN operator) and the use of other approaches for unifying the OWA with the weighted average that have formed the HAN operator, the WOWA operator and the IWAN operator. Several key families of OWAWAN operators have also been studied including the addition OWAWA, the subtraction OWAWA, the multiplication OWAWA, the division OWAWA and the variance OWAWA operator. These aggregation operators have shown the potential for developing a new framework for arithmetic operations.

Acknowledgements. Support from the European Commission (PIEF-GA-2011-300062)is gratefully acknowledged.

References

[1] Beliakov, G., Pradera, A., Calvo, T.: Aggregation functions: A guide for practitioners. STUDFUZZ, vol. 221. Springer, Heidelberg (2007)
[2] Fodor, J., Marichal, J.L., Roubens, M.: Characterization of the ordered weighted averaging operators. IEEE Trans. Fuzzy Syst. 3, 236–240 (1995)
[3] Grabisch, M., Marichal, J.L., Mesiar, R., Pap, E.: Aggregation functions: Means. Inform. Sci. 181, 1–22 (2011)
[4] Hamming, R.W.: Error-detecting and error-correcting codes. Bell Syst. Tech. J. 29, 147–160 (1950)

[5] Karayiannis, N.: Soft learning vector quantization and clustering algorithms based on ordered weighted aggregation operators. IEEE Trans. Neural Networks 11, 1093–1105 (2000)

[6] Merigó, J.M.: A unified model between the weighted average and the induced OWA operator. Expert Syst. Applic. 38, 11560–11572 (2011)

[7] Merigó, J.M., Casanovas, M.: Decision making with distance measures and induced aggregation operators. Computers & Indust. Engin. 60, 66–76 (2011)

[8] Merigó, J.M., Casanovas, M.: A new Minkowski distance based on induced aggregation operators. Int. J. Computational Intelligence Syst. 4, 123–133 (2011)

[9] Merigó, J.M., Gil-Lafuente, A.M.: The induced generalized OWA operator. Inform. Sci. 179, 729–741 (2009)

[10] Merigó, J.M., Gil-Lafuente, A.M.: New decision-making techniques and their application in the selection of financial products. Inform. Sci. 180, 2085–2094 (2010)

[11] Merigó, J.M., Gil-Lafuente, A.M.: Decision making techniques in business and economics based on the OWA operator. SORT - Stat. Oper. Res. Trans. 36, 81–101 (2012)

[12] Merigó, J.M., Xu, Y.J., Zeng, S.Z.: Group decision making with distance measures and probabilistic information. Knowledge-Based Syst. 40, 81–87 (2013)

[13] Shannon, C.E.: A mathematical theory of communication. Bell Syst. Tech. J. 27, 379–423 (1948)

[14] Torra, V.: The weighted OWA operator. Int. J. Intelligent Syst. 12, 153–166 (1997)

[15] Xu, Z.S., Chen, J.: Orderedweighted distance measure. J. Syst. Sci. Syst. Engin. 17, 432–445 (2008)

[16] Xu, Z.S., Da, Q.L.: An overview of operators for aggregating information. Int. J. Intelligent Syst. 18, 953–969 (2003)

[17] Yager, R.R.: On ordered weighted averaging aggregation operators in multi-criteria decision making. IEEE Trans. Syst. Man Cybern. B 18, 183–190 (1988)

[18] Yager, R.R.: Families of OWA operators. Fuzzy Sets Syst. 59, 125–148 (1993)

[19] Yager, R.R.: On the inclusion of variance in decision making under uncertainty. Int. J. Uncert. Fuzz. Knowledge-Based Syst. 4, 401–419 (1996)

[20] Yager, R.R.: Including importances in OWA aggregation using fuzzy systems modelling. IEEE Trans. Fuzzy Syst. 6, 286–294 (1998)

[21] Yager, R.R.: New modes of OWA information fusion. Int. J. Intelligent Syst. 13, 661–681 (1998)

[22] Yager, R.R.: Heavy OWA operators. Fuzzy Optim. Decision Making 1, 379–397 (2002)

[23] Yager, R.R.: Induced aggregation operators. Fuzzy Sets Syst. 137, 59–69 (2003)

[24] Yager, R.R.: Generalizing variance to allow the inclusion of decision attitude in decision making under uncertainty. Int. J. Approximate Reasoning 42, 137–158 (2006)

[25] Yager, R.R.: Norms induced from OWA operators. IEEE Trans. Fuzzy Syst. 18, 57–66 (2010)

[26] Yager, R.R., Engemann, K.J., Filev, D.P.: On the concept of immediate probabilities. Int. J. Intell. Syst. 10, 373–397 (1995)

[27] Yager, R.R., Filev, D.P.: Induced ordered weighted averaging operators. IEEE Trans. Syst. Man Cybern. B 29, 141–150 (1999)

[28] Yager, R.R., Kacprzyk, J., Beliakov, G.: Recent developments on the ordered weighted averaging operators: Theory and practice. Springer, Heidelberg (2011)

[29] Zeng, S.Z., Su, W., Le, A.: Fuzzy generalized ordered weighted averaging distance operator and its application to decision making. Int. J. Fuzzy Syst. 14, 402–412 (2012)

[30] Zhou, L.G., Chen, H.Y., Liu, J.P.: Generalized power aggregation operators and their applications in group decision making. Computers & Indust. Engin. 62, 989–999 (2012)

Part IV
T-norms

Migrativity of Uninorms over T-norms and T-conorms

M. Mas, M. Monserrat, D. Ruiz-Aguilera, and J. Torrens

Abstract. In this paper the notions of α-migrative uninorms over a fixed t-norm T and over a fixed t-conorm S are introduced and studied. All cases when the uninorm U lies in any one of the most usual classes of uninorms are analyzed, characterizing with some assumptions on continuity all solutions of the migrativity equation for all possible combinations of U and T and for all possible combinations of U and S.

1 Introduction

In last decades the study of aggregation functions has been extensively developed mainly because of their great quantity of applications. This is one of the main reasons for the increasing interest in the field of aggregation functions and this interest is endorsed by the publication of some monographs dedicated entirely to aggregation functions ([1], [2], [15]).

One of the main topics in the study of these connectives from the theoretical point of view is directed towards the characterization of those that verify certain properties that may be useful in each context. The study of these properties for certain aggregation functions usually involves the resolution of functional equations. One of these properties is α-*migrativity*, introduced in [9]. For any $\alpha \in [0, 1]$ and a mapping $F : [0, 1]^2 \to [0, 1]$, this property is described as

$$F(\alpha x, y) = F(x, \alpha y) \quad \text{for all } x, y \in [0, 1]. \tag{1}$$

The interest of this property comes from its applications, for example in decision making processes ([5]), when a repeated, partial information needs to be fused in a global result, or in image processing, since in this context migrativity expresses

M. Mas · M. Monserrat · D. Ruiz-Aguilera · J. Torrens
Universitat de les Illes Balears, Carretera de Valldemossa, km 7.5, 07122 Palma, Spain
e-mail: {mmg448,mma112,daniel.ruiz,jts224}@uib.es

H. Bustince et al. (eds.), *Aggregation Functions in Theory and in Practise*,
Advances in Intelligent Systems and Computing 228,
DOI: 10.1007/978-3-642-39165-1_18, © Springer-Verlag Berlin Heidelberg 2013

the invariance of a given property under a proportional rescaling of some part of the image ([21]).

The migrativity property (and successive generalizations) has been studied for t-norms in [11, 12, 13], for semicopulas, quasi-copulas and copulas in [2, 8, 21] and for aggregation functions in general in [4, 5, 22]. Note that in the equation (1) the product αx can be replaced by any t-norm T_0 obtaining the property for t-norms called (α, T_0)-migrativity, that can be written as

$$T(T_0(\alpha, x), y) = T(x, T_0(\alpha, y)) \quad \text{for all } x, y \in [0, 1] \tag{2}$$

being T_0 a t-norm and $\alpha \in [0, 1]$. This generalization of the migrativity for t-norms has been recently studied in [12]. By dualization, a similar definition can be given for t-conorms as it was pointed out in [19]. Moreover, this study has been extended to uninorms with the same neutral element in [20] (the case of representable uninorms was also solved in [4]).

However, all the previous studies have a common point: they always deal with aggregation functions (t-norms, t-conorms or uninorms) having the same neutral element, with the only exception of representable uninorms that were investigated in [4]. This condition is not necessary to find out solutions of the migrativity property as we will see in the present work. As a first step in this direction, we will present in this paper a complete study of those uninorms with neutral element $e \in]0, 1[$ that are α-migrative over t-norms and over t-conorms. We will do it by considering uninorms in any one of the most usual classes of uninorms.

The article is organized into different sections. After this introduction, we include a preliminary section to establish the necessary notation and we recall some basic definitions, specially on uninorms. In Section 3 we introduce the definition of (α, T)-migrative uninorm for a given t-norm T, analyzing some of its initial properties. We continue with the characterization of those (α, T)-migrative uninorms, that lay in each one of the most usual classes of uninorms, i.e., uninorms in $\mathscr{U}_{\min}$ and $\mathscr{U}_{\max}$, idempotent uninorms, representable uninorms and uninorms continuous in the open square $]0, 1[^2$. In Section 4 we dualize these results for t-conorms and we end the paper with a section of conclusions and future work.

2 Preliminaries

We will assume the basic theory of t-norms and t-conorms. The definitions, notations and results on them can be found in [1, 14]. We will just give in this section some basic facts about uninorms. More details can be found in [1, 9, 24].

Definition 1. A binary function $U : [0, 1]^2 \to [0, 1]$ is called a *uninorm* if it is associative, commutative, non-decreasing in each variable and there is a neutral element $e \in [0, 1]$ such that $U(e, x) = x$ for all $x \in [0, 1]$.

Evidently, a uninorm with neutral element $e = 1$ is a t-norm and a uninorm with neutral element $e = 0$ is a t-conorm. For any other value $e \in]0, 1[$ the operation works as a t-norm in $[0,e]^2$, as a t-conorm in $[e,1]^2$ and its values are between minimum and maximum in the set of points $A(e)$ given by

$$A(e) = [0,e[\times]e,1] \cup]e,1] \times [0,e[.$$

We will usually denote a uninorm with neutral element e and underlying t-norm and t-conorm, T and S, by $U \equiv \langle T,e,S \rangle$. For any uninorm it is satisfied that $U(0,1) \in \{0,1\}$ and a uninorm U is called *conjunctive* if $U(1,0) = 0$ and *disjunctive* when $U(1,0) = 1$. On the other hand, the most studied classes of uninorms are:

- Uninorms in $\mathcal{U}_{\min}$ (respectively $\mathcal{U}_{\max}$), those given by minimum (respectively maximum) in $A(e)$.
- *Idempotent* uninorms, those that satisfy $U(x,x) = x$ for all $x \in [0,1]$.
- *Representable* uninorms, those that have an additive generator.
- *Continuous* in the open square $]0,1[^2$.

In what follows we recall the structure of each one of these classes of uninorms.

Theorem 1. ([9]) *Let $U : [0,1]^2 \to [0,1]$ be a uninorm with neutral element $e \in]0,1[$. Then, the sections $x \mapsto U(x,1)$ and $x \mapsto U(x,0)$ are continuous in each point except perhaps for e if and only if U is given by one of the following formulas.*

(a)If $U(0,1) = 0$, then

$$U(x,y) = \begin{cases} eT\left(\frac{x}{e},\frac{y}{e}\right) & \text{if } (x,y) \in [0,e]^2 \\ e+(1-e)S\left(\frac{x-e}{1-e},\frac{y-e}{1-e}\right) & \text{if } (x,y) \in [e,1]^2 \\ \min(x,y) & \text{if } (x,y) \in A(e), \end{cases} \tag{3}$$

where T is a t-norm, and S is a t-conorm.
(b)If $U(0,1) = 1$, then the same structure holds, changing minimum by maximum in $A(e)$.

The set of uninorms as in case (a) will be denoted by $\mathcal{U}_{\min}$ and the set of uninorms as in case (b) by $\mathcal{U}_{\max}$. We will denote a uninorm in $\mathcal{U}_{\min}$ with underlying t-norm T, underlying t-conorm S and neutral element e as $U \equiv \langle T,e,S \rangle_{\min}$ and in a similar way, a uninorm in $\mathcal{U}_{\max}$ as $U \equiv \langle T,e,S \rangle_{\max}$.
Idempotent uninorms were characterized first in [1] for those with a lateral continuity and in [18] for the general case. An improvement of this last result was done in [24] as follows.

Theorem 2. ([24]) *U is an idempotent uninorm with neutral element $e \in [0,1]$ if and only if there exists a non increasing function $g : [0,1] \to [0,1]$, symmetric with respect to the main diagonal, with $g(e) = e$, such that*

$$U(x,y) = \begin{cases} \min(x,y) & \text{if } y < g(x) \text{ or } (y = g(x) \text{ and } x < g(g(x))) \\ \max(x,y) & \text{if } y > g(x) \text{ or } (y = g(x) \text{ and } x > g(g(x))) \\ \min(x,y) \text{ or } \max(x,y) & \text{if } y = g(x) \text{ and } x = g(g(x)), \end{cases}$$

being commutative in the points (x,y) such that $y = g(x)$ with $x = g(g(x))$.

Any idempotent uninorm U with neutral element e and associated function g, will be denoted by $U \equiv \langle g, e \rangle_{\text{ide}}$ and the class of idempotent uninorms will be denoted by $\mathscr{U}_{\text{ide}}$. Obviously, for any of these uninorms the underlying t-norm T is the minimum and the underlying t-conorm S is the maximum.

Definition 2. ([9]) Consider $e \in]0,1[$. A binary operation $U : [0,1]^2 \to [0,1]$ is a *representable uninorm* if and only if there exists a continuous strictly increasing function $h : [0,1] \to [-\infty, +\infty]$ with $h(0) = -\infty$, $h(e) = 0$ and $h(1) = +\infty$ such that

$$U(x,y) = h^{-1}(h(x) + h(y))$$

for all $(x,y) \in [0,1]^2 \setminus \{(0,1),(1,0)\}$ and $U(0,1) = U(1,0) \in \{0,1\}$. The function h is usually called an *additive generator* of U.

Remark 1. Recall that there are no continuous uninorms with neutral element $e \in]0,1[$. In fact, representable uninorms were characterized as those uninorms that are continuous in $[0,1]^2 \setminus \{(1,0),(0,1)\}$ (see [23]) as well as those that are strictly increasing in the open unit square (see [10]). We will denote by $\mathscr{U}_{\text{rep}}$ the class of representable uninorms.

Any representable uninorm U with neutral element e and additive generator h, will be denoted by $U \equiv \langle h, e \rangle_{\text{rep}}$. For any of these uninorms the underlying t-norm T is strict and the underlying t-conorm S is strict as well.

A more general class containing representable uninorms are those continuous in the open unit square $]0,1[^2$, that were characterized in [17] as follows.

Theorem 3. (*[17] and [23] for the current version*) *Suppose U is a uninorm continuous in $]0,1[^2$ with neutral element $e \in]0,1[$. Then either one of the following cases is satisfied:*

(a) There exist $u \in [0,e[$, $\lambda \in [0,u]$, two continuous t-norms T_1 and T_2 and a representable uninorm R such that U can be represented as

$$U(x,y) = \begin{cases} \lambda T_1\left(\frac{x}{\lambda}, \frac{y}{\lambda}\right) & \text{if } x,y \in [0,\lambda] \\ \lambda + (u-\lambda)T_2\left(\frac{x-\lambda}{u-\lambda}, \frac{y-\lambda}{u-\lambda}\right) & \text{if } x,y \in [\lambda,u] \\ u + (1-u)R\left(\frac{x-u}{1-u}, \frac{y-u}{1-u}\right) & \text{if } x,y \in]u,1[\\ 1 & \text{if } \min(x,y) \in]\lambda,1] \text{ and } \max(x,y) = 1 \\ \lambda \text{ or } 1 & \text{if } (x,y) \in \{(\lambda,1),(1,\lambda)\} \\ \min(x,y) & \text{elsewhere.} \end{cases}$$

$$(4)$$

(b) There exist $v \in]e,1]$, $\omega \in [v,1]$, two continuous t-conorms S_1 and S_2 and a representable uninorm R such that U can be represented as

$$
U(x,y) = \begin{cases}
vR\left(\frac{x}{v},\frac{y}{v}\right) & \text{if } x,y \in \,]0,v[\\
v + (\omega - v)S_1\left(\frac{x-v}{\omega-v}, \frac{y-v}{\omega-v}\right) & \text{if } x,y \in [v,\omega] \\
\omega + (1-\omega)S_2\left(\frac{x-\omega}{1-\omega}, \frac{y-\omega}{1-\omega}\right) & \text{if } x,y \in [\omega,1] \\
0 & \text{if } \max(x,y) \in [0,\omega[\text{ and } \min(x,y) = 0 \\
\omega \text{ or } 0 & \text{if } (x,y) \in \{(0,\omega),(\omega,0)\} \\
\max(x,y) & \text{elsewhere.}
\end{cases}
\tag{5}
$$

The class of all uninorms continuous in $]0,1[^2$ will be denoted by $\mathscr{U}_{\cos}$. A uninorm as in (4) will be denoted by $U \equiv \langle T_1, \lambda, T_2, u, (R,e)\rangle_{\cos,\min}$ and the class of all uninorms continuous in the open unit square of this form will be denoted by $\mathscr{U}_{\cos,\min}$. Analogously, a uninorm as in (5) will be denoted by $U \equiv \langle (R,e), v, S_1, \omega, S_2\rangle_{\cos,\max}$ and the class of all uninorms continuous in the open unit square of this form will be denoted by $\mathscr{U}_{\cos,\max}$.

For any uninorm $U \equiv \langle T_1, \lambda, T_2, u, (R,e)\rangle_{\cos,\min}$, the underlying t-norm of U is given by an ordinal sum[1] of three t-norms, T_1, T_2 and a strict t-norm, whereas the underlying t-conorm is strict. For a uninorm $U \equiv \langle (R,e), v, S_1, \omega, S_2\rangle_{\cos,\max}$, the underlying t-norm of U is strict, whereas the underlying t-conorm is given by an ordinal sum of three t-conorms: a strict t-conorm, S_1, and S_2.

A detailed study of continuous t-norms that are α-migrative over the minimum, the product and the Łukasiewicz t-norms can be found in [12]. An extension of this work has been made in [13] where continuous t-norms α-migrative over any continuous t-norm T_0, including ordinal sums, are studied. A similar study can be done for t-conorms just by dualizing the results for t-norms (see for instance [19]). Thus we refer to theses papers for the results concerning migrativity of t-norms and t-conorms, that will be used in this work in order to give the results for uninorms.

3 Migrative Uninorms Over T-norms

Now we will introduce the definition of migrativity of a uninorm U over a t-norm T.

Definition 3. Given a t-norm T and $\alpha \in [0,1]$, a uninorm U is said to be α-*migrative* over T or $(\alpha,T)-$*migrative* if

$$
U(T(\alpha,x),y) = U(x,T(\alpha,y)) \quad \text{for all } x,y \in [0,1].
\tag{6}
$$

Since the cases of t-norms and t-conorms are already known we will consider only uninorms with neutral element $e \in \,]0,1[$. First, note that for the extreme values of α, the migrativity of uninorms over t-norms become trivial, as follows.

---

[1] For details about ordinal sums of t-norms and t-conorms, see [14], definition 3.44.

Lemma 1. *Let U be a uninorm with neutral element $e \in]0,1[$ and T a t-norm. Then,*

i) U is 1-migrative over T.

ii) U is 0-migrative over T if and only if U is conjunctive.

As a consequence of the previous lemma, from now on we will consider $\alpha \in]0,1[$. In this case, we can easily derive the next general result on migrativity.

Proposition 1. *Let U be a uninorm with neutral element $e \in]0,1[$, T a t-norm and $\alpha \in]0,1[$. Suppose that U is α-migrative over T, then the following items hold.*

i) U must be a conjunctive uninorm.

ii) $U(\alpha,y) = U(1,T(\alpha,y))$ for all $y \in [0,1]$.

From the previous result, we will only consider conjunctive uninorms. Moreover, from point *ii)* in the previous proposition, it is clear that the values that can take a conjunctive uninorm on the boundary, i.e. the values $U(1,y)$ with $y \in [0,1]$, will be important. Thus, let us devote the following subsection to this end.

3.1 Uninorms on the Boundary

In this section we will study the values that take a uninorm U on the boundary, $B = [0,1]^2 \backslash]0,1[^2$. It is clear that any conjunctive uninorm satisfies $U(x,0) = U(0,y) = 0$ for all $x,y \in [0,1]$ and $U(x,1) = U(1,y) = 1$ for all $x,y \in [e,1]$, whereas any disjunctive uninorm satisfies $U(x,1) = U(1,y) = 1$ for all $x,y \in [0,1]$ and $U(x,0) = U(0,y) = 0$ for all $x,y \in [0,e]$. Now we investigate the values of the uninorm on the remaining points.

Note that in all classes of uninorms recalled in the preliminaries, it also holds that

$$U(1,y) \in \{1,y\} \qquad \text{for all} \quad y \in [0,1], \tag{7}$$

for conjunctive uninorms and

$$U(0,y) \in \{0,y\} \qquad \text{for all} \quad y \in [0,1]. \tag{8}$$

for disjunctive uninorms. In fact, up to our knowledge, there is no conjunctive uninorms failing property (7) nor disjunctive uninorms failing property (8), and we claim that this boundary property holds in general for all conjunctive and disjunctive uninorms, respectively. To deal with this problem, let us begin by the following definition.

Definition 4. Let U be a conjunctive uninorm. We will say that U is *locally internal on the boundary* if it satisfies property (7). Similarly, if U is a disjunctive uninorm, we will say that U is *locally internal on the boundary* if it satisfies property (8).

Remark 2. Note that uninorms in $\mathscr{U}_{\min}$ and $\mathscr{U}_{\max}$, idempotent, representable or continuous in the open square uninorms, are all of them locally internal on the boundary.

With this definition the previously mentioned claim can be written as follows:

Claim: Any uninorm is locally internal on the boundary.

We have no proof of this fact, but we can give the following partial results on conjunctive uninorms, that can be easily dualized to disjunctive ones.

Lemma 2. *Let U be a conjunctive uninorm. Then, for any $x \in]0, e[$, $U(1, x) < e$ or $U(1, x) = 1$.*

Proposition 2. *Any conjunctive uninorm $U \equiv \langle T, e, S \rangle$ with underlying t-norm T continuous is locally internal on the boundary.*

It is clear that for disjunctive uninorms we obtain similar results just by dualyzing the reasonings. Specifically, we have,

Proposition 3. *Let $U \equiv \langle T, e, S \rangle$ be any disjunctive uninorm. Then, the following items hold:*

 i) *For any $x \in]e, 1[$, $U(0, x) > e$ or $U(0, x) = 0$.*
 ii) *If the underlying t-conorm S is continuous, then U is locally internal on the boundary.*

3.2 Migrativity of Uninorms That Are Locally Internal on the Boundary

Let us now return to the migrative property, in this case for uninorms that are locally internal on the boundary.

Proposition 4. *Let U be a conjunctive uninorm with neutral element $e \in]0, 1[$ that is locally internal on the boundary. Let T be a t-norm and $\alpha \in]0, 1[$. If U is α-migrative over T then $T(\alpha, e) = \alpha$ and $\alpha < e$.*

Thus, from now on we will consider only $\alpha \in]0, e[$. Now, we can prove that for uninorms locally internal on the boundary, the migrativity of U over T is equivalent to have both U and T the same α-section.

Proposition 5. *Let U be a conjunctive uninorm with neutral element $e \in]0, 1[$ locally internal on the boundary, T a t-norm and $\alpha \in]0, e[$. The following items are equivalent:*

 i) *U is (α, T)-migrative.*
 ii) *$U(\alpha, y) = T(\alpha, y)$ for all $y \in [0, 1]$.*

Since uninorms are compensatory in the region $A(e)$ and t-norms are always under the minimum, the result above proves in particular that there are no uninorms (locally internal on the boundary) α-migrative over any Archimedean t-norm. Thus, if we restrict from now on to continuous t-norms, to find (α, T)-migrative uninorms, T must be the minimum or a nontrivial ordinal sum.

Again from the Proposition 5, to characterize uninorms U, that are α-migrative over a t-norm T, we only need to look for uninorms with the same α-section than T. We will distinguish cases depending on the class of uninorms considered. We start with uninorms in $\mathscr{U}_{\min}$.

Theorem 4. *Let $U \equiv \langle T_U, e, S_U \rangle$ be a uninorm in $\mathscr{U}_{\min}$ with neutral element $e \in]0,1[$ and T_U continuous. Let T be a continuous t-norm and $\alpha \in]0,e[$.*

a) If $T(\alpha,\alpha) = \alpha$ then U is α-migrative over T if and only if $U(\alpha,\alpha) = \alpha$. In this case, both T_U and T must be ordinal sums $T_U = (\langle 0, \frac{\alpha}{e}, T_{1U} \rangle, \langle \frac{\alpha}{e}, 1, T_{2U} \rangle)$ and $T = (\langle 0, \alpha, T_1 \rangle, \langle \alpha, 1, T_2 \rangle)$ for some continuous t-norms T_{1U}, T_{2U}, T_1 and T_2.

b) If $T(\alpha,\alpha) < \alpha$ then U is α-migrative over T if and only if both T and T_U are ordinal sums of the form $T = (\ldots, \langle a, b, T_A \rangle, \ldots)$ and $T_U = (\ldots, \langle \frac{a}{e}, \frac{b}{e}, T_{UA} \rangle, \ldots)$, where $a < \alpha < b \le e$, T_A and T_{UA} are Archimedean with T_{UA} being $\frac{\alpha-a}{b-a}$-migrative over T_A.

For representable uninorms the result is negative and there are no representable uninorms α-migrative over t-norms, as it is stated in the following theorem.

Theorem 5. *Let U be a representable uninorm and $\alpha \in]0,1[$. Then U is never (α, T)-migrative for any t-norm T.*

The case of idempotent uninorms is solved in the next theorem.

Theorem 6. *Let $U \equiv \langle g, e \rangle_{\text{ide}}$ be a conjunctive idempotent uninorm with $e \in]0,1[$, T a t-norm and $\alpha \in]0,e[$. Then the following items are equivalent*

i) U is (α, T)-migrative.
ii) $U(\alpha,y) = T(\alpha,y) = \min(\alpha,y)$ for all $y \in [0,1]$.

Note that when T is continuous the α-section of T is given by the minimum if and only if α is an idempotent element of T. In this case, taking into account the structure of idempotent uninorms, the previous theorem can be written as follows.

Theorem 7. *Let $U \equiv \langle g, e \rangle_{\text{ide}}$ be a conjunctive idempotent uninorm with $e \in]0,1[$, T a continuous t-norm and $\alpha \in]0,e[$. Then the following items are equivalent*

i) U is (α, T)-migrative.
ii) α is an idempotent element of T, $g(1) \ge \alpha$ and $U(\alpha,1) = \alpha$.

Now we deal with uninorms continuous in $]0,1[^2$.

Proposition 6. *Let $U \equiv \langle T_1, \lambda, T_2, u, (R,e) \rangle_{\cos,\min}$ be a uninorm in $\mathscr{U}_{\cos,\min}$, $\alpha \in]0,1[$ such that $\alpha > \lambda$, and T a t-norm. Then U is not α-migrative over T.*

Proposition 7. *Let $U \equiv \langle T_1, \lambda, T_2, u, (R,e) \rangle_{\cos,\min}$ be a uninorm in $\mathscr{U}_{\cos,\min}$, $\alpha \in]0,1[$ such that $\alpha = \lambda$ and T a continuous t-norm. Then U is α-migrative over T if and only if $T(\lambda,\lambda) = \lambda$ and $U(\lambda,1) = \lambda$.*

Theorem 8. *Let* $U \equiv \langle T_1, \lambda, T_2, u, (R, e) \rangle_{\cos,\min}$ *be a uninorm in* $\mathcal{U}_{\cos,\min}$, $\alpha \in]0, 1[$ *such that* $\alpha < \lambda$ *and* T *a continuous t-norm.*

a) If $T(\alpha, \alpha) = \alpha$ *then* U *is* α-*migrative over* T *if and only if* $U(\alpha, \alpha) = \alpha$.
Moreover, in this case $T_1 = \left(\langle 0, \frac{\alpha}{\lambda}, T_1' \rangle, \langle \frac{\alpha}{\lambda}, 1, T_1'' \rangle \right)$ *for some continuous t-norms* T_1' *and* T_1''.

b) If $T(\alpha, \alpha) < \alpha$ *then* U *is* α-*migrative over* T *if and only if in the structure of* T *as ordinal sum of archimedean summands* $T = (\ldots, \langle a, b, T_A \rangle, \ldots)$, *where* $\alpha \in]a, b[$, $b \leq \lambda$, *and the structure of* T_1 *as ordinal sum of archimedean summands has the form* $T_1 = (\ldots, \langle \frac{a}{\lambda}, \frac{b}{\lambda}, T_{1A} \rangle, \ldots)$ *with* T_{1A} *being* $\frac{\alpha - a}{b - a}$-*migrative over* T_A.

4 Migrative Uninorms over T-conorms

Similarly for the case of t-norms, it can be introduced the definition of migrativity of a uninorm U over a t-conorm S.

Definition 5. *Given a t-conorm* S *and* $\alpha \in [0, 1]$, *a uninorm* U *is said to be* α-*migrative over* S *or* (α, S)-*migrative if*

$$U(S(\alpha, x), y) = U(x, S(\alpha, y)) \quad \text{for all } x, y \in [0, 1]. \tag{9}$$

Since the cases of t-norms and t-conorms are already known we will consider only uninorms with neutral element $e \in]0, 1[$. We can derive similar results than for the case of t-norms just by duality. Then, for instance we obtain,

Lemma 3. *Let* U *be a uninorm with neutral element* $e \in]0, 1[$ *and* S *a t-conorm. Then,*

i) U is 0-migrative over S.
ii) U is 1-migrative over S if and only if U is disjunctive.

Proposition 8. *Let* U *be a uninorm with neutral element* $e \in]0, 1[$, S *a t-conorm and* $\alpha \in]0, 1[$. *Suppose that* U *is* α-*migrative over* S, *then the following items hold.*

i) U must be a disjunctive uninorm.
ii) $U(\alpha, y) = U(0, S(\alpha, y))$ *for all* $y \in [0, 1]$.
iii) If U is locally internal on the boundary then $S(\alpha, e) = \alpha$ *and* $\alpha > e$.

Proposition 9. *Let* U *be a disjunctive uninorm locally internal on the boundary with neutral element* $e \in]0, 1[$, S *a t-conorm and* $\alpha \in]e, 1[$. *The following items are equivalent:*

i) U is (α, S)-*migrative.*
ii) $U(\alpha, y) = S(\alpha, y)$ *for all* $y \in [0, 1]$.

Clearly, the last result derives into the fact that there are no α-migrative uninorms (locally internal on the boundary) over Archimedean t-conorms. Thus, if we restrict to the case of continuous t-conorms, to find (α, S)-migrative uninorms it must be S

the maximum or a nontrivial ordinal sum. To find out all these solutions we only need to find uninorms with the same α-section than the t-conorm S.

We can do the same study than for t-norms, between the classes of uninorms recalled in the preliminaries. First, we have the following result for uninorms in $\mathscr{U}_{\max}$.

Theorem 9. Let $U \equiv \langle T_U, e, S_U \rangle$ be a uninorm in $\mathscr{U}_{\max}$ with neutral element $e \in]0,1[$ and S_U continuous. Let S be a continuous t-conorm and $\alpha \in]e,1[$.

a) If $S(\alpha,\alpha) = \alpha$ then U is α-migrative over S if and only if $U(\alpha,\alpha) = \alpha$. In this case, both S_U and S must be ordinal sums $S_U = (\langle 0, \frac{\alpha-e}{1-e}, S_{1U} \rangle, \langle \frac{\alpha-e}{1-e}, 1, S_{2U} \rangle)$ and $S = (\langle 0, \alpha, S_1 \rangle, \langle \alpha, 1, S_2 \rangle)$ for some continuous t-conorms S_{1U}, S_{2U}, S_1 and S_2.

b) If $S(\alpha,\alpha) > \alpha$ then U is α-migrative over S if and only if both S and S_U are ordinal sums of the form $S = (\ldots, \langle a,b,S_A \rangle, \ldots)$ and $S_U = (\ldots, \langle \frac{a-e}{1-e}, \frac{b-e}{1-e}, S_{UA} \rangle, \ldots)$, where $e \leq a < \alpha < b$, S_A and S_{UA} are Archimedean with S_{UA} being $\frac{\alpha-a}{b-a}$-migrative over S_A.

For the case of representable uninorms we have no solutions.

Theorem 10. Let U be a representable uninorm and $\alpha \in]0,1[$. Then U is never (α, S)-migrative for any t-conorm S.

The results for idempotent uninorms are as follows.

Theorem 11. Let $U \equiv \langle g, e \rangle_{\text{ide}}$ be a disjunctive idempotent uninorm with $e \in]0,1[$, S a t-conorm and $\alpha \in]e,1[$. Then the following items are equivalent

i) U is (α, S)-migrative.
ii) $U(\alpha, y) = S(\alpha, y) = \max(\alpha, y)$ for all $y \in [0,1]$.

Theorem 12. Let $U \equiv \langle g, e \rangle_{\text{ide}}$ be a disjunctive idempotent uninorm with $e \in]0,1[$, S a continuous t-conorm and $\alpha \in]e,1[$. Then the following items are equivalent

i) U is (α, S)-migrative.
ii) α is an idempotent element of S, $g(0) \leq \alpha$ and $U(0, \alpha) = \alpha$.

Finally, the results for uninorms continuous in $]0,1[^2$ are as follows.

Proposition 10. Let $U \equiv \langle (R,e), v, S_1, \omega, S_2 \rangle_{\cos,\max}$ be a uninorm in $\mathscr{U}_{\cos,\max}$, $\alpha \in]0,1[$ such that $\alpha < \omega$, and S a t-conorm. Then U is not α-migrative over S.

Proposition 11. Let $U \equiv \langle (R,e), v, S_1, \omega, S_2 \rangle_{\cos,\max}$ be a uninorm in $\mathscr{U}_{\cos,\max}$, $\alpha \in]0,1[$ such that $\alpha = \omega$ and S a continuous t-conorm. Then U is α-migrative over S if and only if $S(\omega, \omega) = \omega$ and $U(\omega, 0) = \omega$.

Theorem 13. Let $U \equiv \langle (R,e), v, S_1, \omega, S_2 \rangle_{\cos,\max}$ be a uninorm in $\mathscr{U}_{\cos,\max}$, $\alpha \in]0,1[$ such that $\alpha > \omega$ and S a continuous t-conorm.

a) If $S(\alpha,\alpha) = \alpha$ then U is α-migrative over S if and only if $U(\alpha,\alpha) = \alpha$. Moreover, in this case $S_2 = (\langle 0, \frac{\alpha-\omega}{1-\omega}, S_2' \rangle, \langle \frac{\alpha-\omega}{1-\omega}, 1, S_2'' \rangle)$ for some continuous t-conorms S_2' and S_2''.

b) If $S(\alpha,\alpha) > \alpha$ then U is α-migrative over S if and only if in the structure of S as ordinal sum of Archimedean summands $S = (\ldots, \langle a,b,S_A \rangle, \ldots)$, where $\alpha \in]a,b[$, $a \geq \omega$, and the structure of S_2 as ordinal sum of Archimedean summands has the form $S_2 = (\ldots, \langle \frac{a-\omega}{1-\omega}, \frac{b-\omega}{1-\omega}, S_{2A} \rangle, \ldots)$ with S_{2A} being $\frac{\alpha-a}{b-a}$-migrative over S_A.

5 Conclusions and Future Work

Uninorms that are migrative over uninorms with the same neutral element were studied in [20]. However, the hypothesis of having the same neutral element is not necessary to find out solutions of the migrative property. In this line, a first step in the study of migrative uninorms with different neutral element has been presented in this work. Specifically, migrativity of uninorms with neutral element $e \in]0, 1[$ over t-norms and t-conorms has been investigated in this paper. As a result, it has been proved that there are no uninorms migrative over the product (that is the original migrative equation) nor over any Archimedean t-norm (and similarly for t-conorms). On the contrary, many solutions appear when we deal with the minimum t-norm (maximum t-conorm), or with any ordinal sum t-norms (t-conorms), and all these solutions varying among the most known classes of uninorms have been reached and characterized.

Our future work on this topic will be directed to the study of uninorms that are migrative over uninorms with different neutral element and to the study of nullnorms that are migrative over t-norms, t-conorms and other nullnorms with the same or different absorbing element. We believe that the results obtained in this study will be useful in our work, because the reasoning to do is probably similar to that used here.

Acknowledgements. This work has been supported by the Spanish grant MTM2009-10320 (with FEDER support).

References

[1] Alsina, C., Frank, M.J., Schweizer, B.: Associative Functions. Triangular Norms and Copulas. World Scientific, New Jersey (2006)

[2] Beliakov, G., Calvo, T.: On migrative means and copulas. In: Proceedings of Fifth International Summer School on Aggregation Operators, AGOP 2009, Palma de Mallorca, pp. 107–110 (2009)

[3] Beliakov, G., Pradera, A., Calvo, T.: Aggregation Functions: A Guide for Practicioners. Springer, Heidelberg (2007)

[4] Bustince, H., De Baets, B., Fernandez, J., Mesiar, R., Montero, J.: A generalization of the migrativity property of aggregation functions. Information Sciences 191, 76–85 (2012)

[5] Bustince, H., Montero, J., Mesiar, R.: Migrativity of aggregation functions. Fuzzy Sets and Systems 160, 766–777 (2009)

[6] Calvo, T., Mayor, G., Mesiar, R. (eds.): Aggregation operators. New trends and applications. STUDFUZZ, vol. 97. Physica-Verlag, Heidelberg (2002)

[7] De Baets, B.: Idempotent uninorms. European Journal of Operational Research 118, 631–642 (1999)

[8] Durante, F., Ricci, R.G.: Supermigrative semi-copulas and triangular norms. Information Sciences 179, 2689–2694 (2009)

[9] Durante, F., Sarkoci, P.: A note on the convex combination of triangular norms. Fuzzy Sets and Systems 159, 77–80 (2008)

[10] Fodor, J., De Baets, B.: A single-point characterization of representable uninorms. Fuzzy Sets and Systems 202, 89–99 (2012)

[11] Fodor, J., Rudas, I.J.: On continuous triangular norms that are migrative. Fuzzy Sets and Systems 158, 1692–1697 (2007)

[12] Fodor, J., Rudas, I.J.: An extension of the migrative property for triangular norms. Fuzzy Sets and Systems 168, 70–80 (2011)

[13] Fodor, J., Rudas, I.J.: Migrative t-norms with respect to continuous ordinal sums. Information Sciences 181, 4860–4866 (2011)

[14] Fodor, J., Yager, R.R., Rybalov, A.: Structure of uninorms. International Journal of Uncertainty, Fuzziness and Knowledge-Based Systems 5, 411–427 (1997)

[15] Grabisch, M., Marichal, J.L., Mesiar, R., Pap, E.: Aggregation functions. In: Encyclopedia of Mathematics and its Applications, vol. 127. Cambridge University Press (2009)

[16] Klement, E.P., Mesiar, R., Pap, E.: Triangular norms. Kluwer Academic Publishers, Dordrecht (2000)

[17] Hu, S.K., Li, Z.F.: The structure of continuous uninorms. Fuzzy Sets and Systems 124, 43–52 (2001)

[18] Martín, J., Mayor, G., Torrens, J.: On locally internal monotonic operations. Fuzzy Sets and Systems 137, 27–42 (2003)

[19] Mas, M., Monserrat, M., Ruiz-Aguilera, D., Torrens, J.: On migrative t-conorms and uninorms. In: Greco, S., Bouchon-Meunier, B., Coletti, G., Fedrizzi, M., Matarazzo, B., Yager, R.R. (eds.) IPMU 2012, Part III. CCIS, vol. 299, pp. 286–295. Springer, Heidelberg (2012)

[20] Mas, M., Monserrat, M., Ruiz-Aguilera, D., Torrens, J.: An extension of the migrative property for uninorms. Information Sciences (2012) (submitted)

[21] Mesiar, R., Bustince, H., Fernandez, J.: On the α-migrativity of semicopulas, quasi-copulas and copulas. Information Sciences 180, 1967–1976 (2010)

[22] Ricci, R.G.: Supermigrative aggregation functions. In: Proceedings of the 6th International Summer School on Aggregation Operators, AGOP 2011, Benevento, Italy, pp. 145–150 (2011)

[23] Ruiz, D., Torrens, J.: Distributivity and conditional distributivity of a uninorm and a continuous t-conorm. IEEE Transactions on Fuzzy Systems 14, 180–190 (2006)

[24] Ruiz-Aguilera, D., Torrens, J., De Baets, B., Fodor, J.: Some remarks on the characterization of idempotent uninorms. In: Hüllermeier, E., Kruse, R., Hoffmann, F. (eds.) IPMU 2010. LNCS (LNAI), vol. 6178, pp. 425–434. Springer, Heidelberg (2010)

Additive Generators of Overlap Functions

Graçaliz Pereira Dimuro and Benjamín Bedregal

Abstract. Overlap functions are a particular instance of aggregation functions, consisting by non-decreasing continuous commutative bivariate functions defined over the unit square, satisfying appropriate boundary conditions. Overlap functions play an important role in classification problems, image processing and in some problems of decision making based on fuzzy preference relations. The concepts of indifference and incomparability defined in terms of overlap functions may allow the application in several different contexts. The aim of this papers is to introduce the notion of additive generators of overlap functions, allowing the definition of overlap functions (as two-place functions) by means of one-place functions, which is important since it can reduce the computational complexity in applications. Also, some properties of an overlap function presenting a generator can be related to properties of its generator, pointing to a more systematic methodology for their selection for the various applications.

1 Introduction

Overlap functions are non-decreasing continuous commutative bivariate functions defined over the unit square $[0,1] \times [0,1]$, satisfying appropriate boundary conditions (see Def. 4), constituting a particular instance of *aggregation functions* playing an important role in classification problems and image processing, when the

Graçaliz Pereira Dimuro
Programa de Pós-Graduação em Computação, Centro de Ciências Computacionais,
Universidade Federal do Rio Grande, Av. Itália km 08, Campus Carreiros,
96201-900 Rio Grande, Brazil
e-mail: gracaliz@gmail.com

Benjamín Bedregal
Departamento de Informática e Matemática Aplicada,
Universidade Federal do Rio Grande do Norte, Campus Universitário,
59072-970 Natal, Brazil
e-mail: bedregal@dimap.ufrn.br

H. Bustince et al. (eds.), *Aggregation Functions in Theory and in Practise*,
Advances in Intelligent Systems and Computing 228,
DOI: 10.1007/978-3-642-39165-1_19, © Springer-Verlag Berlin Heidelberg 2013

classifying concepts presents overlapping. In this case, overlapping must be evaluated whenever the evaluated classes are not crisp, so one can check if a certain classification structure is in accordance with the reality, being also adequate for the desired decision making process [9].

Bustince et al. [9] presented the basic properties that must be fulfilled by overlap functions when applied to the problem of object recognition, where the best classification with respect to background is the one with less overlapping between the class object and the class background. More recently, Jurio et al. [16] analyzed several properties of overlap functions, in particular, the convex combination of overlap functions in a new overlap function, presenting an application to image processing.

Overlap functions can also be applied in decision making based on fuzzy preference relations, where the associativity property is not strongly required and the use of t-norms or t-conorms as the combination/separation operators is not necessary. In [10], Bustince et al. applied overlap functions in fuzzy preference modeling and decision making, presenting an algorithm to elaborate on an alternative preference ranking that penalizes the alternatives for which the expert is not sure about his/her preference.

The concepts of indifference and incomparability defined in terms of overlap functions may allow the application in several different contexts. Then, considering their potencial for practical applications, the main properties of overlap functions (for example, migrativity, homogeneity, idempotency, Lipschitzianity condition) have been studied in different aspects (see, e.g., the works by Bustince et al. [8, 9, 10, 16] and Bedregal et al. [3]). Nevertheless, there are also different definitions of overlap functions (see, e.g., [2, 14, 33]).

The most well-known aggregation functions are *t-norms* and *t-conorms*, introduced by Schweizer and Sklar [36, 39], which are used to model conjunction and disjunction in fuzzy logic. An important approach that allows the definition of t-norms and t-conorms (as two-place functions) by means of one-place functions is the use of *additive generators*, which appear explicitly for the first time in [37, 38], with several posterior generalizations (e.g., [15, 17, 18, 21, 22, 26, 27, 29, 31, 32, 40, 41, 42, 43, 44]). Dimuro et al. [12, 13] introduced interval additive generators of interval t-norm and t-conorms, which is approach based on interval fuzzy logic [4, 5, 34, 35]

In the development of applications, the definition of an aggregation function in terms of an additive generator is very important, since it reduces the computational complexity [27]. Moreover, some properties of an aggregation function that has a generator can be related to properties of its generator. Then, the study of aggregation functions in terms of their additive generators can lead to a fresh view of those operators and a more systematic methodology for their selection for the various applications [20].

The aim of this papers is to introduce the notion of additive generator of overlap functions, presenting some preliminary results. The paper is organized as follows. Section 2 summarizes the main concepts related to aggregation functions, t-norms and additive generators of t-norms. In Sect. 3, we briefly present overlap functions, introducing some example. Section 4 introduces the additive generators of overlap functions with some initial results. Section 5 is the Conclusion.

2 T-norms and Their Additive Generators

One important class of fuzzy operators are the *aggregation operators* [3, 23], which were proposed to aggregate several values in just one. A function $A : [0,1]^n \rightarrow [0,1]$ is said to be an aggregation operator if it satisfies the following two conditions:

(A1) A is increasing in each argument: for each $i \in \{1,\ldots,n\}$, if $x_i \leq y$, then
$A(x_1,\ldots,x_n) \leq A(x_1,\ldots,x_{i-1},y,x_{i+1},\ldots,x_n)$;
(A2)   Boundary conditions: $A(0,\ldots,0) = 0$ and $A(1,\ldots,1) = 1$.

In this paper, we are interested in the *overlap functions*al. [8, 9, 10, 16], one important class of aggregation functions that are related in some sense to *triangular norms* (t-norms).

Triangular norms were introduced by Menger [24] to model distance in probabilistic metric spaces. Schweizer and Sklar [37, 38] redefined the t-norm axioms proposed by Menger into the form used today. Nowadays, t-norms are often applied as a generalization of the classical conjunction.

Definition 1. A bivariate aggregation operator $T : [0,1]^2 \rightarrow [0,1]$ is said to be a t-norm if, for all $x,y,z \in [0,1]$, it satisfies the following conditions:
(T1)   Commutativity: $T(x,y) = T(y,x)$;
(T2)   Associativity: $T(x,T(y,z)) = T(T(x,y),z)$;
(T3)   Boundary condition: $T(x,1) = x$.

Some other properties may be required for t-norms, such as the following:
(T4)   Continuity: T is continuous in both arguments at the same time;
(T5)   Idempotency: $T(x,x) = x$, for all $x \in [0,1]$;
(T6)   Positiveness: if $T(x,y) = 0$ then either $x = 0$ or $y = 0$.

Example 1. Typical examples of t-norms are:
(i)   *Gödel* or *minimum*: $G(x,y) = \min\{x,y\}$;
(ii)   *Product*: $P(x,y) = x \cdot y$;
(iii)   *Łukasiewicz*: $L(x,y) = \max\{x+y-1,0\}$.

Notice that $L \leq P \leq G$, that is, for each $x,y \in [0,1]$, it holds that $L(x,y) \leq P(x,y) \leq G(x,y)$. In fact G is the greatest t-norm. G and P are *continuous* and *positive* t-norms. In particular, G is the unique *idempotent* t-norm.

An element $x \in]0,1]$ is said to be a *non-trivial zero divisor* of a t-norm T, if there exists $y \in]0,1]$ such that $T(x,y) = 0$. Clearly, a t-norm is *positive* if and only if it has no non-trivial zero divisor.[1]

Definition 2. [41] Let $f : [a,b] \rightarrow [c,d]$ be an increasing or decreasing function. The function $f^{(-1)} : [c,d] \rightarrow [a,b]$ defined by

---

[1] See [18], for more details about t-norms.

$$f^{(-1)}(y) = \begin{cases} \sup\{x \in [a,b] \mid f(x) < y\} & \text{if } f(a) < f(b), \\ \sup\{x \in [a,b] \mid f(x) > y\} & \text{if } f(a) > f(b), \\ a & \text{if } f(a) = f(b) \end{cases} \quad (1)$$

is called the *pseudo-inverse* of f.

Notice that when f is an increasing (but not constant) function then

$$f^{(-1)}(y) = \sup\{x \in [a,b] \mid f(x) < y\}. \quad (2)$$

Analogously, when f is a decreasing (but not constant) function then

$$f^{(-1)}(y) = \sup\{x \in [a,b] \mid f(x) > y\}. \quad (3)$$

In the following, we denote the *range* or *image* of a function $f : A \to B$ by $Ran(f)$.

Remark 1. [42, page 2] If a function $f : [a,b] \to [c,d]$ is increasing (decreasing) then $f^{(-1)}$ is also increasing (decreasing). If f is strictly increasing (decreasing) then $f^{(-1)}$ is continuous, $f^{(-1)} \circ f = Id_{[a,b]}$ and $f \circ f^{(-1)}(x) = x$ if and only if $x \in Ran(f)$.

Example 2. [12] The functions $p, l : [0,1] \to [0,\infty]$, defined, respectively, by

$$p(x) = \begin{cases} -\ln x & \text{if } x \neq 0 \\ \infty & \text{if } x = 0 \end{cases}$$

and $l(x) = 1 - x$, are strictly decreasing and $p(1) = l(1) = 0$. In this case, for each $y \in [0,\infty]$, it holds that

$$p^{(-1)}(y) = \begin{cases} e^{-y} & \text{if } y \neq \infty \\ 0 & \text{if } y = \infty \end{cases}$$

and

$$l^{(-1)}(y) = \begin{cases} 1 - y & \text{if } y \in [0,1] \\ 0 & \text{otherwise} \end{cases}$$

that is, $l^{(-1)}(y) = \max\{1 - y, 0\}$.

In the literature [6, 11, 17, 18, 19, 20, 22, 25, 27, 28, 30, 37, 38, 31, 32, 40, 42, 43, 44], there exist several definitions for additive generators. In this paper, we adopt the one extracted from [41], which was also applied in [12, 13].

Definition 3. Consider the functions $f : [0,1] \to [0,\infty]$ and $F : [0,1]^2 \to [0,1]$. The function f is said to be an *additive generator* of F if f is a strictly decreasing (increasing) function such that, for all $x, y \in [0,1]$,

$$F(x,y) = f^{(-1)}(f(x) + f(y)), \quad (4)$$

where $f^{(-1)}$ is the pseudo-inverse of f.

A function $F : [0,1]^2 \to [0,1]$ is said to be an *additively generated function* if there exists an additive generator of F. A function F is said to be *additively generated by* $f : [0,1] \to [0,\infty]$ if f is an additive generator of F.

Theorem 1. *[19, Theorem 2.6.3] Let $t : [0,1] \to [0,\infty]$ be a strictly decreasing function with $t(1) = 0$, such that t is right-continuous in 0 and, for each $(x,y) \in [0,1]^2$, it holds that*

$$t(x) + t(y) \in Ran(t) \cup [t(0), \infty]. \tag{5}$$

Then, the function $T_t : [0,1]^2 \to [0,1]$, defined by

$$T_t(x,y) = t^{(-1)}(t(x) + t(y)), \tag{6}$$

is a t-norm.

The t-norm T_t given in Eq. (6) is said to be additively generated by the function t. In this case, t is an additive generator of the t-norm T_t.

Example 3. [12] Let $p,l : [0,1] \to [0,\infty]$ be the functions defined in Example 2 and consider $x,y \in [0,1]$. Then, defining

$$T_l(x,y) = l^{(-1)}(l(x) + l(y)) = l^{(-1)}((1-x) + (1-y)),$$

one has that:

$$T_l(x,y) = \begin{cases} x+y-1 & \text{if } 1 \leq x+y \\ 0 & \text{otherwise} \end{cases}$$

and, then, it holds that $T_l(x,y) = \max\{x+y-1, 0\}$ and so $T_l = L$ (the Łukasiewicz t-norm given in Example 1). Analogously, we have that if $x \neq 0$ and $y \neq 0$ then

$$T_p(x,y) = p^{(-1)}(p(x) + p(y)) = p^{(-1)}((-\ln x) + (-\ln y)) = e^{(\ln x \cdot y)} = x \cdot y.$$

Otherwise (suppose that $x = 0$), one has that

$$T_p(x,y) = p^{(-1)}(p(x) + p(y)) = p^{(-1)}(\infty + p(y)) = 0.$$

It follows that $T_p = P$ (the Product t-norm given in Example 1).

3   Overlap Functions

In this section, we consider the concept of *overlap functions* introduced by Bustince et al. [8, 9, 10]. Some properties of overlap functions (e.g., migrativity, homogeneity, idempotency, convex combination) were also studied by Bedregal et al. [3] and Jurio et al. [16].

Definition 4. A bivariate function $O : [0,1]^2 \to [0,1]$ is said to be an overlap function if it satisfies the following conditions:
(O1) O is commutative;
(O2) $O(x,y) = 0$ if and only if $xy = 0$;
(O3) $O(x,y) = 1$ if and only if $xy = 1$;
(O4) O is non-decreasing;
(O5) O is continuous.

Example 4. It is possible to find several examples of overlap functions, such as any continuous t-norm with no zero divisors (property (O2)). On the other hand, the function $O_{mM} : [0,1]^2 \to [0,1]$, given by

$$O_{mM}(x,y) = \min(x,y) \max(x^2, y^2),$$

is a non associative overlap functions having 1 as neutral element, and, thus, is not a t-norm. This is the same case of the overlap functions $O_{DB}, O_2 : [0,1]^2 \to [0,1]$, defined by

$$O_{DB}(x,y) = \begin{cases} \frac{2xy}{x+y} & \text{if } x+y \neq 0 \\ 0 & \text{if } x+y = 0; \end{cases}$$

and

$$O_2(x,y) = x^2 y^2 \text{ or, more generally, } O_p(x,y) = x^p y^p, \text{ with } p > 1,$$

which are neither associative nor have 1 as neutral element.

Notice that whenever an overlap function has a neutral element, then, by **(O3)**, this element is necessarily equal to 1. Moreover, the following proposition proves that associative overlap functions always have 1 as neutral element and, thus, they are continuous t-norms with no zero divisors.

Proposition 1. *[1, Lemma 2.1.1],[18, Proposition 2.41] Let $F : [0,1]^2 \to [0,1]$ be a continuous bivariate function. If F is associative, $F(1,1) = 1$ and $F(0,1) = F(1,0) = 0$, then 1 is a neutral element of F, that is, $F(x,1) = F(1,x) = x$, for each $x \in [0,1]$.*

As a corollary, we have the Theorem 6 in [9]:

Corollary 1. *[9] $O : [0,1]^2 \to [0,1]$ is an associative overlap function if and only if O is a continuous and positive t-norm.*

Nevertheless, the reverse of Proposition 1 does not hold, since, as shown in Example 4, the overlap function $O(x,y) = \min(x,y)\max(x^2,y^2)$ has 1 as neutral element but is not associative.

4 Additive Generators of Overlap Functions

In this section, we introduce the concept of additive generator of an overlap function, allowing the definition of overlap functions (as two-place functions) by means of one-place functions (their additive generators).[2]

Lemma 1. *Let $\theta : [0,1] \to [0,\infty]$ be a decreasing function such that*
  1. *$\theta(x) + \theta(y) \in Ran(\theta)$, for $x,y \in [0,1]$ and*
  2. *if $\theta(x) = \theta(0)$ then $x = 0$.*

Then $\theta(x) + \theta(y) \geq \theta(0)$ if and only if $x = 0$ or $y = 0$.

Proof. ($\Rightarrow$) Since θ is decreasing and $\theta(x) + \theta(y) \in \text{Ran}(\theta)$ for each $x,y \in [0,1]$, then onde has that $\theta(x) + \theta(y) \leq \theta(0)$. Therefore, if $\theta(x) + \theta(y) \geq \theta(0)$ then it holds that $\theta(x) + \theta(y) = \theta(0)$. Suppose that $\theta(0) = 0$. Then, since θ is decreasing, one has that $\theta(x) = 0$, for each $x \in [0,1]$, which is contradiction with condition 2, and, therefore, it holds that $\theta(0) > 0$. Now, suppose that $\theta(0) \neq \infty$. Then, since $\theta(0) \neq 0$, one has that $\theta(0) + \theta(0) > \theta(0)$, which is also a contradiction. So, it follows that $\theta(0) = \infty$ and, therefore, since $\theta(x) + \theta(y) = \theta(0)$, we have that $\theta(x) = \infty$ or $\theta(y) = \infty$. Hence, by condition 2, one has that $x = 0$ or $y = 0$. ($\Leftarrow$) It is straightforward. $\square$

---

[2] Observe that any migrative overlap function can be naturally defined as an one-place function (see, e.g., [3, 9]).

Lemma 2. *Let* $\theta : [0,1] \to [0,\infty]$ *and* $\vartheta : [0,\infty] \to [0,1]$ *be bivariate functions such that*

$$\vartheta(\theta(x)) = 0 \text{ if and only if } x = 0. \tag{7}$$

Then $\theta(x) = \theta(0)$ *if and only if* $x = 0$

Proof. ($\Rightarrow$) If $\theta(x) = \theta(0)$, then one has that $\vartheta(\theta(x)) = \vartheta(\theta(0))$. By the left side of (7), we have that $\vartheta(\theta(0)) = 0$. Therefore, $\vartheta(\theta(x)) = 0$ and so, by the right side of (7), we have that $x = 0$. ($\Leftarrow$) It is straightforward. $\square$

Lemma 3. *Let* $\theta : [0,1] \to [0,\infty]$ *and* $\vartheta : [0,\infty] \to [0,1]$ *be bivariate functions such that*

$$\vartheta(\theta(x)) = 1 \text{ if and only if } x = 1. \tag{8}$$

Then $\theta(x) = \theta(1)$ *if and only if* $x = 1$

Proof. Analogous to Lemma 2. $\square$

Theorem 2. *Let* $\theta : [0,1] \to [0,\infty]$ *and* $\vartheta : [0,\infty] \to [0,1]$ *be continuous and decreasing functions such that*
1. $\theta(x) + \theta(y) \in Ran(\theta)$;
2. $\vartheta(\theta(x)) = 0$ *if and only* $x = 0$;
3. $\vartheta(\theta(x)) = 1$ *if and only* $x = 1$;
4. $\theta(x) + \theta(y) = \theta(1)$ *if and only* $x = 1$ *and* $y = 1$.

Then, the function $O_{\theta,\vartheta} : [0,1]^2 \to [0,1]$, *defined by*

$$O_{\theta,\vartheta}(x,y) = \vartheta(\theta(x) + \theta(y)), \tag{9}$$

is an overlap function.

Proof. We show that the conditions of Definition 4 holds. The proofs of the commutativity property (condition **(O1)**) and the continuity property (condition **(O5)**) are immediate. Moreover, it follows that

$$O_{\theta,\vartheta}(x,y) = 0 \Leftrightarrow \vartheta(\theta(x) + \theta(y)) = 0 \quad \text{by Eq. (10)}$$
$$\Leftrightarrow \vartheta(\theta(z)) = 0 \text{ for some } z \in [0,1] \quad \text{by condition 1};$$
$$\Leftrightarrow z = 0 \quad \text{by condition 2};$$
$$\Leftrightarrow \theta(x) + \theta(y) = \theta(0) \quad \text{by Lemma 2};$$
$$\Leftrightarrow x = 0 \text{ or } y = 0 \quad \text{by Lemma 1.}$$

which proves the condition **(O2)**. Also, one has that:

$$O_{\theta,\vartheta}(x,y) = 1 \Leftrightarrow \vartheta(\theta(x) + \theta(y)) = 1 \text{ by Eq. (10)}$$
$$\Leftrightarrow \vartheta(\theta(z)) = 1 \text{ for some } z \in [0,1] \quad \text{by condition 1};$$
$$\Leftrightarrow z = 1 \quad \text{by condition 3};$$
$$\Leftrightarrow \theta(x) + \theta(y) = \theta(1) \quad \text{by Lemma 3}$$
$$\Leftrightarrow x = 1 \text{ and } y = 1 \quad \text{by condition 4,}$$

which proves the condition **(O3)**. Finally, to prove the condition **(O4)**, considering $z \in [0,1]$ with $y \le z$, then $\theta(y) \ge \theta(z)$. It follows that

$$O_{\theta,\vartheta}(x,y) = \vartheta(\theta(x) + \theta(y)) \leq \vartheta(\theta(x) + \theta(z)) = O_{\theta,\vartheta}(x,z),$$

since ϑ and θ are decreasing.   □

Corollary 2. *Let* $\theta : [0,1] \to [0,\infty]$ *and* $\vartheta : [0,\infty] \to [0,1]$ *be continuous and strictly decreasing functions such that*
  1. $\theta(0) = \infty$ *and* $\theta(1) = 0$;
  2. $\vartheta(0) = 1$ *and* $\vartheta(\infty) = 0$.

Then, the function $O_{\theta,\vartheta} : [0,1]^2 \to [0,1]$, *defined by*

$$O_{\theta,\vartheta}(x,y) = \vartheta(\theta(x) + \theta(y)), \tag{10}$$

is an overlap function.

Proof. It follows from Theorem 2.   □

(θ, ϑ) is called an *additive generator pair* of the overlap function $O_{\theta,\vartheta}$, and $O_{\theta,\vartheta}$ is said to be additively generated by the pair (θ, ϑ).

Example 5. Consider the functions $\theta : [0,1] \to [0,\infty]$ and $\vartheta : [0,\infty] \to [0,1]$, defined, respectively by:

$$\theta(x) = \begin{cases} -2\ln x & \text{if } x \neq 0 \\ \infty & \text{if } x = 0 \end{cases}$$

and

$$\vartheta(x) = \begin{cases} e^{-y} & \text{if } y \neq \infty \\ 0 & \text{if } y = \infty, \end{cases}$$

which are continuous and strictly decreasing functions, satisfying the conditions 1-3 of Theorem 2. Then, whenever $x \neq 0$ and $y \neq 0$, one has that:

$$O_{\theta,\vartheta}(x,y) = \vartheta(\theta(x) + \theta(y)) = e^{-(-2\ln x - 2\ln y)} = e^{\ln x^2 y^2} = x^2 y^2.$$

Otherwise, if $x = 0$, it holds that

$$O_{\theta,\vartheta}(0,y) = \vartheta(\theta(0) + \theta(y)) = \vartheta(\infty + \theta(y)) = 0,$$

and, similarly, if $y = 0$, then $O_{\theta,\vartheta}(x,0) = 0$. It follows that

$$O_{\theta,\vartheta}(x,y) = O_2(x,y) = x^2 y^2,$$

the non associative overlap function for which 1 is not a neutral element, given in Example 4.

Corollary 3. *Considering the same conditions of Theorem 2, whenever* $\vartheta = \theta^{(-1)}$ *then* $O_{\theta,\vartheta}$ *is a positive t-norm.*

Proof. By Theorem 1, $O_{\theta,\vartheta}$ is a t-norm (additively generated by θ), and by Theorem 2, $O_{\theta,\vartheta}$ is positive.   □

Theorem 3. *Let $O = [0,1] \to [0,1]$ be an overlap function having 1 as neutral element. Then, whenever O is additively generated by a pair (θ, ϑ), with $\theta : [0,1] \to [0,\infty]$ and $\vartheta : [0,\infty] \to [0,1]$ satisfying the conditions of Theorem 2, then O is associative.*

Proof. *If 1 is the neutral element of O, then, since $\theta(1) = 0$, one has that:*

$$y = O(1,y) = \vartheta(\theta(1) + \theta(y)) = \vartheta(0 + \theta(y)) = \vartheta(\theta(y)),$$

which implies that ϑ is the pseudo-inverse of θ, that is,

$$\vartheta = \theta^{(-1)}. \tag{11}$$

It follows that:

$$
\begin{aligned}
O(x, O(y,z)) &= \vartheta(\theta(x) + \theta(O(y,z))) \text{ by Equation (4)} \\
&= \vartheta(\theta(x) + \theta(\vartheta(\theta(y) + \theta(z)))) \text{ by Equation (4)} \\
&= \vartheta(\theta(x) + \theta(\theta^{(-1)}(\theta(y) + \theta(z)))) \text{ by Equation (11)} \\
&= \vartheta(\theta(x) + (\theta(y) + \theta(z))) \\
&= \vartheta((\theta(x) + \theta(y)) + \theta(z)) \text{ by the associativity of the addition} \\
&= \vartheta(\theta(\theta^{(-1)}((\theta(x) + \theta(y)) + \theta(z)))) \\
&= \vartheta(\theta(\vartheta((\theta(x) + \theta(y)) + \theta(z)))) \text{ by Equation (11)} \\
&= \vartheta(\theta(\vartheta(\theta(O(x,y)) + \theta(z)))) \text{ by Equation (4)} \\
&= O(O(x,y),z) \text{ by Equation (4)},
\end{aligned}
$$

which proves that O is associative. □

The following result is immediate:

Corollary 4. *Let $O = [0,1] \to [0,1]$ be an overlap function additively generated by a pair (θ, ϑ). O is a t-norm if and only if 1 is a neutral element of O.*

Notice that whenever T is a positive continuous t-norm (that is, an overlap function) that is additively generated by a function $t : [0,1] \to [0,\infty]$ in the sense of Theorem 1, then it is also additively generated by a pair (θ, ϑ) in the sense of Theorem 2, where $\theta = t$ and $\vartheta = t^{(-1)}$, and vice-versa.

5 Conclusion

In this paper, we introduced the notion of additive generators of overlap functions, presenting some preliminary results. Due to the applicability of overlap functions (e.g., in classification problems, image processing and in some problems of decision making based on fuzzy preference relations), the fresh view of (two-place) overlap functions by means of their (one-place) additive generators can reduce the computational complexity and make easier the analysis of properties, providing a more systematic methodology for their selection for the various applications.

Future work is concerned with the definition of additive generators of grouping functions [3, 10, 16], and with an in-depth study about related properties aiming at the applications on image processing.

Acknowledgements. This work was partially supported by the following Brazilian funding agencies: CNPq (Conselho Nacional de Desenvolvimento Científico e Tecnológico), under the Proc. No. 305131/10-9, 560118/10-4, 476234/2011-5, 480832/2011-0, 307681/2012-2 and FAPERGS (Fundação de Amparo à Pesquisa do Rio Grande do Sul), under the Proc. No. 11/0872-3.

References

[1] Alsina, C., Frank, M.J., Schweizer, B.: Associative Functions: Triangular Norms and Copulas. World Scientific Publishing Company, Singapore (2006)

[2] Beattie, A.R., Landsberg, P.T.: One-dimensional overlap functions and their application to auger recombination in semiconductors. Proceedings of the Royal Society of London. Series A, Mathematical and Physical Sciences 258(1295), 486–495 (1960)

[3] Bedregal, B.C., Dimuro, G.P., Bustince, H., Barrenechea, E.: New results on overlap and grouping functions (to appear, 2013)

[4] Bedregal, B.C., Dimuro, G.P., Reiser, R.H.S.: An approach to interval-valued R-implications and automorphisms. In: Carvalho, J.P., Dubois, D., Kaymak, U., da Costa Sousa, J.M. (eds.) Proceedings of the Joint 2009 International Fuzzy Systems Association World Congress and 2009 European Society of Fuzzy Logic and Technology Conference, IFSA/EUSFLAT, pp. 1–6 (2009)

[5] Bedregal, B.C., Dimuro, G.P., Santiago, R.H.N., Reiser, R.H.S.: On interval fuzzy S-implications. Information Sciences 180(8), 1373–1389 (2010)

[6] Beliakov, G., Bustince, H., Goswami, D.P., Mukherjee, U.K., Pal, N.R.: On averaging operators for Atanassov's intuitionistic fuzzy sets. Information Sciences 181(6), 1116–1124 (2011)

[7] Beliakov, G., Pradera, A., Calvo, T.: Aggregation Functions: A Guide for Practitioners. STUDFUZZ, vol. 221. Springer, Heidelberg (2007)

[8] Bustince, H., Fernández, J., Mesiar, R., Montero, J., Orduna, R.: Overlap index, overlap functions and migrativity. In: Proceedings of IFSA/EUSFLAT Conference, pp. 300–305 (2009)

[9] Bustince, H., Fernandez, J., Mesiar, R., Montero, J., Orduna, R.: Overlap functions. Nonlinear Analysis 72(3-4), 1488–1499 (2010)

[10] Bustince, H., Pagola, M., Mesiar, R., Hüllermeier, E., Herrera, F.: Grouping, overlaps, and generalized bientropic functions for fuzzy modeling of pairwise comparisons. IEEE Transactions on Fuzzy Systems 20(3), 405–415 (2012)

[11] Deschrijver, G.: Additive and multiplicative generators in interval-valued fuzzy set theory. IEEE Transactions on Fuzzy Systems 15(2), 222–237 (2007)

[12] Dimuro, G.P., Bedregal, B.C., Santiago, R.H.N., Reiser, R.H.S.: Interval additive generators of interval t-norms and interval t-conorms. Information Sciences 181(18), 3898–3916 (2011)

[13] Dimuro, G.P., Bedregal, B.R.C., Reiser, R.H.S., Santiago, R.H.N.: Interval additive generators of interval T-norms. In: Hodges, W., de Queiroz, R. (eds.) WoLLIC 2008. LNCS (LNAI), vol. 5110, pp. 123–135. Springer, Heidelberg (2008)

[14] Eskola, K.J., Vogt, R., Wang, X.N.: Nuclear overlap functions. International Journal of Modern Physics A 10(20n21), 3087–3090 (1995)

[15] Faucett, W.M.: Compact semigroups irreducibly connected between two idempotents. Proceedings of the American Mathematical Society 6, 741–747 (1955)

[16] Jurio, A., Bustince, H., Pagola, M., Pradera, A., Yager, R.R.: Some properties of overlap and grouping functions and their application to image thresholding. In: Fuzzy Sets and Systems (in press, corrected proof, 2013) (available online in January 2013)

[17] Klement, E.P., Mesiar, R., Pap, E.: Quasi- and pseudo-inverses of monotone functions, and the construction of t-norms. Fuzzy Sets and Systems 104(1), 3–13 (1999)

[18] Klement, E.P., Mesiar, R., Pap, E.: Triangular Norms. Kluwer Academic Publisher, Dordrecht (2000)

[19] Klement, E.P., Mesiar, R., Pap, E.: Triangular norms: Basic notions and properties. In: Klement, E.P., Mesiar, R. (eds.) Logical, Algebraic, Analytic, and Probabilistic Aspects of Triangular Norms, pp. 17–60. Elsevier, Amsterdam (2005)

[20] Leventides, J., Bounas, A.: An approach to the selection of fuzzy connectives in terms of their additive generators. Fuzzy Sets and Systems 126(2), 219–224 (2002)

[21] Ling, C.H.: Representation of associative functions. Publicationes Mathematicae Debrecen 12, 189–212 (1965)

[22] Mayor, G., Monreal, J.: Additive generators of discrete conjunctive aggregation operations. IEEE Transactions on Fuzzy Systems 15(6), 1046–1052 (2007)

[23] Mayor, G., Trillas, E.: On the representation of some aggregation functions. In: Proceedings of IEEE International Symposium on Multiple-Valued Logic, pp. 111–114. IEEE, Los Alamitos (1986)

[24] Menger, K.: Statistical metrics. Proceedings of the National Academic of Sciences 28(12), 535–537 (1942)

[25] Mesiarová, A.: Generators of triangular norms. In: Klement, E.P., Mesiar, R. (eds.) Logical, Algebraic, Analytic, and Probabilistic Aspects of Triangular Norms, pp. 95–111. Elsevier, Amsterdam (2005)

[26] Mesiarová, A.: H-transformation of t-norms. Information Sciences 176(11), 1531–1545 (2006)

[27] Mesiarová-Zemánková, A.: Ranks of additive generators. Fuzzy Sets and Systems 160(14), 2032–2048 (2009)

[28] Monreal, E.P., Mesiar, R., Pap, E.: Additive generators of t-norms which are not necessarily continuous. In: Proceddings of the Fourth European Congress on Intelligent Techniques and Soft Computing – EUFIT 1996, vol. 1, pp. 60–73. ELITE-Foundation, Aachen (1996)

[29] Mostert, P.S., Shields, A.L.: On the structure of semigroups on a compact manifold with boundary. Annals of Mathematics 65(1), 117–143 (1957)

[30] Nguyen, H.T., Walker, E.A.: A First Course in Fuzzy Logic. Chapman & Hall/CRC, Boca Raton (2006)

[31] Ouyang, Y.: On the construction of boundary weak triangular norms through additive generators. Nonlinear Analysis 66(1), 125–130 (2007)

[32] Ouyang, Y., Fang, J., Zhao, Z.: A generalization of additive generator of triangular norms. International Journal of Approximate Reasoning 49(2), 417–421 (2008)

[33] Povey, A.C., Grainger, R.G., Peters, D.M., Agnew, J.L., Rees, D.: Estimation of a lidar's overlap function and its calibration by nonlinear regression. Applied Optics 51(21), 5130–5143 (2012)

[34] Reiser, R.H.S., Bedregal, B.C., Santiago, R.N., Dimuro, G.P.: Analyzing the relationship between interval-valued D-implications and interval-valued QL-implications. TEMA – Tendencies in Computational and Applied Mathematics 11(1), 89–100 (2010)

[35] Reiser, R.H.S., Dimuro, G.P., Bedregal, B.R.C., Santiago, R.H.N.: Interval valued QL-implications. In: Leivant, D., de Queiroz, R. (eds.) WoLLIC 2007. LNCS, vol. 4576, pp. 307–321. Springer, Heidelberg (2007)

[36] Schweizer, B., Sklar, A.: Statistical metric spaces. Pacific Journal of Mathematics 10(1), 313–334 (1960)

[37] Schweizer, B., Sklar, A.: Associative functions and statistical triangle inequalities. Publicationes Mathematicae Debrecen 8, 168–186 (1961)

[38] Schweizer, B., Sklar, A.: Associative functions and abstract semigroups. Publicationes Mathematicae Debrecen 10, 69–81 (1963)

[39] Schweizer, B., Sklar, A.: Probabilistic Metric Spaces. North-Holland, New York (1983)

[40] Viceník, P.: Additive generators of non-continuous triangular norms. In: Rodabaugh, S.E., Klement, E.P. (eds.) Topological and Algebraic Structures in Fuzzy Sets, pp. 441–454. Kluwer, Dordrecht (2003)

[41] Viceník, P.: Additive generators of associative functions. Fuzzy Sets and Systems 153(2), 137–160 (2005)

[42] Viceník, P.: Additive generators of border-continuous triangular norms. Fuzzy Sets and Systems 159(13), 1631–1645 (2008)

[43] Viceník, P.: Intersections of ranges of additive generators of associative functions. Tatra Mountains Mathematical Publications 40, 117–131 (2008)

[44] Viceník, P.: On a class of generated triangular norms and their isomorphisms. Fuzzy Sets and Systems 161(10), 1448–1458 (2010)

Continuous T-norms and T-conorms Satisfying the Principle of Inclusion and Exclusion

Mária Kuková and Mirko Navara

Abstract. The classical principle of inclusion and exclusion is formulated for set-theoretic union and intersection. It is natural to ask if it can be extended to fuzzy sets. The answer depends on the choice of fuzzy logical operations (which belong to the larger class of aggregation operators). Further, the principle can be generalized to interval-valued fuzzy sets, resp. IF-sets (Atanassov's intuitionistic fuzzy sets). The principle of inclusion and exclusion uses cardinality of sets (which has a natural extension to fuzzy sets, interval-valued fuzzy sets and IF sets) or, more generally, a *measure*, which can be defined in different ways. We also point up the question of the *domain* of the measure which has been neglected so far.

1 Classical Form of the Principle of Inclusion and Exclusion and Its Analogues

Finite sets $A_1, \ldots, A_n$ satisfy the equality

$$m\left(\bigcup_{i=1}^{n} A_i\right) = \sum_{i=1}^{n} m(A_i) - \sum_{i=1}^{n-1}\sum_{j=i+1}^{n} m(A_i \cap A_j) + \cdots + (-1)^{n+1} m\left(\bigcap_{i=1}^{n} A_i\right), \quad (1)$$

where m is the cardinality of sets. More generally, we can take for m any (finitely additive) measure defined on an algebra of subsets containing all $A_1, \ldots, A_n$.

For $n, k \in \mathbb{N}$ we use the notation

Mária Kuková
Department of Mathematics, Faculty of Natural Sciences, Matej Bel University,
Tajovského 40, Banská Bystrica, Slovakia
e-mail: maja.kukova@gmail.com

Mirko Navara
Center for Machine Perception, Department of Cybernetics,
Faculty of Electrical Engineering, Czech Technical University in Prague, Czech Republic
e-mail: navara@cmp.felk.cvut.cz

H. Bustince et al. (eds.), *Aggregation Functions in Theory and in Practise*,
Advances in Intelligent Systems and Computing 228,
DOI: 10.1007/978-3-642-39165-1_20, © Springer-Verlag Berlin Heidelberg 2013

$$\mathbb{N}_n = \{1,2,\ldots,n\}$$

and we denote by $\mathcal{N}_{n,k}$ the set of all k-element subsets of $\mathbb{N}_n$. Then (1) can be written as

$$m\left(\bigcup_{i\in\mathbb{N}_n} A_i\right) = \sum_{k\in\mathbb{N}_n} (-1)^{k+1} \sum_{M\in\mathcal{N}_{n,k}} m\left(\bigcap_{i\in M} A_i\right) \tag{2}$$

and, dually,

$$m\left(\bigcap_{i\in\mathbb{N}_n} A_i\right) = \sum_{k\in\mathbb{N}_n} (-1)^{k+1} \sum_{M\in\mathcal{N}_{n,k}} m\left(\bigcup_{i\in M} A_i\right). \tag{3}$$

An analogous formula holds for numbers instead of sets and the operations of maximum and minimum:

$$\max\{A_1,\ldots,A_n\} =$$

$$= \sum_{i=1}^{n} A_i - \sum_{i=1}^{n-1}\sum_{j=i+1}^{n} \min\{A_i,A_j\} + \ldots + (-1)^{n+1} \min\{A_1,\ldots,A_n\}, \tag{4}$$

equivalently,

$$\max_{i\in\mathbb{N}_n} A_i = \sum_{k\in\mathbb{N}_n} (-1)^{k+1} \sum_{M\in\mathcal{N}_{n,k}} \min_{i\in M} A_i \tag{5}$$

and, dually,

$$\min_{i\in\mathbb{N}_n} A_i = \sum_{k\in\mathbb{N}_n} (-1)^{k+1} \sum_{M\in\mathcal{N}_{n,k}} \max_{i\in M} A_i. \tag{6}$$

The probabilistic sum (product t-conorm)

$$x \boxplus y = x \overset{\mathrm{P}}{\vee} y = x + y - x\cdot y$$

and the (ordinary) product satisfy a similar formula:

$$\boxplus_{i\in\mathbb{N}_n} A_i = \sum_{k\in\mathbb{N}_n} (-1)^{k+1} \sum_{M\in\mathcal{N}_{n,k}} \prod_{i\in M} A_i \tag{7}$$

and, dually,

$$\prod_{i\in\mathbb{N}_n} A_i = \sum_{k\in\mathbb{N}_n} (-1)^{k+1} \sum_{M\in\mathcal{N}_{n,k}} \boxplus_{i\in M} A_i. \tag{8}$$

It is natural to ask which forms of the principle of inclusion and exclusion can be generalized in fuzzy sets theory. After an introductory Section 2, we investigate generalizations of the principle of inclusion and exclusion to fuzzy sets in Section 3. In Section 4, we study further generalizations to IF-sets (also called Atanassov's intuitionistic fuzzy sets in [1] and essentially equivalent to interval-valued fuzzy sets [21]).[1]

---

[1] The term "Atanassov's intuitionistic fuzzy sets" has no relation to intuitionistic logic, thus this terminology is misleading and it was criticized in [7]; instead of it, we prefer the term "IF-sets".

2 Fuzzy Set Operations and Measures on Systems of Fuzzy Sets

For more information on the operations used here, we refer to [10, 14, 19].

We consider fuzzy subsets of a fixed non-empty set (universe) X and denote by $\mu_A : X \to [0,1]$ the membership function of a fuzzy set A. By $\wedge$ (resp. $\dot{\vee}$) we denote a general t-norm (resp. t-conorm), i.e., a binary operation on $[0,1]$ which is commutative, associative, non-decreasing, and has the neutral element 1 (resp. 0). Particular types of t-norms and t-conorms will be distinguished by indices in place of dots. We shall use mainly the following t-norms:

$$x \underset{G}{\wedge} y = \min(x,y), \quad \text{(Gödel (standard, min, Zadeh) t-norm)}$$

$$x \underset{P}{\wedge} y = x \cdot y, \quad \text{(product t-norm)}$$

$$x \underset{L}{\wedge} y = \max(x+y-1,0). \quad \text{(Łukasiewicz t-norm)}$$

Their dual t-conorms are obtained by duality with respect to the standard fuzzy negation, $\neg x = 1 - x$,

$$x \dot{\vee} y = \neg(\neg x \wedge \neg y),$$

the explicit formulas are:

$$x \overset{G}{\vee} y = \max(x,y), \quad \text{(Gödel (standard, min, Zadeh) t-conorm)}$$

$$x \overset{P}{\vee} y = x+y-x \cdot y, \quad \text{(product t-conorm)}$$

$$x \overset{L}{\vee} y = \min(x+y,1). \quad \text{(Łukasiewicz t-conorm)}$$

Fuzzy intersections and unions are defined by

$$\mu_{A \cap B}(x) = \mu_A(x) \wedge \mu_B(x), \tag{9}$$

$$\mu_{A \dot{\cup} B}(x) = \mu_A(x) \dot{\vee} \mu_B(x). \tag{10}$$

The notation $\dot{\cap}$ and $\dot{\cup}$ is assigned to a general intersection and union. Particular types will be distinguished by the same indices as the corresponding t-norms and t-conorms.

The sign $\nearrow$ will denote the limit of an increasing sequence of reals, fuzzy sets, later also of IF-sets.

We fix a σ-algebra S of subsets of X. We denote by $\mathscr{T}$ the family of all S-measurable fuzzy subsets of X. In particular $1_X, 0_X \in \mathscr{T}$ are the constant functions on X with values $1, 0$, respectively.

Definition 1. A mapping $m : \mathscr{T} \to [0,1]$ is called a *state* if the following properties are satisfied:

1. $m(1_X) = 1$, $m(0_X) = 0$,
2. $m(A \overset{L}{\cup} B) = m(A) + m(B) - m(A \underset{L}{\cap} B)$,

3. $A_n \nearrow A \Rightarrow m(A_n) \nearrow m(A)$.

States were characterized by D. Butnariu and E. P. Klement in [2]:

Theorem 1. *Every state m on $\mathscr{T}$ is of the form*

$$m(A) = \int \mu_A \, dP, \tag{11}$$

where the state (probability measure) P is the restriction of m to the Boolean σ-algebra $\mathscr{T} \cap \{0,1\}^X$ of sharp elements of $\mathscr{T}$.

A state of the form (11) is also called an *integral state*. It has been used in many previous studies, even in the pioneering work by Zadeh [20]. However, it was usually introduced without any deeper motivation. The axiomatic approach of Butnariu and Klement proves that, in the most important cases, all states are integral states [13].

3   The Principle of Inclusion and Exclusion for Fuzzy Sets

We investigare the following formulation of the principle of inclusion and exclusion:

$$m\left(\dot{\bigcup}_{i \in \mathbb{N}_n} A_i\right) = \sum_{k \in \mathbb{N}_n} (-1)^{k+1} \sum_{M \in \mathscr{N}_{n,k}} m\left(\dot{\bigcap}_{i \in M} A_i\right), \tag{12}$$

where fuzzy intersection $\dot{\cap}$ and fuzzy union $\dot{\cup}$ are based on a t-norm $\wedge$ and a t-conorm $\dot{\vee}$, respectively, and m is a state on $\mathscr{T}$. In previous papers [3, 4, 5, 9, 17], only Gödel, product, or Łukasiewicz operations were considered. The result was that Gödel and product operations satisfy the principle of inclusion and exclusion [9], Łukasiewicz operations violate it [9, 11]. We asked about the validity of the principle of inclusion and exclusion for all continuous t-norms and t-conorms. Using the results of [3], we have proved in [12] that such operations are quite rare:

Theorem 2. *[12] Suppose that a fuzzy intersection $\dot{\cap}$ and a fuzzy union $\dot{\cup}$ are based on a continuous t-norm $\wedge$ and a continuous t-conorm $\dot{\vee}$, respectively. Let m be a state on $\mathscr{T}$. Then $\dot{\cap}, \dot{\cup}$ satisfy the principle of inclusion and exclusion (12) iff $\wedge$ and $\dot{\vee}$ can be expressed as the following ordinal sums:*

$$\wedge = (\langle a_\alpha, b_\alpha, \underset{P}{\wedge} \rangle)_{\alpha \in I}.$$

$$\dot{\vee} = (\langle a_\alpha, b_\alpha, \overset{P}{\dot{\vee}} \rangle)_{\alpha \in I}$$

for some collection of disjoint intervals $((a_\alpha, b_\alpha))_{\alpha \in I}$ in $[0, 1]$, i.e.,

$$x \,\dot{\wedge}\, y = \begin{cases} a_\alpha + (b_\alpha - a_\alpha) \cdot \left(\frac{x-a_\alpha}{b_\alpha-a_\alpha} \,\dot{\wedge}_P\, \frac{y-a_\alpha}{b_\alpha-a_\alpha} \right) & \text{if } (x,y) \in [a_\alpha, b_\alpha]^2, \\ \min(x,y) & \text{otherwise,} \end{cases}$$

$$x \,\dot{\vee}\, y = \begin{cases} a_\alpha + (b_\alpha - a_\alpha) \cdot \left(\frac{x-a_\alpha}{b_\alpha-a_\alpha} \,\dot{\vee}^P\, \frac{y-a_\alpha}{b_\alpha-a_\alpha} \right) & \text{if } (x,y) \in [a_\alpha, b_\alpha]^2, \\ \max(x,y) & \text{otherwise.} \end{cases}$$

4 The Principle of Inclusion and Exclusion for IF-Sets

The principle of inclusion and exclusion can be generalized to t-representable operations on IF-sets. An interval-valued fuzzy set A on the universe X is described by two functions $\mu_A, \rho_A : X \to [0,1]$, where $\mu_A(x) \le \rho_A(x)$ for all $x \in X$. Another approach is the concept of IF-sets (derived from Atanassov's *intuitionistic fuzzy* sets), introduced by Atanassov (see e.g. [1]). An IF-set is given by

$$A = \{(x, \mu_A(x), \nu_A(x)) : x \in X\},$$

where $\mu_A, \nu_A : X \to [0,1]$ such that

$$\mu_A(x) + \nu_A(x) \le 1 \tag{13}$$

for all $x \in X$. We shall use the notation

$$A = (\mu_A, \nu_A).$$

The set of all IF-sets will be denoted by $\mathscr{IF}$.

The partial ordering on the set $\mathscr{IF}$ is defined by the formula

$$A \le B \iff (\mu_A \le \mu_B \text{ and } \nu_A \ge \nu_B).$$

Evidently $(0_X, 1_X)$ is the least element of $(\mathscr{IF}, \le)$, $(1_X, 0_X)$ is the greatest element of $(\mathscr{IF}, \le)$. Ordinary fuzzy sets can be embedded into IF-sets by the following homomorphism:

$$\mu_A \mapsto (\mu_A, 1 - \mu_A).$$

The operations with IF-sets A, B are defined by the following formulas:

$$A \cap B = (\mu_A \wedge \mu_B, \nu_A \,\dot{\vee}\, \nu_B),$$

$$A \,\dot{\cup}\, B = (\mu_A \,\dot{\vee}\, \mu_B, \nu_A \wedge \nu_B),$$

where $\dot{\vee}$ is the t-conorm dual to the t-norm $\wedge$ (the indices distinguish the used fuzzy operations). These operations are so-called *t-representable* t-norms and t-conorms. It is known that not all t-norms and t-conorms on IF-sets are of this type, see [6] for details.

The following definition comes from Riečan [15].

Definition 2. A mapping $m : \mathscr{IF} \to [0,1]$ is called an *L-state* if the following properties are satisfied:

1. $m((1_X, 0_X)) = 1$, $m((0_X, 1_X)) = 0$,
2. $A \underset{L}{\cap} B = (0_X, 1_X) \Rightarrow m(A \overset{L}{\cup} B) = m(A) + m(B)$,
3. $A_n \nearrow A \Rightarrow m(A_n) \nearrow m(A)$,

$\forall A, B, A_n \in \mathscr{IF}$ $(n = 1, \dots)$.

Grzegorzewski and Mrówka [9] defined an L-probability $\mathscr{P}$ of an IF-set A by the interval

$$\mathscr{P}(A) = \left[\int_X \mu_A \, dP, 1 - \int_X \nu_A \, dP \right], \tag{14}$$

where P is a probability measure over X. More generally, an axiomatic approach to probability on IF-events was proposed by Riečan [16]. Let us have a set

$$\mathscr{J} = \{[a,b] : a, b \in [0,1], a \le b\}$$

with an ordering given by the formula

$$[a_1, a_2] \le [b_1, b_2] \iff (a_1 \le b_i \text{ and } a_2 \le b_2).$$

Definition 3. A mapping $\mathscr{P} : \mathscr{IF} \to \mathscr{J}$ is called an *L-probability* if the following conditions hold:

1. $\mathscr{P}(1_X, 0_X) = [1,1]$, $\mathscr{P}((0_X, 1_X)) = [0,0]$,
2. $A \underset{L}{\cap} B = (0_X, 1_X) \Rightarrow \mathscr{P}(A \overset{L}{\cup} B) = \mathscr{P}(A) + \mathscr{P}(B)$,
3. $A_n \nearrow A \Rightarrow \mathscr{P}(A_n) \nearrow \mathscr{P}(A)$.

We use the notation

$$\mathscr{P}(A) = \left[\mathscr{P}^\flat(A), \mathscr{P}^\sharp(A) \right].$$

It is easy to see that the following proposition holds:

Proposition 1. *Let $\mathscr{P} : \mathscr{IF} \to \mathscr{J}$. Then $\mathscr{P}$ is an L-probability if and only if $\mathscr{P}^\flat, \mathscr{P}^\sharp : \mathscr{F} \to [0,1]$ are L-states and $\mathscr{P}^\flat \le \mathscr{P}^\sharp$.*

In [4], Ciungu and Riečan have proved the following theorem which implies that the notion of an L-state is a generalization of that of L-probability by Grzegorzewski and Mrówka (see also [18, 5]):

Theorem 3. *[4] For any L-state $m : \mathscr{IF} \to [0,1]$ there exist probability measures $P, Q : 2^X \to [0,1]$ and $\alpha \in [0,1]$ such that $\forall A \in \mathscr{IF}$*

$$m(A) = \int_X \mu_A \, dP + \alpha \left(1 - \int_X (\mu_A + \nu_A) \, dQ \right). \tag{15}$$

We obtained the following generalization of the principle of inclusion and exclusion to IF-sets:

Theorem 4. *[11, 12] Let A_i be IF-sets, $A_i = (\mu_{A_i}, \nu_{A_i})$, $i = 1, \ldots, n$. Let m be an L-state. Then the the principle of inclusion and exclusion holds for the Gödel and product operations, i.e.*

$$m\left(\dot{\bigcup}_{i=1}^{n} A_i\right) = \sum_{i=1}^{n} m(A_i) - \sum_{i=1}^{n-1} \sum_{j=i+1}^{n} m(A_i \cap A_j) + \ldots + (-1)^{n+1} m\left(\bigcap_{i=1}^{n} A_i\right),$$

where the pair of operations $(\dot{\cup}, \cap)$ can be chosen from the possibilities $(\overset{G}{\cup}, \overset{}{\underset{G}{\cap}})$ or $(\overset{P}{\cup}, \underset{P}{\cap})$. The product operations are the only ones which are based on continuous Archimedean t-norms and satisfy the principle of inclusion and exclusion.

As a consequence of Theorem 4 and Proposition 1, we obtain the following result for an L-probability $\mathscr{P} : \mathscr{IF} \to \mathscr{J}$:

Theorem 5. *[11, 12] Let A_i be IF-sets, $A_i = (\mu_{A_i}, \nu_{A_i})$, $i = 1, \ldots, n$. Let $\mathscr{P}$ be an L-probability, $\mathscr{P}(A) = [\mathscr{P}^{\flat}(A), \mathscr{P}^{\sharp}(A)]$. Then the the principle of inclusion and exclusion holds for the Gödel and product operations, i.e.*

$$\mathscr{P}^{\flat}\left(\dot{\bigcup}_{i \in \mathbb{N}_n} A_i\right) = \sum_{k \in \mathbb{N}_n} (-1)^{k+1} \sum_{M \in \mathscr{N}_{n,k}} \mathscr{P}^{\flat}\left(\bigcap_{i \in M} A_i\right),$$

$$\mathscr{P}^{\sharp}\left(\dot{\bigcup}_{i \in \mathbb{N}_n} A_i\right) = \sum_{k \in \mathbb{N}_n} (-1)^{k+1} \sum_{M \in \mathscr{N}_{n,k}} \mathscr{P}^{\sharp}\left(\bigcap_{i \in M} A_i\right),$$

where the pair of operations $(\dot{\cup}, \cap)$ can be chosen from the possibilities $(\overset{G}{\cup}, \underset{G}{\cap})$ or $(\overset{P}{\cup}, \underset{P}{\cap})$. The product operations are the only ones which are based on continuous Archimedean t-norms and satisfy the principle of inclusion and exclusion.

5 Conclusions

We studied generalizations of the principle of inclusion and exclusion for fuzzy sets, interval-valued fuzzy sets, and IF-sets (Atanassov's intuitionistic fuzzy sets). The conclusion is that it is satisfied only for the Gödel and product operations and some of their ordinal sums. Future work could concentrate on measures which are not integral measures and which are defined on domains which do not contain all fuzzy sets measurable with respect to a given σ-algebra.

Acknowledgements. The second author was supported by the Czech Technical University in Prague under project SGS12/187/OHK3/3T/13.

References

[1] Atanassov, K.: Intuitionistic Fuzzy Sets: Theory and Applications. Springer, Heidelberg (1999)

[2] Butnariu, D., Klement, E.P.: Triangular Norm-Based Measures and Games with Fuzzy Coalitions. Kluwer Academic Publishers, Dordrecht (1993)

[3] Ciungu, L.C., Kelemenová, J., Riečan, B.: A New Point of View to the Inclusion - Exclusion Principle. In: 6th IEEE International Conference on Intelligent Systems IS 2012, Varna, Bulgaria, pp. 142–144 (2012)

[4] Ciungu, L., Riečan, B.: General form of probabilities on IF-sets. In: Di Gesù, V., Pal, S.K., Petrosino, A. (eds.) WILF 2009. LNCS (LNAI), vol. 5571, pp. 101–107. Springer, Heidelberg (2009)

[5] Ciungu, L., Riečan, B.: Representation theorem for probabilities on IFS-events. Information Sciences 180, 793–798 (2010)

[6] Deschrijver, G., Cornelis, C., Kerre, E.E.: On the representation of intuitionistic fuzzy t-norms and t-conorms. IEEE Trans. Fuzzy Syst. 12(1), 45–61 (2004)

[7] Dubois, D., Gottwald, S., Hájek, P., Kacprzyk, J., Prade, H.: Terminological difficulties in fuzzy set theory—The case of "Intuitionistic Fuzzy Sets". Fuzzy Sets and Systems 156(3), 485–491 (2005)

[8] Dubois, D., Prade, H.: Gradualness, uncertainty and bipolarity: Making sense of fuzzy sets. Fuzzy Sets and Systems 192, 3–24 (2012)

[9] Grzegorzewski, P., Mrówka, E.: Probability of intuitionistic fuzzy events. In: Grzegorzewski, P., et al. (eds.) Soft Methods in Probability, Statistics and Data Analysis, pp. 105–115. Springer, New York (2002)

[10] Klement, E.P., Mesiar, R., Pap, E.: Triangular Norms. Kluwer Academic Publishers, Dordrecht (2000)

[11] Kuková, M.: The inclusion-exclusion principle for L-states and IF-events. Information Sciences 224, 165–169 (2013),
http://dx.doi.org/10.1016/j.ins.2012.10.029

[12] Kuková, M., Navara, M.: Principles of inclusion and exclusion for fuzzy sets. Fuzzy Sets Syst. (accepted)

[13] Navara, M.: Triangular norms and measures of fuzzy sets. In: Klement, E.P., Mesiar, R. (eds.) Logical, Algebraic, Analytic, and Probabilistic Aspects of Triangular Norms, pp. 345–390. Elsevier (2005)

[14] Nguyen, H.T., Walker, E.: A First Course in Fuzzy Logic, 2nd edn. Chapman & Hall/CRC, Boca Raton (2000)

[15] Riečan, B.: On some contributions to quantum structures by fuzzy sets. Kybernetika 43, 481–490 (2007)

[16] Riečan, B.: Probability theory on IF events. In: Aguzzoli, S., Ciabattoni, A., Gerla, B., Manara, C., Marra, V. (eds.) ManyVal 2006. LNCS (LNAI), vol. 4460, pp. 290–308. Springer, Heidelberg (2007)

[17] Riečan, B., Mundici, D.: Probability on MV-algebras. In: Pap, E. (ed.) Handbook of Measure Theory, ch. 21, pp. 869–910. Elsevier Science, Amsterdam (2002)

[18] Riečan, B., Petrovičová, J.: On the Łukasiewicz Probability Theory on IF-sets. Tatra Mt. Math. Publ. 46, 125–146 (2010)

[19] Schweizer, B., Sklar, A.: Probabilistic Metric Spaces. North-Holland, New York (1983)

[20] Zadeh, L.A.: Probability measures of fuzzy sets. J. Math. Anal. Appl. 23, 421–427 (1968)

[21] Zadeh, L.A.: The concept of a linguistic variable and its application to approximate reasoning I. Inform. Sci. 8, 199–249 (1975)

On Mulholland Inequality and Dominance of Strict Triangular Norms

Milan Petrík

Abstract. Mulholland inequality and its consequences for the dominance relation of strict triangular norms are studied and new results made in this area are presented. As a main result, it is presented that the dominance relation on the set of strict triangular norms is not transitive and thus not an order relation.

1 Mulholland Inequality and Mulholland's Condition

We denote by $\mathbb{R}_0^+$ the set of positive real numbers with zero and by $\mathbb{R}^+$ the set of positive real numbers without zero. An increasing bijection $f \colon \mathbb{R}_0^+ \to \mathbb{R}_0^+$ is said to solve *Mulholland inequality* if

$$f^{-1}(f(x+u) + f(y+v)) \leq f^{-1}(f(x) + f(y)) + f^{-1}(f(u) + f(v)) \qquad (1)$$

holds for all $x, y, u, v \in \mathbb{R}_0^+$. By *MI* we denote the set of all increasing bijections of $\mathbb{R}_0^+$ that solve Mulholland inequality.

Mulholland inequality has attracted the attention of the researches studying triangular norms mainly because it is closely related, as it will be shown in the sequel, to the dominance relation on the set of strict triangular norms. Nevertheless, it has been originally introduced by H. P. Mulholland in his paper [12] from 1950 as a generalization of Minkowski inequality which represents the triangular inequality for the *p*-norms.

Every function of the type $x \mapsto x^p$, $p \geq 1$, solves Mulholland inequality; this way we actually obtain Minkowski inequality. The set of solutions is, however, larger and Mulholland in his paper has provided a sufficient condition for the fulfillment of his inequality:

Milan Petrík
Dept. of Mathematics and Statistics, Faculty of Science, Masaryk University. Brno, Czech Republic. Institute of Computer Science, Academy of Sciences, Prague, Czech Republic
e-mail: petrik@cs.cas.cz

H. Bustince et al. (eds.), *Aggregation Functions in Theory and in Practise*,
Advances in Intelligent Systems and Computing 228,
DOI: 10.1007/978-3-642-39165-1_21, © Springer-Verlag Berlin Heidelberg 2013

Theorem 1. *Let* $f \colon \mathbb{R}_0^+ \to \mathbb{R}_0^+$ *be an increasing bijection. If both* f *and* $\log \circ f \circ \exp$ *are convex then* $f \in MI$.

If $\log \circ f \circ \exp$ is convex then we say, following the terminology of Matkowski [10], that f is *geometrically convex* or, shortly, *geo-convex*. An example of an increasing bijection on $\mathbb{R}_0^+$ that is convex and geo-convex while not of the type $x \mapsto x^p$, $p \geq 1$, is $x \mapsto \exp(x) - 1$.

In 1984, Tardiff has shown that Mulholland inequality is closely related to the relation of dominance between strict triangular norms and has provided a different sufficient condition [18]. In 1999 Schweizer posed a question (see Sklar [17] for more details) on comparing the Mulholland's and Tardiff's condition. This question has been answered in 2002 by Jarczyk and Matkowski who demonstrated [7] that the Tardiff's condition implies the one of Mulholland. An alternative proof has been also given by Baricz [3] in 2010.

2   Counter-Example for Mulholland's Condition

By MC we denote the set of all increasing bijections of $\mathbb{R}_0^+$ that are convex and geo-convex, i.e., that comply with Mulholland's condition. Theorem 1 states that $MC \subseteq MI$. It has remained an open question whether also $MI \subseteq MC$, i.e., whether Mulholland's condition is also necessary for the solutions to Mulholland inequality, or not. The condition of f being convex is necessary [12], however, the condition of f being geo-convex is not necessary as it has been shown in a recent paper [13]. This has been done by presenting a new sufficient condition, stronger than the Mulholland's one (see Theorem 2), and by presenting a function that complies with this new condition but not with the Mulholland's one. In this section, we are going to give a brief presentation of the result.

Definition 1. *Let* $f \colon \mathbb{R}_0^+ \to \mathbb{R}_0^+$ *be an increasing bijection.*
    Then f is said to be *k-subscalable* for some given $k \in \mathbb{R}_0^+$ if

$$\forall a, b, x \in \mathbb{R}_0^+, b - a \geq k, x \leq 1 \colon \quad \frac{f(bx)}{f(b)} \leq \frac{f(ax)}{f(a)}. \tag{2}$$

We denote the set of all k-subscalable bijections by S_k.
    Further, f is said to be *k-linear* for some given $k \in \mathbb{R}_0^+$ if it is positive-linear of the interval $[0, k]$, i.e., if there exists $r \in \mathbb{R}^+$ such that $f(x) = rx$ for all $x \in [0, k]$. Notice that also the extreme case $k = 0$ requiring no partial linearity of f is considered. We denote the set of all k-linear bijections by L_k.
    Finally, for a given $k \in \mathbb{R}_0^+$ we introduce the set $LS_k = L_k \cap S_k$ of all bijections that are k-subscalable and k-linear and the set

$$LS = \bigcup_{k \in \mathbb{R}_0^+} LS_k.$$

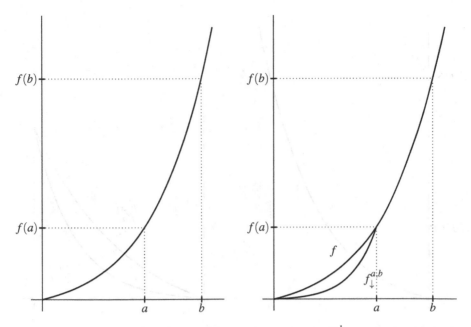

Fig. 1 Graph of the bijection f (left) and graph of the a,b-scale $f_{\downarrow}^{a,b}$ compared to the graph of f (right)

Let us introduce another presentation of the formula (2). The inequality (2) can be rewritten as:

$$\frac{f(a)}{f(b)} f(bx) \le f(ax).$$

Now, making the substitution $y = ax$, we obtain the formula:

$$\forall a,b,y \in \mathbb{R}_0^+, b - a \ge k, y \le a: \quad \frac{f(a)}{f(b)} f\left(\frac{b}{a}y\right) \le f(y). \tag{3}$$

Denote

$$f_{\downarrow}^{a,b}(y) = \frac{f(a)}{f(b)} f\left(\frac{b}{a}y\right).$$

We can see the graph of $f_{\downarrow}^{a,b}$ on $[0,a]$ as a scaling of the graph of f on $[0,b]$ to $[0,a] \times [0,f(a)]$. The notion of "subscalability" then comes from the requirement on this scaled graph to be "under" the original graph of f. See an illustration in Figure 1. The inequality (2) is illustrated in Figure 2.

If $k = 0$ then (2), as well as (3), is equivalent to the condition of geo-convexity. Therefore, $LS_0 = MC$ and, thus, $MC \subseteq LS$.

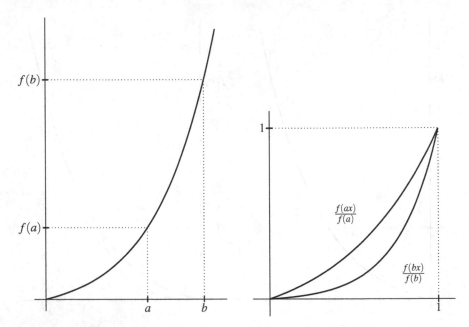

Fig. 2 Graph of the bijection f (left) and graphs of the functions $f(ax)/f(a)$ and $f(bx)/f(b)$ compared (right). The latter graphs are identical to the graph of f on $[0,a]$ and $[0,b]$, respectively, scaled to the unit square $[0,1] \times [0,1]$.

Moreover, it can be proven that

Theorem 2. *[13] Let $f: \mathbb{R}_0^+ \to \mathbb{R}_0^+$ be a convex increasing bijection which, for some $k \in \mathbb{R}_0^+$, is k-subscalable and k-linear. Then f solves Mulholland inequality.*

Thus, $MC \subseteq LS \subseteq MI$. To show that $MC \subsetneq LS$ the following counter-example has been provided in the same paper [13]:

Example 1. The increasing bijection $g: \mathbb{R}_0^+ \to \mathbb{R}_0^+$ is defined, for all $x \in \mathbb{R}_0^+$, as

$$g(x) = \begin{cases} \frac{5}{3}x & \text{if } x \in [0,1], \\ \frac{7}{3}x - \frac{2}{3} & \text{if } x \in]1,2], \\ x^2 & \text{if } x \in]2,\infty[. \end{cases}$$

It can be shown [13] that $g \in LS$ (particularly, $g \in LS_1$) but $g \notin MC$. See the graph of g in Figure 3-left.

3 Mulholland Inequality and Dominance of Strict Triangular Norms

A *triangular norm* (or a *t-norm* for short) [2, 8] is a commutative, associative, and non-decreasing binary operation $*: [0,1] \times [0,1] \to [0,1]$ with neutral element 1. A

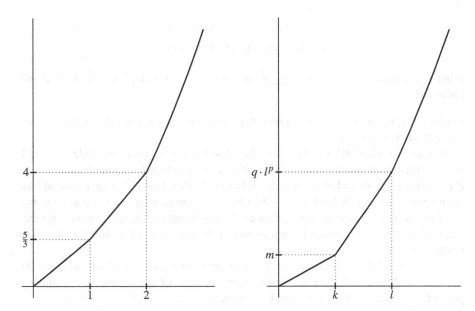

Fig. 3 Left: Graph of the bijection g given by Example 1 which is 1-subscalable and 1-linear but not geo-convex. Right: Graph of a general convex increasing bijection that is linear on $[0,k]$, affine on $]k,l]$, and power on $]l,\infty[$.

t-norm is said to be *strict* if there exists a decreasing bijection $\varphi \colon [0,1] \to [0,\infty]$ such that $x * y = \varphi^{-1}(\varphi(x) + \varphi(y))$ for all $x,y \in [0,1]$; the bijection φ is then called the *generator* of $*$. T-norms are studied nowadays mainly in the framework of the basic logic [5, 6] and the monoidal t-norm based logic [4] which are both prototypical fuzzy logics; particularly, the real unit interval $[0,1]$ endowed with a strict t-norm is isomorphic to the standard semantics of the product logic which is a special case of the basic logic. Nevertheless, originally the t-norms have been introduced within the framework of probabilistic metric spaces [9, 11] where they establish the triangular inequality of the probabilistic metrics.

Dominance is a binary relation on a set of n-ary operations; particularly, a t-norm $*$ *dominates* a t-norm $\diamond$ ($* \gg \diamond$) if, and only if,

$$\forall x,y,u,v \in [0,1]: \qquad (x \diamond y) * (u \diamond v) \quad \geq \quad (x * u) \diamond (y * v).$$

The motivation to study dominance of t-norms comes from Tardiff [18] who recognized that dominance plays an important role when constructing Cartesian products of probabilistic metric spaces. The dominance of strict t-norms is closely related to Mulholland inequality:

Theorem 3. *[18, Theorem 3] Let* $*_1 \colon [0,1] \times [0,1] \to [0,1]$ *and* $*_2 \colon [0,1] \times [0,1] \to [0,1]$ *be two strict t-norms given by the generators* $\varphi_1 \colon [0,1] \to [0,\infty]$ *and* $\varphi_2 \colon [0,1] \to [0,\infty]$, *respectively, i.e.,*

$$x *_1 y = \varphi_1^{-1}(\varphi_1(x) + \varphi_1(y)),$$
$$x *_2 y = \varphi_2^{-1}(\varphi_2(x) + \varphi_2(y)).$$

Then $*_2$ dominates $*_1$, i.e., $*_1 \gg *_2$, if, and only if, $f = \varphi_2 \circ \varphi_1^{-1}$ solves Mulholland inequality.

Remark that recently, this correspondence has been enlarged to all continuous Archimedean triangular norms [14].

It can be checked easily that the dominance relation is reflexive and anti-symmetric. Nevertheless, it remained an open question for a long time whether it is also transitive, and thus an order relation [1, Problem 17]. This question has been answered recently by Sarkoci [16] who has given an example of three continuous t-norms that violate the transitivity of the dominance relation. However, for the class of strict t-norms, which form a subset of the continuous t-norms, the question is still open.

Due to the relation between dominance of strict t-norms and Mulholland inequality, we can do the following reasoning. Suppose three strict t-norms $*_1$, $*_2$, and $*_3$ given by the generators φ_1, φ_2, and φ_3, respectively, i.e., for all $x, y \in [0, 1]$:

$$x *_1 y = \varphi_1^{-1}(\varphi_1(x) + \varphi_1(y)),$$
$$x *_2 y = \varphi_2^{-1}(\varphi_2(x) + \varphi_2(y)),$$
$$x *_3 y = \varphi_3^{-1}(\varphi_3(x) + \varphi_3(y)).$$

We define further the following three increasing bijections of $\mathbb{R}_0^+$:

$$f_{12} = \varphi_1 \circ \varphi_2^{-1},$$
$$f_{23} = \varphi_2 \circ \varphi_3^{-1},$$
$$f_{13} = \varphi_1 \circ \varphi_3^{-1}.$$

Now, the transitivity formula of the dominance relation

$$*_1 \ll *_2 \quad \text{and} \quad *_2 \ll *_3 \quad \Rightarrow \quad *_1 \ll *_3 \tag{4}$$

can be rewritten according to Theorem 3 as

$$f_{12} \in \mathrm{MI} \quad \text{and} \quad f_{23} \in \mathrm{MI} \quad \Rightarrow \quad f_{13} \in \mathrm{MI}.$$

It can be checked easily that $f_{13} = f_{12} \circ f_{23}$; therefore (4) holds if, and only if,

$$f_{12} \in \mathrm{MI} \quad \text{and} \quad f_{23} \in \mathrm{MI} \quad \Rightarrow \quad f_{12} \circ f_{23} \in \mathrm{MI}.$$

Thus we have:

Lemma 1. *The dominance relation on the set of strict t-norms is transitive if, and only if, the set MI is closed with respect to compositions meaning that if $f_1 \in MI$ and $f_2 \in MI$ then also $f_1 \circ f_2 \in MI$.*

It can be proven that the set MI is not closed with respect to compositions. Take the function from Example 1 and denote it by f_{12}. As stated in Example 1, $f_{12} \in MI$. Further, denote $f_{23} \colon x \mapsto x^2$. By Mulholland's condition, also $f_{23} \in MI$. Now, denoting the composition of these two functions by $f_{13} = f_{12} \circ f_{23}$, it can be shown that $f_{13} \notin MI$.

This implies that also transitivity of dominance of the strict t-norms is violated. Let $\varphi_2 \colon [0,1] \to [0,\infty]$ be any decreasing bijection, $\varphi_1 = f_{12} \circ \varphi_2$, $\varphi_3 = f_{23}^{-1} \circ \varphi_2$ and let $*_1$, $*_2$, and $*_3$ be strict t-norms generated by φ_1, φ_2, and φ_3, respectively. Then it can be shown (confer with the proof of Lemma 1) that $*_1 \ll *_2$, $*_2 \ll *_3$, but $*_1 \not\ll *_3$. We introduce this as the result of the paper:

Theorem 4. *The dominance relation on the set of strict t-norms is not transitive and thus not an order.*

4 Open Problems

There has been given a nice characterization for the bijections that are elements of MC; particularly the following one:

Proposition 1. *[15, Proposition 13] Let $f \colon \mathbb{R}_0^+ \to \mathbb{R}_0^+$ be a convex increasing bijection. Then it is geo-convex (i.e. $f \in MC$) if, and only if, there exists a sequence $(g_i \colon \mathbb{R}_0^+ \to \mathbb{R}_0^+)_{i \in \mathbb{N}}$ of power functions $g_i \colon x \mapsto q_i x^{p_i}$, $p_i, q_i \in \mathbb{R}_0^+$, $p_i \geq 1$, such that $f = \bigvee_{i \in \mathbb{N}} g_i$.*

According to this characterization, it can be easily seen that the function g from Example 1 is not an element of MC and, also, that the set MC is closed with respect to compositions.

The structure of LS, however, has remained unexplored as well as remains an open question whether $LS = MI$. We present the following questions:

Problem 1. The function in Example 1 could be defined, more generally, as

$$f[k,l,m,p](x) = \begin{cases} \frac{m}{k} x & \text{if } x \in [0,k] , \\ \frac{lp - m}{l - k} x - \frac{klp - lm}{l - k} & \text{if } x \in \,]k,l] , \\ x^p & \text{if } x \in \,]l,\infty[\end{cases}$$

for real parameters $k,l,m,p \in \mathbb{R}_0^+$ such that $k \leq l$ and $p \geq 1$. (See Figure 3-right for an illustration.) Give some sufficient conditions on the parameters k,l,m,p in order to ensure that f is k-subscalable. Clearly, it is necessary that $l \leq 2k$ and that $m \in \left[k^p, kl^{p-1} \right]$.

Problem 2. More generally, what is the structure of the set LS?

Problem 3. We conjecture that $LS \subsetneq MI$ but, so far, there is no proof.

Problem 4. What is the maximal subset of MI that is closed with respect to compositions?

Acknowledgements. The author would like to thank the referees for a careful reading of the paper and for their comments.

This work was partially supported by the Czech Science Foundation under Project P201/12/P055 and partially by ESF Project CZ.1.07/2.3.00/20.0051 Algebraic methods in Quantum Logic of the Masaryk University. Supported by

INVESTMENTS IN EDUCATION DEVELOPMENT

References

[1] Alsina, C., Frank, M.J., Schweizer, B.: Problems on associative functions. Aequationes Mathematicae 66(1-2), 128–140 (2003)

[2] Alsina, C., Frank, M.J., Schweizer, B.: Associative Functions: Triangular Norms and Copulas, World Scientific, Singapore (2006)

[3] Baricz, Á.: Geometrically concave univariate distributions. Journal of Mathematical Analysis and Applications 363(1), 182–196 (2010), doi:10.1016/j.jmaa.2009.08.029

[4] Esteva, F., Godo, L.: Monoidal t-norm based logic: towards a logic for left-continuous t-norms. Fuzzy Sets and Systems 124, 271–288 (2001)

[5] Hájek, P.: Basic fuzzy logic and BL-algebras. Soft Computing 2, 124–128 (1998)

[6] Hájek, P.: Metamathematics of Fuzzy Logic. Kluwer, Dordrecht (1998)

[7] Jarczyk, W., Matkowski, J.: On Mulholland's inequality. Proceedings of the American Mathematical Society 130(11), 3243–3247 (2002)

[8] Klement, E.P., Mesiar, R., Pap, E.: Triangular Norms. Trends in Logic, vol. 8. Kluwer Academic Publishers, Dordrecht (2000)

[9] Schweizer, B., Sklar, A.: Probabilistic Metric Spaces. Dover Publications, Mineola (2005)

[10] Matkowski, J.: L^p-like paranorms. In: Selected Topics in Functional Equations and Iteration Theory, Proceedings of the Austrian-Polish Seminar, Graz., 1991. Grazer Math. Ber., vol. 316, pp. 103–138 (1992)

[11] Menger, K.: Statistical metrics. Proc. Nat. Acad. Sci. U.S.A. 8, 535–537 (1942)

[12] Mulholland, H.P.: On generalizations of Minkowski's inequality in the form of a triangle inequality. Proc. London Math. Soc. 51(2), 294–307 (1950)

[13] Petrík, M., Navara, M., Sarkoci, P.: Alternative proof of Mulholland's theorem and new solutions to Mulholland inequality. In: ISMVL 2013: IEEE 43rd International Symposium on Multiple-Valued Logic, Toyama, Japan, May 21-24 (2013)

[14] Saminger-Platz, S., De Baets, B., De Meyer, H.: A generalization of the Mulholland inequality for continuous Archimedean t-norms. J. Math. Anal. Appl. 345, 607–614 (2008)

[15] Sarkoci, P.: Dominance Relation for Conjunctors in Fuzzy Logic. Ph.D. thesis (January 2007)

[16] Sarkoci, P.: Dominance is not transitive on continuous triangular norms. Aequationes Math. 75, 201–207 (2008)

[17] Sklar, A.: Remark and Problem. In: Report of Meeting, 37th International Symposium on Functional Equations, Huntington, 1999, Aequationes Math. 60, 187–188 (2000)

[18] Tardiff, R.M.: On a generalized Minkowski inequality and its relation to dominates for t-norms. Aequationes Math. 27(3), 308–316 (1984)

Distributivity Equation in the Class of Noncommutative T-Operators

Paweł Drygaś

Abstract. Recently the distributivity equation was discussed in families of certain operations (e.g. triangular norms, conorms, uninorms and nullnorms). In this paper we describe the solutions of distributivity equation in the class of noncommutative t-operators. Previous results about distributivity between nullnorms can be obtained as simple corollaries.

1 Introduction

The problem of distributivity has been posed many years ago (cf. Aczel [1], pp. 318-319). A new direction of investigations is mainly concerned of distributivity between triangular norms and triangular conorms ([9] p.17). Since a short time many authors deal with solution of distributivity equation for aggregation functions ([4]), fuzzy implications ([2]), uninorms and nullnorms ([14], [20]), which are generalization of triangular norms and conorms.

Our consideration was motivated by intention of getting algebraic structures which have weaker assumptions than nullnorms. A characterization of such binary operations is interesting not only from a theoretical point of view, but also for their applications, since they have proved to be useful in several fields like fuzzy logic framework ([11]), expert system ([13]), neural networks ([13]) or fuzzy quantifiers ([11]).

First, we introduce weak algebraic structures (section 2). Then, the distributivity equations are recalled (section 3). Next, solutions of distributivity equations from

Paweł Drygaś

Institute of Mathematics, University of Rzeszów, Rejtana 16a, 35-310 Rzeszów, Poland

e-mail: paweldr@univ.rzeszow.pl

H. Bustince et al. (eds.), *Aggregation Functions in Theory and in Practise*,

Advances in Intelligent Systems and Computing 228,

DOI: 10.1007/978-3-642-39165-1_22, © Springer-Verlag Berlin Heidelberg 2013

described families are characterized (section 4). Finally, our results are applied to nulnorms, which can be compared with results from [14] and [8] (section 5).

2 Associative, Monotonic Binary Operations

We start with basic definitions and facts.

Definition 1 ([10]). A semi triangular norm T is an increasing, associative operation $T : [0,1]^2 \to [0,1]$ with neutral element 1.
A semi triangular conorm S is an increasing, commutative, associative operation $S : [0,1]^2 \to [0,1]$ with neutral element 0.
A triangular norm T is a commutative semi triangular norm
A triangular conorm S is a commutative semi triangular conorm

Example 1 ([10]). Well-known t-norms and t-conorms are:
$$T_M(x,y) = \min(x,y), \qquad S_M(x,y) = \max(x,y),$$
$$T_P(x,y) = x \cdot y, \qquad S_P(x,y) = x + y - xy,$$
$$T_L(x,y) = \max(x+y-1,0), \qquad S_L(x,y) = \min(x+y,1),$$

Definition 2 ([3]). Operation $V : [0,1]^2 \to [0,1]$ is called nullnorm if it is commutative, associative, increasing, has a zero element $z \in [0,1]$, and that satisfies

$$V(0,x) = x \quad \text{for all } x \leq z, \tag{1}$$

$$V(1,x) = x \quad \text{for all } x \geq z. \tag{2}$$

By definition, the case $z = 0$ leads back to t-norms, while the case $z = 1$ leads back to t-conorms (cf. [10]). The next theorem show that it is built up from a t-norm, a t-conorm and the zero element.

Theorem 1 ([3]). *Let $z \in (0,1)$. A binary operation V is a nullnorm with zero element z if and only if*

$$V(x,y) = \begin{cases} S^*(x,y) & \text{if } x,y \in [0,z] \\ T^*(x,y) & \text{if } x,y \in [z,1] \,, \\ z & \text{otherwise} \end{cases} \tag{3}$$

where

$$\begin{cases} S^*(x,y) = \varphi^{-1}(S(\varphi(x),\varphi(y))), \ \varphi(x) = x/z, & x,y \in [0,z] \\ T^*(x,y) = \psi^{-1}(T(\psi(x),\psi(y))), \ \psi(x) = (x-z)/(1-z), \ x,y \in [z,1] \end{cases} \tag{4}$$

S is triangular conorm and T is triangular norm.

If in definition of nullnorm we omit assumptions (1) and (2), it cannot be shown that a commutative, associative, increasing binary operator V with zero element $z = 0$ or $z = 1$ behaves as a t-norm and t-conorm.
    In Definition 2 the existence of zero element z follows from (1) and (2):

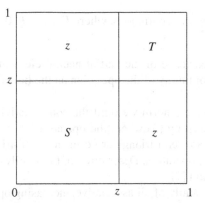

Fig. 1 Structure of nullnorm

Lemma 1. *Let V be a monotonic, binary operation and exist $z \in [0,1]$ such that*

$$V(0,x) = V(x,0) = x \quad \text{for all } x \leq z \tag{5}$$

$$V(1,x) = V(x,1) = x \quad \text{for all } x \geq z \tag{6}$$

then $V(x,y) = z$ for $(x,y) \in [0,z] \times [z,1] \cup [z,1] \times [0,z]$, $V|_{[0,z]}$ is monotonic, binary operation with neutral element 0 and zero element z. $V|_{[z,1]}$ is monotonic, binary operation with neutral element 1 and zero element z.
Moreover V is associative (commutative, idempotent) if and only if $V|_{[0,z]}$ and $V|_{[z,1]}$ are associative (commutative, idempotent).

Definition 3. Element $s \in [0,1]$ is called idempotent element of operation $G :$ $[0,1]^2 \to [0,1]$ if $G(s,s) = s$. Operation G is called idempotent if all elements from $[0,1]$ are idempotent.

Theorem 2 (cf. [6]). *Operation $V : [0,1]^2 \to [0,1]$ is idempotent nullnorm with zero element z if and only if it is given by*

$$V(x,y) = \begin{cases} \max(x,y) & \text{if } x,y \in [0,z] \\ \min(x,y) & \text{if } x,y \in [z,1] \\ z & \text{otherwise} \end{cases} \tag{7}$$

More general families of operations with zero element are examined in [16].
As we known, the structure of nullnorms is the same as the structure of t-operators

Definition 4 ([12]). Operation $F : [0,1]^2 \to [0,1]$ is called t-operator if it is commutative, associative, increasing and such that

$$F(0,0) = 0, \quad F(1,1) = 1, \tag{8}$$

the functions F_0 and F_1 are continuous, where $F_0(x) = F(0,x)$, $F_1(x) = F(1,x)$.

$$(9)$$

In this definition the existence of the partial neutral elements (conditions (1) and (2)) follows from the continuity of the operation on the boundary of the unit square ((8), (9)).

If in the definition of nullnorm we omit the commutativity condition, then we obtain the operation given by (3) where the operations T and S are not necessary commutative (i.e. they are semi-triangular norm and semi-triangular conorm). It is different in the case of t-operators. Description of the family of such operations we may find in [21], [17] and [7]

Let $\mathscr{F}_{a,b}$ denote the family of all associative, increasing operations $F : [0,1]^2 \to [0,1]$, such that $F(0,1) = a$, $F(1,0) = b$ and functions F_0, F_1, F^0, F^1 are continuous, where $F_0(x) = F(0,x)$, $F_1 = F(1,x)$, $F^0(x) = F(x,0)$, $F^1(x) = F(x,1)$, $x \in [0,1]$.

Theorem 3 ([7]). *Let* $F : [0,1]^2 \to [0,1]$, $F(0,1) = a$, $F(1,0) = b$. *Operation* $F \in \mathscr{F}_{a,b}$ *if and only if there exists semi-triangular norm and semi-triangular conorm such that*

$$F(x,y) = \begin{cases} aS\left(\frac{x}{a}, \frac{y}{a}\right) & \text{if } x,y \in [0,a], \\ b+(1-b)T\left(\frac{x-b}{1-b}, \frac{y-b}{1-b}\right) & \text{if } x,y \in [b,1], \\ a & \text{if } x \le a \le y, \\ b & \text{if } y \le b \le x, \\ x & \text{otherwise,} \end{cases} \quad (10)$$

for $a \le b$ and

$$F(x,y) = \begin{cases} bS\left(\frac{x}{b}, \frac{y}{b}\right) & \text{if } x,y \in [0,b], \\ a+(1-a)T\left(\frac{x-a}{1-a}, \frac{y-a}{1-a}\right) & \text{if } x,y \in [a,1], \\ a & \text{if } x \le a \le y, \\ b & \text{if } y \le b \le x, \\ y & \text{otherwise,} \end{cases} \quad (11)$$

for $b \le a$.

Remark 1. The class $\mathscr{F}_{z,z}$ is a class of t-operators with zero element z and noncommutative components T and S.

Lets denote $\mathscr{F}_z := \mathscr{F}_{z,z}$.

3 Distributivity Equations

Now, we consider the distributivity equation (cf. [1], p. 318).

Definition 5. Let $F, G : [0,1]^2 \to [0,1]$. Operation F is distributive over G, if they fulfil the distributivity conditions:

$$\forall_{x,y,z \in [0,1]} \ F(x, G(y,z)) = G(F(x,y), F(x,z)), \quad (12)$$

$$\forall_{x,y,z\in[0,1]} \ F(G(y,z),x) = G(F(y,x),F(z,x)). \tag{13}$$

Lemma 2 (cf. [18]). *Let* $F : X^2 \to X$ *have right (left) neutral element* e *in a subset* $\emptyset \neq Y \subset X$ *(i.e.* $\forall_{x\in Y} \ F(x,e) = x \ (F(e,x)=x))$. *If operation* F *is distributive over operation* $G : X^2 \to X$ *fulfilling* $G(e,e) = e$, *then* G *is idempotent in* Y.

Proof. Let $x \in Y \subset X$, $y, z = e \in Y \subset X$. If F is left distributive over G, then $x = F(x,e) = F(x,G(e,e)) = G(F(x,e),F(x,e)) = G(x,x)$. In the case where operation F has left neutral element the proof is similar.

Corollary 1 ([5]). *If operation* $F : [0,1]^2 \to [0,1]$ *with neutral element* $e \in [0,1]$ *is distributive over operation* $G : [0,1]^2 \to [0,1]$ *fulfilling* $G(e,e) = e$, *then* G *is idempotent.*

Lemma 3 ([18]). *Every increasing operation* $F : [0,1]^2 \to [0,1]$ *is distributive over* max *and* min.

Now we present solutions of distributivity equations (12) and (13) in the family $\mathscr{F} = \bigcup_{a,b\in[0,1]} \mathscr{F}_{a,b}$.

4 Distributivity of $F \in \mathscr{F}_{a,b}$ Over $G \in \mathscr{F}_z$

Now our consideration will concern the distributivity of $F \in \mathscr{F}_{a,b}$ over G which is noncommutative nullnorm. We distinguish here three different cases depending on the inequality between the elements a, b of operation F and the zero element of operation G. If $a = b$ then we obtain case considered in [18].

Theorem 4. *Let* $a,b,z \in [0,1]$, $z < a < b$. $F \in \mathscr{F}_{a,b}$ *is distributive over* $G \in \mathscr{F}_z$ *if and only if* G *is the idempotent t-operator (7) and* F *has the following form:*

$$F(x,y) = \begin{cases} zS_1\left(\frac{x}{z},\frac{y}{z}\right) & \text{if } x,y \in [0,z], \\ z+(a-z)S_2\left(\frac{x-z}{a-z},\frac{y-z}{a-z}\right) & \text{if } x,y \in [z,a], \\ \max(x,y) & \text{if } \min(x,y) \leq z \leq \max(x,y) \leq a, \\ b+(1-b)T\left(\frac{x-b}{1-b},\frac{y-b}{1-b}\right) & \text{if } x,y \in [b,1], \\ a & \text{if } x \leq a \leq y, \\ b & \text{if } y \leq b \leq x, \\ x & \text{otherwise,} \end{cases} \tag{14}$$

where S_1, S_2 *are semi-triangular conorm,* T *is semi-triangular norm. (see Fig. 2).*

Proof. Let $a,b,z \in [0,1]$, $z < a < b$, $F \in \mathscr{F}_{a,b}$ be distributive over $G \in \mathscr{F}_z$. Directly from Lemma 2 operation G is an idempotent t-operator and it is given by (7).

Using (12) and (10) for F we have for $x \in [0,a]$

$$F(x,z) = F(x,G(0,1)) = G(F(x,0),F(x,1)) = G(x,a) = \begin{cases} z & for \ x \in [0,z] \\ x & for \ x \in [z,a] \end{cases},$$

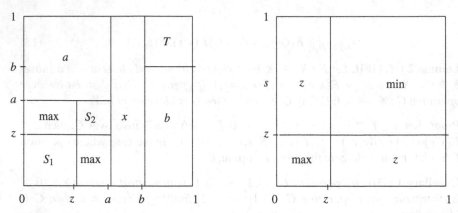

Fig. 2 Structure of operations F and G from Theorem 4

$$F(z,x) = F(G(0,1),x) = G(F(0,x),F(1,x)) = G(x,b) = \begin{cases} z & \text{for } x \in [0,z] \\ x & \text{for } x \in [z,a] \end{cases}.$$

So, S is an ordinal sum of semi-triangular conorm S_1 and S_2.

Conversely, let F be given by (14) and G by (7). Since F is increasing in suitable rectangular domains with common boundaries, then it is increasing in $[0,1]^2$. To prove (12) we have to consider 64 cases. Moreover, directly from Lemma 3 we can omit cases with distributivity over max or min

In a similar way we obtain

Theorem 5. *Let $a,b,z \in [0,1]$, $a \leq z \leq b$. $F \in \mathscr{F}_{a,b}$ is distributive over $G \in \mathscr{F}_z$ if and only if G is the idempotent t-operator* (7).

Theorem 6. *Let $a,b,z \in [0,1]$, $a < b < z$. $F \in \mathscr{F}_{a,b}$ is distributive over $G \in \mathscr{F}_z$ if and only if G is the idempotent t-operator* (7) *and F has the following form:*

$$F(x,y) = \begin{cases} aS\left(\frac{x}{a},\frac{y}{a}\right) & \text{if } x,y \in [0,a], \\ b+(z-b)T_1\left(\frac{x-z}{z-b},\frac{y-z}{z-b}\right) & \text{if } x,y \in [b,z], \\ z+(1-z)T_2\left(\frac{x-z}{1-z},\frac{y-z}{1-z}\right) & \text{if } x,y \in [z,1], \\ \min(x,y) & \text{if } b \leq \min(x,y) \leq z \leq \max(x,y), \quad (15) \\ a & \text{if } x \leq a \leq y, \\ b & \text{if } y \leq b \leq x, \\ x & \text{otherwise}, \end{cases}$$

where T_1, T_2 are semi-triangular norm, S is semi-triangular conorm.

5 Application to Nullnorms

From the above results we get corollaries about distributivity between nullnorms.

From the above results we get corollaries about distributivity between two null-norms.

Corollary 2 (cf.[14], Preposition 3.2). *Let $z_1, z_2 \in [0, 1]$ and operations F and G be nullnorms with $z_1 \leq z_2$. Then F is distributive over G, if and only if G is idempotent and F has the following form:*

$$F = \begin{cases} S & in \ [0, z_1]^2 \\ T_1 & in \ [z_1, z_2]^2 \\ T_2 & in \ [z_2, 1]^2 \\ min & in \ [z_1, z_2] \times [z_2, 1] \cup [z_2, 1] \times [z_1, z_2] \\ z_1 & otherwise \end{cases},$$

where S is triangular conorm, T_1 and T_2 are triangular norms.

By duality

Corollary 3 (cf.[14], Preposition 3.3). *Let $z_1, z_2 \in [0, 1]$ and operations F and G be nullnorms with $z_2 \leq z_1$. Then F is distributive over G, if and only if G is idempotent and F has the following form:*

$$F = \begin{cases} S_1 & in \ [0, z_2]^2 \\ S_2 & in \ [z_2, z_1]^2 \\ max & in \ [0, z_2] \times [z_2, z_1] \cup [z_2, z_1] \times [0, z_2] \\ T & in \ [z_1, 1]^2 \\ z_1 & otherwise \end{cases},$$

where S_1, S_2 are triangular conorms and B is triangular norm.

6 Conclusion

In this paper we give the partial characterization of distributivity equations (12) and (13) for noncommutative t-operators. In future work we will be present solutions of distributivity equations (12) and (13) in the family $\mathscr{F}$, i.e we will be consider the distributivity of $F \in \mathscr{F}_{a,b}$ over $G \in \mathscr{F}_{c,d}$ in all cases, depending on ordering of a, b, c, d.

References

[1] Aczél, J.: Lectures on Functional Equations and their Applications. Acad. Press, New York (1966)

[2] Baczyński, M.: On a class of distributive fuzzy implications. Internat. J. Uncertainty, Fuzzines Knowledge-Based Syst. 9, 229–238 (2001)

[3] Calvo, T., De Baets, B., Fodor, J.: The functional equations of Frank and Alsina for uninorms and nullnorms. Fuzzy Sets and Systems 120, 385–394 (2001)

[4] Calvo, T.: On some solutions of the distributivity equation. Fuzzy Sets and Systems 104, 85–96 (1999)

[5] Drewniak, J.: Binary operations on fuzzy sets. BUSEFAL 14, 69–74 (1983)

[6] Drygaś, P.: A characterization of idempotent nullnorms. Fuzzy Sets and Systems 145, 455–461 (2004)

[7] Drygaś, P.: Noncommutative t-operators (submitted)

[8] Drewniak, J., Drygaś, P., Rak, E.: Distributivity between uninorms and nullnorms. Fuzzy Sets Syst. 159, 1646–1657 (2008)

[9] Fodor, J., Roubens, M.: Fuzzy Preference Modeling and Multicriteria Decision Support. Kluwer Acad. Publ., New York (1994)

[10] Klement, E.P., Mesiar, R., Pap, E.: Triangular norms. Kluwer Acad. Publ., Dordrecht (2000)

[11] Klir, G.J., Yuan, B.: Fuzzy Sets and Fuzzy Logic, Theory and Application. Prentice Hall PTR, Upper Saddle River (1995)

[12] Mas, M., Mayor, G., Torrens, J.: t-operators. Internat. J. Uncertainty, Fuzzines Knowledge-Based Syst. 7, 31–50 (1999)

[13] Mas, M., Mayor, G., Torrens, J.: The modularity condition for uninorms and t-operators. Fuzzy Sets and Systems 126, 207–218 (2002)

[14] Mas, M., Mayor, G., Torrens, J.: The distributivity condition for uninorms and t-operators. Fuzzy Sets and Systems 128, 209–225 (2002)

[15] Mas, M., Mayor, G., Torrens, J.: Corrigendum to "The distributivity condition for uninorms and t-operators" [Fuzzy Sets and Systems 128 (2002) 209–225]. Fuzzy Sets and Systems 153, 297–299 (2002)

[16] Mas, M., Mesiar, R., Monserat, M., Torrens, J.: Aggregation operations with annihilator. Internat. J. Gen. Syst. 34, 1–22 (2005)

[17] Marichal, J.-L.: On the associativity functional equation. Fuzzy Sets Syst. 114, 381–389 (2000)

[18] Rak, E.: Distributivity equation for nullnorms. J. Electrical Engin. 56(12/s), 53–55 (2005)

[19] Rak, E., Drygaś, P.: Distributivity between uninorms. J. Electrical Engin. 57(7/s), 35–38 (2006)

[20] Ruiz, D., Torrens, J.: Distributive idempotent uninorms. Internat. J. Uncertainty, Fuzzines Knowledge-Based Syst. 11, 413–428 (2003)

[21] Sander, W.: Associative aggregation operators. In: Calvo, T., Mayor, G., Mesiar, R. (eds.) Aggregation Operators, pp. 124–158. Physica-Verlag, Heidelberg (2002)

On Some Classes of Discrete Additive Generators

G. Mayor and J. Monreal

Abstract. This work is developed in the field of the additive generation of discrete aggregation operators. Specifically, this article deals with the study and applicability of disjunctions that are additively generated by several special types of generators.

Keywords: discrete binary operation, disjunction, t–conorm, additive generator, symmetric additive generator, S–implication function.

1 Introduction

Fuzzy logic is one of the tools for management of uncertainty. In Fuzzy logic we usually work with a continuous scale of certainty values, the real unit interval $[0, 1]$, however implementation restrictions in applications force us to use a finite scale of truth degrees instead of the mentioned continuous one. In this paper we deal with the class of finitely valued disjunction-like operations that contains, in particular, the family of finitely valued t–conorms. In full analogy to the representation theorem of continuous t–conorms, there exists a characterization of smooth (divisible) discrete t–conorms as ordinal sums of Łukasiewicz discrete t–conorms [6, 7]. Other references on smooth discrete associative operations are [1, 6, 3]. Here our goal is the study of different aspects of the additive generation of a class of discrete binary operations that we call disjunctions; in particular the additive generation of t–conorms (associative disjunctions). Some results related to discrete t–conorms differ substantially from those obtained for ordinary t–conorms defined on $[0, 1]$. In this sense, for instance, we know that a t–conorm with nontrivial idempotent elements has not an additive generator; this is not true for discrete t–conorms as we recall in this paper. It seems clear that to develop a theory focused on the additive generation of dis-

G. Mayor · J. Monreal

Department of Mathematics and Computer Science, University of the Balearic Islands, E-07122 Palma de Mallorca, Spain

e-mail: {gmayor,jaume.monreal}@uib.es

H. Bustince et al. (eds.), *Aggregation Functions in Theory and in Practise*,
Advances in Intelligent Systems and Computing 228,
DOI: 10.1007/978-3-642-39165-1_23, © Springer-Verlag Berlin Heidelberg 2013

crete (associative) disjunctions is useful. Thus, a t–conorm S defined on the scale $\{0,1,2,\ldots,n\}$ is determined by $\frac{n(n-1)}{2}$ entries. If S admits an additive generator $(0,a_1,\ldots,a_n)$ then it can be managed by only n integer values [4, 5, 8]. In this paper we also point out the usefulness of having discrete additive generators when we have to describe and manage properties of discrete binary operations.

2 Preliminaries

Consider $L = \{0,1,\ldots,n\}$, $n \geq 1$, equipped with the usual ordering. We begin recalling basic definitions, examples and properties of finitely-valued t–conorms. A complete exposition of this topic can be found in [7].

2.1 Disjunctions and T–Conorms on a Finite Totally Ordered Set

Definition 1. A *disjunction* on L is a binary operation $D : L \times L \to L$ such that for all $i,i',j,j' \in L$ the following axioms are satisfied:

1. $D(i,j) = D(j,i)$ (commutativity)
2. $D(i,0) = i$ (boundary condition)
3. $D(i,j) \leq D(i',j')$ whenever $i \leq i'$, $j \leq j'$ (monotonicity)

Definition 2. A t–conorm on L is an associative disjunction $(D(D(i,j),k) = D(i,D(k,k)), \forall i,j,k \in L)$

Example 1. We can consider as basic t–conorms:

i) the drastic,
$$S_D(i,j) = \begin{cases} i \text{ if } j = 0, \\ j \text{ if } i = 0, \\ n \text{ otherwise;} \end{cases}$$

ii) the maximum $S_M(i,j) = \max\{i,j\}$,
iii) the bounded sum, or Łukasiewicz t–conorm, $S_L(i,j) = \min\{i+j,n\}$.

Remark 1. The mapping $N(i) = n - i$ is the only strong negation on L ($N : L \to L$ decreasing and involutive).

Given a t-conorm S, the binary operation $T : L \times L \to L$ defined by $T(i,j) = NS(N(i),N(j))$ is a t-norm on L (commutative, associative, increasing in each variable with n as neutral element) called the N-dual of S.

Definition 3. A disjunction D on L is *smooth* if
$$0 \leq D(i+1,j) - D(i,j) \leq 1, \ \forall i,j \in L, i < n.$$

Definition 4. A t–conorm S is Archimedean if for all $i,j \in L \backslash \{0,n\}$ there exists $m \in N$ satisfying $i_S^{(m)} > j$ where $i_S^{(m)} = S(S(\ldots S(i,i)\ldots))$.

It can be easily proved that a t-conorm on L is Archimedean if and only if the only idempotent elements are 0 and n.

Proposition 1. S_L *is the only smooth Archimedean t–conorm on L.*

Now, we recall a well-known method for constructing a new t–conorm from two given t–conorms.

Proposition 2. *Let S_1 be a t–conorm on $\{0,1,\ldots,m\}$ and S_2 a t–conorm on $L = \{0,1,\ldots,n\}$, with $m,n \geq 1$. Consider the binary operation S defined on $\{0,1,\ldots,m+n\}$ as follows:*

$$S(i,j) = \begin{cases} S_1(i,j) & \text{if } (i,j) \in \{0,1,\ldots,m\}^2, \\ m+S_2(i-m,j-m) & \text{if } (i,j) \in \{m,m+1,\ldots,m+n\}^2, \\ \max(i,j) & \text{otherwise.} \end{cases}$$

Then S is a t–conorm on $\{0,1,\ldots,m+n\}$, called the ordinal sum of S_1 and S_2 and denoted by $S = \langle S_1, S_2 \rangle$.

Next, we are going to characterize the class of smooth t–conorms as ordinal sums of Łukasiewicz t–conorms.

Proposition 3. *A t–conorm S on $L = \{0,1,\ldots,n\}$ is smooth if and only if there exists a set $I = \{0 = a_0 < a_1 < \ldots < a_r < a_{r+1} = n\}$, $0 \leq r \leq n-1$, of elements of L, such that:*

$$S(i,j) = \begin{cases} \min\{a_{l+1}, i+j-a_l\} & \text{if } (i,j) \in (a_l, a_{l+1})^2, \ 0 \leq l \leq r, \\ \max\{i,j\} & \text{otherwise.} \end{cases}$$

Remark 2.

i) $S_M \leq S \leq S_L$ for any smooth t–conorm S.

ii) There are exactly 2^{n-1} smooth t–conorms on L.

Example 2. The table below shows the cardinality of different classes of t–conorms on L for distinct values of n.

n	t-conorms	smooth	Archimedean	ordinal sums	others
1	1	1	1	0	0
2	2	2	1	1	0
3	6	4	2	3	1
4	22	8	6	11	5
5	94	16	22	45	27
6	451	32	95	205	151
7	2386	64	471	1021	894
8	13775	128	2670	5512	5593
9	86417	256	17387	32095	36935
10	590489	512	131753	201367	257369

Next, we introduce the pseudo-inverse of appropriate monotone functions from L to R^+, and we consider a construction similar to that given in case of ordinary t–conorms. Thus, we state a general method to construct disjunctive operations on L involving only a one-place real function and the usual addition.

2.2 Additive Generation of Disjunctions and T–Conorms

Let R^+ be the set of non-negative real numbers, and consider the class F of strictly increasing functions $f : L \to R^+$ with $f(0) = 0$. We can represent such functions by ordered strictly increasing lists of length $n + 1$, $(a_0 = 0, a_1, ..., a_n)$ where $a_i = f(i)$, $i \in L$.

Given $f \in F$, we define the pseudo–inverse of f as the function $f^{(-1)} : R^+ \to L$ defined by $f^{(-1)}(t) = \max\{i \in L; f(i) \le t\}$.

Proposition 4. *Given $f \in F$, the binary operation on L, $D : L \times L \to L$, defined by*

$$D(i, j) = f^{(-1)}(f(i) + f(j)) \qquad i, j \in L \tag{1}$$

is a disjunction on L. We write $D = \langle f \rangle$ to indicate that the disjunction D is defined from f via (1). In this case we say that D is additively generated by f and we also say that f is an additive generator of D.

For brevity, the functions in F will be called additive generators.

Proposition 5. *If $D = \langle (0, a_1, ..., a_n) \rangle$ then we have the following equivalences:*

1. *$D(i, j) = k < n$ if and only if $a_k \le a_i + a_j < a_{k+1}$, $k < n$.*
2. *$D(i, j) = n$ if and only if $a_n \le a_i + a_j$.*

Example 3. If $D = \langle (0, 1, 2, 3, 14, 21, 28, 35, 42) \rangle$ then we have $S(4, 5) = 7$ because $a_7 \le a_4 + a_5 < a_8$.

D	0	1	2	3	4	5	6	7	8
0	0	1	2	3	4	5	6	7	8
1	1	2	3	3	4	5	6	7	8
2	2	3	3	3	4	5	6	7	8
3	3	3	3	3	4	5	6	7	8
4	4	4	4	4	6	7	8	8	8
5	5	5	5	5	7	8	8	8	8
6	6	6	6	6	8	8	8	8	8
7	7	7	7	7	8	8	8	8	8
8	8	8	8	8	8	8	8	8	8

We say that an additive generator f is associative if the disjunction $D = \langle f \rangle$ is associative (a t–conorm). The function $f = (0, 2, 3, 4, 6)$ is associative and it generates an Archimedean non smooth t-conorm on $\{0, 1, 2, 3, 4\}$.

Proposition 6. *Consider $Ran f + Ran f = \{f(i) + f(j) ; i, j \in L\}$. If $f \in F$ is such that $Ran f + Ran f \subset Ran f \cup [f(n), +\infty)$ then the disjunction generated by f is an Archimedean t-conorm.*

Remark 3. There are disjunctions which are not additively generables. There are t–conorms which are not additively generables.

Here we show the three t–conorms on $\{0,1,2,3,4,5,6,7,8\}$ without additive generator.

S_1	0 1 2 3 4 5 6 7 8
0	0 1 2 3 4 5 6 7 8
1	1 4 5 6 6 8 8 8 8
2	2 5 5 7 8 8 8 8 8
3	3 6 7 7 8 8 8 8 8
4	4 6 8 8 8 8 8 8 8
5	5 8 8 8 8 8 8 8 8
6	6 8 8 8 8 8 8 8 8
7	7 8 8 8 8 8 8 8 8
8	8 8 8 8 8 8 8 8 8

S_2	0 1 2 3 4 5 6 7 8
0	0 1 2 3 4 5 6 7 8
1	1 5 5 5 7 8 8 8 8
2	2 5 6 6 7 8 8 8 8
3	3 5 6 8 8 8 8 8 8
4	4 7 7 8 8 8 8 8 8
5	5 8 8 8 8 8 8 8 8
6	6 8 8 8 8 8 8 8 8
7	7 8 8 8 8 8 8 8 8
8	8 8 8 8 8 8 8 8 8

S_3	0 1 2 3 4 5 6 7 8
0	0 1 2 3 4 5 6 7 8
1	1 5 5 6 6 8 8 8 8
2	2 5 5 7 8 8 8 8 8
3	3 6 7 7 8 8 8 8 8
4	4 6 8 8 8 8 8 8 8
5	5 8 8 8 8 8 8 8 8
6	6 8 8 8 8 8 8 8 8
7	7 8 8 8 8 8 8 8 8
8	8 8 8 8 8 8 8 8 8

Example 4. The three basic t–conorms quoted above have the following additive generators:

  i. $S_{\text{Ł}} = \langle (0,1,\ldots,n-1,n) \rangle$
  ii. $S_M = \langle (0,1,3,7,\ldots,2^{n-1}-1,2^n-1) \rangle$
  iii. $S_D = \langle (0,n-1,n,\ldots,2n-3,2n-2) \rangle$

Proposition 7. Let $f = (a_0,a_1,\ldots,a_n)$ and $g = (b_0,b_1,\ldots,b_n)$ in F. Then $\langle f \rangle = \langle g \rangle$ if and only if the following conditions hold:

  1. $a_k \leq a_i + a_j < a_{k+1}$ implies $b_k \leq b_i + b_j < b_{k+1}$
  2. $a_i + a_j \geq a_n$ implies $b_i + b_j \geq b_n$

for all $i,j,k \in L$ with $k < n$.

As a consequence of the previous proposition, if D is a disjunction additively generable then it has a generator f with $Ran\, f \subset Z^+$, where Z^+ is the set of non-negative integers.

Proposition 8. *Any smooth t–conorm is additively generable.*

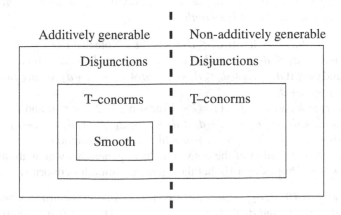

3   Some Classes of Additive Generators

Given an additive generator f, we will study the relationship between its structure and the types of t–conorms that they generate.

3.1   Concave Additive Generators

Definition 5. An additive generator $f = (0, a_1, \ldots, a_n)$ is called concave when

$$a_1 \geq a_2 - a_1 \geq \ldots \geq a_{n-1} - a_{n-2} \geq a_n - a_{n-1}$$

In other words, $f \in F$ is concave when there exists a decreasing list $d = (d_1, \ldots, d_n)$ of positive real numbers such that $a_i = d_1 + \ldots + d_i \ \forall i \in L$.

From this kind of additive generators, we can obtain disjunctions without non-trivial idempotent elements.

Proposition 9. *Let f a concave additive generator. Then, $D = \langle f \rangle$ is a disjuntion on L having only 0 and n as idempotent elements.*

Proof. If we suppose $0 < i < n$ and $D(i, i) = i$ then $2a_i < a_{i+1}$, that leads to to $2(d_1 + \ldots + d_i) < d_1 + \ldots + d_{i+1}$, that is, $d_1 + \ldots + d_i < d_{i+1}$, which is impossible because $d_i \geq d_{i+1}$. Thus, $D(i, i) > i$.

3.2   Convex Additive Generators

Definition 6. An additive generator $f = (0, a_1, \ldots, a_n)$ is called concave when

$$a_1 \leq a_2 - a_1 \leq \ldots \leq a_{n-1} - a_{n-2} \leq a_n - a_{n-1}$$

In other words, $f \in F$ is concave when there exists an increasing list $d = (d_1, \ldots, d_n)$ of positive real numbers such that $a_i = d_1 + \ldots + d_i \ \forall i \in L$.

Proposition 10. *Let $f = (0, a_1, a_2, \ldots, a_n)$ be a convex additive generator. The disjunction D additively generated by f is smooth.*

Proof. It is $a_i = d_1 + \ldots + d_i \ \forall i \geq 1$ with $d_1 \leq d_2 \leq \ldots \leq d_n$. Note that $D(i, j) = k < n$ if and only if $d_1 + \ldots + d_k \leq d_1 + \ldots + d_i + d_1 + \ldots + d_j < d_1 + \ldots + d_{k+1}$, and $D(i, j) = k = n$ if and only if $d_1 + \ldots + d_n \leq d_1 + \ldots + d_i + d_1 + \ldots + d_j$. In any case, $d_{j+1} + \ldots + d_k \leq d_1 + \ldots + d_i$.
If we suppose $D(i, j) = k$ and $D(i, j - 1) = k - l$ for some k, $2 < k \leq n$ and $l \geq 2$ then we obtain $d_j + \ldots + d_{k-l} \leq d_1 + \ldots + d_i < d_j + \ldots + d_{k-l+1} < d_j + \ldots + d_{k-1}$. But $d_k \geq d_j$, thus $d_1 + \ldots + d_i < d_{j+1} + \ldots + d_k$ which is a contradiction.

Next we give a characterization of the convex additive generators which are associative. We know from Proposition 10 that they generate smooth t-conorms.

Proposition 11. *Let $f = (0, a_1, a_2, \ldots, a_n)$ be a convex additive generator where $a_i = d_1 + \ldots + d_i$, $1 \leq i \leq n$, and $d_1 \leq d_2 \leq \ldots \leq d_n$. Then, $D = \langle f \rangle$ is a t–conorm* *if and only if there exist* $0 = i_0 < i_1 < \ldots < i_r = n$, *with* $r \geq 1$, $i_1, \ldots, i_{r-1} \in L$ *such that* $\forall k : 0 \leq k \leq r - 1$:

$$a_{i_k} < d_{i_k+1} \leq \ldots \leq d_{i_{k+1}} < a_{i_k+1},$$
$$a_{i_k} + d_{i_k+1} + \ldots + d_{\lfloor \frac{i_k+i_{k+1}}{2} \rfloor} \geq d_{i_{k+1} - \lfloor \frac{i_k+i_{k+1}}{2} \rfloor} + \ldots + d_{i_{k+1}}$$

In this case, we can obtain an additive generator with the following structure

$$0 < \overbrace{d_1 = \ldots = d_{i_1}} < a_{i_1} < \overbrace{d_{i_1+1} \leq \ldots \leq d_{i_2}} < a_{i_1+1} < \ldots < a_{i_{r-1}} < \overbrace{d_{i_{r-1}+1} \leq \ldots \leq d_{i_r}} < a_{i_{r-1}+1}$$

Example 5.

Associative generator
$d = ($ 1, **3**, 3, 3, **11**, 12, 13, 14, **65**, 70, 75, 80, 85, 90, 95$)$
$f = (0,$ **1**, 4, 7, **10**, 21, 33, 46, **60**, 125, 195, 270, 350, 435, 525, 620$)$

The list of differences of this additive generator has four blocks ($r = 4$, $i_0 = 0$, $i_1 = 1$; $i_2 = 4$, $i_3 = 8$, $i_4 = 15$) having one, three, four and seven elements respectively.

A simple method to obtain a convex (concave) additive generator from one that is concave (convex) is described below.

Proposition 12. *Let* $f = (0, a_1, \ldots, a_n)$ *be a concave (convex) additive generator. Then the additive generator* $f^* = (b_0, b_1, \ldots, b_n)$ *where* $b_i = a_n - a_{n-i}$, $i = 0, 1, \ldots, n$, *is convex (concave).*

Proof. If $f = (0, a_1, \ldots, a_n)$ is concave (convex) then $\forall i = 1, \ldots, n$,

$$d_i^* = b_i - b_{i-1} = a_n - a_{n-i} - (a_n - a_{n-i+1}) = a_{n-i+1} - a_{n-i} = d_{n-i}$$

Thus, when d_i decrease (increase), the differences $d_i^* = b_i - b_{i-1}$ increase (decrease).

3.3 Symmetric Additive Generators

Definition 7. We say that an additive generator $f = (0, a_1, \ldots, a_n)$ is symmetric if $a_i + a_{n-i} = a_n \ \forall i \in L$.

Next proposition shows the structure of such additive generators.

Proposition 13. *An additive generator* $f = (0, a_1, \ldots, a_n)$ *is symmetric if and only if there exists* k, $0 < k < n$ *such that:*

$$a_n = 2a_k \text{ and } a_{k+r} = a_n - a_{k-r}, \ r = 1, \ldots, k, \text{ if } n \text{ is even.}$$
$$a_n = a_{k-1} + a_k \text{ and } a_{k+r} = a_n - a_{k-1-r}, \ r = 1, \ldots, k-1, \text{ if } n \text{ is odd.}$$

Next we treat the construction of associative symmetric additive generators. To do that, we take a basic t–conorm and, using the first half part of its standard additive generator, we complete it in order to obtain a symmetric additive generator.

We only develop here the construction from the maximum t–conorm. For more details on this construction from the other basic t–conorms see [8].

3.3.1 Construction of Associative Symmetric Additive Generators ($n = 2k$)

From the standard additive generator of S_M on $L_k = \{0,1,...,k\}$, that is $(0,1,3,...,2^k-1)$, we consider the symmetric additive generator on L_{2k} with its first part $(0,1,3,...,2^k-1)$ and the second one defined by the condition in Definition 7.

Note that the first part is convex and the second cancave. We call these additive generators convex-concave. In the table below are written these additive generators for different values of k.

k	symmetric additive generator
1	$(0,1,2)$
2	$(0,1,3,5,6)$
3	$(0,1,3,7,11,13,14)$
4	$(0,1,3,7,15,23,27,29,30)$
5	$(0,1,3,7,15,31,47,55,59,61,62)$
6	$(0,1,3,7,15,31,63,95,111,119,123,125,126)$
7	$(0,1,3,7,15,31,63,127,191,223,239,247,251,253,254)$
8	$(0,1,3,7,15,31,63,127,255,383,447,479,495,503,507,509,510)$

Here we draw the symmetric additive generator in case $k = 6$ ($n = 12$). It generates the t–conorm given below.

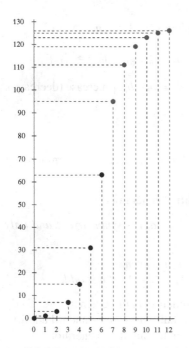

S	0	1	2	3	4	5	6	7	8	9	10	11	12
0	0	1	2	3	4	5	6	7	8	9	10	11	12
1	1	1	2	3	4	5	6	7	8	9	10	12	12
2	2	2	2	3	4	5	6	7	8	9	12	12	12
3	3	3	3	3	4	5	6	7	8	12	12	12	12
4	4	4	4	4	4	5	6	7	12	12	12	12	12
5	5	5	5	5	5	5	6	12	12	12	12	12	12
6	6	6	6	6	6	6	6	12	12	12	12	12	12
7	7	7	7	7	7	7	12	12	12	12	12	12	12
8	8	8	8	8	12	12	12	12	12	12	12	12	12
9	9	9	9	12	12	12	12	12	12	12	12	12	12
10	10	10	12	12	12	12	12	12	12	12	12	12	12
11	11	12	12	12	12	12	12	12	12	12	12	12	12
12	12	12	12	12	12	12	12	12	12	12	12	12	12

$f = (0,1,3,7,15,31,63,95,111,119,123,125,126)$

Next proposition shows that the above additive generators are associative.

Proposition 14. *The symmetric additive generator* $f = (0, a_1, \ldots, a_k, a_{k+1}, \ldots, a_{2k})$
given by

$$\begin{aligned} a_i &= 2^i - 1 \\ a_{k+i} &= 2 \cdot (2^k - 1) - a_{k-i} \end{aligned} \quad i = 1, \ldots, k$$

is associative.

Proof. We have to proof $S(S(i, i'), i'') = S(i, S(i', i''))$, $\forall 1 \leq i, i', i'' \leq 2k$. Consider only one of the possible cases: $i \leq i' < k < i''$. Taking $j = i'' - k$, we have $S(S(i, i'), i'') = S(i', k + j)$. Taking into account the Proposition 5, we distinguish two subcases:

- If $i' < k - j$ then $a_{k+j} \leq a_{i'} + a_{k+j} = 2^{i'} - 1 + 2 \cdot (2^k - 1) - (2^{k-j} - 1) = 2 \cdot (2^k - 1) + 2^{i'} - 2^{k-j} \leq 2 \cdot (2^k - 1) + 2^{k-j-1} - 2^{k-j} = 2 \cdot (2^k - 1) - 2^{k-j-1} < 2 \cdot (2^k - 1) - (2^{k-j-1} - 1) = a_{k+j+1}$ and thus $S(i', k + j) = k + j$. On the other hand, $S(i, S(i', i'')) = S(i, k + j)$ and from a similar computation as above, we obtain $S(i, k + j) = k + j$
- If $i' \geq k - j$ then $a_{i'} + a_{k+j} = 2^{i'} - 1 + 2 \cdot (2^k - 1) - (2^{k-j} - 1) = 2 \cdot (2^k - 1) + (2^{i'} - 2^{k-j}) \geq 2 \cdot (2^k - 1) = a_n$ and thus $S(i', k + j) = n$. On the other hand, $S(i, S(i', k + j)) = S(i, n) = n$.

Now to highlight the advantage of having additive generators for aggregation operators, we show below how to use them in the study of properties of S-implication functions.

4 Additive Generation and the Study of Discrete Implication Functions

Definition 8. A binary operation $I : L \times L \to L$ is an implication on L if the following conditions hold $\forall i, j, k \in L$:

(I1) $I(i, k) \geq I(j, k)$ whenever $i \leq j$ (decreasing in the first place),
(I2) $I(i, j) \leq I(i, k)$ whenever $j \leq k$ (increasing in the second place),
(I3) $I(0, 0) = I(1, 1) = 1, I(1, 0) = 0$ (boundary conditions).

Proposition 15. *Let S be a t-conorm on L and N the only strong negation on L. The binary operation I_S on L defined by*

$$I_S(i, j) = S(N(i), j) \quad i, j \in L. \tag{2}$$

is an implication on L that we call S-implication.

We will study three properties related to implication functions:

P1 Identity principle: $I(i, i) = n, \forall i \in L$.
P2 Ordering principle: $I(i, j) = n \Leftrightarrow i \leq j, \forall i, j \in L$.

P3 Generalized Modus Ponens principle: $T(i, I(i, j)) \leq j$, $\forall i, j \in L$, where T is a t-norm and I an implication with n as left-neutral element.

From now on we suppose that I is an S–implication function, T the N-dual of S and $f = (0, a_1, \ldots, a_n)$ an additive generator of S.

First, our aim is to describe the properties above in terms of the given additive generator of S.

Proposition 16. *Let I be an S–implication function as above. Then:*

i) *The implication I satisfies the Identity principle if and only if $a_{n-i} + a_i \geq a_n$ $\forall i \in L$.*

ii) *The implication I satisfies the Ordering principle if and only if $a_i + a_j \geq a_n \Leftrightarrow i + j \geq n$.*

iii) *The implication I satisfies the Generalized Modus Ponens principle if and only if*

1) $a_i + a_j \geq a_n \implies i + j \geq n$.

2) $a_i + a_j < a_n \implies \exists k < n$ *tal que* $\begin{cases} a_k \leq a_i + a_j < a_{k+1} \\ a_{n-j} \leq a_i + a_{n-k} \end{cases}$

Next proposition states that the symmetry of an additive generator is a sufficient condition to assure that the all above mentioned properties are satisfied.

Proposition 17. *If the t–conorm S admits a symmetric additive generator then the corresponding S–implication function satisfies the properties P1, P2 and P3.*

According to this result, the S–implication function obtained from the additive generator of subsection 3.3.1, $f = (0, 1, 3, 7, 15, 31, 63, 95, 111, 119, 123, 125, 126)$, satisfies the properties P1, P2 and P3.

5 Conclusions

This article deals with the study and applicability of disjunctions that are additively generated by several special types of generators. From one hand, some classes of additive generator are introduced, then the disjunctions generated by them are studied. In particular, we show the advantages of having symmetric additive generators when S–implication functions are considered.

Acknowledgements. The authors acknowledge the support by the Spanish Government Grant MTM2009-10962.

References

[1] Fodor, J.: Smooth associative operations on finite ordinal scales. IEEE Transactions on Fuzzy Systems 8, 791–795 (2000)

[2] Godo, L., Sierra, C.: A new approach to connective generation in the framework of expert systems using fuzzy logic. In: Proc. 18th IEEE Int. Symposium on Multiple-Valued Logic, pp. 157–162 (1988)

[3] Klement, E.P., Mesiar, R., Pap, E.: Triangular Norms. Kluwer Academic Publishers (2000)

[4] Martín, J., Mayor, G., Monreal, J.: The problem of the additive generation of finitely-valued t–conorms. Mathware & Soft Computing 16, 17–27 (2009)

[5] Mayor, G., Monreal, J.: Additive generators of discrete conjunctive aggregation operations. IEEE Transactions on Fuzzy Systems 15(6), 1046–1052 (2007)

[6] Mayor, G., Torrens, J.: On a class of operators for expert systems. International Journal of Intellingent Systems 8, 771–778 (1993)

[7] Mayor, G., Torrens, J.: Triangular Norms on Discrete Settings. In: Logical, Algebraic, Analytic and Probabilistic Aspects of Triangular Norms, pp. 189–230. Elsevier (2005)

[8] Monreal, J.: Generació additiva de funcions d'agregació conjuntives i disjuntives discretes. PhD thesis, University of the Balearic Islands (2012)

Part V
Multi-criteria Decision Making

The Consensus Functional Equation in Agreement Theory

Juan Carlos Candeal, Esteban Induráin, and José Alberto Molina

Abstract. We introduce the concept of the consensus functional equation, for a bivariate map defined on an abstract choice set. This equation is motivated by miscellaneous examples coming from different contexts. In particular, it appears in the analysis of sufficiently robust agreements arising in Social Choice. We study the solutions of this equation, relating them to the notion of a rationalizable agreement rule. Specific functional forms of the solutions of the consensus functional equation are also considered when the choice sets have particular common features. Some extension of the consensus equation to a multivariate context are also explored.

Keywords: Functional equations in two variables, Agreement rules in Social Choice.

Mathematics Subject Classification (2010): 91B16, 91B14.

1 Introduction

Assume that a research team, working individually, can reach a best output, say x. In the same way, a second (different) team, working individually, can reach their best performance, say y. However, if both teams collaborate and work together, they could reach an even better achievement, say $F(x,y)$. In this situation we may think that if one of the research teams could be able to get, working individually, the best possible output $F(x,y)$, the collaboration with the other team would not lead beyond that best attainment $F(x,y)$. In other words, the following *functional equation* arises:

Juan Carlos Candeal · José Alberto Molina
Departamento de Análisis Económico. Universidad de Zaragoza, Spain
e-mail: {candeal,jamolina}@unizar.es

Esteban Induráin
Departamento de Matemáticas. Universidad Pública de Navarra. Pamplona, Spain
e-mail: steiner@unavarra.es

H. Bustince et al. (eds.), *Aggregation Functions in Theory and in Practise*,
Advances in Intelligent Systems and Computing 228,
DOI: 10.1007/978-3-642-39165-1_24, © Springer-Verlag Berlin Heidelberg 2013

$$F(x,y) = F(F(x,y),y) = F(x,F(x,y)).$$

Here $x,y \in X$, where X denotes he set of all possible goals of any research team.

This functional equation was already introduced in [11] and [13], where it was named the *consensus functional equation*.

The same equation arises when we study some particular kinds of agreements between two individuals, encountered in Social Choice. In this direction, let us assume that X is a nonempty set, that represents the collection of possible choices of each individual. (i.e.: X can be interpreted as the *choice set*, which is the same for both individuals). Suppose that $F : X \times X \to X$ is the map or that expresses the agreement between them. (i.e.: F can be interpreted as an *agreement rule*). In other words, if the first individual chooses the alternative $x \in X$ while the second individual chooses the alternative $y \in X$, then $F(x,y) \in X$ is another alternative, where the two agents agree, and so-to-say represents a *consensus option* for both agents.

The consensus equation is based upon the following idea: Suppose that we consider a situation in which the agreement is so *robust* that, if either of the individiuals changes her/his initial position on the one agreed by both of them, then the former achieved agreement should not vary, and remains the same. In formula, when analyzing this kind of agreements we should study the functional equation

$$F(F(x,y),y) = F(x,F(x,y)) = F(x,y), \ (x,y \in X).$$

In this second example, we may notice that the last equality of the formula is exactly the *unanimity principle* over the alternatives that are in the codomain of the map F (i.e., $F(z,z) = z$, for every $z \in F(X \times X) \subseteq X$). Thus, if F satisfies *non-imposition* condition (i.e., F is surjective, so that $X = F(X \times X)$), then the consensus equation implies the unanimity principle for the whole X.

The consensus equation transpires a property that, in some sense, reminds us the Nash equilibrium concept coming from Game Theory (see e.g. [26, 20]). As a matter of fact, if, for a given $x,y \in X$, we interpret $F(x,y)$ as the "best social agreement" (provided that the first agent chooses x whereas the second one chooses y), then the "best choice" for the first agent in order to reach that "best collective agreement", provided that the second agent keeps at her/his choice y, is to single out $F(x,y)$. The same argument applies for the second individual.

Here we furnish two further *examples*, that may also constitute a *motivation* to study the functional equation of consensus by its own merit:

1. *Suppose that several different tasks must be done to achieve a goal. Let T be the set of those tasks to be done. We may identify an individual with the subset of tasks that she/he is able to do. In this case, we may interpret X as the power set $\mathscr{P}(T)$, and consider the union "$\cup$" of subsets of T as a binary operation defined on $\mathscr{P}(T)$. Then it is clear that $(x \cup y) \cup y = x \cup (x \cup y) = x \cup y$ for every $x,y \in \mathscr{P}(T)$. Therefore, if we change the notation, setting $F(x,y) = x \cup y \ (x,y \in X)$ we immediately get*

$$F(F(x,y),y) = F(x,F(x,y)) = F(x,y) \ (x,y \in X).$$

2. *Suppose that a nonempty set X is given a total order "$\preceq$". Given two elements $x, y \in X$, let $F(x,y) = y$ if $x \preceq y$, otherwise let $F(x,y) = x$ $(x, y \in X)$. It is obvious that this bivariate map $F : X \times X \rightarrow X$ satisfies, in particular, the functional equation*

$$F(F(x,y),y) = F(x, F(x,y)) = F(x,y) \quad (x, y \in X).$$

The paper is organized as follows:

Section 2 contains the basic background.

In Section 3 we study the consensus equation from an abstract point of view. To that end, we introduce a key concept; namely, that of a *rationalizable bivariate map*. Rationalizability is a notion that resembles the one already introduced in the literature of single-valued choice functions (see [4, 5, 28] or, more recently, [25]). Here, it means that a particular binary relation, that we call the *revealed relation*, describes F. (Thus, a map $F : X \times X$, on a choice set X, is said to be *rationalizable* if it can be expressed in terms of a suitable binary relation defined on X, as stated in Definition 5 in Section 2 below). We characterize bivariate maps that satisfy the consensus equation plus the *anonymity principle* as those that are rationalizable.

Associativity (i.e., $F(F(x,y),z) = F(x, F(y,z))$, for every $x, y, z \in X$) is a slightly more demanding property than the fulfilment of the consensus equation (see [11, 13]). When an agreement rule $F : X \times X \rightarrow X$ is associative, we get a more appealing result: namely, in this case there is a partial order, say $\preceq$, defined on X such that $(X, \preceq)$ is a *semi-lattice* and $F(x,y)$ turns out to be the supremum, with respect to $\preceq$, of $\{x,y\}$, for every $x, y \in X$ (see [11]). Notice that associativity can be viewed as an *extension* property (see e.g. [21, 16, 12, 22, 1, 29]). That is, if F is associative then, for any finite number of agents, we can induce agreement rules based on it. For instance, if the number of agents is three we may induce a trivariate rule $G : X \times X \times X \rightarrow X$ by declaring that $G(x,y,z) = F(F(x,y),z) = F(x, F(y,z))$ $(x, y, z \in X)$. In other words, associativity invites everyone "to join the party".

We also pay attention to the case where F is a *selector* (see [23, 18]), i.e., $F(x,y) \in \{x,y\}$, for every $x, y \in X$. In this case, and for obvious reasons, we will say that F satisfies the *independence of irrelevant alternatives condition*. Quite surprisingly, the independence of irrelevant alternatives condition is proved to be more restrictive than consensus.

In Section 4 we study several aspects of the solutions of the consensus equation in concrete scenarios.

In particular, we pay attention to the case in which X can be identified to a real interval. In this case, we add some extra conditions on F, namely monotonicity (Paretian properties) and continuity. Then we get some impossibility as well as some possibility results about the existence of agreement rules. On the one hand, we prove that there is no strongly Paretian bivariate map which satisfies consensus. On the other hand, we show that the only continuous agreement rules that satisfy the independence of irrelevant alternatives condition are the max and the min (i.e., those based upon the most and the least favoured individuals, respectively).

In Section 5 we explore the extended consensus equation, considering n-variate maps that correspond to general models of consensus where n-individuals are involved, obviously with $n \geq 2$.

A final Section 6 of further comments closes the paper.

Remark 1. Throughout the paper we will focus on the consensus equation involving only *two variables*. The generalization of this equation for more than two variables could constitute the raw material to build future pieces of research.

2 Preliminaries

In what follows, X will denote a nonempty set, that we interpret as the choice set (or the set of alternatives). Moreover, $F : X \times X \to X$ will be a bivariate map defined on X.

Definition 1. The map F is said to satisfy:

(1) the *unanimity principle* if $F(x,x) = x$ for every $x \in X$,
(2) the *anonymity principle* if $F(x,y) = F(y,x)$ for every $x,y \in X$,
(3) the *consensus functional equation (short, consensus)* if it holds that
   $F(F(x,y),y) = F(x,F(x,y)) = F(x,y)$ for every $x,y \in X$,
(4) the *associativity equation* if $F(x,F(y,z)) = F(F(x,y),z)$ for every $x,y,z \in X$,
(5) the *independence of irrelevant alternatives condition (shortly denoted by IIA))*
   if $F(x,y) \in \{x,y\}$, for every $x,y \in X$.

Remark 2. Needless to say that the word "*consensus*" is encountered in many branches of mathematical Social Choice theory, under a wide sort of scopes and approaches (see e.g. [15, 27, 30, 10, 14, 8, 19, 24, 6, 9, 20, 2]). All these contexts, as the equation introduced throughout the present manuscript, transpire the idea of buying models to interpret situations in which a "social agreement" between individuals is reached, following some "rules" or procedures.

Definition 2. A bivariate map $F : X \times X \to X$ is said to be an *agreement rule* if it satisfies the conditions (1) to (3) of Definition 1 above.

Now we recall some basic concepts on binary relations. A binary relation $\preceq$ defined on X is said to be a *partial order* if it is reflexive (i.e., $x \preceq x$ holds for every $x \in X$), antisymmetric (i.e., $(x \preceq y) \wedge (y \preceq x) \Rightarrow x = y$, for every $x,y \in X$) and transitive (i.e., $(x \preceq y) \wedge (y \preceq z) \Rightarrow x \preceq z$, for every $x,y,z \in X$). If, in addition, $\preceq$ is total (i.e., $(x \preceq y) \vee (y \preceq x)$ holds for every $x,y \in X$), then $\preceq$ is said to be a *total order*.

A binary relation $\mathscr{R}$ defined on X is said to have the *supremum property* if, for every $x,y \in X$, there is a unique $z \in X$ such that the following two conditions are met: (i) $(x\mathscr{R}z) \wedge (y\mathscr{R}z)$ holds; (ii) if there is $u \in X$ such that $x\mathscr{R}u$ and $y\mathscr{R}u$ hold, then $z\mathscr{R}u$ also holds. The unique element z that satisfies conditions (i) and (ii) is called the *supremum* of x and y and it is denoted by $sup_{\mathscr{R}}\{x,y\}$. Whenever $sup_{\mathscr{R}}\{x,y\} \in \{x,y\}$, then it is called *maximum* of x and y and it is denoted by $max_{\mathscr{R}}\{x,y\}$.

Definition 3. Let $\preceq$ be a partial order defined on X. Then $(X, \preceq)$ is said to be a *semi-lattice* if $\preceq$ has the supremum property.[1]

3 Consensus vs. Rationalizable Bivariate Maps

The main purpose of this section is to provide a description of the agreement rules defined on a nonempty choice set X in terms of certain binary relations on X with special features. To that end, the following concept will play an important role.

Definition 4. Let X be a nonempty set. Let F be a bivariate map defined on X. Associated with F we consider on X a new binary relation, denoted by $\mathscr{R}_r$ and defined as follows: $x\mathscr{R}_r y \iff F(x,y) = y$, for every $x,y \in X$. The binary relation $\mathscr{R}_r$ is said to be the *revealed relation* of F.

Before introducing the notion of a rationalizable bivariate map, a notational convention is needed.

Notation. Let $\mathscr{R}$ be a binary relation defined on X. Then, for each $x \in X$, $G_{\mathscr{R}}(x)$ will denote the *upper contour set* of x, i.e., $G_{\mathscr{R}}(x) = \{z \in X : x\mathscr{R}z\}$.

Definition 5. A bivariate map F on X is said to be *rationalizable* if $F(x,y) \in G_{\mathscr{R}_r}(x) \cap G_{\mathscr{R}_r}(y)$, for every $x,y \in X$.

Remark 3. That is, the concept of rationalizability intends to describe a bivariate map by means of the upper contour sets of its corresponding revealed relation.

Next Theorem 1 characterizes the bivariate anonymous maps that satisfy the consensus equation in terms of those which are rationalizable.

Theorem 1. *Let F be a unanimous and anonymous bivariate map defined on X. Then F is rationalizable if and only if it satisfies consensus.*

Proof. Suppose that F is an anonymous bivariate map defined on X which satisfies the consensus equation. Let $x,y \in X$ be fixed. In order to show that F is rationalizable, notice that $F(x,F(x,y)) = F(x,y)$ since F satisfies the consensus equation. Thus, by definition of $\mathscr{R}_r$, $F(x,y) \in G_{\mathscr{R}_r}(x)$. Moreover, by anonymity together with consensus, it holds that $F(y,F(x,y)) = F(F(x,y),y) = F(x,y)$. Therefore, $F(x,y) \in G_{\mathscr{R}_r}(y)$. So, $F(x,y) \in G_{\mathscr{R}_r}(x) \cap G_{\mathscr{R}_r}(y)$. Since x,y are arbitrary elements of X, it follows that F is rationalizable.

For the converse, suppose that F is an anonymous rationalizable bivariate map defined on X. We want to see that F satisfies consensus. To that end, let $x,y \in X$ be fixed. Since F is rationalizable, it holds that $x\mathscr{R}_r F(x,y)$ and $y\mathscr{R}_r F(x,y)$. But, by definition of the revealed consensus relation, this means that $F(x,F(x,y)) = F(x,y)$ and $F(y,F(x,y)) = F(x,y)$. Now, by anonymity, $F(y,F(x,y)) = F(F(x,y),y)$ and

---

[1] For an excellent account of the material related to latticial or semi-latticial structures, see e.g., [7, 17].

therefore $F(F(x,y),y) = F(x,y)$. The fact that $F(F(x,y),F(x,y)) = F(x,y)$ follows directly from unanimity. Since x,y are arbitrary elements of X, we have shown that F satisfies consensus.   □

Remark 4. A unanimous and anonymous bivariate map F defined on X may fail to be rationalizable, even when $\mathscr{R}_r$ is transitive. Indeed, let $X = \{x,y,z\}$ and define F : $X \times X \to X$ as follows: $F(x,x) = F(y,z) = F(z,y) = x$, $F(y,y) = F(x,z) = F(z,x) = y$ and $F(z,z) = F(x,y) = F(y,x) = z$. It is clear that this map F is unanimous and anonymous. In addition, an easy calculation gives: $\mathscr{R}_r = \{(x,x),(y,y),(z,z)\}$. In other words, $x\mathscr{R}_r x$, $y\mathscr{R}_r y$ and $z\mathscr{R}_r z$ are the only possible relationships, according to $\mathscr{R}_r$, among the three elements of X. Thus, in addition to being reflexive and anti-symmetric, $\mathscr{R}_r$ is transitive too. However, F is not rationalizable since, for example, $z = F(x,y) \notin G_{\mathscr{R}_r}(x) \cap G_{\mathscr{R}_r}(y)$.

In general, as the next Proposition 1 shows, for a (unanimous) bivariate map F, consensus is a less restrictive condition than associativity or independence of irrelevant alternatives.

Proposition 1. *Let F be a bivariate map defined on X.*

(i) If F is unanimous and associative, then it safisfies consensus.
(ii) If F satisfies IIA, then it safisfies consensus.

Proof. (i) Let $x,y \in X$ be fixed. Then, by associativity and unanimity, it holds that $F(F(x,y),y) = F(x,F(y,y)) = F(x,y)$. The other equality of consensus is proved similarly. So, since x,y are arbitrary points of X, F satisfies consensus.

(ii) Let $x,y \in X$ be fixed. Since F satisfies IIA, either $F(x,y) = x$ or $F(x,y) = y$. If $F(x,y) = x$, then we have that $F(F(x,y),y) = F(x,y) = x = F(x,x) = F(x,F(x,y))$. Now, if $F(x,y) = y$, then $F(F(x,y),y) = F(y,y) = y = F(x,y) = F(x,F(x,y))$. So, in any of the two cases, we have that $F(F(x,y),y) = F(x,F(y,y)) = F(x,y)$. Since x,y are arbitrary points of X, F satisfies consensus.   □

As a direct consequence of Theorem 1 and Proposition 1 we obtain the following corollary.

Corollary 1. *(i) Every unanimous, anonymous and associative bivariate map defined on X is rationalizable.*

(ii) Every bivariate map defined on X which satisfies IIA is rationalizable.

Remark 5. It is easy to see that, for a unanimous and anonymous bivariate map F, associativity and IIA are independent conditions. Moreover, there are agreement rules (hence rationalizable bivariate maps) other than associative maps or those that satisfy IIA. For a thorough description of the links that can be established among the mentioned properties of bivariate maps, see [13].

We now focus on associative agreement rules. As we have just seen, associativity is more restrictive than consensus. Indeed, associativity reinforces in a significant manner the scope of Theorem 1, as next Theorem 2 states.

Theorem 2. *Let F be an associative agreement rule defined on X. Then, $(X, \mathscr{R}_r)$ is a semi-lattice and $F(x,y) = \sup_{\mathscr{R}_r}\{x,y\}$, for every $x, y \in X$.*

Proof. Let us first prove that $\mathscr{R}_r$ is a partial order on X. Indeed, reflexivity follows directly from unanimity of F. To see that $\mathscr{R}_r$ is antisymmetric, let $x, y \in X$ be such that $x\mathscr{R}_r y$ and $y\mathscr{R}_r x$ hold. Then, by definition of $\mathscr{R}_r$, we have that $F(x,y) = y$ and $F(y,x) = x$. So, by anonymity, $x = y$ and therefore $\mathscr{R}_r$ is antisymmetric. To prove transitivity of $\mathscr{R}_r$, let $x, y, z \in X$ be such that $x\mathscr{R}_r y$ and $y\mathscr{R}_r z$ hold. Then, by definition of $\mathscr{R}_r$ again, we have that $F(x,y) = y$ and $F(y,z) = z$. Let us see that $F(x,z) = z$, which would mean that $x\mathscr{R}_r z$. Indeed, $F(x,z) = F(x,F(y,z)) = F(F(x,y),z) = F(y,z) = z$, the second equality being true since F is associative. Thus, $\mathscr{R}_r$ is transitive too.

Let us show now that $(X, \mathscr{R}_r)$ is a semi-lattice. To that end, we must prove that, for given arbitrary elements $x, y \in X$, the supremum $\sup_{\mathscr{R}_r}\{x,y\}$ exists. Notice that, since F is asociative, by Proposition 1 (i), it satisfies consensus too. So, $F(x,F(x,y)) = F(x,y)$ and therefore, by definition of $\mathscr{R}_r$, we have that $x\mathscr{R}_r F(x,y)$. In a similar way, now using anonymity and consensus, we get $F(y,F(x,y)) = F(F(x,y),y) = F(x,y)$. That is, $y\mathscr{R}_r F(x,y)$. So $F(x,y)$ is an upper bound, with respect to $\mathscr{R}_r$, of x and y. Let us see that it is the least upper bound. To see this, let $z \in X$ such that $x\mathscr{R}_r z$ and $y\mathscr{R}_r z$. Then, by definition of $\mathscr{R}_r$ again, we have that $F(x,z) = F(y,z) = z$. Hence $F(F(x,y),z)) = F(x,F(y,z)),y) = F(x,z) = z$, the first equality being true by associativity. Therefore, it follows that $F(F(x,y),z)) = z$ which means that $F(x,y)\mathscr{R}_r z$. So, we have shown that $F(x,y) = \sup_{\mathscr{R}_r}\{x,y\}$, which proves the second claim of the statement of Theorem 2. This finishes the proof. □

We now present some illuminating observations about the concepts introduced above.

Remark 6. (i) It should be observed that, if $\mathscr{R}$ is a binary relation on X for which $(X, \mathscr{R})$ is a semi-lattice, then the bivariate map $F_{\mathscr{R}}$ defined on X as $F_{\mathscr{R}}(x,y) = \sup_{\mathscr{R}}\{x,y\} \in X$ $(x, y \in X)$ is an associative agreement rule. Moreover, in this case, it can be easily proved that $\mathscr{R}$ and $\mathscr{R}_r$ coincide. So, associative agreement rules are characterized as those that can be rationalized by means of semi-latticial structures.

(ii) An agreement rule that satisfies IIA need not be associative. Moreover, and unlike the associative case, the revealed relation $\mathscr{R}_r$ in this situation can exhibit intransitivities. To see an example, consider the set $X = \{x, y, z\}$ and the bivariate map $F : X \times X \to X$ given by $F(x,x) = F(x,z) = F(z,x) = x$; $F(x,y) = F(y,x) = F(y,y) = y$; $F(y,z) = F(z,y) = F(z,z) = z$. It is clear that F is anonymous and satisfies IIA. However, it is not associative since $F(x,F(y,z)) = F(x,z) = x$, whereas $F(F(x,y),z) = F(y,z) = y$. In terms of the revealed relation $\mathscr{R}_r$ we have that $x\mathscr{R}_r y$, $y\mathscr{R}_r z$ and $z\mathscr{R}_r x$. So, there is a "cycle", with respect to $\mathscr{R}_r$, for the three-element set $\{x, y, z\}$.

(iii) If an agreement rule F satisfies IIA, then the revealed consensus relation $\mathscr{R}_r$ becomes a total order on X. Moreover, if an agreement rule F which satisfies IIA is also associative, then $F(x,y) = \max_{\mathscr{R}_r}\{x,y\}$, for every $x, y \in X$. So, associative agreement rules that satisfy IIA are characterized as those that can be rationalized by means of totally ordered structures.

(iv) Associative agreement rules have an interesting property that we call the *extension property*. The extension property means that an associative (bivariate) agreement rule generates associative, unanimous and anonymous n-variate rules, for any finite number of agents $n \in \mathbb{N}$. In other words, if a (unanimous and anonymous) map involving just two individuals is associative then it is possible that more and more individuals can "join the party" and enjoy a "stable" agreement. So, from a behavioural perspective, associativity is an appealing property. Indeed, let F_2 be an associative (bivariate) agreement rule. Then, by Theorem 2, $F_2(x,y) = sup_{\mathscr{R}_r}\{x,y\}$, for every $x,y \in X$. Now, for any $n \geq 3$, define $F_n : X^n = X \times \ldots (n\text{-times}) \ldots \times X \to X$ as follows: $F_n(x_1,\ldots,x_n) = sup_{\mathscr{R}_r}\{x_1,\ldots,x_n\}$, for every $x_1,\ldots,x_n \in X$. It is then straightforward to see that, for every $n \geq 3$, the n-variate map F_n so-defined is associative, unanimous and anonymous.

(v) It should be noted that Theorem 2 can be applied to scenarios in which the choice set X is, on its own, a space of preferences. Indeed, let X denote the collection of all the total preorders (i.e.: transitive and total binary relations) that can be defined on a finite set Z. Let $F : X \times X \to X$ be the *Borda rule* (see [25]). Then, it is straightforward to see that F is an associative agreement rule. Thus, Theorem 2 states that the Borda rule is entirely described by the revealed relation on X. Actually, it is simple to prove that, in this case, $\mathscr{R}_r$ is given as follows: $\precsim_1 \mathscr{R}_r \precsim_2$ if and only if $\precsim_2 \subseteq \precsim_1$ and $\prec_1 \subset \prec_2$, $(\precsim_1, \precsim_2 \in X)$. Here, $\prec$ stands for the asymmetric part of $\precsim$ (i.e., $x \prec y$ if and only if $\neg(y \precsim x)$, for every $x,y \in X$).

As seen in the proof of Theorem 2, an associative, unanimous and anonymous bivariate map defined on X has the property that its revealed relation turns out to be transitive. The converse is not true even though the bivariate map is rationalizable (or, equivalently by Theorem 1, it satisfies consensus). Nevertheless, for a unanimous bivariate map that satisfies IIA, transitivity of its revealed relation implies associativity. These two facts are proved through the next Proposition 2, which closes this section.

Proposition 2. *(i) An agreement rule such that its associated revealed relation is transitive may fail to be associative.*
(ii) Every agreement rule that satisfies IIA is associative.

Proof. (i) Let $X = \{x,y,z,u\}$. Let $F : X \times X \to X$ be the bivariate map given by $F(x,x) = x$; $F(y,y) = y$; $F(x,y) = F(x,z) = F(y,z) = F(z,x) = F(z,y) = F(z,z) = F(z,u) = F(u,z) = z$; $F(x,u) = F(y,u) = F(u,x) = F(u,y) = F(u,u) = u$. It is clear that F satisfies unanimity and anonymity. Let us see that it is an agreement rule (i.e., it satisfies consensus) by showing that it is rationalizable (see Theorem 1 above). To that end, let $\mathscr{R}_r$ be its revealed relation. A direct calculation proves that $\mathscr{R}_r$ is given by: $x\mathscr{R}_r x$, $x\mathscr{R}_r z$, $x\mathscr{R}_r u$, $y\mathscr{R}_r y$; $y\mathscr{R}_r z$, $y\mathscr{R}_r u$, $z\mathscr{R}_r z$, $u\mathscr{R}_r z$, $u\mathscr{R}_r u$. Let us observe that $\mathscr{R}_r$ is transitive. Now, by checking the upper contour sets of $\mathscr{R}_r$, we obtain: $G_{\mathscr{R}_r}(x) = \{x,z,u\}$; $G_{\mathscr{R}_r}(y) = \{y,z,u\}$; $G_{\mathscr{R}_r}(z) = \{z\}$; $G_{\mathscr{R}_r}(u) = \{z,u\}$. Thus $F(x,x) = x \in G_{\mathscr{R}_r}(x)$; $F(x,y) = z \in G_{\mathscr{R}_r}(x) \cap G_{\mathscr{R}_r}(y)$; $F(x,z) = z \in G_{\mathscr{R}_r}(x) \cap G_{\mathscr{R}_r}(z)$; $F(x,u) = u \in G_{\mathscr{R}_r}(x) \cap G_{\mathscr{R}_r}(u)$; $F(y,y) = y \in G_{\mathscr{R}_r}(y)$; $F(y,z) = z \in G_{\mathscr{R}_r}(y) \cap G_{\mathscr{R}_r}(z)$; $F(y,u) = u \in G_{\mathscr{R}_r}(y) \cap G_{\mathscr{R}_r}(u)$; $F(z,z) = z \in G_{\mathscr{R}_r}(z)$;

$F(z,u) = z \in G_{\mathcal{R}_r}(z) \cap G_{\mathcal{R}_r}(u)$. Therefore F is rationalizable. Finally, observe that F is not associative since $F(F(x,y),u) = F(z,u) = z \neq u = F(x,u) = F(x,F(y,u))$.
(ii) Let $x,y,z \in X$ be fixed. We must show that $F(F(x,y),z) = F(x,F(y,z))$. Since F satisfies IIA, it follows that $F(x,y) \in \{x,y\}$, $F(x,z) \in \{x,z\}$ and $F(y,z) \in \{y,z\}$. So we distinguish among eight possibilities:

(1) $F(x,y) = x$, $F(x,z) = x$ and $F(y,z) = y$. In this case, $F(F(x,y),z) = F(x,z) = x = F(x,y) = F(x,F(y,z))$ and we are done.
(2) $F(x,y) = x$, $F(x,z) = x$ and $F(y,z) = z$. In this case, $F(F(x,y),z) = F(x,z) = x = F(x,z) = F(x,F(y,z))$ and we are done again.
(3) $F(x,y) = x$, $F(x,z) = z$ and $F(y,z) = y$. Now, since F is anonymous, $F(y,x) = F(x,y) = x$ and $F(z,y) = F(y,z) = y$. So we get $y\mathcal{R}_r x$ and $x\mathcal{R}_r z$. Thus, by transitivity of $\mathcal{R}_r$, it follows that $y\mathcal{R}_r z$. But $F(z,y) = F(y,z) = y$ means that $z\mathcal{R}_r y$ too. In addition, $\mathcal{R}_r$ is antisymmetric since F is anonymous. Therefore, $y = z$. Now, if $y = z$, $F(F(x,y),z) = F(x,F(y,z))$ becomes $F(F(x,y),y) = F(x,F(y,y))$ or, equivalently, $F(F(x,y),y) = x = F(x,y) = F(x,F(y,y))$ and we are done.
(4) $F(x,y) = x$, $F(x,z) = z$ and $F(y,z) = z$. In this case, $F(F(x,y),z) = F(x,z) = z = F(x,F(y,z))$ and we are done.
(5) $F(x,y) = y$, $F(x,z) = x$ and $F(y,z) = y$. In this case, $F(F(x,y),z) = F(y,z) = y = F(x,y) = F(x,F(y,z))$ and we are done.
(6) $F(x,y) = y$, $F(x,z) = x$ and $F(y,z) = z$. In this case, and arguing in the same way as in case (3) above, we have that $x\mathcal{R}_r y$ and $z\mathcal{R}_r x$ which, by transitivity, implies that $z\mathcal{R}_r y$. This, together with $y\mathcal{R}_r z$, implies that $y = z$. Then, $F(F(x,y),z) = F(x,F(y,z))$ becomes $F(F(x,y),y) = F(x,F(y,y))$ or, equivalently, $F(F(x,y),y) = y = F(x,y) = F(x,F(y,y))$ and we are done again.
(7) $F(x,y) = y$, $F(x,z) = z$ and $F(y,z) = y$. In this case, $F(F(x,y),z) = F(y,z) = y = F(x,y) = F(x,F(y,z))$ and we are done. Finally,
(8) $F(x,y) = y$, $F(x,z) = z$ and $F(y,z) = z$. In this case, $F(F(x,y),z) = F(y,z) = z = F(x,z) = F(x,F(y,z))$ which concludes the proof. $\qquad\square$

Remark 7. It can be shown that if X is a three-elements set (i.e, $X = \{x,y,z\}$), then any agreement rule defined on X for which $\mathcal{R}_r$ is transitive is, in fact, associative.

4 Possibility vs. Impossibility Results on the Existence of Agreement Rules in Continuum Spaces

In this section, we study the consensus equation in particular contexts. In general, the solutions of this equation cannot be described in an easy way (see [13] for details). However, in some special cases, and imposing also some natural extra conditions on the map F, it is indeed possible to entirely describe its solutions. (See e.g. [23, 13] for further results in this direction).

Throughout this section, we assume that the choice set X is a real interval.

Both impossibility as well as possibility results arise. On the one hand, we prove that there is no strongly Paretian bivariate map which satisfies consensus. On the other hand, we show that the only continuous agreement rules that satisfy IIA are the max and the min.

In what follows, $\mathscr{I}$ will represent an interval of the real line $\mathbb{R}$.

Remark 8. At this stage, we point out that real intervals naturally arise to represent the set of alternatives in several contexts of Social Choice. Thus, X could be a set of monetary payoffs or, in a probabilistic scenario, X could represent the space of lotteries between two outcomes. In the first situation X can be identified as the real interval $[0, \infty)$. And in the second case, X can be identified as $[0, 1]$.

Notation. Let $F : \mathscr{I} \times \mathscr{I} \to \mathscr{I}$ be a bivariate map. For every $x \in \mathscr{I}$, F_x (respectively, F^x) stands for the *vertical* (respectively, *horizontal*) *restriction* of F, that is, $F_x(y) = F(x,y) \in \mathscr{I}$ (respectively, $F^x(y) = F(y,x) \in \mathscr{I}$).

We recall the concept of an idempotent function defined on $\mathscr{I}$. This concept will play a significant role in the sequel, in particular in the next Proposition 3.

Definition 6. A function $f : \mathscr{I} \to \mathscr{I}$ is said to be *idempotent* if $f(f(x)) = f(x)$, for every $x \in \mathscr{I}$.

Proposition 3. *Let $F : \mathscr{I} \times \mathscr{I} \to \mathscr{I}$ be a bivariate map.*

(i) F is unanimous if and only if $F_x(x) = F^x(x) = x$ for every $x \in \mathscr{I}$.
(ii) F is anonymous if and only if $F_x(y) = F^x(y)$ for every $x, y \in \mathscr{I}$.
(iii) F satisfies consensus if and only if, for every $x \in X$, both restrictions F_x and F^x are idempotent functions and, for each $z \in F(\mathscr{I} \times \mathscr{I})$, it holds that $F_z(z) = F^z(z) = z$.

Proof. Parts (i) and (ii) follow directly. So we prove only part (iii). Suppose that F satisfies consensus and let $x \in X$ be fixed. Then, we have that $F_x(F_x(y)) = F(x, F_x(y)) = F(x, F(x,y)) = F(x,y) = F_x(y)$ for every $y \in \mathscr{I}$. Also, we have that $F^x(F^x(y)) = F(F^x(y), x) = F(F(y,x), x) = F(y,x) = F^x(y)$, for every $y \in \mathscr{I}$. Therefore, F_x and F^x are both idempotent functions. Since x is an arbitrary element of $\mathscr{I}$, we have proved that F_x and F^x are both idempotent functions for every $x \in \mathscr{I}$. The fact that, for each $z \in F(\mathscr{I} \times \mathscr{I})$, $F_z(z) = F^z(z) = z$ follows directly from consensus.

Conversely, suppose that, for every $x \in \mathscr{I}$, F_x and F^x are both idempotent functions. Let $x, y \in \mathscr{I}$ be fixed. Then, we have that $F(x, F(x,y)) = F_x(F_x(y)) = F_x(y) = F(x,y)$, and also we have that $F(x,y) = F^y(x) = F^y(F^y(x)) = F(F^y(x), y) = F(F(x,y), y)$. Moreover, $F(F(x,y), F(x,y)) = F(x,y)$ since, by hypothesis, $F_z(z) = F^z(z) = z$, for every $z \in F(\mathscr{I} \times \mathscr{I})$. Therefore, F satisfies consensus.

Remark 9. It should be noted that the concepts introduced above can indeed be defined in a more abstract setting. As a matter of fact, Proposition 3 remains true if $\mathscr{I}$ is replaced by a nonempty choice set X.

Before presenting a basic definition of the most familiar notions involving monotonicity properties of real-valued bivariate functions, we recall that given $(x,y), (u,v) \in \mathscr{I} \times \mathscr{I}$, the notation $(x,y) \leq (u,v)$ means that both $x \leq u$ and $y \leq v$ hold. Similarly, $(x,y) < (u,v)$ means that both $(x,y) \leq (u,v)$ and $(x,y) \neq (u,v)$ hold. Finally, $(x,y) \ll (u,v)$ means that both $x < u$ and $y < v$ hold.

Definition 7. A bivariate map $F : \mathscr{I} \times \mathscr{I} \to \mathscr{I}$ is said to be:

(i) *Paretian (or non-decreasing)* if $(x,y) \leq (u,v)$ implies $F(x,y) \leq F(u,v)$, for every $x,y,u,v \in \mathscr{I}$.

(ii) *weakly Paretian* if $(x,y) \ll (u,v)$ implies $F(x,y) < F(u,v)$, for every $x,y,u,v \in \mathscr{I}$.

(iii) *strongly Paretian* if $(x,y) < (u,v)$ implies $F(x,y) < F(u,v)$, for every $x,y,u,v \in \mathscr{I}$.

(iv) *dictatorial* if either $F(x,y) = x$ for every $x,y \in \mathscr{I}$ holds, or $F(x,y) = y$ for every $x,y \in \mathscr{I}$ holds.

(v) *dichotomic* if for every $x \in \mathscr{I}$ the functions F_x and F^x are either constant or strictly increasing.

(vi) *continuous* if the inverse image of every Euclidean open subset of $\mathscr{I}$ is an open subset of $\mathscr{I} \times \mathscr{I}$, where $\mathscr{I} \times \mathscr{I}$ is endowed with the usual product (Euclidean) topology.

Remark 10. Notice that the monotonicity properties that appear in Definition 7 above are meaningful in the case that the choice set X is a set of monetary pay-offs.

Now we present a general theorem that allows us to derive certain impossibility results. Before a simple and useful lemma concerning strictly increasing real-valued idempotent functions is shown.

Lemma 1. *Let* $f : \mathscr{I} \to \mathscr{I}$ *be a strictly increasing idempotent function. Then,* $f(x) = x$, *for every* $x \in \mathscr{I}$. *(In other words, the identity function is the only one strictly increasing real-valued function that is idempotent.)*

Proof. Let $x \in \mathscr{I}$ arbitrarily be given. Let us see that $f(x) = x$. If, on the contrary, $f(x) \neq x$ then either $f(x) < x$ or $x < f(x)$. Assume that $f(x) \neq x$. Then, since f is strictly increasing, we have that $f(f(x)) < f(x)$. But, since f is idempotent, $f(f(x)) = f(x)$. So, we get $f(x) < f(x)$, which is a contradiction. The case $x < f(x)$ is handled in a similar way.

Theorem 3. *Let* $F : \mathscr{I} \times \mathscr{I} \to \mathscr{I}$ *be a bivariate map. Then the following assertions are equivalent:*

i) *F is dichotomic, unanimous and satisfies consensus.*
ii) *F is dictatorial.*

Proof. (ii) implies (i) is routine. So, we will concentrate on (i) implies (ii). Assume then that F is dichotomic, unanimous and satisfies consensus. Let us prove first that there cannot exist $x \in \mathscr{I}$ such that both F_x and F^x are constant functions. Indeed, suppose, by way of contradiction, that there is $x_0 \in \mathscr{I}$ for which both F_{x_0} and F^{x_0} are constant functions. Then, since F is unanimous and therefore $F_{x_0}(x_0) = x_0$, it holds that $F_{x_0}(y) = x_0 = F^{x_0}(y)$, for all $y \in \mathscr{I}$. Now, let $x_1 \in \mathscr{I}$ so that $x_0 < x_1$ (if any). The case $x_1 < x_0$ (if any) is similar. Then, $F_{x_1}(x_0) = F^{x_0}(x_1) = x_0 < x_1 = F_{x_1}(x_1)$. So, since, by hypothesis, F is dichotomic, it follows that F_{x_1} is a strictly increasing function. Now, since F satisfies consensus, by Proposition 3(iii), F_{x_1} is idempotent. Thus, by the previous lemma, F_{x_1} is the identity map (i.e., $F_{x_1}(y) = y$, for all $y \in \mathscr{I}$). In a similar way, we can prove that F^{x_1} is the identity map. Therefore, we have shown, in fact, that, for every $z \in \mathscr{I}$ such that $x_0 < z$ (if any), both functions F_z as well as F^z are the identity map. Let now consider three points $x_0, x_1, x_2 \in \mathscr{I}$ such that $x_0 < x_1 < x_2$ (if any). Then, since F_{x_1} is the identity map, it follows that $F(x_2, x_1) = F_{x_2}(x_1) = x_1$. Now, by definition, we have that $F(x_2, x_1) = F^{x_1}(x_2)$. So, $F^{x_1}(x_2) = x_1 \neq x_2$, which contradicts the fact, shown above, that F^{x_1} is the identity function. Therefore, there cannot exist $x \in \mathscr{I}$ such that both F_x and F^x are constant functions.

Using a similar argument to that employed above we can prove that there cannot exist $x \in \mathscr{I}$ such that both F_x and F^x are strictly increasing functions. (The proof of this assertion is left to the reader).

So, we have proved that, for every $x \in \mathscr{I}$, if F_x is a constant (respectively, the identity) function, then F^x is the identity (respectively, a constant) function. Suppose now that, for some $x_0 \in \mathscr{I}$, F_{x_0} is constant and F^{x_0} is the identity. Let us show that this situation leads to the conclusion $F(x, y) = x$, for all $x, y \in \mathscr{I}$ (in other words, F is dictatorial, the first individual acting as a dictator). Indeed, let $x_1 \in \mathscr{I}$ so that $x_0 < x_1$ (if any). Then $F_{x_1}(x_0) = F(x_1, x_0) = F^{x_0}(x_1) = x_1$, the last equality being true since F^{x_0} is the identity. Now, by unanimity, $F_{x_1}(x_1) = x_1$. So, $F_{x_1}(x_0) = x_1 = F_{x_1}(x_1)$, hence, since F is dichotomic, it follows that $F_{x_1}(y) = x_1$, for all $y \in \mathscr{I}$. The case $x_1 < x_0$ (if any) is similar leading to the same conclusion (i.e., $F_{x_1}(y) = x_1$, for all $y \in \mathscr{I}$). Thus, $F(x, y) = F_x(y) = x$, for all $x, y \in \mathscr{I}$.

Suppose now that, for some $x_0 \in \mathscr{I}$, F_{x_0} is the identity and F^{x_0} is constant. Arguing in a similar manner as above, it can be seen now that $F(x, y) = F_x(y) = y$, for all $x, y \in \mathscr{I}$. This ends the proof.

Theorem 3 immediately gives rise to the following corollaries.

Corollary 2. *There is no dichotomic bivariate map* $F : \mathscr{I} \times \mathscr{I} \to \mathscr{I}$ *that satisfies unanimity, anonymity and consensus.*

Proof. Just observe that dictatorial bivariate maps on $\mathscr{I}$ are not anonymous.

Corollary 3. *There is no strongly Paretian bivariate map* $F : \mathscr{I} \times \mathscr{I} \to \mathscr{I}$ *that satisfies unanimity and consensus.*

Proof. It is also a straightforward consequence of Theorem 3. Indeed, suppose that there is a bivariate map, say F, that is strongly Paretian, unanimous and satisfies consensus. Then, since, clearly, strongly Paretian implies dichotomic, it follows, by Theorem 3, that F is dictatorial. But neither of the two dictatorial bivarite maps are strongly Paretian. This contradiction provides the result.

Remark 11. (i) A careful glance at the proof of Theorem 3 above shows that the only bivariate map on $\mathscr{I}$ which satisfies consensus and has the additional property that all of its vertical restrictions are strictly increasing functions (respectively, all of its horizontal restrictions are strictly increasing functions) is dictatorial over the second (respectively, first) coordinate. That is, $F(x,y) = y$ for every $x, y \in \mathscr{I}$ (respectively, $F(x,y) = x$ for every $x, y \in \mathscr{I}$).

(ii) If strongly Paretian is relaxed to Paretian (or weakly Paretian) then the impossibility result does not hold true. For example, consider the dictatorial bivariate maps or the max/min functions.

It is interesting to search for some possibilities results based on certain natural properties, in addition to consensus, of the bivarite map. In [13], it was offered a characterization of the maximum rule (i.e., $F(x,y) = \max\{x,y\}$), for the case $\mathscr{I} = \mathbb{R}$, in terms of five properties; namely, continuity, unanimity, anonimity, consensus and upper-Pareto. A bivarite map $F : \mathbb{R} \times \mathbb{R} \to \mathbb{R}$ is said to be *upper-Paretian* if it is non-decreasing and for every $x, y \in \mathbb{R}$ there exists $u \in \mathbb{R}$ such that $y < u$ and $F(x,y) < F(x,u)$. It is not difficult to show that this latter characterization result remains true if $\mathbb{R}$ is replaced by a real interval $\mathscr{I}$. A much more easy result can be obtained if a more demanding property than consensus is required; namely the fulfilment of IIA. We now state this possibility result. Actually, we establish that the only continuous bivariate maps $F : \mathscr{I} \times \mathscr{I} \to \mathscr{I}$ that satisfy IIA are the max, the min and the dictatorial functions. In particular, we have that the only continuous agreement rules that satisfy IIA are the max and the min functions.

Theorem 4. *Let $F : \mathscr{I} \times \mathscr{I} \to \mathscr{I}$ be a bivariate map. Then the following conditions are equivalent:*

(i) F is continuous and satisfies IIA.
(ii) F is of one of the following forms:
    (1) $F(x,y) = x$, for every $x, y \in \mathscr{I}$.
    (2) $F(x,y) = y$, for every $x, y \in \mathscr{I}$.
    (3) $F(x,y) = \max\{x,y\}$, for every $x, y \in \mathscr{I}$.
    (4) $F(x,y) = \min\{x,y\}$, for every $x, y \in \mathscr{I}$.

Proof. It is straightforward to see that (ii) implies (i).

To prove the converse implication, (i) implies (ii), let $F : \mathscr{I} \times \mathscr{I} \to \mathscr{I}$ be a continuous bivariate map which satisfies IIA. Let $x \in \mathscr{I}$ be fixed and consider the vertical restriction F_x. Since F satisfies IIA, $F_x(y) \in \{x,y\}$ for all $y \in \mathscr{I}$. The continuity of F_x, together with IIA, clearly implies that F_x must be of one of the following types:

(1) $F_x(y) = x$, for every $y \in \mathscr{I}$.
(2) $F_x(y) = y$, for every $y \in \mathscr{I}$.
(3) $F_x(y) = y$, if $y \geq x$ and $F_x(y) = x$, if $y < x$.
(4) $F_x(y) = x$, if $y \geq x$ and $F_x(y) = y$, if $y < x$.

Now, the continuity of F (in two variables) clearly implies that if for some $x_0 \in \mathscr{I}$, F_{x_0} is of the type (i), i= 1 to 4, then F_x is of the type (i), for all $x \in \mathscr{I}$. Finally, it is straightforward to see that the situation for each of the four cases leads to the corresponding functional form given in the statement of the theorem.

As a direct consequence of Theorem 4 we obtain the following corollary.

Corollary 4. *Let* $F : \mathscr{I} \times \mathscr{I} \to \mathscr{I}$ *be a bivariate map. Then the following conditions are equivalent:*

(i) F is continuous, anonymous and satisfies IIA.

(ii) Either $F(x,y) = max\{x,y\}$ *(for all* $x,y \in \mathscr{I}$*), or* $F(x,y) = min\{x,y\}$ *(for all* $x,y \in \mathscr{I}$*).*

Remark 12. Theorem 4 strongly depends on the independence of irrelevant alternatives (IIA) condition imposed to F, since there are continuous bivariate maps $F : \mathscr{I} \times \mathscr{I} \to \mathscr{I}$ which satisfy consensus other than those belonging to the four types that appear in the statement of the theorem (for details, see [13]).

5   Extending the Consensus Equation to a Multivariate Context

Until now we have studied the consensus equation that involves only two factors in its definition. We now explore the extended consensus equation which means that we are going to consider the n-factors (or n-individuals) case. To that end, the next notation and definition are in order.

Notation. Let X be a set and let $n \in \mathbb{N}$. Let us denote by X^n the n-fold Cartesian product of X and consider a n-variate map $F : X^n \to X$. Let $\mathbf{x} = (x_j)_{j \in N} \in X^n$. In order to make the notation as simple as possible, let us denote by $\mathbf{x}^F \in X^n$ any of the 2^n elements of X^n derived from $\mathbf{x}$ in the following way: For every $j \in N$, $\mathbf{x}_j^F = x_j$, or $F(\mathbf{x})$.

Definition 8. A n-variate map $F : X^n \to X$ is said to satisfy the *extended consensus equation* if $F(\mathbf{x}^F) = F(\mathbf{x})$, for every $\mathbf{x} \in X^n$.

The following result states that if a n-variate unanimous map satisfies extended consensus then it can be fully described by a family of $(n-1)$-variate maps that satisfy extended consensus too, together with a kind of (weak) unanimity. So, the entire description of the class of bivariate maps that satisfy consensus is important since it allows us to also describe those that satisfy extended consensus for any number of factors (agents). Before presenting the result let us introduce the following notation.

Notation. Let $\mathbf{x} \in X^n$, $j \in N$ and $z \in X$ be given. Then by $\mathbf{x}_{+j}(z)$ we mean the following element of X^{n+1}: $\mathbf{x}_{+j}(z) = (\mathbf{x}_{+j}(z))_k = x_k$, if $k < j$, or z, if $k = j$, or x_{k-1}, if $k > j$. In words, the $j-1$ first components of $\mathbf{x}_{+j}(z)$ are the same as those of $\mathbf{x}$, the j-th component is z and the remaining components are those of $\mathbf{x}$ shifted one place on the right. Let now $\mathbf{x} \in X^n$ and $j \in N$ be given. Then $\mathbf{x}_{+j}$ will denote the element of X^{n-1} obtained by removing from $\mathbf{x}$ the j-th component while keeping the remaining components equal to those of $\mathbf{x}$.

In addition, 1_n will denote the vector of $\mathbb{R}^n$ with all the coordinates equal to one. Similarly, for any $x \in X$ given, $x1_n$ will stand for the element of X^n with all the components equal to x. Let $F : X^n \to X$ be a n-variate map. For every $x \in X$ and $j \in N$, denote by F_x^j the $(n-1)$-variate map defined as follows: $F_x^j(\mathbf{z}) = \mathbf{F}(\mathbf{z}_{+j}(x))$, for every $\mathbf{z} \in X^{n-1}$.

Once the above tedious notation has been introduced we are ready to offer the main result of this section.

Theorem 5. *A n-variate unanimous map $F : X^n \to X$ satisfies the extended consensus equation if and only if F_x^j does (for every $x \in X$, $j \in N$).*

Proof. Suppose first that F fulfils consensus. Let $x \in X$ and $j \in N$ be given and consider the $(n-1)$-variate map F_x^j. Assume, without loss of generality that $j = 1$. Then, for every $\mathbf{z} \in X^{n-1}$, it follows that $F_x^j(\mathbf{z}^{F_x^j}) = F(\mathbf{z}_{+j}^{F_x^j}(x)) = F(\mathbf{z}_{+j}(x)) = F_x^j(\mathbf{z})$, since F satisfies consensus. So, F_x^j fulfils consensus.

Conversely, assume now that, for each $x \in X$ and each $j \in N$, F_x^j satisfies consensus and let us prove that so F does. To that end, let $\mathbf{x} \in X^n$ be fixed and consider any of the 2^n elements $\mathbf{x}^F \in X^n$, as defined above. We distinguish between the two following cases: (i) There is at least one component of $\mathbf{x}^F$, say $\mathbf{x}_j^F \in N$, which is different from $F(\mathbf{x})$, or (ii) All the components of $\mathbf{x}^F$ are $F(\mathbf{x})$. If (i) occurs, then $F(\mathbf{x}^F) = F_{x_j}^j(\mathbf{x}_{-j}^F)$. Now observe that, for each $k \in N \setminus \{1\}$, the k-th component of $\mathbf{x}_{-j}^F$ is equal to x_k or equal to $F(\mathbf{x})$. So, since $F(\mathbf{x}) = F_{x_j}^j(\mathbf{x}_{-j})$ and, by hypothesis, $F_{x_j}^j$ satisfies consensus, it turns out that $F(\mathbf{x}^F) = F_{x_j}^j(\mathbf{x}_{-j}^F)^{F_{x_j}^j} = F(\mathbf{x})$. If (ii) happens, then the fact that $F(F(\mathbf{x}1_n)) = F(\mathbf{x})$ follows from the unanimity of F. So, the proof is ended.

Remark 13. It is interesting to study the functional form the of the unanimous n-variate maps that, in addition to fulfil the consensus equation, also satisfy natural conditions like anonymity or continuity. For the particular case $X = \mathbb{R}$, the class of unanimous n-variate maps that fulfil consensus plus continuity is closely related to the class of lattice polynomial functions (see [23] for a thorough discussion of these functions). Indeed, it is not difficult to see that a lattice polynomial function in $\mathbb{R}^n$ satisfies consensus, unanimity and continuity. Nevertheless, the class of real-valued functions defined on $\mathbb{R}^n$ that satisfy consensus, unanimity and continuity is larger than the class of lattice polynomial functions as the next example shows. Let $F : \mathbb{R}^2 \to \mathbb{R}$ be the function given by: $F(x,y) = x$, if $y \le 1$ and $x \ge y$, or $F(x,y) = x$, if $y \le 1$ and $x \le y$, or $F(x,y) = x$, if $y \ge 1$ and $x \ge 1$, or $F(x,y) = 1$, if $y \ge 1$ and $x \le 1$. Then it is straightforward to see that F so-defined satisfies consensus, unanimity and continuity. Actually, it can be shown that $F(x,y) = \max\{x, \min\{y, 1\}\}$.

If anonymity is added to the previous discussion then the class of lattice polynomial functions reduces to the so-called order statistics functions (for a discussion of this latter family, see also [23]). We conjecture that the class of real-valued functions defined on $\mathbb{R}^n$ that satisfy consensus, unanimity, anonymity and continuity agrees with the family of order statistics functions.

6 Further Comments

One of the achievements in [13] is showing that under unanimity plus anonymity, a new functional equation for bivariate maps (namely, the so-called equation of consensus, also analyzed in the present manuscript), is indeed equivalent to a weaker version of the associativity equation.

Throughout the present paper, we have not intended to solve the functional equation of consensus in the general case of bivariate maps F defined on a nonempty set X. Indeed, we may observe that the even more restrictive condition of associativity leads to a too wide set of possible solutions. In this direction, a glance at [3] may give us an idea of how large could be the set of solutions, even in relevant particular cases (e.g. : $X = \mathbb{R}$ or $X = [0,1]$).

In what concerns the consensus equation, it is important to point out that, under unanimity plus anonymity, any finite sequence of applications of F in which only the elements $x, y \in X$ are involved[2] always leads to $F(x,y)$. Viewing $F(x,y)$ as an agreement rule defined by means of a binary operation $*_F$ on X (i.e. $F(x,y) = x *_F y$, for every $x, y \in X$), the algebraic structure $(X, *_F)$ could be understood as being a weakening of the notion of a semigroup, that is called a magma in the specialized literature. This magma, namely the set X matched with the operation $*_F$, has the aforementioned property of simplification for finite sequences. (See [13] for further details).

In particular cases the operation $*_F$ is semi-latticial. In other special cases, it is a selector. And in some more restrictive cases, it corresponds to the idea of taking a maximum as proved in Corollary 4 above. Obviously, this fact of "taking a maximum" strongly agrees with the underlying idea of "reaching the best possible agreement" or "selecting the best possible option" commonly encountered in any process of aggregation of individual alternatives into a social one, typical of a wide variety of Social Choice contexts.

Acknowledgements. This work has been supported by the research projects ECO2008-01297 and MTM2009-12872-C02-02 (Spain).

References

[1] Aczél, J.: The associativity equation re-revisited. In: AIP Conf. Proc., vol. 707(1), pp. 195–203 (2004)

[2] Alcantud, J.C.R., de Andrés Calle, R., Cascón, J.M.: A unifying model to measure consensus solutions in a society. Math. Comput. Modelling (2012) (to appear), doi 10.1016/j.mcm.2011.12.020

[3] Alsina, C., Frank, M.J., Schweizer, B.: Associative functions: triangular norms and copulas. World Scientific Publishing Co. Pte. Ltd., Singapore (2006)

[4] Arrow, K.J.: Rational choice functions and orderings. Econometrica 26, 121–127 (1959)

[5] Arrow, K.J.: Social choice and individual values. John Wiley and Sons, New York (1963)

[6] Beliakov, G., Calvo, T., James, S.: On penalty-based aggregation functions and consensus. In: Herrera-Viedma, E., García-Lapresta, J.L., Kacprzyk, J., Fedrizzi, M., Nurmi, H., Zadrożny, S. (eds.) Consensual Processes. STUDFUZZ, vol. 267, pp. 23–40. Springer, Heidelberg (2011)

[7] Birkhoff, G.: Lattice theory, 3rd edn. American Mathematical Society, Providence (1967)

---

[2] An example could be $(((y *_F ((x *_F y) *_F x)) *_F (y *_F x)) *_F (y *_F y)) *_F x.$

[8] Bradley, R.: Reaching a consensus. Soc. Choice Welf. 29(4), 609–632 (2007)

[9] Bustince, H., Barrenechea, E., Calvo, T., James, S., Beliakov, G.: Consensus in multiexpert decision making problems using penalty functions defined over a Cartesian product of lattices. Information Fusion (2011), doi:10.1016/j.inffus.2011.10.002

[10] Calvo, T., Mesiar, R., Yager, R.R.: Quantitative weights and aggregation. IEEE Trans. on Fuzzy Systems 12, 62–69 (2004)

[11] Campión, M.J., Candeal, J.C., Catalán, R.G., De Miguel, J.R., Induráin, E., Molina, J.A.: Aggregation of preferences in crisp and fuzzy settings: functional equations leading to possibility results. Internat. J. Uncertain. Fuzziness Knowledge-Based Systems 19(1), 89–114 (2011)

[12] Candeal, J.C., De Miguel, J.R., Induráin, E., Olóriz, E.: Associativity equation revisited. Publ. Math. Debrecen 51(1-2), 133–144 (1997)

[13] Candeal, J.C., Induráin, E.: Bivariate functional equations around associativity. Aequat. Math. 84, 137–155 (2012)

[14] Castillo, E., Iglesias, A., Ruiz-Cobo, R.: Functional Equations in Applied Sciences. Elsevier, Amsterdam (2005)

[15] Cook, W.D., Seiford, L.M.: Priority ranking and consensus formation. Management Science 28, 1721–1732 (1978)

[16] Craigen, R., Páles, Z.: The associativity equation revisited. Aequat. Math. 37, 306–312 (1989)

[17] Davey, B.A., Priestley, H.A.: Introduction to lattices and order. Cambridge University Press, Cambridge (1990)

[18] García-Ferreira, S., Gutev, V., Nogura, T.: Extensions of 2-point selections. New Zealand J. Math. 38, 1–8 (2008)

[19] García-Lapresta, J.L.: Favoring consensus and penalizing disagreement in group decision making. Journal of Advanced Computational Intelligence and Intelligent Informatics 12, 416–421 (2008)

[20] Heitzing, J., Simmons, F.W.: Some chance for consensus: voting methods for which consensus is an equilibrium. Soc. Choice Welf. 38(1), 43–57 (2012)

[21] Ling, C.H.: Representation of associative functions. Publ. Math. Debrecen 12, 189–212 (1965)

[22] Marichal, J.L.: On the associativity functional equation. Fuzzy Sets and Systems 114(3), 381–389 (2000)

[23] Marichal, J.L., Mathonet, P.: On comparison meaningfulness of aggregation functions. J. Math. Psych. 45, 213–223 (2001)

[24] Mc Morris, F.R., Powers, R.C.: Consensus rules based on decisive families: The case of hierarchies. Math. Social Sci. 57, 333–338 (2009)

[25] Moulin, H.: Axioms for cooperative decision making. Econometric Society Monographs. Cambridge University Press, Cambridge (1988)

[26] Nash, J.: Equilibrium points in n-person games. Proc. Natl. Acad. Sci. USA 36(1), 48–49 (1950)

[27] Rescher, N.: Some observations on social consensus methodology. Theory and Decision 3(2), 175–179 (1972)

[28] Sen, A.K.: Collective choice and social welfare. Holden-Day, San Francisco (1970)

[29] Stupňanová, A., Kolesárová: Associative n-dimensional copulas. Kybernetika 47(1), 93–99 (2011)

[30] Wagner, C.G.: Consensus for belief functions and related uncertainty measures. Theory and Decision 26(3), 295–304 (1989)

A Discriminative Dynamic Index Based on Bipolar Aggregation Operators for Supporting Dynamic Multi-criteria Decision Making

Yeleny Zulueta, Juan Martínez-Moreno, Luis Martínez, and Macarena Espinilla

Abstract. While Multi-Criteria Decision Making (MCDM) models are focus on selecting the best alternative from a finite number of feasible solutions according to a set of criteria, in Dynamic Multi-Criteria Decision Making (DMCDM) the selection process also takes into account the temporal performance of such alternatives during different time periods. In this contribution is proposed a new discriminative dynamic index to handling differences in temporal behavior of alternatives, which are not discriminated in preceding dynamic approaches. An example is provided to illustrate the feasibility and effectiveness of the proposed index.

1 Introduction

A Multi-Criteria Decision Making (MCDM) problem consists of selecting the most desirable alternative from a given feasible set according to a set of criteria [12, 16]. As a matter of fact, MCDM problems could involve the current and past performance of alternatives, they are called Dynamic Multi-Criteria Decision Making (DMCDM) problems because the time dimension is considered [4, 8, 14].

DMCDM approaches are commonly focused on problems in which the final decision is performed based on all information collected at multiple time periods [8, 15, 18, 19, 24, 25]. However, they are not effective in handling situations including large sets of alternatives or criteria and changes of such sets over the time. Recently in [4] was introduced a framework for DMCDM that allows to overcome this weakness by means of a dynamic feedback mechanism. The crucial phase in the DMCDM framework is the selection of an appropriate associative aggregation

Yeleny Zulueta
University of Informatics Science, Carretera San Antonio de los Baños, Havana, Cuba
e-mail: yeleny@uci.cu

Juan Martínez-Moreno · Luis Martínez · Macarena Espinilla
University of Jaen, Campus Las Lagunillas, Jaen, Spain
e-mail: {jmmoreno,luis.martinez,mestevez}@ujaen.es

H. Bustince et al. (eds.), *Aggregation Functions in Theory and in Practise*,
Advances in Intelligent Systems and Computing 228,
DOI: 10.1007/978-3-642-39165-1_25, © Springer-Verlag Berlin Heidelberg 2013

operator for the computation of dynamic ratings due to its properties can highly modify the computing cost (e.g.: associativity) and obtain very different results attending to the type of reinforcement supported by the aggregation operator [13, 22]. However using any associative aggregation operator there are situations in which equal dynamic ratings are generated independently from the temporal performance of the alternatives. While the associativity property of the aggregation operator avoids the storing of all past alternatives (dynamic and non-dynamic) rating values, the lack of such information prevents a final decision based on temporal evolution of alternatives.

Therefore, this contribution proposes a novel *discriminative dynamic index* to extend the general approach in [4] such that the use of this index in the framework provides a temporal behavior differentiation of alternatives throughout time. The remaining of this paper is organized as follows. Sect. 2 reviews DMCDM approaches with special attention to the framework presented in [4]. In Sect. 3 it is introduced the new discriminative dynamic index to extend the initial approach. Sect. 4 shows an illustrative example and Sect. 5 concludes the paper.

2 Dynamic Multi-criteria Decision Making Approaches

In [3] are stated three common characteristics for a DMCDM problem: alternatives are not fixed, criteria are not fixed and the temporal profile of an alternative matters for comparison with other ones. To deal with decision making in dynamic environments, some authors have proposed different approaches [8, 15, 18, 19, 24, 25] that commonly model the problem as a three-dimensional decision matrix which is firstly transformed into conventional two-dimensional decision matrix by aggregating the time dimension and next is solving the problem through traditional MCDM models (or viceversa).

As in MCDM, an important issue in DMCDM is the selection of the aggregation operator (see [2] for a formal definition) because it directly impacts output values as well as the final ranking of alternatives. Some proposals have presented time dependent aggregation operators to deal with the information provided at different periods. Xu developed in [18] the concept of dynamic weighted averaging operator, and introduced some methods to obtain the associated weights, while in [19] the dynamic intuitionistic fuzzy weighted averaging operator and the uncertain dynamic intuitionistic fuzzy weighted averaging operator is defined.

Previous studies are focused on decision making problems in which the original decision information is usually collected at different time periods and a final decision is needed. Therefore, they are dynamic because the temporal profile of alternatives is considered for such final decision. However, there are other MCDM problems in which different, separated and interlinked decisions are taken either frequently, or just at the end of the process. In such context, it is remarkable the framework for DMCDM recently introduced in [4].

While most of the revised approaches provide solutions based on specific MCDM techniques oriented to problems dealing with specific types of information and where the final decision is performed using all that information collected at multiple periods; in [4] it is properly formalized the DMCDM, by extending the classic MCDM model, in a general framework operating without the need of storing all past information. Such framework is suitable for any dynamic problem, including consensus problems or situations requiring several steps before reaching a final decision. It is revised in further detail below.

2.1 The General Framework

Some basic notations from the original framework [4] are reviewed in the following.

Let $T = \{1, 2, \ldots\}$ be the (possibly infinite) set of discrete decision moments, and A_t the set of available alternatives at each decision moment $t \in T$.

At each time period $t \in T$, for each available alternative $a \in A_t$, a *non-dynamic rating* $R_t(a) \in [0, 1]$ is computed. It is usually obtained by using an aggregation operator $Agg_1 : [0, 1]^n \to [0, 1]$, that combines the assessments of all criteria, $M_t = \{m_1, \ldots, m_n\}$ according to their weights $w_t \in [0, 1]^n, \sum_{w \in w_t} w = 1, \forall t \in T$.

The information about the set of alternatives over time is carried out from one iteration to another in the historical set. Depending on the specific characteristics of each dynamic problem we may fix a *retention policy* that is the rule for selecting alternatives to be remembered in the H_t, which is defined as:

$$H_0 = \emptyset, \qquad H_t = \bigcup_{t' \leq t} A_{t'}, \quad t, t' \in T. \tag{1}$$

The dynamic nature of the decision process is supported by an evaluation function $E_t(a)$ it is defined for each $t \in T$ as:

$$E_t : A_t \cup H_{t-1} \to [0, 1]$$

$$E_t(a) = \begin{cases} R_t(a), & a \in A_t \setminus H_{t-1} \\ Agg_2(E_{t-1}(a), R_t(a)), & a \in A_t \cap H_{t-1} \\ E_{t-1}(a), & a \in H_{t-1} \setminus A_t \end{cases} \tag{2}$$

Being $Agg_2 : [0, 1]^n \to [0, 1]$ an associative aggregation operator that can apply different types of reinforcements to the alternatives according to the attitudinal character of the decision making problem.

Aggregation operator for scoring alternatives in the non-dynamic part (Agg_1) is completely independent from one used in evaluation function of the dynamic part (Agg_2). It is worth noting that the dynamic rating computation requires the associativity property for the aggregation operator Agg_2, to ensure that repeated application of the aggregation function will generate, at every particular decision moment, the same result as application over the whole set of past non-dynamic ratings. Furthermore it is suggested that Agg_2 should fulfill the reinforcement property [13, 22] in order to strength high or low ratings in the dynamic context.

2.2 Drawbacks on Dynamic Evaluation Function Performance

The associativity property of Agg_2 avoids the storing of all past alternatives (dynamic and non-dynamic) rating values and it is simple to calculate the effect of adding new arguments to the aggregation. As stated in [21] this can be seen as a kind of Markovian property in which the new aggregated value just depends on the previous aggregated value and the new argument. However, this advantage brings out that the original framework outputs equal dynamic ratings for different alternatives without a discrimination about their temporal profile because associativity property does not allow to distinguish the order of such previous and new aggregated values.

Remark: This drawback arises from the associativity property of the aggregation operator therefore it appears using any associative aggregation operator.

Without loss of generality and for the sake of simplicity, this problem is illustrated in the following situation in which a decision maker wants to select the best option from alternatives $a1$ and $a2$ considering the retention policy of accumulating all alternatives in historical set. The dynamic ratings are calculated using the probabilistic sum operator (which exhibits an upward reinforcement) in order to corroborate the tendency of previous high non-dynamic ratings. Table 1 shows the results during five decision periods.

Table 1 Results obtained for alternatives with different temporal profile

Alternative	$R_1 = E_1$	R_2	E_2	R_3	E_3	R_4	E_4	R_5	E_5
$a1$	0.100	0.800	0.820	0.900	0.982	0.200	0.996	0.910	0.999
$a2$	0.900	0.800	0.980	0.100	0.982	0.200	0.996	0.910	0.999

At $t = 3$, $a1$ increases its rating while $a2$ decreases it, however both obtain the same dynamic rating $(E_3(a1) = E_3(a2))$. At $t = 4$, the rating of $a1$ decreases and the rating of $a2$ increases, but still both obtain the same dynamic rating $(E_4(a1) = E_4(a2))$. Eventually at $t = 5$ the rating of both alternatives performances the same increment and the dynamic rating is also the same $(E_5(a1) = E_5(a2))$. At independent decision periods $t = 3, 4, 5$, the decision maker cannot choose the best alternative just based on the dynamic rating because:

1. Alternatives obtain equal dynamic rating although they perform different rating evolution.
2. Alternatives obtain equal dynamic rating despite they perform opposed rating evolution.
3. Alternatives obtain equal dynamic rating though all of them perform an increasing evolution or decreasing rating evolution.

Different perspectives to solve the problem can be assumed. From a *static perspective*, the decision maker can select the alternative with highest rating at the current period but this contradictorily implies to *loss the dynamic perspective* of the DM-CDM problem.

To overcome this drawback, our aim in this contribution is to extend the original framework formalizing a new dynamic index that allows the decision maker to discriminate the best alternative according to the rating changes behavior throughout time.

3   A Discriminative Dynamic Index for DMCDM

To keep the *dynamic perspective* of the decision making problem when the situations pointed out in Sect. 2.2 arises, seems logic and suitable to find a solution in which *the temporal profile of an alternative matters for comparison with other alternatives*, as stated in Sect. 2.

To that end, we improve the resolution procedure for DMCDM, as can be seen in Figure 1, by performing a new aggregation process for computing a discriminative dynamic index that allows to distinguish alternatives and consequently obtain rankings for supporting dynamic decisions.

In this general resolution procedure the first step is essentially carried out through MCDM traditional methods. The second step lies on the DMCDM approach previously reviewed. The third step consists of computing the discriminative dynamic index and is performed just if equal dynamic ratings values are generated in the second step. These tree steps will finally enable to obtain a final ranking of alternatives.

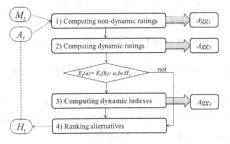

Fig. 1 Improved DMCDM resolution procedure

3.1   Computation of the Discriminative Dynamic Index

In this subsection we present in detail how to compute the discriminative dynamic index to perform steps 3) and 4) from the improved DMCDM resolution procedure.

Definition 1. The change in rating, $D_t(a)$, is the difference between the ratings at the current and previous period and is defined as:

$$D_t(a) = \begin{cases} 0, & t = 1 \\ R_t(a) - R_{t-1}(a), & t > 1. \end{cases} \tag{3}$$

Since $R_t(a), R_{t-1}(a) \in [0,1]$, the rating increment/decrement, $D_t(a)$, at each decision period is assessed in a *bipolar scale* $D_t(a) \in [-1,1]$ [5]. In which 0 is so-called the *neutral* element that represents no change in rating from period $t-1$ to t.

The rating change $D_t(a)$, just encloses the rating behavior from $t-1$ to t, hence it is necessary to formalize a dynamic mechanism that encloses all rating changes during all considered periods.

The benefits of computing final results without storing all previous values (through the associativity) and additionally modulating the importance of these values in such final results (through the reinforcements) are also used in the discriminative dynamic index proposal, $I_t(\cdot)$, due to its features:

- *Dynamic*: it must represent the rating change over time without storing all of them.
- *Customizable*: it should be able to model different behaviors regarding alternative rating decrements or increments over different periods.

Definition 2. Let $D_t(a)$ be the change in rating of an alternative a at a decision period t and $Agg_3 : [-1,1]^2 \to [-1,1]$ be a bipolar aggregation operator, the discriminative dynamic index, which represents the rating behavior of the alternative until t, is defined as:

$$I_t : A_t \cup H_{t-1} \to [-1,1]$$

$$I_t(a) = \begin{cases} D_t(a), & a \in A_t \setminus H_{t-1} \\ Agg_3(I_{t-1}(a), D_t(a)), & a \in A_t \cap H_{t-1} \\ I_{t-1}(a), & a \in H_{t-1} \setminus A_t. \end{cases} \tag{4}$$

The index $I_t(a)$ performance depends on the alternative, a as:

- if $a \in A_t \setminus H_{t-1}$ then its discriminative dynamic index $I_t(a)$ is the rating change $D_t(a)$,
- if $a \in A_t \cap H_{t-1}$, then its discriminative dynamic index is computed by Agg_3 that aggregates the discriminative dynamic index in the previous iteration with the current rating change, $Agg_3(I_{t-1}(a), D_t(a))$,
- if $a \in H_{t-1} \setminus A_t$, then its discriminative dynamic index is obtained from previous iteration, $I_t(a) = I_{t-1}(a)$.

Therefore, if different alternatives obtain equal dynamic rating, $E_t(.)$ at a period t, the final ranking will be generated considering the discriminative dynamic index values $I_t(.)$ that will reflect a *dynamic perspective*.

The choice of the aggregation operator Agg_3 will depend on the decision makers' attitude regarding the dynamic rating change but independent of the others aggregation operators as Agg_1 and Agg_2.

Table 2 summarizes the key features of aggregation operators used in the three aggregation processes illustrated in Figure 1 which are applied in the resolution procedure of the DMCDM improved approach. It is noteworthy to point out that

Table 2 Characterization of the aggregation operators to be used in DMCDM

Feature	Agg_1	Agg_2	Agg_3
Definition	$[0,1]^n \to [0,1]$	$[0,1]^2 \to [0,1]$	$[-1,1]^2 \to [-1,1]$
Required Property		Associativity	Associativity, Bipolarity
Desired Property		Reinforcement	Reinforcement

there is a key difference between the characterization of Agg_2 and Agg_3: Agg_3 must deal with values in a *bipolar scale* $[-1,1]$ meanwhile Agg_2 operates in $[0,1]$.

Consequently it is necessary to extend the latter to the bipolar scale [6] in $[-1,1]$ in which a remarkable point, e, of the interval plays an specific role as a neutral or an absorbant element. This fact leads to a *bipolar aggregation* in which the key feature is the different effects of arguments above and below e on the aggregated value [23].

Uninorms [17] satisfies this characterization but in $[0,1]$. A uninorm U is a commutative, associative and increasing binary operator with a neutral element $e \in [0,1]$.

In [9, 10, 11] the authors development the topic of "pseudo-operations". Pseudo-addition and pseudo-multiplication are examples of them. In [6] was proposed a rescaling to consider $[-1,1]$, such that, given a continuous $S : [0,1]^2 \to [0,1]$ t-conorm, the *symmetric pseudo-addition* $\oplus$ is a binary operation on $[-1,1]$ defined by:

R1 For $x,y \geq 0$: $x \oplus y = S(x,y)$.
R2 For $x,y \leq 0$: $x \oplus y = -S(-x,-y)$.
R3 For $x \in [0,1[, y \in]-1,0]$: $x \oplus y = x \ominus_S (-y)$. Moreover, $1 \oplus (-1) = 1$ or -1.
R4 For $x \leq 0, y \geq 0$: just reverse x and y.

The structure of the binary operation $\oplus$ is closely related to uninorms. From the point of view of bipolar scales, the interval $[-1,1]$ is viewed as the union of two unipolar scales.

Proposition 1. $T = \oplus_{/[-1,0]^2}$ is a t-norm on $[-1,0]$ (i.e., in particular $T(x,0) = x$, for every $x \in [-1,0]$), $S = \oplus_{/[0,1]^2}$ is a t-conorm on $[0,1]$ and H is an average function $H = \oplus_{/[-1,0] \times [0,1] \cup [0,1] \times [-1,0]}$. They have the following properties:

- If $x,y \in [0,1]$, then $x \oplus y = S(x,y) \geq \max\{x,y\}$.
- If $x,y \in [-1,0]$, then $x \oplus y = T(x,y) \leq \min\{x,y\}$.
- If $-1 \leq y \leq 0 \leq x \leq 1$, then $y \leq x \oplus y = H(x,y) \leq x$.

The previous proposition provides a performance that can be interpreted as *attitudes to deal with the ratings changes*:

- Optimistic: when both values are positive the aggregation acts as an upward reinforcement.
- Pessimistic: when both values are negative, it acts as a downward reinforcement.
- Averaging: when one value is negative and the another positive, it acts as an averaging operator.

The aggregation function $\oplus$ exhibits conjunctive behavior on $[-1,0]$ and disjunctive behavior on $[0,1]$. On the rest of the domain the behavior is averaging.

Let S be a strict t-conorm S with additive generator $s : [0,1] \to [0,\infty]$ and $g : [-1,1] \to [0,\infty]$ the symmetric extension of s, i.e.,

$$g(x) = \begin{cases} s(x), & x \geq 0 \\ -s(-x), & x < 0 \end{cases} \tag{5}$$

It is possible to rescal $\oplus$ to a binary operator U on $[0,1]$ such that U is a generated uninorm operator. Then, $x \oplus y = g^{-1}(g(x) + g(y))$ for any $x,y \in [-1,1]$. We also introduce another function $u : [0,1] \to [-\infty,\infty]$ defined by $u(x) = g(2x-1)$ that is strictly increasing and satisfies $u(\frac{1}{2}) = 0$. Then $U(z,t) = u^{-1}(u(z) + u(t))$ for any $z,t \in [0,1]$. U is an uninorm that is continuous (except in $(0,1)$ and $(1,0)$), is strictly increasing on $]0,1[^2$ and has neutral element $\frac{1}{2}$. Moreover, the induced t-norm T_U is the dual of S.

Such an operator should be used if the decision maker's attitude is influenced by the number of increment or decrement ratings received. Particularly, when all the attributes' ratings are positive, the more these there are, the more positive the agent becomes in its aggregation. That is similar for negative values and when conflict occurs, the ratings are aggregated in a risk-neutral way.

4 Illustrative Example

A high-technology manufacturing company desires to select at five different periods, suitable material supplier to purchase the key components of products. There are six candidates for initial evaluation but at successive periods, there will be additional suppliers while others will be unavailable due to market conditions. The company is interested about supplier's evolution and considers the following elements:

- *Criteria*: quality (m_1), delivery performance (m_2), price (m_3) and technological capability (m_4).
- *Retention policy* keeps all alternatives from A_t to H_t.
- *Non-dynamic rating* is computed with the weighted sum operator, using the weighting vector $w_t = (0.15, 0.20, 0.25, 0.40), \forall t \in T$.
- *Dynamic rating* is computed with the probabilistic sum operator .
- *Discriminative dynamic index* is computed with the Van Melle's combining function $C : [-1,1]^2 \to [-1,1]$ modified in [17] as:

$$C(x,y) = \begin{cases} S(x,y) = x+y-xy, & \text{if } \min\{x,y\} \geq 0 \\ T(x,y) = x+y+xy, & \text{if } \max\{x,y\} \leq 0 \\ H(x,y) = \frac{x+y}{1-\min\{|x|,|y|\}} & \text{otherwise} \end{cases} \tag{6}$$

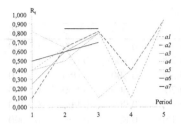

Fig. 2 Rating behavior of suppliers

4.1 The Resolution Procedure and Index Performance

In order to clarify the suppliers behavior Figure 2 depicts their ratings over the five periods. The difference between time events is not a variable neither for the index nor for the original approach.

As illustrated in Figure 1, at each period the non-dynamic and dynamic ratings are computed. These results are shown in Table 3, columns "R_t" and "E_t" respectively. To better understand the resolution procedure, following we focus on the discriminative dynamic index computations and performance. Index values are shown in Table 3, column "I_t".

Period t=1: Here it is not necessary to compute the dynamic ratings.

Period t=2: Dynamic ratings for a_3 and a_4 are equal therefore we have to compute their discriminative dynamic index, for instance:
$I_2(a_3) = C(D_2(a_3), I_1(a_3))$
$D_2(a_3) = R_2(a_3) - R_1(a_3) = 0.3500$ and $I_1(a_3) = D_1(a_3) = 0$ then
$I_2(a_3) = 0.35000$.
Note that for both alternatives the index evidences an *optimistic attitude*.

Period t=3: The discriminative dynamic index is computed for: a_2 and a_5 that present opposed rating evolution; and for a_3, a_4 and a_6 that present increasing rating evolution. The index attitude for a_2 is *optimistic* while for a_5 is *pessimistic*.

Period t=4: All suppliers obtain same dynamic ratings. The discriminative dynamic index shows an *averaging attitude* because a_1 and a_2 present increasing ratings in temporal profile but decreasing one at the current period while a_5 presents the inverse situation. This *averaging attitude* compensates both values but does not allow ignore rating decrements at current or previous periods.

Period t=5: All the suppliers not only obtain equal dynamic rating but also a_2 and a_5 present equal improvement from $t = 4$ to $t = 5$: $D_5(a_2) = D_5(a_5)$. Despite $D_5(a_1) > D_5(a_2)$, their indexes are $I_5(a_1) < I_5(a_2)$ because $I_4(a_1)$ was negative. Furthermore, as $D_5(a_2) = D_5(a_5)$ and $I_4(a_5) < 0$, $I_4(a_2) > 0$ then $I_5(a_2)$ is better than $I_5(a_5)$. Therefore the dynamic index provides for a_1 and a_5 an *average attitude* while for a_2 presents an *optimistic attitude*.

Table 3 Results at each period

Period	A_t	m_1	m_2	m_3	m_4	R_t	E_t	I_t
	a_1	0.700	0.900	0.100	0.300	0.4000	0.40000	-
	a_2	0.100	0.200	0.100	0.050	0.1000	0.10000	-
$t = 1$	a_3	0.200	0.300	0.500	0.050	0.2500	0.25000	-
	a_4	0.100	0.200	0.500	0.500	0.4000	0.40000	-
	a_5	0.900	0.950	0.900	0.900	0.8200	0.82000	-
	a_6	0.450	0.150	0.850	0.150	0.5000	0.50000	-
	a_1	0.300	0.600	0.750	0.600	0.6450	0.78700	-
	a_2	0.150	0.200	0.900	0.800	0.6450	0.68050	0.350000
	a_3	0.900	0.550	0.400	0.700	0.6000	0.68050	0.100000
$t = 2$	a_4	0.450	0.150	0.850	0.150	0.5000	0.70000	-
	a_5	0.900	0.800	0.410	0.400	0.6450	0.93610	-
	a_6	0.800	0.700	0.200	0.800	0.6000	0.80000	-
	a_7	0.800	0.700	0.900	0.900	0.8500	0.85000	-
	a_1	0.700	0.650	0.800	0.950	0.8200	0.96166	-
	a_2	0.800	0.700	0.800	0.900	0.8200	0.94249	0.624625
	a_3	0.500	0.800	0.900	0.800	0.8000	0.94000	0.480000
$t = 3$	a_4	0.900	0.900	0.900	0.650	0.8000	0.94000	0.370000
	a_5	0.100	0.200	0.100	0.050	0.1000	0.94249	-0.624625
	a_6	1.000	0.500	1.000	0.500	0.7000	0.94000	0.190000
	a_7	0.800	0.700	0.900	0.900	0.8500	0.97750	-
	a_1	0.200	0.100	0.050	0.100	0.1000	0.96500	-0.550472
$t = 4$	a_2	0.100	0.200	0.500	0.500	0.4000	0.96500	0.352802
	a_5	0.800	0.550	0.300	0.300	0.4000	0.96500	-0.463750
	a_1	0.950	0.950	1.000	0.900	0.9450	0.99800	0.655194
$t = 5$	a_2	0.950	0.850	1.000	0.950	0.9450	0.99800	0.705525
	a_5	0.850	0.950	0.900	1.000	0.9450	0.99800	0.151515

4.2 Ranking Alternatives and Results Analysis

In Table 4 is depicted the summary of rankings obtained with the original DMCDM framework as well as with the improved one using the new index.

Table 4 Suppliers rankings

Period	Original framework	Discriminative dynamic index
$t = 1$	$a_5 \succ a_6 \succ a_1 = a_4 \succ a_3 \succ a_2$	-
$t = 2$	$a_5 \succ a_7 \succ a_6 \succ a_1 \succ a_3 = a_4 \succ a_2$	$a_5 \succ a_7 \succ a_6 \succ a_1 \succ a_3 \succ a_4 \succ a_2$
$t = 3$	$a_7 \succ a_1 \succ a_2 = a_5 \succ a_3 = a_4 = a_6$	$a_7 \succ a_1 \succ a_2 \succ a_5 \succ a_3 \succ a_4 \succ a_6$
$t = 4$	$a_1 = a_2 = a_5$	$a_2 \succ a_5 \succ a_1$
$t = 5$	$a_1 = a_2 = a_5$	$a_2 \succ a_1 \succ a_5$

In Sect. 2.2 were summarized situations in which the original framework can not discriminate alternatives consequently the *main objective of the DMCDM was not accomplished* since the *most desirable alternatives can not be selected considering their current and past performance.*

However it is remarkable that in all circumstances the discriminative dynamic index ranks the alternatives taking into account the desired *pessimistic, averaging* or *optimistic* attitude. Consequently our proposal support the DMCDM by improving the original framework in such way that the crucial purpose of DMCDM is achieved.

5 Conclusion

In this contribution, we focused on the DMCDM problems. To support consistent decisions in cases in which the framework in [4] is not effective, we introduced a novel discriminative dynamic index in a general resolution procedure for DMCDM. It uses an aggregation process based on associative bipolar operators. This features allows to exploit their associativity property to represent the rating behavior of alternatives over different periods as well as to model effects of rating changes above and below neutral element on the final aggregated value.

Acknowledgements. This work is partially supported by the Research Project TIN-2012-31263 and ERDF.

References

[1] De Baets, B., Fodor, J.: Van melle's combining function in mycin is a representable uninorm: An alternative proof. Fuzzy Sets and Systems 104, 133–136 (1998)

[2] Beliakov, G., Pradera, A., Calvo, T.: Aggregation Functions: A Guide for Practitioners. STUDFUZZ, vol. 221. Springer, Heidelberg (2008)

[3] Campanella, G., Pereira, A., Ribeiro, R.A., Varela, M.L.R.: Collaborative dynamic decision making: A case study from B2B supplier selection. In: Hernández, J.E., Zarate, P., Dargam, F., Delibašić, B., Liu, S., Ribeiro, R. (eds.) EWG-DSS 2011. LNBIP, vol. 121, pp. 88–102. Springer, Heidelberg (2012)

[4] Campanella, G., Ribeiro, R.: A framework for dynamic multiple-criteria decision making. Decision Support Systems 52(1), 52–60 (2011)

[5] Dubois, D., Prade, H.: An introduction to bipolar representations of information and preference. Int. J. Intell. Syst. 23(8), 866–877 (2008)

[6] Grabisch, M., De Baets, B., Fodor, J.: The quest for rings on bipolar scales. International Journal of Uncertainty, Fuzziness, and Knowledge-Based Systems 12(4), 499–512 (2004)

[7] Grabisch, M.: Aggregation on bipolar scales. In: de Swart, H.C.M., Orlowska, E., Schmidt, G., Roubens, M. (eds.) Theory and Applications of Relational Structures as Knowledge Instruments II, pp. 355–371 (2006)

[8] Lin, Y., Lee, P., Ting, H.: Dynamic multi-attribute decision making model with grey number evaluations. Expert Systems with Applications 35, 1638–1644 (2008)

[9] Pap, E.: g-calculus. Rad. Prirod.-Mat. Fak. Ser. Mat. 23(1), 145–156 (1993)

[10] Pap, E.: Pseudo-additive measures and their applications. In: Handbook of Measure Theory. Elsevier Science (2002)

[11] Pap, E.: Applications of the generated pseudo-analysis to nonlinear partial differential equations. Contemporary Mathematics 377, 239–259 (2005)

[12] Pedrycz, W., Ekel, P., Parreiras, R.: Models and algorithms of fuzzy multicriteria decisionmaking and their applications. Wiley, West Sussex (2011)

[13] Ribeiro, R.A., Pais, T.C., Simões, L.F.: Benefits of Full-Reinforcement Operators for Spacecraft Target Landing. In: Greco, S., Pereira, R.A.M., Squillante, M., Yager, R.R., Kacprzyk, J. (eds.) Preferences and Decisions. STUDFUZZ, vol. 257, pp. 353–367. Springer, Heidelberg (2010)

[14] Saaty, T.: Time dependent decision-making; dynamic priorities in the ahp/anp: Generalizing from points to functions and from real to complex variables. Mathematical and Computer Modelling 46(78), 860–891 (2007)

[15] Teng, D.: Topsis method for dynamic evaluation of hi-tech enterprises strategic performance with intuitionistic fuzzy information. Advances in Information Sciences and Service Sciences 3(11), 443–449 (2011)

[16] Triantaphyllou, E.: Multi-criteria decision making methods: A comparative study. In: Applied Optimization. Applied Optimization Series, vol. 44. Kluwer Academic (2000)

[17] Tsadiras, A., Margaritis, K.: The mycin certainty factor handling function as uninorm operator and its use as a threshold function in artificial neurons. Fuzzy Sets and Systems 93, 263–274 (1998)

[18] Xu, Z.: On multi-period multi-attribute decision making. Knowledge-Based Systems 21(2), 164–171 (2008)

[19] Xu, Z., Yager, R.: Dynamic intuitionistic fuzzy multi-attribute decision making. International Journal of Approximate Reasoning 48(1), 246–262 (2008)

[20] Yager, R., Rybalov, A.: Uninorm aggregation operators. Fuzzy Sets and Systems 80, 111–120 (1996)

[21] Yager, R., Rybalov, A.: A note on the incompatibility of openness and associativity. Fuzzy Sets and Systems 89, 125–127 (1997)

[22] Yager, R., Rybalov, A.: Full reinforcement operators in aggregation techniques. IEEE Transactions on Systems, Man, and Cybernetics, Part B: Cybernetics 28(6), 757–769 (1998)

[23] Yager, R., Rybalov, A.: Bipolar aggregation using the Uninorms. Fuzzy Optimi Decis Making 10(1), 59–70 (2010)

[24] Yao, S.: A distance method for multi-period fuzzy multi-attribute decision making, pp. 1–4 (2010)

[25] Zhang, L., Zou, H., Yang, F.: A dynamic web service composition algorithm based on TOPSIS. Journal of Networks 6(9), 1296–1304 (2011)

Social Choice Voting with Linguistic Preferences and Difference in Support

Patrizia Pérez-Asurmendi and Francisco Chiclana

Abstract. A new aggregation rule in social choice voting with linguistic intensities of preferences between pairs of alternatives is introduced. This new aggregation rule leads to the definition of linguistic majorities with difference in support under which an alternative defeats another if the first one reaches a concrete collective preference support fixed before the election process. This new rule extends existing rules from the context of crisp and reciprocal [0,1]-valued preferences to the framework of linguistic preferences. Both possible representation formats of linguistic information are addressed: fuzzy sets and 2–tuples. The linguistic majorities constitute a class of majority rules because they generalize all the possible majority rules by adjusting the required threshold of support.

1 Introduction

Decision making problems deal with the social choice of the best alternative from a set of feasible alternatives taking into account the individual preferences of the voters of a social group. Once the individual preferences are collected, these problems could be tackled following a direct or an indirect approach. The indirect approach is based on the application of a rule to aggregate individual preferences into a collective preference, followed by a selection process to choose the final social decision. The aggregation rule is superfluous in the direct approach. The type of aggrega-

Patrizia Pérez-Asurmendi
Universidad de Valladolid, Spain
e-mail: patrizia.perez@eco.uva.es

Francisco Chiclana
De Montfort Universitiy, Leicester, UK
e-mail: chiclana@dmu.ac.uk

H. Bustince et al. (eds.), *Aggregation Functions in Theory and in Practise*,
Advances in Intelligent Systems and Computing 228,
DOI: 10.1007/978-3-642-39165-1_26, © Springer-Verlag Berlin Heidelberg 2013

tion rule to apply is therefore crucial in the final choice of indirect decision making processes, an issue that is central to the present paper.

Pairwise individual preferences could be expressed using either numerical or linguistic values. Classical voting systems assume crisp preferences, i.e. preferences are represented using numerical information values of either $\{-1, 0, -1\}$ or $\{0, 0.5, 1\}$. Thus, a voter reports indifference between the alternatives in comparison or the preferred alternative, without quantifying the intensity of preference. A natural extension consists in allowing voters to provide intensities of preference [20, 18] by means of numerical values in the unit interval, i.e. providing values that belong to $[0, 1]$. Alternatively, voters are allowed to express their preferences using a set of finite linguistic values or labels. Voters could feel more comfortable using linguistic assessments because they are not pressed to quantify their opinions precisely with numerical values. Such facility could also increase the satisfaction of the voters during the voting process and also their concern with the final outcome. Additionally, it is argued that the vagueness of the linguistic labels represents better the human capacity of making decisions without using complex calculus process than numerical values [1, 5, 10, 16]. In this paper, it is assumed that preferences between pair of alternatives will be represented by means of linguistic labels.

This paper focuses on majority voting systems that aggregate individual preferences. In the case of crisp preferences, simple majority rule [17] stands out amongst the different majority rules. Under this rule, an alternative defeats another one when the votes cast for the first one exceed the votes cast for the second one. Simple majority rule can be seen as the most decisive aggregation rule because the requirement for the indifference state is quite strong given that both alternatives have to reach exactly the same number of votes. However, it is also very unstable because the requirement to be the winner is quite weak because the social choice could be reverted with the change of few votes; in some cases even the change of a single vote would be sufficient to revert the social choice. To overcome this drawback, different rules are defined via the strengthening of the requirement to declare an alternative as the winner. Among these rules are: unanimous majority, absolute majority, qualified majorities, and majorities based on difference of votes [4, 3, 19, 12, 15, 9].

Majorities based on difference of votes allow to control the support required for an alternative to be the winner by fixing, in advance of the election process, the difference of votes. This allows for the indifference between two alternatives to be declared in a higher number of cases than under simple majority rule. In fact, the indifference state could be enlarged as much as desired, and therefore provides a greater level of flexibility in decision making processes. These majorities constitute a class of voting rules because they generalize any other majority by changing the difference of votes required to an alternative with respect to another one to be the winner [12].

The extension of these rules to the context of $[0,1]$-valued preferences is known as majorities based on difference in support [13]. Under these rules, a specific difference in support is fixed in advanced for an alternative to be declared as the winner.

The aim of this paper is to fill the gap between majorities based on difference of votes and majorities based on difference in support by providing a new rule when preferences are of a linguistic nature. Linguistic majorities with difference in support will be developed for the two different methodologies possible for representing linguistic preferences: fuzzy sets [23] and 2–tuples [7]. Consequently, two different linguistic rules are presented: linguistic fuzzy majorities and 2–tuples linguistic majorities. In Figure 1, the new linguistic majorities are set forth in relation with the corresponding numerical ones. As in the case of the latest ones, voters' role under these linguistic majorities remains the same, i.e. they are required to compare pairs of alternatives with independence of the nature of the voting issue. That contributes to simplify and clarify elections to the voters. But, in fact the nature of the voting issue is taken into account to establish the concrete required difference in support. That permits the modification of the requirement of support from one voting situation to other ones, which seems to be reasonable in practical applications.

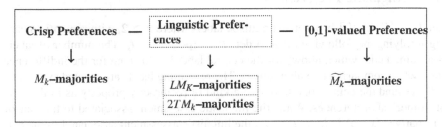

Fig. 1 Individual preferences and majorities based on differences

The rest of the paper is organized as follows: Section 2 provides some basic concepts needed for the development of the paper. In Section 3, linguistic majorities with difference in support are introduced. Conclusions and future research questions are provided in Section 4.

2 Preliminaries

Consider m voters provide their preferences over pairs of alternatives of a set $X = \{x_1, \ldots, x_n\}$. The preferences of each voter can be represented in matrix form: $R^p = \left(r_{ij}^p \right)$ where r_{ij}^p stands for the degree of preference of alternative x_i over x_j for voter p. The coefficients of that matrix could be of numerical or linguistic nature.

2.1 Numerical Preferences

If the values of r_{ij}^p are restricted to $\{0, \frac{1}{2}, 1\}$, we face the crisp case where voters declare only their preference for an alternative or the indifference between two alternatives. When $r_{ij}^p = 1$ the voter p prefers alternative x_i to alternative x_j; alternative x_j is preferred to alternative x_i when $r_{ij}^p = 0$; whilst the value $r_{ij}^p = \frac{1}{2}$ represents the voter indifference between both alternatives. The implied reciprocity property of crisp preferences assures that the asymmetric property usually required to a weak order relation is verified. Let us recall that a binary preference relation represented by $\succ_p$ is asymmetric if given two alternatives x_i and x_j, $x_i \succ_p x_j$ implies that $x_j \nsucc_p x_i$.

If the values r_{ij}^p belong to $[0,1]$ then voters are allowed to express their intensity of preference between pairs of alternatives. If $r_{ij}^p > 0.5$, the individual p prefers the alternative x_i to the alternative x_j; the nearer the value of r_{ij} to 1 the greater is the preference for x_i with respect to x_j. As in the crisp case, $r_{ij}^p = 0.5$ means that the voter p is indifferent between x_i and x_j. The $[0,1]$-valued preferences are assumed to fulfil the following reciprocity property, $r_{ij}^p + r_{ji}^p = 1$, that extends the crisp preferences asymmetry property.

2.2 Linguistic Preferences

Let $L = \{l_0, \ldots, l_s\}$ be a set of linguistic labels where $s \geq 2$, with semantic meaning implying the following linear order: $l_0 < l_1 < \ldots < l_s$. The number of labels is assumed odd, which allows for the central label $l_{s/2}$ to stand for the indifference state when comparing two alternatives. The remaining labels are located symmetrically around the central one to derive a kind of reciprocity property as in the case of numerical preferences. When the linguistic assessment associated to the pair of alternatives (x_i, x_j) is $l_{ij} = l_h \in \mathscr{L}$, the linguistic assessment associated to the pair (x_j, x_i) would be $l_{ji} = l_{s-h}$. The following negation operator is defined $N(l_h) = l_g$ with $|g - h| = s$. An possible set of seven linguistic labels with their corresponding semantic meanings is given in Table 1.

Table 1 Seven linguistic labels

Label	Meaning
l_0	x_j is absolutely preferred to x_i
l_1	x_j is highly preferred to x_i
l_2	x_j is slightly preferred to x_i
l_3	x_i and x_j are indifferent
l_4	x_i is slightly preferred to x_j
l_5	x_i is highly preferred to x_j
l_6	x_i is absolutely preferred to x_j

In the following, the linguistic preference of individual p is represented by the matrix $R^p = \left(l_{ij}^p\right)$ with $l_{ij}^p \in \mathscr{L}$. A profile of linguistic preferences for the pair of alternatives (x_i, x_j) is a vector $(l^1, \ldots, l^m) \in \mathscr{L}^m$ of the associated linguistic preferences provided by the m voters for that pair of alternatives.

The main two representations formats of linguistic information [8] are the cardinal and the ordinal one. The first one is based on the use of fuzzy set with associated membership function that are mathematically processed via Zadeh's *Extension*

Principle [24]. The second one is based on the use of *2-Tuples symbolic methodology* [7].

2.2.1 Fuzzy Set Representation of Linguistic Values

Convex normal fuzzy subsets of the real line, also known as fuzzy numbers, are commonly used to represent linguistic terms. By doing this, each linguistic assessment is represented using a fuzzy number that is characterized by a membership function, with base variable the unit interval $[0,1]$, describing its semantic meaning. The membership function maps each value in $[0,1]$ to a degree of performance which represents its compatibility with the linguistic assessment [24].

It is worth mentioning that some authors consider trapezoidal fuzzy numbers as the most appropriate to represent linguistic preferences [2, 14] because they are more general than triangular and interval fuzzy numbers. A representation of the set of seven balanced linguistic terms of Table 1 using trapezoidal fuzzy numbers is given in Figure 2.

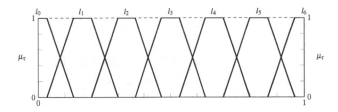

Fig. 2 Representation of seven linguistic terms with fuzzy trapezoidal membership functions

2.2.2 2–Tuple Representation of Linguistic Values and Their Computational Procedure

In Herrera and Martinez [7] linguistic values are modelled by means of linguistic 2–tuples: (l_b, λ_b) where $l_b \in \mathcal{L}$ and λ_b is a numeric value representing the symbolic translation. This representation structure allows, on the one hand, to obtain the same information than with the symbolic representation model based on indexes without losing information in the aggregation phase. On the other hand, the result of the aggregation is expressed on the same domain as the one of the initial linguistic labels and therefore, the well-known re-translation problem of the above methods is avoided.

Definition 1. Let $a \in [0,s]$ be the result of a symbolic aggregation of the indexes of a set of labels assessed in a linguistic term set $\mathcal{L} = \{l_0, \ldots, l_s\}$. Let $b = round(a) \in \{0, \ldots, s\}$. The value $\lambda_b = a - b \in [-0.5, 0.5)$ is called a symbolic translation, and the pair of values (l_b, λ_b) is called the 2–tuple linguistic representation of the symbolic aggregation a.

The 2–tuple linguistic representation of symbolic aggregation can be mathematically formalized with the following mapping:

$$\phi : [0,s] \rightarrow \mathscr{L} \times [-0.5,0.5)$$
$$\phi(a) = (l_b, \lambda_b). \tag{1}$$

Based on the linear order of the linguistic term set and the complete ordering of the set $[-0.5, 0.5)$, it is easy to prove that ϕ is strictly increasing and continuous and, therefore its inverse function exists:

$$\phi^{-1} : \mathscr{L} \times [-0.5,0.5) \rightarrow [0,s]$$
$$\phi^{-1}(l_b, \lambda_b) = b + \lambda_b = a.$$

The following negation operator is defined: $N(\phi(a)) = \phi(s - a)$.

Figure 3 illustrates the application of the 2-tuple function ϕ and its inverse for a linguistic term set of cardinality seven. The value of the symbolic translation is assumed to be 3.7, which means that $round(3.7) = 4$ and therefore it can be represented with the 2-tuple $(l_4, -0,3)$.

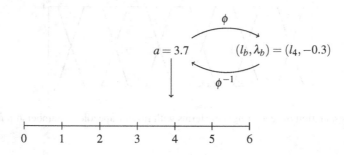

Fig. 3 Subindexes, symbolic translation and 2-tuples

2.3 Majorities Based on Differences

In 2001, García-Lapresta and Llamazares [12] introduce the *majorities based on difference of votes* or M_k*–majorities* with the aim of reducing the support problems commonly attached to simple Majority rule (see for the axiomatic characterization of such rules [15, 9]). Under these rules, given a difference of votes k, an alternative, x_i, defeats another alternative, x_j, by k votes when the difference between the votes cast for the alternative x_i and the votes cast for the alternative x_j is greater than k. M_k–majorities generalize other majority rules.

García-Lapresta and Llamazares extend M_k–majorities to the framework of $[0,1]$-valued preferences [13] with the *majorities based on difference in support* or $\overline{M}_k$*–majorities*. Under these majorities, an alternative, x_i, defeats another one, x_j, by a threshold of support k, when the sum of the intensities of preference of x_i over x_j,

r_{ij}, of the m voters exceeds the sum of the intensities of preference of x_j over x_i, r_{ji}, in a quantity greater than k. The reciprocity property of $[0,1]$-valued preferences allow to define $\widetilde{M_k}$–majorities via the average of individual intensities of preference [13]:

Definition 2. Given a threshold $k \in [0,m)$ and a profile of reciprocal preference relations $R(X) = (R^1, \ldots, R^m)$, the $\widetilde{M_k}$–*majority* is a mapping from $[0,1]^m$ to $\{1, \frac{1}{2}, 0\}$ defined by:

$$\widetilde{M_k}(x_i, x_j) = \begin{cases} 1 & \text{if } \frac{1}{m} \sum\limits_{p=1}^{m} r_{ij}^p > \frac{m+k}{2m} \\ 0 & \text{if } \frac{1}{m} \sum\limits_{p=1}^{m} r_{ij}^p < \frac{m-k}{2m} \\ \frac{1}{2} & \text{otherwise,} \end{cases} \tag{2}$$

where $\widetilde{M_k}(x_i, x_j) = 1$ means that x_i defeats x_j by a threshold of support greater than k, $\widetilde{M_k}(x_i, x_j) = 0$ symbolizes that x_j defeats x_i by a threshold of support greater than k and $\widetilde{M_k}(x_i, x_j) = \frac{1}{2}$ stands for a tie of both alternatives meaning that the difference in between the support for both alternatives in absolute value is lower than or equal to k.

Under these rules, two alternatives are socially indifferent if the collective preference belongs to a numerical interval, specifically $\left[0.5 - \frac{k}{2m}, 0.5 + \frac{k}{2m}\right]$, and not just when the collective preference equals the value 0.5. When the collective preference is greater than the upper bound of such interval, the first alternative is preferred to the second one. On the contrary, when the collective preference is lower than the lower bound of that interval, the second alternative is preferred to the first one.

3 Linguistic Majorities with Difference in Support

To define majority rules with difference in support in the context of linguistic preferences, we need first to introduce the concept of linguistic decision rule:

Definition 3. A linguistic decision rule is a mapping

$$F : \mathscr{L}^m \to \{0, 0.5, 1\},$$

that associates to a profile of linguistic preferences, $(l^1, \ldots, l^m) \in \mathscr{L}^m$, the following values:

- $F(l^1, \ldots, l^m) = 1$ when x_i defeats x_j;
- $F(l^1, \ldots, l^m) = 0$ when x_j defeats x_i; and
- $F(l^1, \ldots, l^m) = 0.5$ when x_i and x_j tie.

The extension of the $\widetilde{M_k}$–majority (2) from the context of the numerical preferences to the linguistic ones, involves the computation of the voters' average linguistic evaluation for a pair of alternatives and the comparison between two linguistic assessments. In the following, we formalize that for the fuzzy set and the 2-tuple linguistic representation methodologies.

3.1 Linguistic Majority with Difference in Support Represented by Fuzzy Sets

Let $\widetilde{A}_{ij}^p$ a type–1 normal and convex fuzzy set that represents the linguistic preference of the voter p when comparing alternatives x_i against alternative x_j.

The application of the *extension principle* [24] and the *representation theorem* of fuzzy sets [23], via the α–level sets, lead to the following [25]:

$$(\widetilde{A}_1 \oplus \widetilde{A}_2)^\alpha = \widetilde{A}_1^\alpha \oplus \widetilde{A}_2^\alpha \tag{3}$$

The α–level sets of fuzzy numbers are closed intervals, and therefore interval arithmetic yields:

$$(\widetilde{A}_1 \oplus \widetilde{A}_2)^\alpha = \widetilde{A}_1^\alpha \oplus \widetilde{A}_2^\alpha = [u_1^-, u_1^+] + [u_2^-, u_2^+] = [u_1^- + u_2^-, u_1^+ + u_2^+].$$

A problem that needs to be addressed here is the comparison of fuzzy numbers. Yager [22], pointed out that this problem has been extensively studied and that there is no unique best approach. Indeed, the set of fuzzy numbers is not totally ordered and therefore it is not possible to achieve a clear social decision in this case. Thus, we require a method to classify them with respect to the intervals of social preference or social indifference established by $\widetilde{M}_k$–majorities.

A widely used approach to rank fuzzy numbers is to convert them into a representative crisp value and perform the comparison on them [22]. Two methods to develop that reduction are common in the literature, namely, the centre of area method (COA) and the mean of maximum method (MOM). The first one computes the centre of mass of the membership function of the fuzzy set (the centroid), whereas the second one computes the mid-point of the 1-level set of the fuzzy set. For a symmetric trapezoidal number $\widetilde{A}$, we have that $u_{COA}(\widetilde{A}) = u_{MOM}(\widetilde{A})$ because of internal symmetry of linguistic labels. Therefore, we refer these real numbers simply as $\mathbf{u}(\widetilde{A})$. Moreover, given two normal and convex trapezoidal fuzzy numbers, namely $\widetilde{A}_1$ and $\widetilde{A}_2$ it holds that, $\mathbf{u}(\widetilde{A}_1 + \widetilde{A}_2) = \mathbf{u}(\widetilde{A}_1) + \mathbf{u}(\widetilde{A}_2)$. Hence, $\mathbf{u}$ is an additive function.

The range of the function $\mathbf{u}$ is $[\mathbf{u}(l_0), \mathbf{u}(l_s)]$, whilst the range of $\frac{m+k}{2m}$ is $[0, 1]$. To compare the values of these functions, we transform function $\mathbf{u}$ into a function $\mathbf{u}'$ with range $[0, 1]$, as follows:

$$\mathbf{u}'(l_h) = \frac{\mathbf{u}(l_h) - \mathbf{u}(l_0)}{\mathbf{u}(l_s) - \mathbf{u}(l_0)}.$$

Below, we formally define the linguistic majority with difference in support represented by fuzzy sets. Under this rule, an alternative, say x_i, defeats another, say x_j by a threshold of support K, if the defuzzified value attached to the average fuzzy set of the voters' linguistic valuations between x_i and x_j exceeds the value 0.5 in a quantity that depends on the threshold K, fixed before the election process.

Definition 4. Given a set of alternatives X and a profile of individual reciprocal fuzzy linguistic preference relations $R(X) = (R^1, \ldots, R^m)$, the LM_K–**majority with difference in support** is the following linguistic decision rule:

$$
LM_K(x_i, x_j) = \begin{cases} 1 & \text{if } \mathbf{u'}\left(\dfrac{1}{m} \sum_{p=1}^{m} \widetilde{A}_{ij}^p \right) > \dfrac{m+K}{2m} \\[2mm] 0 & \text{if } \mathbf{u'}\left(\dfrac{1}{m} \sum_{p=1}^{m} \widetilde{A}_{ji}^p \right) > \dfrac{m-K}{2m} \\[2mm] 0.5 & \text{otherwise,} \end{cases} \tag{4}
$$

where $\mathbf{u'}\left(\dfrac{1}{m} \sum_{p=1}^{m} \widetilde{A}_{ij}^p \right)$ and $\mathbf{u'}\left(\dfrac{1}{m} \sum_{p=1}^{m} \widetilde{A}_{ij}^p \right)$ are the defuzzified values of the fuzzy average linguistic preference of the profile of fuzzy linguistic preferences of the pairs of alternatives (x_i, x_j) and (x_j, x_i) respectively; and $K \in [0, m)$ represents the threshold of support required to an alternative to be the social winner.

Because $\mathbf{u}$ is additive, then we have that

$$
\mathbf{u'}\left(\frac{1}{m} \sum_{p=1}^{m} A_{ij}^p \right) = \frac{1}{m} \sum_{p=1}^{m} \mathbf{u'}(A_{ij}^p).
$$

i.e. $\mathbf{u'}$ is additive. Therefore expression (4) can be rewritten as:

$$
LM_K(x_i, x_j) = \begin{cases} 1 & \text{if } \dfrac{1}{m} \sum_{p=1}^{m} \mathbf{u'}(\widetilde{A}_{ij}^p) > \dfrac{m+K}{2m} \\[2mm] 0 & \text{if } \dfrac{1}{m} \sum_{p=1}^{m} \mathbf{u'}(\widetilde{A}_{ij}^p) < \dfrac{m-K}{2m} \\[2mm] 0.5 & \text{otherwise,} \end{cases} \tag{5}
$$

where the threshold $K \in [0, m)$ and $\dfrac{1}{m} \sum_{p=1}^{m} \mathbf{u'}(\widetilde{A}_{ij}^p)$ is the average of the defuzzified values associated with the profile of fuzzy linguistic preferences of the pair of alternatives (x_i, x_j) as per the assessment of each individual voter.

When $K = 0$, the rule is equivalent to the *simple majority based on linguistic labels* introduced in [11]. In such a case, no difference of support between the alternatives is required.

3.2 Linguistic Majority with Difference in Support Represented by 2–Tuples

To extend the $\widetilde{M}_k$–majority to the 2–tuple representation, the addition and the rule to compare pairs of 2–tuples are needed.

Let two different 2–tuples, $\phi(a_1) = (l_{b_1}, \lambda_{b_1})$ and $\phi(a_2) = (l_{b_2}, \lambda_{b_2})$, with $a_1, a_2 \in [0, s]$ the results of symbolic aggregations, $b_1 = round(a_1), b_2 = round(a_2), \lambda_{b_1} = a_1 - b_1$ and $\lambda_{b_2} = a_2 - b_2$.

Definition 5. (2-tuple addition [7]).

$$\phi(a_1) + \phi(a_2) = (l_{b_{12}}, \lambda_{b_{12}}),$$

with $b_{12} = round(a_1 + a_2)$, and $\lambda_{b_{12}} = (a_1 + a_2) - b_{12}$.

Definition 6. (2-tuple lexicographic ordering [7]). Given $\phi(a_1) = (l_{b_1}, \lambda_{b_1})$ and $\phi(a_2) = (l_{b_2}, \lambda_{b_2})$,

1. If b_1 is greater than b_2, then $\phi(a_1) > \phi(a_2)$.
2. If b_1 is equal to b_2 and λ_{b_1} is greater than λ_{b_1}, then $\phi(a_1) > \phi(a_2)$.
3. If b_1 is equal to b_2 and λ_{b_1} is equal to λ_{b_1}, then $\phi(a_1) = \phi(a_2)$.

Below, we formally define the linguistic majority with difference in support represented by 2–tuples. Under this rule, an alternative, x_i, defeats another, x_j, by a threshold of support k if the 2–tuple linguistic representation of the average symbolic aggregation of the linguistic preferences of x_i over x_j exceeds the 2–tuple linguistic representation associated of the indifference state in a value that depends on the threshold k, fixed before the election process.

Definition 7. Given a set of alternatives X and a profile of individual reciprocal 2–tuple linguistic preferences relations $R(X) = (R^1, \ldots, R^2)$, the $2TM_k$**–majority with difference in support** is the following linguistic decision rule:

$$2TM_k(x_i, x_j) = \begin{cases} 1 & \text{if } \frac{1}{m} \sum_{p=1}^{m} \phi(a_{ij}^p) > \phi\left(\frac{s \times m + k}{2m}\right) \\ 0 & \text{if } \frac{1}{m} \sum_{p=1}^{m} \phi(a_{ij}^p) < \phi\left(\frac{s \times m - k}{2m}\right) \\ 0.5 & \text{otherwise,} \end{cases} \tag{6}$$

where $\frac{1}{m} \sum_{p=1}^{m} \phi(a_{ij}^p)$ is the average of the 2–tuple representation of the linguistic preferences provided by the voters for the pair of alternatives (x_i, x_j), ϕ is the 2–tuple symbolic aggregation mapping in (1); and k is a real number in $[0, m \times s)$ that represents the threshold of support, fixed before the election process.

Given that in the ordinal representation for linguistic information the addition of linguistic labels is defined as $l_{a_1} + l_{a_2} = l_{a_1 + a_2}$ [21], function ϕ is additive. Therefore expression (6) can be rewritten as:

$$2TM_k(x_i,x_j) = \begin{cases} 1 & \text{if } \phi\left(\frac{1}{m}\sum_{p=1}^{m} a_{ij}^p\right) > \phi\left(\frac{s\times m+k}{2m}\right) \\ 0 & \text{if } \phi\left(\frac{1}{m}\sum_{p=1}^{m} a_{ij}^p\right) < \phi\left(\frac{s\times m-k}{2m}\right) \\ 0.5 & \text{otherwise,} \end{cases} \tag{7}$$

where $\frac{1}{m}\sum_{p=1}^{m} a_{ij}^p$ is the symbolic aggregation, specifically the arithmetic mean, of the linguistic preferences provided by the voters for the pair of alternatives (x_i,x_j).

We note that in the context of the 2–tuples, the threshold k varies between 0 and $s \times m$. That happens because each voter could select a linguistic label from l_0 to l_s and consequently the symbolic translation of each of these labels goes from 0 to s. Therefore, the addition of these individual symbolic values varies in $[0, s \times m]$.

When $k = 0$, the rule could be seen as the simple majority rule in the framework of the 2–tuples. In fact, when we take $k = 0$ in expression (7), we are dealing with the arithmetic mean operator described in [7].

4 Concluding Remarks

We have introduced a new class of aggregation rules in the context of linguistic preferences, namely Linguistic majorities with difference in support that generalize other linguistic majorities. These rules extend Majorities based on difference in support from the interval-value preferences to the context of the linguistic preferences through two different representations, i.e. the fuzzy set representation and the 2–tuple representation. They also reach different majorities by adjusting the support required to the winner alternative, getting the possibility of varying such support depending on the nature of the voting issue.

This research leaves open some interesting extensions. For instance, the extension of this new rule to the context of fuzzy majorities by formalizing linguistic labels as type–2 fuzzy sets and the rule as a type–1 fuzzy set. Also, the relationship between the LM_k–majority with difference in support and the $2TM_k$–majority with difference in support is to be established.

References

[1] Chen, S., Hwan, C.: Fuzzy multiple attribute decision making-methods and applications. Springer, Berlin (1992)
[2] Delgado, M., Verdegay, J.L., Vila, M.A.: Linguistic decision making models. Int. J. Intell. Syst. 7, 479–492 (1992)
[3] Ferejohn, J.A., Grether, D.M.: On a class of rational social decisions procedures. Journal of Economic Theory 8, 471–482 (1974)
[4] Fishburn, P.C.: The Theory of Social Choice. Princeton University Press, Princeton (1973)

260 P. Pérez-Asurmendi and F. Chiclana

[5] Fodor, J., Roubens, M.: Fuzzy preference modelling and multicriteria decission support. Kluwer Academics Publishers, Dordrecht (1994)

[6] Herrera, F., Herrera-Viedma, E., Verdegay, J.L.: A linguistic decision process in group decision making. Group Decision and Negociation 5, 165–176 (1996)

[7] Herrera, F., Martínez, L.: A 2-tuple fuzzy linguistic representation model for computing with words. IEEE Trans. Fuzzy Syst. 8, 746–752 (2000)

[8] Herrera, F., Alonso, S., Chiclana, F., Herrera-Viedma, E.: Computing with words in decision making: foundations, trends and prospects. Fuzzy Optim. Decis. Making 8, 337–364 (2009)

[9] Houy, N.: Some further characterizations for the forgotten voting rules. Mathematical Social Sciences 53, 111–121 (2007)

[10] Kacprzyk, J., Fedrizzi, M.: Multiperson decision making models using fuzzy sets and possibility theory. Kluwer Academic Publishers, Dordrecht (1990)

[11] García-Lapresta, J.L.: A general class simple majority decision rules based on linguistic opinions. Information Sciences 176(4), 352–365 (2006)

[12] García-Lapresta, J.L., Llamazares, B.: Majority decisions based on difference of votes. Journal of Mathematical Economics 35, 463–481 (2001)

[13] García-Lapresta, J.L., Llamazares, B.: Preference intensities and majority decisions based on difference of support between alternatives. Group Decision and Negotiation 19, 527–542 (2010)

[14] Garc'ıa-Lapresta, J.L., Llamazares, B., Martínez-Panero, M.: A social choice analysis of the Borda rule in a general linguistic framework. International Journal of Computational Intelligence Systems 19(8), 527–542 (2010)

[15] Llamazares, B.: The forgotten decision rules: Majority rules based on difference of votes. Mathematical Social Sciences 51, 311–326 (2006)

[16] Lu, J., Zhang, G., Ruan, D., Wu, F.: Multi-objective group decision making. Methods, software and applications with fuzzy set techniques. Imperial College Press, London (2007)

[17] May, K.O.: A set of independent necessary and sufficient conditions for simple majority decisions. Econometrica 20, 680–684 (1952)

[18] Nurmi, H.: Fuzzy social choice: A selective retrospect. Soft Computing 12, 281–288 (2008)

[19] Saari, D.G.: Consistency of decision processes. Annals of Operations Research 23, 103–137 (1990)

[20] Sen, A.K.: Collective Choice and Social Welfare. Holden-Day, San Francisco (1970)

[21] Xu, Z.: A method based on linguistic aggregation operators for group decision making with linguistic preference relations. Information Sciences 166(1-4), 19–30 (2004)

[22] Yager, R.: Owa aggregation over a continuous interval argument with applications to decision making. IEEE Transactions on Systems, Man and Cybernetics, Part B: Cybernetics 34(4), 1952–1963 (2004)

[23] Zadeh, L.A.: Fuzzy sets. Information and Control 8(3), 338–357 (1965)

[24] Zadeh, L.A.: The concept of a linguistic variable and its application to approximate reasoning-I. Information Sciences 8, 199–249 (1975)

[25] Zhou, S.M., Chiclana, F., John, R.I., Garibaldi, J.M.: Alpha-level aggregation: A practical approach to Type-1 OWA operation for aggregating uncertain information with applications to breast cancer treatments. IEEE Transactions on Knowledge and Data Engineering 23(10), 1455–1468 (2011)

Calibration of Utility Function

Jana Špirková

Abstract. Utility theory belongs to a very interesting part of a modern decision making theory. We develop the basic concept of utility theory on a determination of a personal utility function wich was founded empirically by short personal interview. We suggest a calibration of theoretical utility function by expected utility maximization criterion. Moreover, on the basis of calibration of utility functions we determine a model of specific values of a maximal and minimal annual premium acceptable for an insured and for an insurer, too.

1 Introduction

A more modern approach of the utility theory was advanced by John von Neumann and Oskar Morgenstern in 1947 in their book *Theory of Games and Economic Behavior* [4]. There, they proposed that a utility function may be tailored for any individual, provided certain assumptions about the individual's preferences hold. These assumptions provide several valid, basic shapes for the utility function. In 2007 was published 60th-anniversary edition of this book [5]. We can find a very interesting approach about utility functions in [3].

Our work was inspired by the books *Modern Actuarial Risk Theory* [2] and *Actuarial Models - The Mathematics of Insurance* [4] and by my students, which want to know a more information about a generation of a personal utility function. Especially, this paper was inspired by the paper *Some useful optimization problems in portfolio theory* [6], where author described calibration of a utility function according to maximization of the expected utility.

Jana Špirková
Faculty of Economics, Matej Bel University,
Tajovského 10, 975 90 Banská Bystrica, Slovakia
e-mail: jana.spirkova@umb.sk

H. Bustince et al. (eds.), *Aggregation Functions in Theory and in Practise*,
Advances in Intelligent Systems and Computing 228,
DOI: 10.1007/978-3-642-39165-1_27, © Springer-Verlag Berlin Heidelberg 2013

On the basis of the calibration of theoretical utility function and expected utility maximization we offer the model on the determination maximal and minimal premiums acceptable for an insured and for an insurer in a general insurance.

Our paper is organized as follows. In Section 2 we recall a few of utility functions which have a constant relative risk aversion. In Section 3 we introduce basic propeties of the expected utility with respect to concavity and convexity of the utility function. Moreover, we recall expected utility maximization criterion. Section 4 represents the basic part of our paper. It contains a calibration of the utility function (5) according to short questionnaire and the expected utility maximization criterion. Section 5 explains a model on a determination of maximal and minimal premium in general insurance with respect to calibrated personal utility functions. Finally, in Section 6 some conclusions and indications of our next investigation about mentioned topic are included.

2 Preliminaries

D. Bernoulli proceeded from the simple observation that the "degree of satisfaction" of having capital, or in other words, the "utility of capital", depends on the particular amount of capital in a nonlinear way. For example, if we give $1,000$ euros to a person with a wealth of $1,000,000$ euros, and the same $1,000$ euros to a person with zero capital, the former will feel much less satisfied than the latter. To model this phenomenon, D. Bernuolli assumed that the satisfaction of possessing a capital x, or the "utility" of x, may be measured by a function $u(x)$, as a rule, is not linear. Such a function is called a *utility function*, or a *utility of money function*. According to [4] the word "satisfaction" would possibly reflect the significance of the definition better, but the term "utility" has been already accepted.

Utility function may be used as a basis for describing individual approaches to risk. Three basic approaches have been characterized. Opposite cases refer to *risk loving* and *risk averse* who accepts favorable gambles only. There is *risk-neutral* between these two extremes. Risk-neutral behavior is typical of persons who are enormously wealthy. Many people may be both risk averse and loving, depending on the range of monetary values being considered.

In real life people do not behave according to theoretical utility functions. There is a psychological problem rather than a mathematical one. Seriousness and uncertainty of respondent's answers depend on situation, on form of questions asked, on time which respondents have, and on many psychological and social factors. We can find a very interesting approach about utility functions in [3].

In [4] Rotar introduced many types of classical utility functions, for example:

$$u(x) = -A \times x^{1-\alpha} + B \ \ if \ \alpha > 1; \tag{1}$$

$$u(x) = A \times \ln x + B \ \ if \ \alpha = 1; \tag{2}$$

$$u(x) = A \times x^{1-\alpha} + B \ if \ \alpha < 1; \tag{3}$$

$$u(x) = \frac{1}{\alpha} \times \left(1 - exp^{-\alpha \times x}\right) \ if \ \alpha > 0; \tag{4}$$

$$u(x) = \frac{x^{1-\alpha}}{1-\alpha} \ if \ \alpha \neq 1. \tag{5}$$

All above mentioned functions represent a standard class of the constant coefficient of relative risk aversion functions (CRRA functions) with the constant relative risk aversion which is given by

$$R(x) = -x\frac{u''(x)}{u'(x)} = \alpha = const. \tag{6}$$

In our paper we focus on an investigation of the utility function (5). Our aim is to calibrate mentioned utility function by the short questionnaire.

3 Expected Utility

Within the expected utility investigation, the explanation for risk aversion is that the utility function for wealth is concave and non-decreasing, and for risk loving is convex and non-decreasing.

The theorem below describes properties of the utility function and its expected value.

Theorem 1. *(Jensen's inequality) [2], [4] Let X be a random variable (with a finite expectation). Then,*
if u(x) is concave,

$$E\left[u\left(X\right)\right] \leq u\left(E\left[X\right]\right). \tag{7}$$

If u(x) is convex,

$$E\left[u\left(X\right)\right] \geq u\left(E\left[X\right]\right). \tag{8}$$

Equality holds if and only if u(x) is linear according to X or var(X) = 0.

Jensen's inequality follows directly from the properties of concave and convex functions. See Fig. 3.

Remark 1. Expected utility is calculated by the well-known formula

$$E\left[u\left(X\right)\right] = \sum_{i=1}^{n} u(x_i) \cdot p_i, \tag{9}$$

where $X = (x_1, x_2, \ldots, x_n)$ is a vector of the possible alternatives and p_i is the probability of alternative x_i.

Expected utilities can be calculated by linear function too, which is determined uniquely by points on the utility function which represent the worst and the best

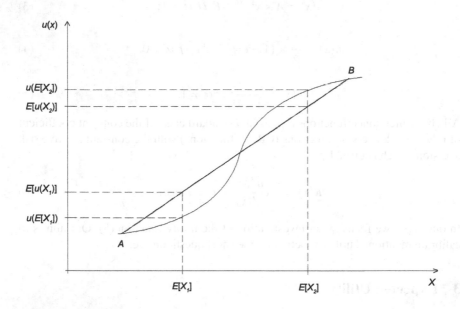

Fig. 1 Jensen's inequalities

case. In both cases we get the same values of the expected utilities. On the Fig. 3 the best and the worst options are represented by the points A, B.

3.1 Expected Utility Maximization

Expected utility maximization criterion corresponds to the preference order $\preceq$ for which

$$X \preceq Y \equiv E\left[u(X)\right] \preceq E\left[u(Y)\right] \tag{10}$$

for a utility function u. The relation (10) means that among two random variables X, Y we prefer the random variable with the larger expected utility. If $u(x)$ is non-decreasing, the rule (10) is monotone. The investor who follows (10) is called an expected utility maximizer.

Remark 2. The preference order (10) does not change if $u(x)$ is replaced by any function

$$u^* = a \times u(x) + b, \tag{11}$$

where a is a positive and b is an arbitrary number. The values of a and b do not matter, because an expression (11) represents an affine transformation.

4 Calibration of Utility Function

As we mentioned in Section (2), we focus on the calibration of the utility function (5) by the next short questionnaire, [6]:

Suppose that you are going to invest 17 000 euros and you have a choice between four different investment strategies for a three-year investment. Which one would you prefer?

- alternative A_1: in the best case profit 1 700 euros (10%), in the worst case profit 550 euros (3.24%);
- alternative A_2: in the best case profit 2 600 euros (15.29%), in the worst case zero profit (but no loss);
- alternative A_3: in the best case profit 4 000 euros (23.53 %), in the worst case loss 1 700 euros (10%);
- alternative A_4: in the best case profit 6 500 euros (38.53 %), in the worst case loss 4 000 euros (23.53 %).

Respondents have chosen one of the previous alternatives. It is apparent that individual alternatives are put in order, so that the first alternative is of the lowest risk and the fourth alternative is of the highest risk. We evaluated expected utilities for all alternatives and for some selected relative risk aversion coefficients α in the utility function (5). On the basis of the equation (9) we evaluated expected utilities and determined maximal expected utilities by

$$E[u(x)] = p \cdot u(x_1) + (1-p) \cdot u(x_2) \to max \qquad (12)$$

for all alternatives on the level of all used α. Our results are written in the Tab. 1. Please, observe, that all values - the best and the worst cases we divided by 10,000 for easier evaluation with respect to suitable decimal places.

You can see from the Tab. 1 a respondent who has chosen alternative A_1 will have the utility function with risk aversion coefficient $\alpha = 11$, hence his/her utility function is given by $u(x) = \frac{-1}{10 \times x^{10}}$; a respondent who has chosen alternative A_2 will have α from 6 to 10 and corresponding utility function (we have chosen for this case $\alpha = 7$) $u(x) = \frac{-1}{6 \times x^6}$; a respondent who has chosen alternative A_3 will have $\alpha = 5$ or $\alpha = 6$ and corresponding utility function for $\alpha = 5$ $u(x) = \frac{-1}{4 \times x^4}$; and at the end, a respondent who has chosen alternative A_4 will have $\alpha = 2$ or $\alpha = 3$ and corresponding utility function with $\alpha = 3$ is given by $u(x) = \frac{-1}{2 \times x^2}$.

5 Expected Utility in Insurance

In this section we introduce a model of the determination of minimal and maximal premium in a general insurance. We use the utility functions in the shape (5) which were calibrated according to the questionnaire from Section 4.

Table 1 Expected utility according to α and individual alternatives

Alternative	A_1	A_2	A_3	A_4
the best case x_1	1.87	1.96	2.10	2.35
the worst case x_2	1.75	1.70	1.53	1.30
probability p	0.99	0.80	0.80	0.80
probability $(1-p)$	0.01	0.20	0.20	0.20
$\alpha = 2$	-0.5351260504	-0.5258103241	-0.5116713352	**-0.4942716858**
$\alpha = 3$	-0.1431866008	-0.1387253581	-0.1334215603	**-0.1316025619**
$\alpha = 4$	-0.0510868286	-0.0489855239	**-0.0474083611**	-0.0508922003
$\alpha = 5$	-0.0205064749	-0.0195385907	**-0.0194081802**	-0.0240641956
$\alpha = 6$	-0.0087806430	**-0.0083486432**	-0.0086885586	-0.0130056075
$\alpha = 7$	-0.0039166661	**-0.0037327835**	-0.0041531599	-0.0076975207
$\alpha = 8$	-0.0017970871	**-0.0017247775**	-0.0020903041	-0.0048420699
$\alpha = 9$	-0.0008417949	**-0.0008175307**	-0.0010969370	-0.0031722489
$\alpha = 10$	-0.0004006034	**-0.0003956201**	-0.0005955985	-0.0021362131
$\alpha = 11$	**-0.0001930420**	-0.0001948224	-0.0003324839	-0.0014663374

Source: own construction

5.1 Utility of Insured

Suppose that our respondent has two alternatives - to buy insurance or not. Suppose that he owns a capital w and that he values wealth by the utility function u. Let's assume he is insured against a loss X for a gross annual premium GP. If he is insured that means a certain alternative. This decision gives us the utility value $u(w - GP)$. If he is not insured that means an uncertain alternative. In this case the expected utility is $E[u(w - X)]$. Based on Jensen's inequality (7) we get

$$E[u(w-X)] \le u(E[w-X]) = u(w - E[X]) \le u(w - GP). \tag{13}$$

Since utility function u is a non-decreasing continuous function, this is equivalent to $GP \le P^{max}$, where P^{max} denotes the maximum premium to be paid. This so-called *zero utility premium* is the solution to the following utility equilibrium equation

$$E[u(w-X)] = u(w - P^{max}). \tag{14}$$

On the basis of individual personal utility functions we can determine maximal premium what our respondent - client of an insurance company will be willing to pay for insurance his/her wealth on the basis of the following model: our client has 17 000 euros and he wants to insure his wealth in the size 12 000 euros. Maximal premium P^{max} is calculated by inverse function u^{-1} to the utility equilibrium equation (14) which is given by

$$P^{max} = w - u^{-1}(E[u(w-X)]). \tag{15}$$

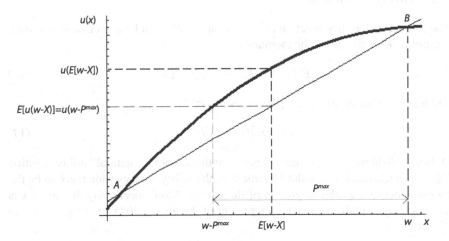

Fig. 2 Expected utility for risk averse approach

Resulting values of maximal premiums with respect to risk aversion coefficients are written in Tab. 2. Moreover, minimal premiums with respect to insurer, are calculated on the basis of (16) and (17).

Table 2 Premiums According to Constant Relative Risk Aversion

probability p	$E[X]$	p^{min}	p^{max} $\alpha = 11$	p^{max} $\alpha = 7$	p^{max} $\alpha = 5$	p^{max} $\alpha = 3$
0.001	12	12.03	7,028.50	2,449.82	521.16	89.06
0.002	24	24.60	7,694.02	3,555.73	971.11	176.73
0.003	36	36.09	8,063.08	4,254.82	1,365.65	263.05
0.004	48	48.11	8,316.16	4,759.20	1,715.96	348.05
0.005	60	60.14	8,507.59	5,150.38	2,030,24	431.78
⋮	⋮	⋮	⋮	⋮	⋮	⋮
0.9	10,800	10,802.43	11,947.04	11,911.49	11,867.62	11,754.69
1.0	12,000	12,000.00	12,000.00	12,000.00	12,000.00	12,000.00

Source: own construction

Remark 3. Observe, that the insured is willing to pay the premium, which is of equal value as the loss. For more information see, for example [1].

5.2   Utility of Insurer

The insurer with utility function $U(x)$ and capital W, with insurance of loss X for a premium GP must satisfy the inequality

$$E\left[U\left(W+GP-X\right)\right] \geq U\left(W\right),\tag{16}$$

and hence for the minimal accepted premium P^{min}

$$U\left(W\right) = E\left[U\left(W+P^{min}-X\right)\right].\tag{17}$$

D. Bernoulli himself suggested as a good candidate for the "natural" utility function $U(x) = lnx$, assuming that the increment of the utility is proportional not to be the absolute but to the relative growth of the capital. More specifically, if capital x is increased by a small dx, then the increment of the utility, $dU(x)$, is proportional to $\frac{dx}{x}$, that is

$$dU(x) = k \times \frac{dx}{x}\tag{18}$$

for a constant k. The solution to this equation is

$$U(x) = k \times lnx + C,\tag{19}$$

where C is another constant. We determine minimal premium by the means of (17) with respect to utility function for insurer $U(x) = lnx$ with his basic capital (Slovak Republic) $W = 2,655,513.51$ euros and loss $X = 12,000$ euros.

Equation (17) can be rewritten as follows:

$$U\left(W\right) = p \cdot U(W + P^{min} - X) + (1-p) \cdot U(W + P^{min}),\tag{20}$$

and hence

$$W = (W + P^{min} - X)^{p} \cdot (W + P^{min})^{(1-p)}.\tag{21}$$

6   Conclusion

In our paper we offered a process of a calibration of a utility function on the basis of expeced utility maximization. Moreover, we used created individual personal utility function on a determination of maximal premium in the case of an insurance of wealth in the size 12,000 euros.

Because, as you can see from Tab. 1, we have more posibilities to choose relative risk aversion coefficients, we want in our next investigation to aggregate risk aversion coefficients for individual alternatives. Our next aim is to aggregate individual personal utility functions for determination of the best (minimal) maximal premium which is the insured willing to pay.

We evaluated expected utilities and correspondig maximal and minimal premiums by MS Office Excel 2010 and Mathematica 8 systems.

Acknowledgements. This work was supported by the Agency of the Ministry of Education, Science, Research and Sport of the Slovak Republic for the Structural Funds of EU, under project ITMS: 26110230082.

References

[1] Brunovský, P.: Mikroeconomics (in Slovak), http://www.iam.fmph.uniba.sk/skripta/brunovsky2/PART0.pdf (cited December 17, 2012)

[2] Kaas, R., Goovaerts, M., Dhaene, J., Denuit, M.: Modern Actuarial Risk Theory. Kluwer Academic Publishers, Boston (2001)

[3] Khurshid, A.: The end of rationality?, Abstracts. In: Tenth International Conference on Fuzzy Set Theory and Applications, FSTA 2010, Slovak Republic, pp. 10–11 (2010)

[4] Neumann, J., Morgenstern, O.: Theory of Games and Economic Behavior. Princeton University Press, US (1947)

[5] Neumann, J., et al.: Theory of Games and Economic Behavior, Sixtieth-Anniversary Edition. Princeton University Press, US (2007)

[6] Melicherčík, I.: Some useful optimization problems in portfolio theory, http://www.defm.fmph.uniba.sk/ludia/melichercik/papers/jarna_skola_2012.pdf (cited September 10, 2012)

[7] Rotar, I.V.: Actuarial Models: The Mathematics of Insurance. Chapman& Hall/CRC Press, Boca Raton, London (2007)

Uncertain Choices: A Comparison of Fuzzy and Probabilistic Approaches

Davide Martinetti, Susana Montes, Susana Díaz, and Bernard De Baets

Abstract. Choices among alternatives in a set can be expressed in three different ways: by means of choice functions, by means of preference relations or using choice probabilities. The connection between the two first formalizations has been widely studied in the literature, both in the crisp or classical context and in the setting of fuzzy relations. However, the connection between probabilistic choice functions and fuzzy choice functions seems to have been forgotten and as far as we know, no literature can be found about it.

In this contribution we focus on the comparison of both types of choice functions. We provide a way to obtain the fuzzy choice function from the probabilistic choice function and the other way around. Moreover, we can prove that under Luce's Choice Axiom the fuzzy choice function derived from the probabilistic choice function is G-normal.

1 Introduction

There is no unique way of representing the act of choice in mathematical language. As pointed out by Fishburn in [7], there exist at least three ways of representing them:

(i) binary relations;
(ii) choice functions;
(iii) choice probabilities.

Davide Martinetti · Susana Montes · Susana Díaz
Dept. of Statistics and Operational Research, University of Oviedo, E-33071, Oviedo, Spain
e-mail: {martinettidavide.uo,montes,diazsusana}@uniovi.es

Bernard De Baets
Dept. of Mathematical Modelling, Statistics and Bioinformatics, Ghent University, Coupure links 653, B-9000, Ghent, Belgium
e-mail: bernard.debaets@ugent.be

H. Bustince et al. (eds.), *Aggregation Functions in Theory and in Practise*,
Advances in Intelligent Systems and Computing 228,
DOI: 10.1007/978-3-642-39165-1_28, © Springer-Verlag Berlin Heidelberg 2013

The relations between the first two formalisms have been widely studied (see [1, 19, 20, 21, 22, 23, 24]), giving birth to an extensive literature that goes under the name of choice theory with revealed preferences. Fishburn already addressed in [7] the lack of results on the connections between choice functions and choice probabilities and hence he proved a set of propositions on the conditions that should be satisfied by the choice probability function in order for the associated choice function to be rational.

The same situation appears in the framework of fuzzy choice and fuzzy preferences. In the last years the results of classical (crisp) revealed choice theory have been recovered in the fuzzy framework, laying bare the connections between fuzzy preference relations and fuzzy choice functions (see, amongst others, [2, 3, 4, 8, 9, 10, 13, 14, 16, 18, 25, 26, 27]). Surprisingly, while the connection between fuzzy preference relations and fuzzy choice functions has been studied in depth, there appears to be no literature on the comparison of fuzzy choice functions and probabilistic choice functions. Furthermore, we consider interesting to compare the two formalisms, since in the first case the uncertainty on the choices is handled using fuzzy set theory, while in the second case, using probability theory.

To find a proper way to translate one formalism into another, we make use of implication operators, trying to maintain the semantic of uncertainty associated to the formalism of probabilistic choice functions, while the semantic of fuzziness associated to the fuzzy choice functions.

2 Preliminary Concepts

2.1 Choice Probability Functions and Reciprocal Relations

Let X be a finite set of alternatives and $\mathscr{B}$ the family of all non-empty subsets of X ($\mathscr{B} = 2^X \setminus \{\emptyset\}$). Imagine that a decision maker is observed to make his choices over the set X, when different bundles of alternatives are presented to him. The set X can be a set of products on the shelf at a supermarket. The decision maker is a consumer, whose purchases are recorded. Not all the possible products in X are always available (seasonal products, out-of-stock, etc.). Hence the choices of the decision maker are recorded also considering these restrictions on the available products. A probability measure P on $\mathscr{B}$ can be defined: for every pair $A, B \in \mathscr{B}$, such that $A \subseteq B$, $P(A, B)$ is the probability that the choice from the set B lies in the subset A. The probability measure nP is completely determined by its values $P(x, A) = P(\{x\}, A)$ (probability of choosing alternative x from the set A), in the sense that $P(A, B) = \sum_{x \in A} P(x, B)$. This probability measure is also called choice probability function and associates to every (x, A) in $X \times \mathscr{B}$ a value $P(x, A)$ that represent the probability that element x is chosen, when A is available. The function P can be approximated using frequency of observations in a dataset. For a matter of convenience, we will denote the choice over the pairs by $p(x, y) = P(\{x\}, \{x, y\})$. This relation p is usually called probabilistic relation or reciprocal relation in the literature and obviously satisfies condition $p(x,y) + p(y,x) = 1$, for any x, y in X. Different definitions of transitivity have been proposed for probabilistic relations:

(i) g-stochastic transitivity, see [5, 6];
(ii) h-isostochastic transitivity, see [5, 6];
(iii) cycle-transitivity, see [5, 6];

In this work we will pay particular interest to a specific kind of h-isostochastic transitivity, also known as multiplicative transitivity: for any x, y, z in X it holds that

$$\frac{p(x,y)}{p(y,x)} \frac{p(y,z)}{p(z,y)} = \frac{p(x,z)}{p(z,x)}. \tag{1}$$

2.2 Luce's Choice Axiom

In probability theory, Luce's Choice Axiom, formulated by R. Duncan Luce in [11, 12], states that the probability of selecting one alternative over another from a set of available alternatives is not affected by the presence or absence of other items in the set. Selection of this kind is said to be independent from irrelevant alternatives (IIA). Formally, it is composed of two parts:

Part 1   If $p(x,y) \neq 0$, for all x, y in X, $x \neq y$, then for all $A \in \mathscr{B}$ and all x in A

$$P(x,X) = P(x,A)P(A,X). \tag{2}$$

Part 2   If $p(x,y) = 0$ for some x, y in X, then for all $A \in \mathscr{B}$, we have

$$P(A,X) = P(A \setminus \{x\}, X \setminus \{x\}). \tag{3}$$

Notice that Luce's Choice Axiom is not always fulfilled. It is easy to find examples of choice probabilities that do not satisfy this axiom. As a consequence of Luce's Choice Axiom, we can guarantee that the probabilistic relation p satisfies the multiplicative transitive property and also the so called, constant ratio rule: for any x, y in X, it holds that

$$\frac{P(x,A)}{P(y,A)} = \frac{P(x,X)}{P(y,X)}. \tag{4}$$

This condition trivially implies that $\frac{p(x,y)}{p(y,x)} = \frac{P(x,A)}{P(y,A)}$, for any $A \in \mathscr{B}$.

2.3 Fuzzy Choice Functions and Fuzzy Preference Relations

A fuzzy choice function C in the sense of Banerjee (see [2]) is defined over $X \times \mathscr{B}$ and associates to every (x,A) a value $C(x,A)$ in the unit interval that represents the degree of choice of alternative x when the set of alternatives A is available. The only condition imposed on C is that, for every $A \in \mathscr{B}$, there exists at least one alternative $x \in A$ such that $C(x,A) > 0$. The fuzzy choice function C satisfies condition H1 if, for every $A \in \mathscr{B}$, there exists one element $x \in A$, such that $C(x,A) = 1$, i.e. the fuzzy set $C(\cdot,A)$ is a normal fuzzy set. It satisfies condition H2 if the family $\mathscr{B}$ of subsets of X is equal to $2^X \setminus \{\emptyset\}$.

A fuzzy preference relation on X is a mapping $Q : X^2 \to [0,1]$, that to every pair of elements x and y of X associates a real value $Q(x,y) \in [0,1]$ that represents the degree of preference of the first element over the second. In this work we consider the following properties of Q:

(i) reflexivity: if $Q(x,x) = 1$, for every $x \in X$;
(ii) moderate completeness: if $Q(x,y) + Q(y,x) \geq 1$, for every $x,y \in X$;
(iii) strong completeness: if $\max(Q(x,y), Q(y,x)) = 1$, for every $x,y \in X$;
(iv) T-transitivity: if $T(Q(x,y), Q(y,z)) \leq Q(x,z)$, for every $x,y,z \in X$ and where T represents a triangular norm;
(v) acyclicity: for all $n \geq 2$ and $(x_1, x_2, \ldots, x_n) \in X$, if

$$\text{if } \begin{cases} Q(x_1, x_2) > Q(x_2, x_1) \\ Q(x_2, x_3) > Q(x_3, x_2) \\ \qquad \cdots \\ Q(x_{n-1}, x_n) > Q(x_n, x_{n-1}) \end{cases}, \text{ then } Q(x_1, x_n) \geq Q(x_n, x_1);$$

Definition 1 ([2]). A fuzzy preference relation R can be revealed from the fuzzy choice function C using the following:

$$R(x,y) = \bigvee_{\{A \in \mathscr{B}|x,y \in A\}} C(x,A). \tag{5}$$

Definition 2 ([2]). A fuzzy choice function C is G-rational if there exists a fuzzy preference relation Q such that:

$$C(x,A) = G(A,Q)(x) = \bigwedge_{y \in A} Q(x,y). \tag{6}$$

Definition 3. A fuzzy choice function is G-normal if it is the G-rationalization of a fuzzy preference Q and its fuzzy revealed preference R is equal to Q.

We recall here, without proof, some previous results on fuzzy choice functions and revealed preference:

(i) If H1 and H2 are verified, then R is strongly complete and reflexive. See [8].
(ii) Every G-normal fuzzy choice function is G-rational. See [8].
(iii) If H1 and H2 are verified, then any G-rational choice function is also G-normal. See [16].
(iv) If Q is acyclic, moderately complete and reflexive, then it generates a G-rational fuzzy choice function through Eq. (6). See [17].

3 From Choice Probabilities to Fuzzy Choice Functions

Choice probabilities are relatively easy to observe and gather. Imagine for example the tracking of the purchases made by a supermarket over its customers. Choice probabilities can be easily calculated using the frequency of choice of a customer over a bundle of products, that can vary over time, according to seasonal availability, sales, etc. The drawback of probability choice functions is that their handling is made difficult by the strong conditions imposed by the probabilistic setting, which impose, for example, that $\Sigma_{x\in A}P(x,A) = 1$. For this reason, it would be interesting to obtain another choice function from the probabilistic one, without the drawback of such strong conditions but that preserves the uncertainty contained in the probabilistic formulation. Fishburn already attempted this approach in [7], considering crisp choice functions $C_{\min}$, $C_{\max}$, but we think that this translation loses important information on the uncertainty described by the probabilistic choice function. For this reason, we propose a formula for computing a fuzzy choice function from a given choice probability.

Definition 4. [7] Given the choice probability P, a fuzzy choice function ρ can be defined in the following way:

$$\rho(x,A) = \frac{P(x,A)}{\max_{a\in A}(P(a,A))}. \tag{7}$$

Notice that this definition was already introduced in [7], but the author considered there the alpha-cuts of ρ, hence he was using a crisp choice function again.

Remark 1. Interested readers can easily prove that Eq. (7) can also be rewritten using the minimum operator and the residual implication operator derived from the product triangular norm $(a \to_P b = b/a \wedge 1)$, i.e.

$$\rho(x,A) = \frac{P(x,A)}{\max_{a\in A}(P(a,A))} = \bigwedge_{a\in A} P(x,A) \to_P P(a,A). \tag{8}$$

Some properties of the function ρ:

  (i) ρ is a fuzzy choice function, since, for every $A \in \mathscr{B}$, there exists at least one x in X, such that $\rho(x,A) > 0$;
 (ii) it verifies hypothesis H1: in fact, for any $A \in \mathscr{B}$, there exists an element x such that $\rho(x,A) = 1$; in particular, the element x is the one in A for which $P(x,A)$ is maximal.
(iii) if the choice probability is defined on $\mathscr{B} = 2^X \setminus \{\emptyset\}$, then ρ is automatically defined on the same set, and hence condition H2 holds.

It would be interesting to find some conditions on the probability choice function P such that the derived fuzzy choice function ρ will show good rationality features. For this reason, we further assume that Luce's Choice Axiom is verified by the probability choice function. Accordingly, we are able to prove the following results.

Proposition 1. *The fuzzy choice function ρ can be computed using only p (the probabilities over the pairs) instead of P:*

$$\rho(x,A) = \bigwedge_{k \in A} \frac{p(x,k)}{p(k,x)}. \tag{9}$$

Sketch of the proof: use the constant ratio rule and Eq. (7).

Proposition 2. *The fuzzy revealed preference relation R from ρ can be computed using p:*

$$R(x,y) = 1 \wedge \frac{p(x,y)}{p(y,x)}. \tag{10}$$

Sketch of the proof: use Eq. (5) and the new formulation of ρ presented in Proposition 1.

Given the fuzzy revealed preference relation R over X, the probabilistic relation p can be computed as:

$$p(x,y) = \frac{R(x,y)}{R(x,y) + R(y,x)}. \tag{11}$$

In fact, using Proposition 2 it is easy to prove that Eq. (11) is equivalent to

$$\frac{R(x,y)}{R(x,y) + R(y,x)} = \frac{1 \wedge \frac{p(x,y)}{p(y,x)}}{\left(1 \wedge \frac{p(x,y)}{p(y,x)}\right) + \left(1 \wedge \frac{p(y,x)}{p(x,y)}\right)} = p(x,y). \tag{12}$$

Then, thanks to Eq. (10) and Eq. (11), we can unveil a way for passing from the probabilistic relation to a fuzzy preference relation and vice versa. On this subject, we already presented different results in [15], where special attention was paid to the transferability of the transitivity property from one relation to another. Talking of transitivity, we have been able to prove the following result:

Proposition 3. *If p is multiplicative transitive, then the fuzzy revealed preference relation R obtained from ρ is T_P-transitive, i.e. $R(x,y)R(y,z) \leq R(x,z)$, for all $x,y,z \in X$.*

Remark 2. Multiplicative transitivity of p does not imply T_M-transitivity of the fuzzy revealed preference relation R, i.e. transitive w.r.t. the minimum t-norm. Consider the following example, where $\operatorname{card}(X) = 3$.

$$p = \begin{pmatrix} \frac{1}{2} & \frac{1}{3} & \frac{2}{23} \\ \frac{2}{3} & \frac{1}{2} & \frac{3}{13} \\ \frac{20}{23} & \frac{10}{13} & \frac{1}{2} \end{pmatrix} \qquad R = \begin{pmatrix} 1 & \frac{1}{2} & \frac{3}{20} \\ 1 & 1 & \frac{3}{10} \\ 1 & 1 & 1 \end{pmatrix}.$$

The probabilistic relation p is multiplicative transitive, while R is not T_M-transitive. In fact,

$$R(a,b) \land R(b,c) = 0.3 > 0.15 = R(a,c).$$

Remark 3. The results of Proposition 3 and Remark 2, namely that multiplicative transitivity of the probabilistic relation p implies T_P-transitivity, but not T_M-transitivity, of the fuzzy revealed preference relation R, were anticipated in the light of the results contained in [5], where it was proved that multiplicative transitivity of p is sufficient condition for T_P-transitivity of p, but it is independent from T_M-transitivity of p.

We conclude with the key result of this contribution:

Theorem 1. *If P satisfies Luce's Choice Axiom, then the fuzzy choice function ρ is G-normal.*

Proof. We know that if a fuzzy preference relation Q is acyclic, moderately complete and reflexive, then it G-rationalizes a fuzzy choice function (see [17]). Consider the fuzzy preference relation R. We have proved that it is strongly complete, reflexive and T_P-transitive. Hence, it is also acyclic, moderately complete and reflexive (all of them are weaker conditions). It thus generates a G-rational fuzzy choice function. It can be the case that the fuzzy choice function generated by R does not coincide with ρ, but we also know that as long as hypotheses H1 and H2 are satisfied, G-rationality implies G-normality (see [16]). Hence R is the fuzzy revealed preference relation of some G-normal fuzzy choice function and we revealed R from ρ, then ρ is G-normal. □

4 Conclusions and Future Works

The results proposed are thought to connect two formalisms that are used for representing human choices in a situation in which uncertainty is involved: probabilistic choices and fuzzy choice functions. Being aware of the semantic differences between the probabilistic and fuzzy approaches, we have been able to define a clear way for passing from one to another. Furthermore, under the additional hypothesis of Luce's Choice Axiom is verified, we also proved that the rationality of the derived fuzzy choice function is ensured. As pointed out in Remark 1, the formula for deriving the fuzzy choice function from a probabilistic choice function involves the use of residual implication operator of the product t-norm T_P. It may just look as a coincidence that multiplicative transitivity gives such a good result in combination with this implication operator, but it should be the key for a more general result, involving other kinds of implication operators.

Acknowledgements. The research reported on in this contribution has been partially supported by Project MTM2010-17844 and the Foundation for the promotion in Asturias of the scientific and technologic research grant BP10-090.

References

[1] Arrow, K.J.: Rational choice functions and orderings. Economica 26, 121–127 (1959)
[2] Banerjee, A.: Fuzzy choice functions, revealed preference and rationality. Fuzzy Sets and Systems 70, 31–43 (1995)
[3] Barrett, R., Pattanaik, P.K., Salles, M.: On choosing rationally when preferences are fuzzy. Fuzzy Sets and Systems 34, 197–212 (1990)
[4] Barrett, R., Pattanaik, P.K., Salles, M.: Rationality and aggregation of preferences in an ordinally fuzzy framework. Fuzzy Sets and Systems 49, 9–13 (1992)
[5] De Baets, B., De Meyer, H., De Loof, K.: On the cycle-transitivity of the mutual rank probability relation of a poset. Fuzzy Sets and Systems 161, 2695–2708 (2010)
[6] De Schuymer, B., De Meyer, H., De Baets, B.: Cycle-transitive comparison of independent random variables. Journal of Multivariate Analysis 96, 352–373 (2005)
[7] Fishburn, P.C.: Choice probabilities and choice functions. Journal of Mathematical Psychology 18, 205–219 (1978)
[8] Georgescu, I.: Fuzzy choice functions, A revealed Preference Approach. Springer, Berlin (2007)
[9] Georgescu, I.: Arrow's axiom and full rationality for fuzzy choice functions. Social Choice and Welfare 28, 303–319 (2007)
[10] Georgescu, I.: Acyclic rationality indicators of fuzzy choice functions. Fuzzy Sets and Systems 160, 2673–2685 (2009)
[11] Luce, R.D.: Individual Choice Behavior: A Theoretical Analysis. Wiley, New York (1959)
[12] Luce, R.D., Suppes, P.: Preferences utility and subject probability. In: Luce, R.D., Bush, R.R., Galanter, E. (eds.) Handbook of Mathematical Psychology III, pp. 249–410. Wiley (1965)
[13] Martinetti, D., De Baets, B., Díaz, S. and Montes, S.: On the role of acyclicity in the study of rationality of fuzzy choice functions. In: International Conference on Intelligent Systems Design and Applications, pp. 350–355 (2011)
[14] Martinetti, D., Montes, I., Díaz, S.: From Preference Relations to Fuzzy Choice Functions. In: Liu, W. (ed.) ECSQARU 2011. LNCS, vol. 6717, pp. 594–605. Springer, Heidelberg (2011)
[15] Martinetti, D., Montes, I., Díaz, S., Montes, S.: A study on the transitivity of probabilistic and fuzzy relations. Fuzzy Sets and Systems 184, 156–170 (2011)
[16] Martinetti, D., Montes, S., Díaz, S., De Baets, B.: Some comments to the fuzzy version of the Arrow-Sen theorem. In: Greco, S., Bouchon-Meunier, B., Coletti, G., Fedrizzi, M., Matarazzo, B., Yager, R.R. (eds.) IPMU 2012, Part IV. CCIS, vol. 300, pp. 286–295. Springer, Heidelberg (2012)
[17] Martinetti, D., De Baets, B., Díaz, S., Montes, S.: On the role of acyclicity in the study of rationality of fuzzy choice functions (under revision)
[18] Richardson, G.: The structure of fuzzy preference: Social choice implications. Social Choice and Welfare 15, 359–369 (1998)
[19] Richter, M.K.: Revealed preference theory. Econometrica 34, 635–645 (1996)
[20] Samuelson, P.A.: Foundation of Economic Analysis, 604 pages. Harvard University Press, Cambridge (1983)
[21] Sen, A.K.: Choice functions and revealed preference. Review of Economic Studies 38, 307–317 (1971)
[22] Sen, A.K.: Collective Choice and Social Welfare. Holden-Day, San Francisco (1970)

[23] Suzumura, K.: Rational choice and revealed preference. Review of Economic Studies 46, 149–158 (1976)
[24] Uzawa, I.: Preference and Rational Choice in the Theory of Consumption. In: Arrow, K.J., Karlin, S., Suppes, P. (eds.) Mathematical Methods in the Social Sciences. Stanford University Press, Stanford (1959)
[25] Wang, X.: A note on congruence conditions of fuzzy choice functions. Fuzzy Sets and Systems 145, 355–358 (2004)
[26] Wang, X., Wu, C., Wu, X.: Choice Functions in Fuzzy Environment: An Overview. In: Cornelis, C., Deschrijver, G., Nachtegael, M., Schockaert, S., Shi, Y. (eds.) 35 Years of Fuzzy Set Theory. STUDFUZZ, vol. 261, pp. 149–169. Springer, Heidelberg (2010)
[27] Wu, C., Wang, X., Hao, Y.: A further study on rationality conditions of fuzzy choice functions. Fuzzy Sets and Systems 176, 1–19 (2011)

Bayes Theorem, Uninorms and Aggregating Expert Opinions

József Dombi

Abstract. In the introduction we examine Dombi aggregative operators, uninorms, strict t-norms and t-conorms. We give a certain class of weighted aggregative operators (weighted representable uninorms). After, we focus on a specific form of the aggregative operator. Using Dombi's generator function, we show that this form is the same as that for the aggregation of expert probability values, and we can get this operator via Bayes' theorem. These two theorems shed new light on the class of aggregative operators.

1 Introduction

The term *uninorm* was first introduced by Yager and Rybalov [17] in 1996. Uninorms are a generalization of t-norms and t-conorms achieved by relaxing the constraint on the identity element in the unit interval $\{0, 1\}$. Since then many articles have focused on uninorms, both from a theoretical [11, 12, 6, 14, 3, 15] and a practical point of view [16]. The paper of Fodor, Yager and Rybalov [5] is notable since it defined a new subclass of uninorms called representable uninorms. This characterization is similar to the representation theorem of strict t-norms and t-conorms, in the sense that both originate from the solution of the associativity functional equation given by Aczél [1].

In this article we will show that the weighted form of the aggregative operator then we use the Dombi's generator function $\left(\frac{1-x}{x}\right)$ clearly related to certain aspects of probability theory and decision making. In the first section, we summarize the basic results of uninorm and aggregation operators.

In section 3, we will show that the aggregative operator is closely related to the aggregation of expert probability values and also show how it can be used to com-

József Dombi
University of Szeged, 6720 Szeged, Árpád tér 2, Hungary
e-mail: dombi@inf.u-szeged.hu

H. Bustince et al. (eds.), *Aggregation Functions in Theory and in Practise*,
Advances in Intelligent Systems and Computing 228,
DOI: 10.1007/978-3-642-39165-1_29, © Springer-Verlag Berlin Heidelberg 2013

pute the aggregative value. In section 4, we show that the aggregation operator can be derived by using Bayes' theorem.

2   Uninorm and Aggregative Operator

The aggregative operators were first introduced in [4] by selecting a set of minimal concepts that must be fulfilled by an evaluation-like operator.

In 1982, Dombi [4] defined the aggregative operator in the following way:

Definition 1. An aggregative operator is a function $a : [0,1]^2 \to [0,1]$ with the properties:

1. Continuous on $[0,1]^2 \backslash \{(0,1),(1,0)\}$
2. $a(x,y) < a(x,y')$ if $y < y', x \neq 0, x \neq 1$
   $a(x,y) < a(x',y)$ if $x < x', y \neq 0, y \neq 1$
3. $a(0,0) = 0$ and $a(1,1) = 1$ (boundary conditions)
4. $a(x,a(y,z)) = a(a(x,y),z)$ (associativity)
5. There exists a strong negation η such that $a(x,y) = \eta(a(\eta(x),\eta(y)))$ (self-DeMorgan identity) if $\{x,y\} \neq \{0,1\}$
6. $a(1,0) = a(0,1) = 0$ or $a(1,0) = a(0,1) = 1$

The definition of uninorms, originally given by Yager and Rybalov [17] in 1996, is the following:

Definition 2. A uninorm U is a mapping $U : [0,1]^2 \to [0,1]$ having the following properties:

- $U(x,y) = U(y,x)$ (commutativity)
- $U(x_1,y_1) \geq U(x_2,y_2)$ if $x_1 \geq x_2$ and $y_1 \geq y_2$ (monotonicity)
- $U(x,U(y,z)) = U(U(x,y),z)$ (associativity)
- $\exists v_* \in [0,1] \; \forall x \in [0,1] \; U(x,v_*) = x$ (neutral element)

A uninorm is a generalization of t-norms and t-conorms. By adjusting its neutral element, a uninorm is a t-norm if $v_* = 1$ and a t-conorm if $v_* = 0$. The main difference in the definition of the uninorms and aggregative operators is that the self-DeMorgan identity requirement does not appear in uninorms, and the neutral element property is not in the definition for the aggregative operators. The following representation theorem of strict, continuous on $[0,1] \times [0,1] \backslash \{(0,1),(1,0)\}$ uninorms (or *representable uninorms*) was given by Fodor et al. [5] (see also Klement et al. [20]).

Theorem 1. *Let* $U : [0,1]^2 \to [0,1]$ *be a function and* $v_* \in]0,1[$. *The following are equivalent:*

1. *U is a uninorm with neutral element* v_* *which is strictly monotone on* $]0,1[^2$ *and continuous on* $[0,1]^2 \backslash \{(0,1),(1,0)\}$.

2. *There exists a strictly increasing bijection* $g_u : [0,1] \to [-\infty, \infty]$ *with* $g_u(v_*) = 0$
such that for all $(x,y) \in [0,1]^2$, *we have*

$$U(x,y) = g_u^{-1}(g_u(x) + g_u(y)),\tag{1}$$

where, in the case of a conjunctive uninorm U, *we use the convention* $\infty + (-\infty) = -\infty$, *while, in the disjunctive case, we use* $\infty + (-\infty) = \infty$ *or there exists a strictly increasing continuous function* $f_u : [0,1] \to [0,\infty]$ *with* $f_u(0) = 0$
, $f(v_*) = 1$ *and* $f_u(1) = \infty$. *The binary operator is defined by*

$$U(x,y) = f_u^{-1}(f_u(x)f_u(y))\tag{2}$$

for all $(x,y) \in [0,1] \times [0,1]/(0,1),(1,0)$ *and either* $U(0,1) = U(1,0) = 0$
or $U(0,1) = U(1,0) = 1$.

If Eq.(1) holds, the function g_u *is uniquely determined by* U *up to a positive multiplicative constant, and it is called an additive generator of the uninorm* U. *Here,* f_u *is called the multiplicative generator function of the operator.*

Such uninorms are called representable uninorms and they were previously introduced as aggregative operators [4].

Definition 3. A representable uninorm is called an aggregative operator. We will denote it by $a(x,y)$.

In the article by János Fodor and Bernard De Baets the authors give a fine characterization of representable uninorms, i. e. if the underlying uninorms are strict and the remaining part of the unit square at a single point of the uninorm lies between the minimum and maximum, then the uninorm is representable [24].

A recent paper of Sándor Jenei also deals with uninorms; namely, it is used to characterize all types of uninorms that have only a few basic properties. He shows that only three general types of uninorms exist. More details can be found in [25], [26], [27].

Since an aggregative operator (i. e. a representable uninorm) is associative it can be extended via associativity to n arguments.

Theorem 2 (Dombi [4])

Let g *be an additive generator of an aggregative operator (i. e. representable uninorm) and consider* $v_* \in (0,1)$, *then* $a_{v_*} : [0,1]^2 \to [0,1]$ *defined by*

$$a_{v_*}(x,y) = g^{-1}(g(x) + g(y) - g(v_*))$$

is an aggregation operator (i. e. representable uninorm) with neutral element v_*.
The extension to n *arguments is given by the formula*

$$a_{v_*}(\mathbf{x}) = g^{-1}(g(v_*)) + \sum_{i=1}^{n}(g(x_i) - g(v_*)).$$

The general form of the weighted operator in the additive representation case is

$$a_{v_*}(\mathbf{w},\mathbf{x}) = g_a^{-1}\left(\sum_{i=1}^{n} w_i g_a(x_i) + g_a(v_*)\left(1 - \sum_{i=1}^{n} w_i\right)\right) \tag{3}$$

In general, a weighted aggregative operator loses associativity and commutativity and is not a representable uninorm.

With this, one can construct an aggregative operator from any given generator function that has the desired neutral value.

There are infinitely many possible neutral values, and with each different neutral value, there is a different aggregative operator.

We will use the transformation defined in $g(x) = ln(f(x))$ to get the multiplicative operator

$$a_{v_*}(x) = f_a^{-1}\left(f_a(v_*)\prod_{i=1}^{n}\frac{f_a(x_i)}{f_a(v_*)}\right) = f_a^{-1}\left(f_a^{1-n}(v_*)\prod_{i=1}^{n}f_a(x_i)\right), \tag{4}$$

where $f_a : [0,1] \to [0,\infty]$.

The multiplicative form of the weighted aggregative operator is

$$a_{v_*}(\mathbf{w},\mathbf{x}) = f_a^{-1}\left(f_a(v_*)\prod_{i=1}^{n}\left(\frac{f_a(x)}{f_a(v_*)}\right)^{w_i}\right) = f_a^{-1}\left(f_a^{1-\sum_{i=1}^{n}w_i}(v_*)\prod_{i=1}^{n}f_a^{w_i}(x_i)\right) \tag{5}$$

In the following we will use the multiplication form of the aggregative operator.

The corresponding negation of the aggregative operator is

$$\eta(x) = f_a^{-1}\left(\frac{f_a^2(v_*)}{f_a(x)}\right)$$

In the Dombi operator case,

$$a_{v_*}(\mathbf{w},\mathbf{x}) = \cfrac{1}{1+\cfrac{1-v_*}{v_*}\prod_{i=1}^{n}\left(\cfrac{1-x_i}{x_i}\cfrac{v_*}{1-v_*}\right)^{w_i}} \tag{6}$$

$$a_{v_*}(\mathbf{w},\mathbf{x}) = \cfrac{v_*(1-v_*)^{\sum_{i=1}^{n}w_i}\prod_{i=1}^{n}x_i^{w_i}}{v_*(1-v_*)^{\sum_{i=1}^{n}w_i}\prod_{i=1}^{n}x_i^{w_i}+(1-v_*)v_*^{\sum_{i=1}^{n}w_i}\prod_{i=1}^{n}(1-x_i)^{w_i}} \tag{7}$$

If $w_i = 1$, then

$$a(\mathbf{w},\mathbf{x}) = \cfrac{\prod_{i=1}^{n}x_i^{w_i}}{\prod_{i=1}^{n}x_i^{w_i}+\prod_{i=1}^{n}(1-x_i)^{w_i}}. \tag{8}$$

$$a_{v_*}(\mathbf{x}) = \frac{(1-v_*)^{n-1}\prod\limits_{i=1}^{n}x_i}{(1-v_*)^{n-1}\prod\limits_{i=1}^{n}x_i + v_*^{n-1}\prod\limits_{i=1}^{n}(1-x_i)} \tag{9}$$

If $v_* = \frac{1}{2}$, then we get

$$a_{\frac{1}{2}}(\mathbf{x}) = \frac{\prod\limits_{i=1}^{n}x_i}{\prod\limits_{i=1}^{n}x_i + \prod\limits_{i=1}^{n}(1-x_i)}. \tag{10}$$

Eq. (10) is called the $3\,\Pi$ operator because it consists of three product operators. This operator was first introduced by Dombi [4].

3 Aggregative Operator and the Aggregation of Expert Probability Values

Here, we will show that the aggregative operator is closely related to the aggregation of expert probability values and also show how it can be used to compute the aggregative value.

A decision maker consults a group of experts labelled 1,2,...,n who individually assess probability values of $p_1, p_2, \ldots, p_n$ for the occurrence of the event E. Based on his feelings of how reliable the experts are and how independent their opinions are, the decision maker would like to form his own probability assessment p for the occurrence of the event E. With these probability estimates, Bordley's paper [2] develops a mathematical formula that relates p to $p_1, p_2, \ldots, p_n$.

A generalization of these problems leads to the Logarithmic Opinion Pool (LOP) model. This class of problem is very important in different fields of economics theory. In the article by C. Genest and J. V. Zidek [7] the procedures for combining probability distribution values are described. This article is virtually the best survey on this topic. In 2002, G. L Gilardoni [8] gave a fine axiomatic characterization of the LOP procedure.

Bordley uses an axiomatic approach. The first part begins by reformulating the problem in terms of statistical odds:

$$\left(o = \frac{p}{1-p}, o_k = \frac{p_k}{1-p_k}, k = 1, 2, \ldots, n \right). \tag{11}$$

Then he states axioms which imply that

$$o = F^E\left(\sum_{k=1}^{n} u_k^E(o_k) \right) \tag{12}$$

for continuous functions $F(u_1,\ldots,u_n)$. In essence, he applied axioms taken from the theory of additive conjoint measurements.

Consider some other event, A, relevant to the occurrence of event, E. Let $o(E/A)$ be the assessment the decision maker makes for the conditional event, (E/A). Then

$$o(E/A) = F^{E/A}\left(\sum_{k=1}^{n} u_k^{E/A}(o_k(E/A))\right), \tag{13}$$

where $o_k(E/A)$ is expert k's assessment of the odds favouring the conditional event, (E/A).

Let us consider the assessment of (not $E = \overline{E}$). The formula for it is

$$o(\overline{E}) = F^{\overline{E}}\left[\sum_{k=1}^{n} u_k^{\overline{E}}(o_k(\overline{E}))\right]. \tag{14}$$

Now we suppose that the following equations are valid, based on conjoint measurement axioms.

$$o(E) = F^E\left(\sum_{i=1}^{n} u_i^E(o_i(E))\right)$$

$$o(E/A) = F^{E/A}\left(\sum_{i=1}^{n} u_i^{E/A}(o_i(E/A))\right) \tag{15}$$

$$o(\overline{E}) = F^{\overline{E}}\left(\sum_{i=1}^{n} u_i^{\overline{E}}(o_i(\overline{E}))\right)$$

Let us introduce the following notation:

$x^* = o(E) \quad F^E(x) = F(x) \quad u^E(x) = f(x) \quad x_i = o_i(E)$

$y^* = o(E/A) \; F^{E/A}(y) = G(y) \; u^{E/A}(y) = g(y) \; y_i = o_i(E/A)$

$z^* = o(\overline{E}) \quad F^{\overline{E}}(z) = H(z) \quad u^{\overline{E}}(z) = h(z) \quad z_i = o_i(\overline{E})$

So the formulas in (15) have the form:

$$x^* = F(\sum_{i=1}^{n} f(x_i)), \quad y^* = G(\sum_{i=1}^{n} g(y_i)), \quad z^* = H(\sum_{i=1}^{n} h(z_i)) \tag{16}$$

Let α be the likelihood ratio, i.e.,

$$\alpha = \frac{p(A|E)}{p(A|\overline{E})}. \tag{17}$$

Bordley in his article supposed this is public knowledge; i.e. all experts and the decision maker agree on α.

The Weak Likelihood Ratio Axiom is

$$o(E|A) = \alpha o(E) \quad o_j(E|A) = \alpha o_j(E), j = 1, \ldots, n,$$

i.e.

$$y^* = \alpha x^* \quad y_j = \alpha x_j$$

The basis of Bordley's proof is that the general solution of the functional equation

$$G\left(\sum_{i=1}^{n} g(\alpha x_i)\right) = \alpha F\left(\sum_{i=1}^{n} f(x_i)\right) \tag{18}$$

is either

$$F(x) = (x + a_1)^{\frac{1}{c}} \qquad G(x) = (\alpha x + a_2)^{\frac{1}{c}}$$

I. $f(x) = wx^c$ $\qquad g(x) = w(\alpha x)^c$

$a_1 = (1 - \sum_{i=1}^{n} w_i)(x_0)^c$ $a_2 = (1 - \sum_{i=1}^{n} w_i)(\alpha x_0)^c$
$F(x) = a_1 e^x$ $\qquad G(x) = a_2 e^{\alpha x}$

II. $f(x) = w \ln(x)$ $\qquad g(x) = w \ln(\alpha x)$

$a_1 = (1 - \sum_{i=1}^{n} w_i) \ln(x_0)$ $a_2 = (1 - \sum_{i=1}^{n} w_i) \ln(\alpha x_0)$

Now another reasonable condition is:

$$\begin{aligned} o(E)o(\overline{E}) &= 1 \\ o_k(E)o_k(\overline{E}) &= 1, \end{aligned} \tag{19}$$

i.e. the product of the odds of E and $\overline{E}$ is 1.

The Normalization Axiom

$$F^E\left(\sum_{k=1}^{n} u_k^E(o_k)\right) F^{\overline{E}}\left(\sum_{k=1}^{n} u_k^E\left(\frac{1}{o_k}\right)\right) = 1. \tag{20}$$

It is not hard to verify that only the second solution fulfils (19), hence

$$x^* = x_0^{(1 - \sum_{i=1}^{n} w_i)} \prod_{i=1}^{n} x_i^{w_i}$$

Let us make the following substitution:

$$x^* = \frac{p}{1-p} \quad x_i = \frac{p_i}{1-p_i} \quad x_0 = \frac{p_0}{1-p_0} \tag{21}$$

Then the solution for p is

$$p = \frac{1}{1 + \frac{1-p_0}{p_0} \prod_{i=1}^{n} \left(\frac{p_0}{1-p_0} \frac{1-p_i}{p_i} \right)^{w_i}}, \tag{22}$$

where p_0 is the prior probability.

This formula is the same as the weighted aggregative operator of pliant logic.

In Bordley's paper, there is a different formula for (22). However, by rearranging it we can get the Bordley formula. That is,

$$p = \frac{\prod_{k=1}^{n}(p_k/p_0)^{w_k} p_0}{\prod_{k=1}^{n}(p_k/p_0)^{w_k} p_0 + \prod_{k=1}^{n}((1-p_k)/(1-p_0))^{w_k}(1-p_0)} \tag{23}$$

Now, we can find the corresponding negation formula for the aggregation of the expert opinion.

Let

$$\eta(p_i) = \frac{1}{1 + \left(\frac{1-p_0}{p_0} \right)^2 \frac{p_i}{1-p_i}}, i = 1, \ldots, n$$

So if p_i is given, it has a probability value of E_i, and p_0 is an apriori probability. Then $\eta(p_i)$ is the possibility of the complementer of the E_i event. Because the self-DeMorgan identity is valid, substituting p_i by its negated value we get the negated value of p.

Example 1. Let $[0.2, 0.8, 0.3, 0.2, 0.3]$ be the experts' opinions of the probability of a certain event and let $(1, 7, 2, 3, 2)$ be the (weighted) importance vector of opinions. The aggregated value (if our probability assessment is 0.5) is

$$\frac{1}{1 + (\frac{1-0.2}{0.2})(\frac{1-0.8}{0.8})^7(\frac{1-0.3}{0.3})^2(\frac{1-0.2}{0.2})^3(\frac{1-0.3}{0.3})^2} = 0.6834 \tag{24}$$

4   Bayes' Theorem and the Aggregative Operator

Here, we show that the aggregation operator can be derived by using Bayes' theorem. This theorem assumes that there is a conditional independence among the events $B_i, i = 1, \ldots, n$ given A. This corresponds to the same assumption made for the Naive-Bayes model in statistical classifications and it is usually depicted as a Bayesian network [23, 18, 19, 20, 21], see Figure 1.

The conditional independence assumption means that there is no connection between any of the attributes. We can write

$$P(B_n|A, B_1 \ldots B_{n-1}) = P(B_n|A)$$
$$P(B_{n-1}|A, B_1 \ldots B_{n-2}) = P(B_{n-1}|A)$$

$$\vdots$$

$$P(B_2|A, B_1) = P(B_2|A)$$

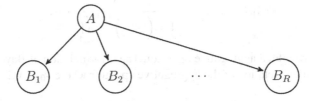

Fig. 1 Bayesian network representing the Naive Bayes classifier

Because the recursive application of Bayes' law is

$$P(A|B_1\ldots B_n) = \frac{P(B_n|A,B_1\ldots B_{n-1})P(B_{n-1}|A,B_1\ldots B_{n-2})\ldots P(B_1|A)P(A)}{P(B_1\ldots B_n)},$$

we get

$$P(A|B_1\ldots B_n) = \frac{P(B_n|A)P(B_{n-1}|A)\ldots P(B_1|A)P(A)}{P(B_1\ldots B_n)} \tag{25}$$

With a similar assumption for $\overline{A}$ (the complement of A),

$$P(\overline{A}|B_1\ldots B_n) = \frac{P(B_n|\overline{A})P(B_{n-1}|\overline{A})\ldots P(B_1|\overline{A})P(\overline{A})}{P(B_1\ldots B_n)} \tag{26}$$

Taking the ratio of (25) and (26)

$$\frac{P(\overline{A}|B_1\ldots B_n)}{P(A|B_1\ldots B_n)} = \frac{P(B_n|\overline{A})P(B_{n-1}|\overline{A})\ldots P(B_1|\overline{A})P(\overline{A})}{P(B_n|A)P(B_{n-1}|A)\ldots P(B_1|A)P(A)}$$

Moreover, since this can be rewritten as

$$\frac{P(\overline{A}|B_1\ldots B_n)}{P(A|B_1\ldots B_n)} = \frac{\frac{P(A|B_n)P(B_n)}{P(A)}\frac{P(A|B_{n-1})P(B_{n-1})}{P(A)}\ldots\frac{P(A|B_1)P(B)}{P(A)}P(A)}{\frac{P(A|B_n)P(B_n)}{P(A)}\frac{P(A|B_{n-1})P(B_{n-1})}{P(A)}\ldots\frac{P(A|B_1)P(B)}{P(A)}P(A)}$$

$$P(B_i|A) = \frac{P(AB_i)}{P(A)} = \frac{P(A|B_i)P(B_i)}{P(A)}.$$

Since the probability of the complement of an event is 1-the probability of that event, we can write

$$P(A|B_1\ldots B_n) = \frac{1}{1 + \left(\frac{P(A)}{P(\overline{A})}\right)^{n-1}\prod_{i=1}^{n}\frac{P(\overline{A}|B_i)}{P(A|B_i)}}$$

Let $P(A) = v_0$, then $P(\overline{A}) = 1 - v_0$ and $P(\overline{A}|B_i) = 1 - x_i$, then $P(A|B_i) = x_i$, so

$$P(A|B_1,\ldots,B_n) = \frac{1}{1 + \left(\frac{v_0}{1-v_0}\right)^{n-1} \prod\limits_{i=1}^{n} \frac{1-x_i}{x_i}}.$$
(27)

and this is the weighted aggregative operator. The reformulation of Bayes' theorem gives the new interpretation of the aggregative operator where $v_0 = v_*$.

5 Conclusions

In this paper it is recalled the equivalence between the concepts of representable uninorm and aggregative operator in the sense of 4 (See for example the paragraph above Theorem 1 in 21). We studied a certain class of weighted aggregative operators (representable uninorms). We showed that there is correspondence between Bayes' theorem and aggregative operators, and also showed that there is a correspondence between the aggregation of expert opinion and aggregative operators.

Acknowledgements. The publication/presentation is supported by the European Union and co-funded by the European Social Fund. Project title: "Telemedicine-focused research activities on the field of Mathematics, Informatics and Medical sciences" Project number: TÁMOP-4.2.2.A-11/1/KONV-2012-0073

References

[1] Aczél, J.: Lectures on functional equations and their applications. Academic Press, New York (1966)
[2] Bordley, R.F.: A multiplicative formula for aggregating probability assessments. Management Science 28, 1137–1148 (1982)
[3] Calvo, T., Baets, B.D., Fodor, J.: The functional equations of frank and alsina for uninorms and nullnorms. Fuzzy Sets and Systems 120, 385–394 (2001)
[4] Dombi, J.: Basic concepts for a theory of evaluation: the aggregative operator. European Journal of Operations Research 10, 282–293 (1982)
[5] Fodor, J., Yager, R.R., Rybalov, A.: Structure of uninorms. International Journal of Uncertainty, Fuzziness and Knowledge-Based Systems 5(4), 411–427 (1997)
[6] Hu, S.-K., Li, Z.-F.: The structure of continuous uni-norms. Fuzzy Sets and Systems 124, 43–52 (2001)
[7] Genest, C., Zidek, J.V.: Combining Probability Distributions: A Critique and an Annotated Bibliography. Statistical Science 1(1), 114–135 (1986)
[8] Gilardoni, G.L.: On irrevelance of alternatives and opinion pooling. Brazilian Journal of Probability and Statistics 16, 87–98
[9] Klement, E.P., Mesiar, R., Pap, E.: Triangular norms. Kluwer (2000)
[10] Klement, E.P., Mesiar, R., Pap, E.: On the relationship of associative compensatory operators to triangular norms and conorms. Uncertainty, Fuzziness and Knowledge-Based Systems 4, 129–144 (1996)
[11] Li, Y.-M., Shi, Z.-K.: Weak uninorm aggregation operators. Information Sciences 124, 317–323 (2000)

[12] Li, Y.M., Shi, Z.K.: Remarks on uninorms aggregation operators. Fuzzy Sets and Systems 114, 377–380 (2000)

[13] Li, J., Mesiar, R., Struk, P.: Pseudo-optimal measures. Information Sciences 180(20), 4015–4021 (2010)

[14] Mas, M., Mayor, G., Torrens, J.: The distributivity condition for uninorms and t-operators. Fuzzy Sets and Systems 128, 209–225 (2002)

[15] Monserrat, M., Torrens, J.: On the reversibility of uninorms and t-operators. Fuzzy Sets and Systems 131, 303–314 (2002)

[16] Yager, R.R.: Uninorms in fuzzy systems modeling. Fuzzy Sets and Systems 122, 167–175 (2001)

[17] Yager, R.R., Rybalov, A.: Uninorm aggregation operators. Fuzzy Sets and Systems 80(1), 111–120 (1996)

[18] Ramoni, M., Sebastiani, P.: Robust Bayes classifiers. Artificial Intelligence 125, 209–226 (2001)

[19] Frank, E., Trigg, L., Holmes, G., Witten, I.H.: Technical note: Naive Bayes for regression. Machine Learning 41, 5–25 (2000)

[20] Friedman, J.H.: On bias, variance, 0/1 loss, and the curse-of-dimensionality. Data Mining and Knowledge Discovery 1, 55–77 (1997)

[21] Friedman, J.H., Geiger, D., Goldszmidt, M.: Bayesian network classifiers. Machine Learning 29, 131–163 (1997)

[22] Triangular Norms. Kluwer, Dordrecht (2000)

[23] Ortiz, J.M., Deutsch, C.V.: Indicator simulationAccounting for Multiple-Point Statistics. Mathematical Geology 36(5), 545–565 (2004)

[24] Fodor, J., De Baets, B.: A single-point characterization of representable uninorms. Fuzzy Sets and Systems 202, 89–99 (2012)

[25] Jenei, S., Montagna, F.: Strongly involutive uninorm algebras. Journal of Logic and Computation (2012), doi:10.1093/logcom/exs019

[26] Jenei, S.: Structural description of a class of involutive uninorms via skew symmetrization. Journal of Logic and Computation 21(5), 729–737 (2011)

[27] Jenei, S., Montagna, F.: Classification of absorbent-continuous, sharp FL_e algebras over weakly real chains (under review process)

Part VI
Applications

On the Induction of New Fuzzy Relations, New Fuzzy Operators and Their Aggregation

Neus Carmona, Jorge Elorza, Jordi Recasens, and Jean Bragard

Abstract. In this paper we generate fuzzy relations and fuzzy operators using different kind of generators and we study the relationship between them. Firstly, we introduce a new fuzzy preorder induced by a fuzzy operator. We generalize this preorder to a fuzzy relation generated by two fuzzy operators and we analyze its properties. Secondly, we introduce and explore two ways of inducing a fuzzy operator, one from a fuzzy operator and a fuzzy relation and the other one from two fuzzy operators. The first one is an extension of the well-known fuzzy operator induced by a fuzzy relation through Zadeh's compositional rule. Finally, we aggregate these operators using the quasi-arithmetic mean associated to a continuous Archimedean t-norm. The aim is to compare the operator induced by the quasi-arithmetic mean of the generators with the quasi-arithmetic mean of the generated operators.

1 Introduction

Fuzzy relations and fuzzy consequence operators are main concepts in fuzzy logic. The fuzzy relation induced by a fuzzy operator and the fuzzy operator induced by a fuzzy relation through Zadeh's compositional rule are notions that have been extensively explored (see for instance [2, 3, 5, 6, 7, 9]).

In Section 2 we recall the main definitions and results that will be used throughout the paper.

Neus Carmona · Jorge Elorza · Jean Bragard
Departamento de Física y Matemática Aplicada, Facultad de Ciencias, Universidad de Navarra, P.O. Box 31008 Pamplona, Spain
e-mail: ncarmona@alumni.unav.es,{jelorza,jbragard}@unav.es

Jordi Recasens
ETS Arquitectura del Vallès, Department of Mathematics and Computer Sciences,
c/ Pere Serra 1-15, P.O. Box 08190 Sant Cugat del Vallès, Spain
e-mail: j.recasens@upc.edu

H. Bustince et al. (eds.), *Aggregation Functions in Theory and in Practise,*
Advances in Intelligent Systems and Computing 228,
DOI: 10.1007/978-3-642-39165-1_30, © Springer-Verlag Berlin Heidelberg 2013

In Section 3 we introduce a fuzzy preorder R_c^c induced by a fuzzy operator c such that collects the information of c over all the fuzzy subsets of the universal set. Recall that the classical relation induced by a fuzzy operator only considers the information over the singletons. We generalize this preorder to a fuzzy relation R_f^g induced by two fuzzy operators f and g and study its properties. From a logical point of view, R_f^g stablishes a crossed relation between the consequences of g and the consequences of f.

In Section 4 we define two new operators C_R^g and C_f^g. The first one is induced by a fuzzy relation and a fuzzy operator and the second one is induced by two fuzzy operators. We explore the properties that are transmitted from the generators. In particular, we show for wich cases the properties of a fuzzy consequence operator (inclusion, monotony and idempotence) and the coherence property hold.

In Section 5 we use the quasi-arithmetic mean associated to a continuous Archimedean t-norm to aggregate these induced fuzzy operators. We study the difference between two cases. In the first one, we consider the operator generated by the quasi-arithmetic mean of some fuzzy operators. In the second one, we aggregate the operators induced by each of fuzzy operators individually.

Finally, in Section 6 we present the conclusions.

2 Preliminaries

Let $\langle L, \wedge, \vee, *, \rightarrow, 0, 1 \rangle$ be a complete commutative residuated lattice in the sense of Bělohlávek [1]. That is, a complete lattice $\langle L, \wedge, \vee, 0, 1 \rangle$, where 0 denotes the least element and 1 denotes the greatest one, such that $(L, *)$ is a commutative monoid i.e. $*$ is associative, commutative and with neutral element 1, and the operations $*$ and $\rightarrow$ satisfy the adjointness property:

$$x * y \leq z \quad \Leftrightarrow \quad y \leq x \rightarrow z$$

where $\leq$ denotes the lattice ordering.

Let us recall in Propositions 1 and 2 the following properties of commutative residuated lattices (residuated lattices for short) [1] that will be used in the paper.

Proposition 1. *Each residuated lattice* $\langle L, \wedge, \vee, *, \rightarrow, 0, 1 \rangle$ *satisfies the following conditions for all* $x, y, z \in X$:

1. $x \rightarrow x = 1$
2. $1 \rightarrow x = x$
3. $x \leq y \Leftrightarrow x \rightarrow y = 1$
4. $x * 0 = 0$
5. $x * (x \rightarrow y) \leq y$
6. $(x \rightarrow y) * (y \rightarrow z) \leq (x \rightarrow z)$

Proposition 2. *Let* $\langle L, \wedge, \vee, *, \rightarrow, 0, 1 \rangle$ *be a residuated lattice. The following conditions hold for each index set I whenever both sides of the (in)equality exist. In the first case, if the left hand side makes sense, so does the right one. For all* $x, y_i \in L$ *with* $i \in I$,

1. $x * \bigvee_{i \in I} y_i = \bigvee_{i \in I} (x * y_i)$
2. $x * \bigwedge_{i \in I} y_i \leq \bigwedge_{i \in I} (x * y_i)$

The frame for our work will be the complete commutative residuated lattice $\langle [0,1], \wedge, \vee, *, \rightarrow, 0, 1 \rangle$ where $\wedge$ and $\vee$ are the usual infimum and supremum, $*$ is a left continuous t-norm and $\rightarrow$ is the residuum of $*$ defined for $\forall a, b \in X$ as $a \rightarrow b = \sup\{\gamma \in [0,1] \mid a * \gamma \leq b\}$. Recall that a t-norm is monotone in both arguments and the residuum is antitone in the first argument and monotone in the second one.

In this paper, X will be a non-empty classical universal set, $[0,1]^X$ will be the set of all fuzzy subsets of X, Γ' will denote the set of all fuzzy relations defined on X and Ω' the set of fuzzy operators defined from $[0,1]^X$ to $[0,1]^X$.

Definition 1. *(Fuzzy Consequence Operator)* A fuzzy operator $C \in \Omega'$ is called a fuzzy consequence operator when it satisfies for all $\mu, \nu \in [0,1]^X$:

(C1) Inclusion $\mu \subseteq C(\mu)$
(C2) Monotony $\mu \subseteq \nu \Rightarrow C(\mu) \subseteq C(\nu)$
(C3) Idempotence $C(C(\mu)) = C(\mu)$

The inclusion of fuzzy subsets is given by the puntual order, i.e. $\mu \subseteq \nu$ if and only if $\mu(x) \leq \nu(x)$ for all $x \in X$.

Definition 2. *(Coherent Fuzzy Operator)* Let $C \in \Omega'$ be a fuzzy operator in Ω'. We say that C is coherent if it satisfies for all $x, a \in X$ and $\mu \in [0,1]^X$

$$\mu(a) * C(\{a\})(x) \leq C(\mu)(x)$$

Let us look back on some properties of fuzzy relations. A fuzzy relation on X is said to be:

(R) Reflexive if $R(x,x) = 1 \quad \forall x \in X$
(S) Symmetric if $R(x,y) = R(y,x) \quad \forall x, y \in X$
(T) $*$-Transitive if $R(x,y) * R(y,z) \leq R(x,z) \quad \forall x, y, z \in X$

A fuzzy relation satisfying (R) and (T) is called a fuzzy preorder. If it also satisfies (S), then it is called a fuzzy similarity or indistinguishability operator. Given R and S fuzzy relations, we say that $R \leq S$ if and only if $R(x,y) \leq S(x,y)$ for all $x, y \in X$.

For a given fuzzy relation R, a fuzzy subset μ of X is called $*$-compatible with R if $\mu(x) * R(x,y) \leq \mu(y)$ for all $x, y \in X$. From its logical implications, these sets are also called true-sets or closed under modus ponens. This notion gets special interest when R is a preorder [3]. When R is not only a preorder but also an indistinguishability operator, these sets are called *extensional sets* and the set of all these subsets has very interesting properties [9].

Every fuzzy operator induces a fuzzy relation in a very natural way and every fuzzy relation also induces a fuzzy operator using Zadeh's compositional product:

Definition 3. Let C be a fuzzy operator in Ω'. The fuzzy relation induced by C is given by

$$R_C(x,y) = C(\{x\})(y) \tag{1}$$

where $\{x\}$ denotes the singleton x.

Definition 4. Let $R \in \Gamma'$ be a fuzzy relation on X. The fuzzy operator induced by R through Zadeh's compositional rule is defined by

$$C_R^*(\mu)(x) = \sup_{w \in X}\{\mu(w) * R(w,x)\} \tag{2}$$

These concepts are strongly connected and they have been extensively explored in several contexts (see for instance [2, 3, 5, 6, 7, 9]).

3 Relation Induced by Fuzzy Operators

Notice that the relation induced by (1) only takes into account the behaviour of C over the singletons and not over more general fuzzy subsets. In order to include this information, we define a new fuzzy relation induced by a fuzzy operator in a different way.

Definition 5. Let c be a fuzzy operator in Ω'. The fuzzy relation R_c^c induced by c is given by

$$R_c^c(x,y) = \inf_{\mu \in [0,1]^X}\{c(\mu)(x) \to c(\mu)(y)\} \tag{3}$$

It is easy to see that this relation is a fuzzy preorder on X since it is the infimum of a family of preorders. From a logical point of view, the crisp interpretation of this relation would be

$x \leq y$ (or related to y) $\Leftrightarrow \forall A \subseteq X$, if x is a consequence of A then y is also a consequence of A

Notice that if c is an inclusive operator, then $R_c^c \leq R_c$. In fact, for all $x,y \in X$ we have $R_c^c(x,y) = \inf_{\mu \in [0,1]^X}\{c(\mu)(x) \to c(\mu)(y)\} \leq c(\{x\})(x) \to c(\{x\})(y) = R_c(x,y)$.

In Definition 6 we generalize the previous definition to the fuzzy relation R_f^g induced by two fuzzy operators f and g. R_f^g is a crossed relation whose logical interpretation in the crisp case would be the following

x is related to y $\Leftrightarrow$ whenever x is a consequence by g of some subset A, then y is a consequence of the same subset by f.

Definition 6. Let f and g be fuzzy operators in Ω'. The fuzzy relation R_f^g induced by f and g is defined by

$$R_f^g(x,y) = \inf_{\mu \in [0,1]^X}\{g(\mu)(x) \to f(\mu)(y)\}$$

g and f will be called the upper and lower generators of R_f^g respectively.

In Propositions 3 and 4 we study the reflexive and $*$-trasitive properties of R_f^g.

Proposition 3. *Let f and g be fuzzy operators in Ω'. Then, R_f^g is reflexive if and only if $g \leq f$, i.e. $g(\mu)(x) \leq f(\mu)(x)$ for all $\mu \in [0,1]^X$ and $x \in X$.*

Proof.

R_f^g is reflexive $\Leftrightarrow$ $R_f^g(x,x) = 1 \quad \forall x \in X$

$$\Leftrightarrow \inf_{\mu \in [0,1]^X} \{g(\mu)(x) \to f(\mu)(x)\} = 1 \quad \forall x \in X$$

$$\Leftrightarrow g(\mu)(x) \to f(\mu)(x) = 1 \quad \forall \mu \in [0,1]^X, \quad \forall x \in X$$

$$\Leftrightarrow g(\mu)(x) \leq f(\mu)(x) \quad \forall \mu \in [0,1]^X, \quad \forall x \in X \quad \Leftrightarrow \quad g \leq f \quad \square$$

Proposition 4. *Let $f, g \in \Omega'$ be fuzzy operators with $f \leq g$. Then, the induced fuzzy relation R_f^g is $*$-transitive.*

Proof.

$$R_f^g(x,y) * R_f^g(y,z) = \inf_{\mu \in [0,1]^X} \{g(\mu)(x) \to f(\mu)(y)\} * \inf_{\mu \in [0,1]^X} \{g(\mu)(y) \to f(\mu)(z)\}$$

$$\leq \inf_{\mu \in [0,1]^X} \{(g(\mu)(x) \to f(\mu)(y)) * (g(\mu)(y) \to f(\mu)(z))\}$$

$$\leq \inf_{\mu \in [0,1]^X} \{(g(\mu)(x) \to g(\mu)(y)) * (g(\mu)(y) \to f(\mu)(z))\}$$

$$\leq \inf_{\mu \in [0,1]^X} \{g(\mu)(x) \to f(\mu)(z)\} = R_f^g(x,z) \qquad \square$$

4 Inducing Fuzzy Operators from Different Generators

In this section we introduce two new operators C_R^g and C_f^g. Their construction is based on Zadeh's compositional rule in a very similar way to the construction given by (2). In this case, it involves either a fuzzy relation R and a fuzzy operator g or two fuzzy operators f, g (generators).

Definition 7. *Let $g \in \Omega'$ be a fuzzy operator and let $R \in \Gamma'$ be a fuzzy relation on X. We define the operator C_R^g induced by g and R as*

$$C_R^g(\mu)(x) = \sup_{w \in X} \{g(\mu)(w) * R(w,x)\} \tag{4}$$

R and g are called the generators of C_R^g.

Notice that C_R^* is a particular case of C_R^g. Taking $g = id$, where id denotes the identity operator on $[0,1]^X$, we obtain $C_R^{id} = C_R^*$.

Definition 8. Let $g, f \in \Omega'$ be fuzzy operators. The operator C_f^g induced by g and f is defined by

$$C_f^g(\mu)(x) = \sup_{w \in X}\{g(\mu)(w) * f(\{w\})(x)\} \tag{5}$$

g and f will be called the upper and lower generators of C_f^g respectively.

The following result shows some basic properties of C_R^g and C_f^g.

Proposition 5. *Given* g_1, g_2, f_1, f_2 *fuzzy operators and* R_1, R_2 *fuzzy relations, the following holds*

1. *If* $g_1 \leq g_2$, *then* $C_R^{g_1} \leq C_R^{g_2}$ $\forall R \in \Gamma'$
2. *If* $R_1 \leq R_2$ *then* $C_{R_1}^g \leq C_{R_2}^g$ $\forall g \in \Omega'$
3. *If* $f_1 \leq f_2$ *then* $C_{f_1}^g \leq C_{f_2}^g$ $\forall g \in \Omega'$
4. *If* $g_1 \leq g_2$ *then* $C_f^{g_1} \leq C_f^{g_1}$ $\forall f \in \Omega'$

Proof. All implications directly follow from the monotony of $*$. To illustrate it, we will prove 4. For any $\mu \in [0,1]^X$ and $x \in X$ we have

$$C_f^{g_1}(\mu)(x) = \sup_{y \in X}\{g_1(\mu)(y) * f(\{y\})(x)\} \leq \sup_{y \in X}\{g_2(\mu)(y) * f(\{y\})(x)\} = C_f^{g_2}(\mu)(x)$$

There exists a close relationship between the operators C_f^g and C_R^g.

Theorem 1. *For every pair* (g, f) *of fuzzy operators, there exists a fuzzy relation* R *such that* $C_R^g = C_f^g$. R *is uniquely determined. Conversely, for every pair* (g, R) *of a fuzzy operator and a fuzzy relation, there exists at least a fuzzy operator* f *such that* $C_f^g = C_R^g$.

Proof. To prove the first statement of the theorem, notice that given (g, f) and using the usual definition $R_f(x, y) = f(\{x\})(y)$, C_f^g coincides with $C_{R_f}^g$. The unicity follows from the construction.

To prove the second statement, notice that for every fuzzy relation $R \in \Gamma'$ we can define a fuzzy operator f_R as follows:

$$f_R(\mu)(y) = \begin{cases} R(x, y) & \text{if } \mu \text{ is the singleton } \{x\} \\ \\ \mu(y) & \text{if } \mu \text{ is not a singleton} \end{cases}$$

Then, for all $\mu \in [0,1]^X$ and $x \in X$,

$$C_{f_R}^g(\mu)(x) = \sup_{w \in X}\{g(\mu)(w) * f_R(\{w\})(x)\} = \sup_{w \in X}\{g(\mu)(w) * R(w, x)\} = C_R^g(\mu)(x) \quad \square$$

Remark 1. Observe that there are infinite choices for the operator f_R since we are only concerned about its effect over the singletons.

Remark 2. From Theorem 1 we can conclude that every property satisfied for C_f^g for arbitrary f will also be satisfied for C_R^g for arbitrary R. Conversely, every property satisfied for C_R^g for arbitrary R will also be satisfied for C_f^g for arbitrary R.

Given f, g two operators and C_f^g the operator that they generate, there is a fuzzy relation R such that $C_f^g = C_R^g$ and it is exactly R_f. Suppose that a property is satisfied for C_R^g for every $R \in \Gamma'$. It will particulary be satisfied for $C_{R_f}^g$. Hence, it will also be satisfied for C_f^g.

On the other hand, for any relation $R \in \Gamma'$ and $g \in \Omega'$ there exist an infinite number of operators f_R for which $C_{f_R}^g$ coincides with C_R^g. Every property satisfied for C_f^g for an arbitrary f will be satisfied for $C_{f_R}^g$ independently of the f_R chosen. Hence, it will also be satisfied for C_R^g.

Let us study which properties of C_f^g and C_R^g are transmitted from the generators. Our main interest is to characterize for which generators we obtain fuzzy consequence operators (FCO).

Lemma 1. *Let $g \in \Omega'$ and $R \in \Gamma'$. If R is reflexive, then $C_R^g \geq g$.*

Proof. $C_R^g(\mu)(x) = \sup_{w \in X}\{g(\mu)(w) * R(w,x)\} \geq g(\mu)(x) * R(x,x) = g(\mu)(x)$

Proposition 6. *Let $g \in \Omega'$ be an inclusive fuzzy operator and $R \in \Gamma'$ a reflexive fuzzy relation. Then, C_R^g is also an inclusive fuzzy operator.*

Proof. From lemma 1 and the inclusion of g, $C_R^g(\mu)(x) \geq g(\mu)(x) \geq \mu(x)$.

We have an equivalent result for the inclusion of C_f^g.

Proposition 7. *Let $g \in \Omega'$ be an inclusive fuzzy operator and $f \in \Omega'$ a fuzzy operator which is inclusive over the singletons. Then, C_f^g is also an inclusive fuzzy operator.*

Proof. Since f is inclusive over the singletons, the relation $R_f(x,y) = f(\{x\})(y)$ is reflexive. From the proof of Theorem 1, we know that $C_f^g = C_{R_f}^g$. Then, it follows from the previous proposition that C_f^g is also inclusive.

Proposition 8. *Let $g \in \Omega'$ be a monotone fuzzy operator. Then, C_R^g is also a monotone fuzzy operator for any $R \in \Gamma'$.*

Proof. Suppose $\mu_1 \subseteq \mu_2$. Then, $g(\mu_1)(x) \leq g(\mu_2)(x)$ for all $x \in X$ and it follows that

$$C_R^g(\mu_1)(x) = \sup_{w \in X}\{g(\mu_1)(w) * R(w,x)\} \leq \sup_{w \in X}\{g(\mu_2)(w) * R(w,x)\} = C_R^g(\mu_2)(x) \quad \square$$

Remark 3. Notice that Proposition 8 and Remark 2 ensure that if g is a monotone fuzzy operator , then C_f^g is also monotone for any $f \in \Omega'$.

Thus, C_f^g and C_R^g inherit the monotony of its upper generator g. This is due to the fact that g has an effect over general fuzzy subsets. Notice that neither the lower generator f nor R do. Hence, the monotony of the lower generator f does not imply the monotony of C_f^g as it is shown in the following simple example.

Example 1. Let f be the identity operator which is trivially monotone. Let g be any operator which is not monotone. Then, $C_f^g(\mu)(x) = \sup_{y \in X}\{g(\mu)(y) * f(\{y\})(x)\} = \{g(\mu)(x) * \{x\}(x)\} = g(\mu)(x)$. Since g is not monotone, neither is C_f^g.

The idempotence does not follow from the idempotence of the generators as directly as the inclusion or the monotony do. In order to generate a FCO from another FCO, we require an additional property. We need the subsets from the image of the upper generator g to be $*$-compatible with the given relation.

Definition 9. Let g be a fuzzy operator and R a fuzzy relation. We will say that g is $*$-*concordant with R* if all the subsets from the image of g are $*$-compatible with R.

Theorem 2. *Let* $R \in \Gamma'$ *be a reflexive fuzzy relation and let* $g \in \Omega'$ *be a FCO. Suppose that* g *is* $*$-*concordant with R. Then, the operator* C_R^g *induced by g and R is also a FCO.*

Proof. Propositions 6 and 8 give us the properties of inclusion and monotony of C_R^g. It only remains to prove the idempotence. To prove the first inclusion notice that, since $g(\mu)$ belongs to $Im(g)$, it is $*$-compatible with R, so $g(\mu)(y) * R(y,x) \leq g(\mu)(x)$ for all $y,x \in X$. Hence, $\sup_{y \in X}\{g(\mu)(y) * R(y,x)\} \leq g(\mu)(x)$ for all $x \in X$. Using this fact, the monotony and idempotence of g and the monotony of $*$ we get

$$
\begin{aligned}
C_R^g(C_R^g(\mu))(x) &= \sup_{w \in X}\{g(C_R^g(\mu))(w) * R(w,x)\} \\
&= \sup_{w \in X}\{g(\sup_{y \in X}\{g(\mu)(y) * R(y,w)\}) * R(w,x)\} \\
&\leq \sup_{w \in X}\{g(g(\mu)(w)) * R(w,x)\} \\
&= \sup_{w \in X}\{g(\mu)(w) * R(w,x)\} = C_R^g(\mu)(x)
\end{aligned}
$$

The other inclusion follows immediately from the inclusion property.

Remark 4. We can state an equivalent result for the operator C_f^g. Let g and f be two fuzzy operators such that g is FCO and f is inclusive over the singletons. If g is $*$-concordant with $R_f(x,y) = f(\{x\})(y)$, then C_f^g is a FCO.

Let us prove that the coherence property is inherited from the upper generator.

Proposition 9. *Let* $g \in \Omega'$ *be a coherent fuzzy operator and R a fuzzy relation in X. Then,* C_R^g *is also a coherent fuzzy operator.*

Proof. Using property 1 from Proposition 2 we have that $\forall a \in X$ and $\forall \mu \in [0,1]^X$,

$$
\mu(a) * C_R^g(\{a\})(x) = \mu(a) * \sup_{y \in X}\{g(\{a\})(y) * R(y,x)\}
$$

$$= \sup_{y \in X} \{\mu(a) * g(\{a\})(y) * R(y,x)\}$$

$$= \sup_{y \in X} \{(\mu(a) * g(\{a\})(y)) * R(y,x)\}$$

$$\leq \sup_{y \in X} \{g(\mu)(y) * R(y,x)\} = C_R^g(\mu)(x)$$

where the inequality holds because of the coherence of g.

Remark 5. From remark 2 we can state the same about the coherence of C_f^g. That is, if g is a coherent fuzzy operator, then C_f^g is also a coherent fuzzy operator.

5 Aggregation of Fuzzy Operators through the Quasi-arithmetic Mean

In this section, we will assume that $*$ is not only a left-continuous t-norm, but also Archimedean and with an additive generator t. Let us recall that a t-norm is Archimedean if for each $x, y \in (0, 1)$ there is an $n \in \mathbb{N}$ with $x^n = x * \overset{n}{\cdots} * x < y$. An additive generator of a t-norm is a strictly decreasing function $t : [0, 1] \to [0, \infty]$, right continuous in 0, with $t(1) = 0$ and satisfying $t(x) + t(y) \in \mathrm{Ran}(t) \cup [t(0), \infty]$ such that

$$x * y = t^{(-1)}(t(x) + t(y))$$

where $t^{(-1)}$ denotes the pseudo-inverse of t defined as:

$$t^{(-1)}(y) = \sup\{x \in [0, 1] | t(x) > y\}$$

The left-continuity of a t-norm $*$ with additive generator t, implies its continuity and therefore, the continuity of its generator. In this case, the pseudo-inverse becomes the usual inverse of t [8].

Given a continuous Archimedean t-norm $*$ with additive generator t, there is a natural way to define the extended quasi-arithmetic mean associated to $*$, $m_t : \bigcup_{n \in \mathbb{N}} [0, 1]^n \longrightarrow [0, 1]$ (see [4]):

$$m_t(x_1, ..., x_n) = t^{-1}\left(\frac{1}{n}\sum_{i=1}^{n} t(x_i)\right) \tag{6}$$

Given a finite family of fuzzy operators, we can aggregate them using the quasi-arithmetic mean associated to $*$ in order to obtain another fuzzy operator.

Definition 10. (*Quasi-arithmetic mean of fuzzy operators*) Let $t : [0, 1] \to [0, \infty]$ be an additive generator of a continuous Archimedean t-norm $*$. Let $\{g_1, .., g_n\}$ be a finite family of fuzzy operators. The *n-ary quasi-arithmetic mean generated by t* is the fuzzy operator given by

$$m_t(g_1,...,g_n) = t^{-1}\left(\frac{1}{n}\sum_{i=1}^{n}t(g_i)\right) \tag{7}$$

such that for every fuzzy subset $\mu \in [0,1]^X$ and every $x \in X$ is

$$m_t(g_1,...,g_n)(\mu)(x) = t^{-1}\left(\frac{1}{n}\sum_{i=1}^{n}t(g_i(\mu)(x))\right) \tag{8}$$

The *extended quasi-arithmetic mean generated by* t is the function $m_t : \bigcup_{n\in\mathbb{N}}(\Omega')^n \to \Omega'$ that maps any finite family of n fuzzy operators to their n-ary quasi arithmetic mean.

The quasi-arithmetic mean can be defined more generally [4]. Indeed, it can be defined for any continuous and strictly increasing or strictlty decreasing function $f : [0,1] \longrightarrow [-\infty,\infty]$. In this case, the expression $\infty - \infty$ needs to be defined (it is often considered $-\infty$). However, we will focus on the natural case where the generator of m_t is the additive generator of the given continuous Archimedean t-norm $*$.

Remark 6. Observe that, if $g_1,..,g_n \in \Omega'$ are fuzzy operators. Then, their arithmetic mean satisfies that

$$min(g_1,...,g_n) \leq m_t(g_1,...,g_n) \leq max(g_1,...,g_n)$$

It is known that the quasi-arithmetic mean m_t generated by t is strictly increasing and idempotent (in the sense that $m_t(g,...,g) = g$) if the generator is continuous and stricly increasing or strictly decreasing [4]. From this fact, the next two propositions follow:

Proposition 10. *Let* $g_1,..,g_n \in \Omega'$ *be inclusive fuzzy operators. Then, its quasi arithmetic mean is also an inclusive fuzzy operator.*

Proposition 11. *Let* $g_1,..,g_n \in \Omega'$ *be monotone fuzzy operators. Then, its quasi arithmetic mean is also a monotone fuzzy operator.*

Remark 7. Observe that the idempotence of the g_i is in general not translated into the idempotence of their quasi-arithmetic mean. Consider for example the quasi arithmetic mean of $g_1 = id$ and $g_2 = \frac{1}{2}id$ with the product t-norm.

Consider the operators C_f^g and C_R^g from the previous section. Given a finite family of fuzzy operators, let us compare two different processes of aggregation through the quasi-arithmetic mean. The first one by aggregating the generators, the second one by aggregating the generated operators.

Theorem 3. *Let* $g_1,..,g_n \in \Omega'$ *be fuzzy operators and* $t : [0,1] \longrightarrow [0,\infty]$ *be an additive generator of the continuous Archimedean t-norm* $*$. *Let* m_t *be the extended quasi-arithmetic mean generated by* t. *Then, for every* $f \in \Omega'$ *and every* $R \in \Gamma'$

$$C_f^{m_t(g_1,...,g_n)} \leq m_t(C_f^{g_1},...,C_f^{g_n}) \quad and \quad C_R^{m_t(g_1,...,g_n)} \leq m_t(C_R^{g_1},...,C_R^{g_n})$$

Proof. We prove the first inequality:

$$C_f^{m_t(g_1,\ldots,g_n)}(\mu)(x) = \sup_{w \in X} \{m_t(g_1,\ldots,g_n)(\mu)(w) * f(\{w\})(x)\}$$

$$= \sup_{w \in X} \left\{ t^{-1} \left(\frac{\sum_{i=1}^n t(g_i(\mu)(w))}{n} \right) * f(\{w\})(x) \right\}$$

$$= \sup_{w \in X} \left\{ t^{-1} \left(t \left[t^{-1} \left(\frac{\sum_{i=1}^n t(g_i(\mu)(w))}{n} \right) \right] + t(f(\{w\})(x)) \right) \right\}$$

$$= \sup_{w \in X} \left\{ t^{-1} \left(\frac{t(g_1(\mu)(w)) + \cdots + t(g_n(\mu)(w))}{n} + t(f(\{w\})(x)) \right) \right\}$$

$$= \sup_{w \in X} \left\{ t^{-1} \left(\frac{t(g_1(\mu)(w)) + \cdots + t(g_n(\mu)(w)) + n \cdot t(f(\{w\})(x))}{n} \right) \right\}$$

$$= \sup_{w \in X} \left\{ t^{-1} \left(\frac{t(g_1(\mu)(w)) + t(f(\{w\})(x))}{n} + \cdots + \frac{t(g_n(\mu)(w)) + t(f(\{w\})(x))}{n} \right) \right\}$$

$$= \sup_{w \in X} \left\{ t^{-1} \left(\frac{t(g_1(\mu)(w) * f(\{w\})(x))}{n} + \cdots + \frac{t(g_n(\mu)(w) * f(\{w\})(x))}{n} \right) \right\}$$

$$\leq t^{-1} \left(\frac{t(\sup_{w \in X}\{g_1(\mu)(w) * f(\{w\})(x)\})}{n} + \cdots + \frac{t(\sup_{w \in X}\{g_n(\mu)(w) * f(\{w\})(x)\})}{n} \right)$$

$$= t^{-1} \left(\frac{1}{n} \sum_{i=1}^n t \left(\sup_{w \in X}\{g_i(\mu)(w) * f(\{w\})(x)\} \right) \right) = m_t(C_f^{g_1},\ldots,C_f^{g_n})(\mu)(x) \qquad \square$$

Finally, we can prove the following Theorem similarly to the previous one:

Theorem 4. *Let* $f_1,\ldots,f_n \in \Omega'$ *be fuzzy operators and* $t : [0,1] \longrightarrow [0,\infty]$ *be an additive generator of the continuous Archimedean t-norm* $*$. *Let* m_t *be the extended quasi-arithmetic mean generated by* t. *Then, for every* $g \in \Omega'$,

$$C_{m_t(f_1,\ldots,f_n)}^g \leq m_t(C_{f_1}^g,\ldots,C_{f_n}^g)$$

6 Conclusions

In this paper we have generated fuzzy relations and fuzzy operators using different kind of generators and we have studied their properties. We have defined a fuzzy relation induced by two operators f,g that uses more information than the behaviour of f and g over the singletons. We have proved that this relation is reflexive if and only if $g \leq f$, $*$-transitive when $f \leq g$ and a preorder when $f = g$.

We have defined two fuzzy operators C_R^g and C_f^g, the first one induced by a fuzzy relation and a fuzzy operator and the second one induced by two fuzzy operators. We have shown that they are equivalent in the following sense: For every C_f^g, there exists R such that $C_f^g = C_R^g$. Conversely, for every C_R^g there exists f such that $C_R^g = C_f^g$.

We have defined the $*$-concordance of a fuzzy operator with a fuzzy relation and we have shown that for a FCO g which is $*$-concordant with a reflexive fuzzy relation R, the generated C_R^g is also a FCO. The same holds for C_f^g if g and the relation R_f induced by f in the classical way satisfy the mentioned conditions. We have also shown that the coherence property is directly transmitted from the upper generator.

We have studied the aggregation of these induced operators using the quasi-arithmetic mean associated to a continuous Archimedean t-norm. On one hand, we have considered the operators generated by the quasi-arithmetic mean of a family of fuzzy operators. On the other hand, we have considered the aggregation of the individually induced fuzzy operators. For a finite family of fuzzy operators, it holds that $C_f^{m_t(g_i)} \leq m_t(C_f^{g_i})$, $C_{m_t(f_i)}^g \leq m_t(C_{f_i}^g)$ and $C_R^{m_t(g_i)} \leq m_t(C_R^{g_i})$.

Acknowledgements. We ackowledge the partial support of the project FIS2011-28820-C02-02 from the Spanish Government and N.C. acknowledges the financial support of the "Asociación de Amigos de la Universidad de Navarra".

References

[1] Bělohlávek, R.: Fuzzy Relational Systems: Foundations and Principles. In: Ifsr. International Series on Systems Science and Engineering, vol. 20. Kluwer Academic/Plenum Publishers, New York (2002)

[2] Castro, J.L., Trillas, E.: Tarski's fuzzy consequences. In: Proceedings of the International Fuzzy Engineering Symposium 1991, vol. 1, pp. 70–81 (1991)

[3] Castro, J.L., Delgado, M., Trillas, E.: Inducing implication relations. International Journal of Approximate Reasoning 10(3), 235–250 (1994)

[4] Grabisch, M., Marichal, J.L., Mesiar, R., Pap, E.: Aggregation Functions (Encyclopedia of Mathematics and its Applications). Cambridge University Press (2009)

[5] Elorza, J., Burillo, P.: On the relation of fuzzy preorders and fuzzy consequence operators. International Journal of Uncertainty, Fuzziness and Knowledge-based Systems 7(3), 219–234 (1999)

[6] Elorza, J., Burillo, P.: Connecting fuzzy preorders, fuzzy consequence operators and fuzzy closure and co-closure systems. Fuzzy Sets and Systems 139(3), 601–613 (2003)

[7] Elorza, J., et al.: On the relation between fuzzy closing morphological operators, fuzzy consequence operators induced by fuzzy preorders and fuzzy closure and co-closure systems. Fuzzy Sets and Systems (2012),
http://dx.doi.org/10.1016/j.fss.2012.08.010

[8] Klement, E.P., Mesiar, R., Pap, E.: Triangular Norms. Trends in logic, Studia logica library. Springer (2000)

[9] Recasens, J.: Indistinguishability Operators. STUDFUZZ, vol. 260. Springer, Heidelberg (2010)

Comparison of Different Algorithms of Approximation by Extensional Fuzzy Subsets

Gabriel Mattioli and Jordi Recasens

Abstract. How to approximate an arbitrary fuzzy subset by an adequate extensional one is a key question within the theory of Extensional Fuzzy Subsets. In a recent paper by the authors [19] different methods were provided to find good approximations. In this work these methods are compared in order to understand better the performance and improvement they give.

1 Introduction

Indistinguishability operators were introduced to fuzzify the concept of crisp equivalence relations. These operators allow to model the idea of "similarity" between elements, which is key to understand how we "identify" objects. The operation of identification is the mainstone to simplify the representation we have of the environment and understand the information given by our perception. Being able to identify objects enables us to store less quantity of information if favour of being able to extract a qualitative analysis of it.

An eye without a mechanism to identify objects is nothing but a sensor of outern reality. An eye with this mechanism becomes a perceptive system that can "understand" the environment.

Under an indistinguishability operator the observable fuzzy sets are the extensional ones. These sets correspond to the fuzzification of classical equivalence classes. Within the theory of Fuzzy Logic the first researcher to point the relevance of these sets was Zadeh when he discussed the concept of granularity [24].

If we assume that indistinguishability operators are a good model to understand similarities between objects (and there is evidence to think so), then a very

Gabriel Mattioli · Jordi Recasens
Department of Mathematics and Computer Science, ETS Arquitectura del Valles,
Universitat Politècnica de Catalunya
C/ Pere Serra 1-15. 08190 Sant Cugat del Vallès. Spain
e-mail: {gabriel.mattioli,j.recasens}@upc.edu

H. Bustince et al. (eds.), *Aggregation Functions in Theory and in Practise*,
Advances in Intelligent Systems and Computing 228,
DOI: 10.1007/978-3-642-39165-1_31, © Springer-Verlag Berlin Heidelberg 2013

interesting problem is how an arbitrary fuzzy subset can be approximated by an extensional one with the minimum loss of accuracy.

In a previous work by the authors [19] this problem was faced and 3 methods were derived for Archimedean t-norms ($T = $ Łukasievicz and $T = \Pi$ product) and one for the Minimum t-norm.

Restricting to Archimedean t-norms, the first method was based on finding an adequate mean between two operators that provide the best upper and lower approximation by extensional fuzzy subsets of a given fuzzy subset μ. The second one computed an adequate power of the lower approximation of μ, and the last one found the solution solving a Quadratic Programming problem.

Big differences can be found between the first two and the last method. The QP-based one guarantees that the solution found is optimal while the first two do not. On the other hand, the last method suffers drastically the curse of dimensionality and becomes computationally unaffordable for large cardinalities of the universe of discourse X. The first two do not have this problem and work even when X is non-finite.

The aim of this work is to compare in depth the mean-based and the power-based methods. In order to reduce the scope of this comparison we will restrict to the Łukasievicz t-norm and to finite sets. This has been done because in [19] explicit formulas were provided to find the best approximations when $T = $ Ł, while the best approximation for $T = \Pi$ had to be found by numerical methods.

The work is structured as follows:

In Section 2 the Preliminaries to this work are given. In this section the definition and main properties of indistinguishability operators and extensional sets will be recalled.

Section 3 will show how the mean-based method can be built. Natural weighted means will be introduced first and explicit formulas will be provided to find the extensional fuzzy subset that better approximates μ following this method.

In Section 4 the power-based method will be given. First of all it will be shown how powers can be defined with respect to a t-norm T and further how this can be used to find good approximations by extensional fuzzy subsets.

In Section 5 a comparison between these two methods will be provided. Fixed an indistinguishability operator E we will study the output and error committed by each of the methods when approximating different fuzzy subsets and some conclusions will be extracted.

Finally, the Concluding Remarks of this work will be given in Section 6.

2 Preliminaries

In this section the main concepts and results used in this work will be given. The definition of indistinguishability operator will be recalled as well as the main properties of the extensional fuzzy subsets related to an indistinguishability operator.

First of all let us recall the well known Ling's Theorem which introduces the concept of additive generator t of a continuous Archimedean t-norm. Additive generators will prove to be very useful further in this work.

Theorem 1. *[15] A continuous t-norm T is Archimedean if and only if there exists a continuous and strictly decreasing function $t : [0,1] \to [0,\infty]$ with $t(1) = 0$ such that*

$$T(x,y) = t^{[-1]}(t(x) + t(y))$$

where $t^{[-1]}$ is the pseudo inverse of t defined by

$$t^{[-1]}(x) = \begin{cases} 1 & \text{if } x \leq 0 \\ t^{-1}(x) & \text{if } 0 \leq x \leq t(0) \\ 0 & \text{if } t(0) \leq x. \end{cases}$$

The function t will be called an additive generator of the t-norm and two generators of the t-norm T differ only by a positive multiplicative constant.

If $T = \text{Ł}$ is the Łukasievicz t-norm, then an additive generator is $t(x) = 1 - x$.
    If $T = \Pi$ is the Product t-norm, then $t(x) = -log(x)$.

Definition 1. Let T be a t-norm.

- The residuation $\overrightarrow{T}$ of T is defined for all $x, y \in [0,1]$ by

$$\overrightarrow{T}(x|y) = \sup\{\alpha \in [0,1] | T(\alpha, x) \leq y\}.$$

- The birresiduation $\overleftrightarrow{T}$ of T is defined for all $x, y \in [0,1]$ by

$$\overleftrightarrow{T}(x,y) = \min\{\overrightarrow{T}(x|y), \overrightarrow{T}(y|x)\} = T(\overrightarrow{T}(x|y), \overrightarrow{T}(y|x)).$$

When the t-norm T is continuous Archimedean, these operations can be rewritten in terms of the additive generator t.

Proposition 1. *Let T be a continuous Archimedean t-norm generated by an additive generator t. Then:*

- $T(x,y) = t^{[-1]}(t(x) + t(y))$
- $\overrightarrow{T}(x|y) = t^{[-1]}(t(y) - t(x))$
- $\overleftrightarrow{T}(x,y) = t^{[-1]}(|t(x) - t(y)|).$

Indistinguishability operators are the fuzzification of classical equivalence relations and model the intuitive idea of "similarity" between objects. For a more detailed explanation on this operators readers are referred to [4], [21].

Definition 2. Let T be a t-norm. A fuzzy relation E on a set X is a T-indistinguishability operator if and only if for all $x, y, z \in X$

a) $E(x,x) = 1$ (Reflexivity)
b) $E(x,y) = E(y,x)$ (Symmetry)
c) $T(E(x,y), E(y,z)) \leq E(x,z)$ (T-transitivity).

Whereas indistinguishability operators represent the fuzzification of equivalence relations, extensional fuzzy subsets play the role of fuzzy equivalence classes altogether with their intersections and unions. Extensional fuzzy subsets are a key concept in the comprehension of the universe of discourse X under the effect of an indistinguishability operator E as they correspond with the observable sets or granules of X.

Definition 3. Let X be a set and E a T-indistinguishability operator on X. A fuzzy subset μ of X is called extensional with respect to E if and only if:

$$\forall x, y \in X \quad T(E(x,y), \mu(y)) \leq \mu(x).$$

We will denote H_E the set of all extensional fuzzy subsets of X with respect to E.

Extensional fuzzy subsets have been widely studied in the literature [7], [11], [12].
    If the t-norm T is continuous Archimedean then the condition of extensionality can be rewritten in terms of additive generators. This result will be recalled several times along this paper.

Lemma 1. *Let E be a T-indistinguishability operator on a set X. $\mu \in H_E$ if and only if $\forall x, y \in X$:*
$$t(E(x,y)) + t(\mu(y)) \geq t(\mu(x)).$$

Proof.
    $\mu \in H_E \Leftrightarrow T(E(x,y), \mu(y)) \leq \mu(x)$
$\Leftrightarrow t^{-1}(t(E(x,y)) + t(\mu(y))) \leq \mu(x).$
    And as t is a monotone decreasing function this is equivalent to

$$t(E(x,y)) + t(\mu(y)) \geq t(\mu(x)).$$

3 Approximation Using Means

In this section we will propose a method to approximate an arbitrary fuzzy subset by an extensional one. First we will introduce two approximation operators, $\phi_E(\mu)$ and $\psi_E(\mu)$, that provide the best upper and lower approximation respectively by extensional fuzzy subsets of μ given an indistinguishability operator E. The method will consist in computing an adequate weight in order to minimize an error function between μ and the natural weighted mean of $\phi_E(\mu)$ and $\psi_E(\mu)$.

Definition 4. Let X be a set and E a T-indistinguishability operator on X. The maps $\phi_E : [0,1]^X \to [0,1]^X$ and $\psi_E : [0,1]^X \to [0,1]^X$ are defined $\forall x \in X$ by:

$$\phi_E(\mu)(x) = \sup_{y \in X} T(E(x,y), \mu(y)),$$

$$\psi_E(\mu)(x) = \inf_{y \in X} \overrightarrow{T}(E(x,y)|\mu(y)).$$

$\phi_E(\mu)$ is the smallest extensional fuzzy subset greater than or equal to μ; hence it is its best upper approximation by extensional fuzzy subsets. Analogously, $\psi_E(\mu)$ provides the best approximation by extensional fuzzy subsets smaller than or equal to μ. From a topological viewpoint these operators can be seen as closure and interior operators on the set $[0,1]^X$ [11]. It is remarkable that these operators also appear in a natural way in fields such as fuzzy rough sets [20], fuzzy modal logic [6], [5], fuzzy mathematical morphology [8] and fuzzy contexts [3] among many others.

Though $\phi_E(\mu)$ and $\psi_E(\mu)$ provide extensional fuzzy subsets that approximate μ there is no guarantee in general that there are no better approximations of μ by extensional fuzzy subsets. In [19] the authors faced this problem and provided three methods to find approximations for Archimedean t-norms and one for the Minimum t-norm. The two methods compared in this paper were introduced there.

Definition 5. [1] Let $t : [0,1] \to [-\infty, \infty]$ be a non-increasing monotonic map, $x, y \in [0,1]$ and $r \in [0,1]$. The weighted quasi-arithmetic mean m_t of x and y is defined as:

$$m_t^r(x,y) = t^{-1}(r \cdot t(x) + (1-r) \cdot t(y))$$

m_t is continuous if and only if $\{-\infty, \infty\} \nsubseteq Ran(t)$.

There is a bijection between the set of continuous Archimedean t-norms and the set of quasi-arithmetic means by taking as map the additive generator t of the t-norm [14]. Under this interpretation in the literature quasi-arithmetic means are sometimes called natural means [17], as we will recall them from now on.

We want to approximate μ by $m_t^r(\phi_E(\mu), \psi_E(\mu))$. Below we prove that this mean is extensional for any value of r.

Proposition 2. *[19] Let X be a set and μ, ν extensional fuzzy subsets of X with respect to an indistinguishability operator E on X. Then:*

$$m^r(\mu, \nu) \in H_E.$$

Corollary 1. *Let μ be a fuzzy subset on a set X and E an indistinguishability operator. Then:*

$$m^r(\phi_E(\mu), \psi_E(\mu)) \in H_E.$$

It is straightforward that for the limit values $r = 0, 1$ this mean is equal to $\phi_E(\mu)$ and $\psi_E(\mu)$ respectively . The question that arises here is for what value of r the error made in this approximation is the lowest one. In mathematical terms, this problem reduces to finding the minimum value of the following function:

$$F(r) = ||\mu - m^r(\phi_E(\mu), \psi_E(\mu))||$$

Considering the Euclidean distance, without loss of generalization, minimizing the previous expression is equivalent to minimize the square of the norm.

$$F(r) = ||\mu - m^r(\phi_E, \psi_E)||^2$$

For the Łukasievicz t-norm the result below provides an explicit formula to find this optimal weight r.

In order to simplify the notation we will denote $\mu_i = \mu(x_i)$, $\phi_i = \phi_E(\mu)(x_i)$ and $\psi_i = \psi_E(\mu)(x_i)$.

Theorem 2. *[19] Let μ be a fuzzy subset of a finite set $X = \{x_1,...,x_n\}$ and $T = Ł$ the Łukasiewicz t-norm. Then the expression $F(r) = ||\mu - m^r(\phi_E(\mu), \psi_E(\mu))||^2$ is minimized when:*

$$r = \frac{\sum \mu_i \cdot \phi_i - \sum \mu_i \cdot \psi_i - \sum \phi_i \cdot \psi_i + \sum \psi_i^2}{\sum \phi_i^2 + \sum \psi_i^2 - 2\sum \phi_i \cdot \psi_i}.$$

4   Approximation Using Powers

In this section we will provide another method to find an approximation of an arbitrary fuzzy subset μ by an extensional one. This method will be based on approximating μ by an adequate power $\psi_E(\mu)^r$ of its lower approximation operator with respect to the t-norm. It will be shown how, for values $r < 1$, the fuzzy subset $\psi_E(\mu)^r$ is extensional and that a global minimum of the error made can be obtained.

Let us recall the definition of power with respect to a t-norm T.

Definition 6. Let T be a t-norm and n a natural number. We will call the nth power of X with respect to T to:

$$T^n(x) = T(\overbrace{x, x, ..., x}^{n}).$$

To simplify notation we will denote $T^n(x) = x^n$.

It is possible to extend this definition to all positive rational numbers as follows.

Definition 7. Let T be a t-norm and n a natural number. We will define x to the power of $1/n$ with respect to T as:

$$x^{1/n} = \sup_{z \in [0,1]} \{T^n(z) \leq x\}$$

and for p, q natural numbers,

$$x^{p/q}(x) = (x^{1/q})^p.$$

Passing to the limit it is possible to define x^r for all $r \in \mathbb{R}^+$ for continuous t-norms.

The following result allows us to calculate powers by using an additive generator t of T.

Proposition 3. *Let T be an Archimedean t-norm with additive generator t and $r \in \mathbb{R}^+$. Then:*

$$x^r = t^{[-1]}(r \cdot t(x))$$

It is straightforward to observe from the previous proposition that $r \leq s \Rightarrow x^r \geq x^s$. Besides, the continuity of t assures continuity of powers when we let the exponent vary.

The key idea of this method follows from the next Corollary 2.

Proposition 4. *[19] Let E be an indistinguishability operator of a set X, μ an extensional fuzzy subset of E and $r \leq 1$. Then*

$$\mu^r \in H_E.$$

Corollary 2. *$\psi(\mu)^r$ is extensional for $r \leq 1$.*

The problem of approximating a fuzzy subset μ by an adequate power $\psi(\mu)^r$ reduces then to compute the value of r for which the following function is minimized.

$$F(r) = ||\mu - \psi_E(\mu)^r||^2$$

For the Łukasievicz t-norm we have the following result.

As it was done in the previous section we will denote $\mu_i = \mu(x_i)$, $\phi_i = \phi_E(\mu)(x_i)$ and $\psi_i = \psi_E(\mu)(x_i)$.

Theorem 3. *[19] Let μ be a fuzzy subset of a finite set $X = \{x_1, ..., x_n\}$ and $T = Ł$ the Łukasievicz t-norm. Then $F(r) = ||\mu - \psi_E(\mu)^r||^2$ is minimized when*

$$r = \frac{\sum \mu_i + \sum \psi_i - \sum \mu_i \cdot \psi_i - n}{2 \sum \psi_i - \sum \psi_i^2 - n}$$

It would be expectable to find an analogous method to find good approximations of μ by extensional fuzzy subsets using powers of ϕ_E. In [19] it is discussed and illustrated with a counterexample that this is not possible in general.

5 Comparative Analysis of the Two Methods Proposed

In this section we will compare the two methods proposed in this paper to approximate arbitrary fuzzy subsets by extensional ones with respect to a given indistinguishability operator. We will compare the output and error committed by each of the methods applied to four fuzzy subsets of a finite set X.

Let us consider the following Ł-indistinguishability operator

$$E = \begin{pmatrix} 1 & 0.9 & 0.7 & 0.4 & 0.2 \\ 0.9 & 1 & 0.7 & 0.4 & 0.2 \\ 0.7 & 0.7 & 1 & 0.4 & 0.2 \\ 0.4 & 0.4 & 0.4 & 1 & 0.2 \\ 0.2 & 0.2 & 0.2 & 0.2 & 1 \end{pmatrix}$$

and the fuzzy subsets

$$\mu_1 = \begin{pmatrix} 0.9 \\ 0.5 \\ 0.6 \\ 0.8 \\ 0.3 \end{pmatrix} \quad \mu_2 = \begin{pmatrix} 0.2 \\ 0 \\ 0.2 \\ 0.6 \\ 0.9 \end{pmatrix} \quad \mu_3 = \begin{pmatrix} 0.7 \\ 0.5 \\ 0.1 \\ 0.8 \\ 0.6 \end{pmatrix} \quad \mu_4 = \begin{pmatrix} 0.1 \\ 0.9 \\ 0.2 \\ 0.8 \\ 0.5 \end{pmatrix}.$$

It is straightforward to observe that none of these sets is extensional with respect to E.

The corresponding upper and lower approximations by extensional fuzzy subsets of these sets are

$$\phi(\mu_1) = \begin{pmatrix} 0.9 \\ 0.8 \\ 0.6 \\ 0.8 \\ 0.3 \end{pmatrix} \quad \phi(\mu_2) = \begin{pmatrix} 0.2 \\ 0.1 \\ 0.2 \\ 0.6 \\ 0.9 \end{pmatrix} \quad \phi(\mu_3) = \begin{pmatrix} 0.7 \\ 0.6 \\ 0.4 \\ 0.8 \\ 0.6 \end{pmatrix} \quad \phi(\mu_4) = \begin{pmatrix} 0.8 \\ 0.9 \\ 0.6 \\ 0.8 \\ 0.5 \end{pmatrix}.$$

$$\psi(\mu_1) = \begin{pmatrix} 0.6 \\ 0.5 \\ 0.6 \\ 0.8 \\ 0.3 \end{pmatrix} \quad \psi(\mu_2) = \begin{pmatrix} 0.1 \\ 0 \\ 0.2 \\ 0.6 \\ 0.8 \end{pmatrix} \quad \psi(\mu_3) = \begin{pmatrix} 0.4 \\ 0.4 \\ 0.1 \\ 0.7 \\ 0.6 \end{pmatrix} \quad \psi(\mu_4) = \begin{pmatrix} 0.1 \\ 0.2 \\ 0.2 \\ 0.7 \\ 0.5 \end{pmatrix}.$$

Let us denote m_i and p_i the fuzzy extensional subsets obtained following the mean-based and power-based methods respectively for each of the sets μ_i ($i = 1, ..., 4$). The output fuzzy extensional subsets are given are given below.

$$m_1 = \begin{pmatrix} 0.75 \\ 0.65 \\ 0.6 \\ 0.8 \\ 0.3 \end{pmatrix} \quad m_2 = \begin{pmatrix} 0.1667 \\ 0.0667 \\ 0.2 \\ 0.6 \\ 0.8667 \end{pmatrix} \quad m_3 = \begin{pmatrix} 0.5565 \\ 0.5043 \\ 0.2565 \\ 0.7522 \\ 0.6 \end{pmatrix} \quad m_4 = \begin{pmatrix} 0.4043 \\ 0.5043 \\ 0.3739 \\ 0.7434 \\ 0.5 \end{pmatrix}.$$

$$p_1 = \begin{pmatrix} 0.6436 \\ 0.5545 \\ 0.6436 \\ 0.8218 \\ 0.3763 \end{pmatrix} \quad p_2 = \begin{pmatrix} 0.1374 \\ 0.0415 \\ 0.2332 \\ 0.6166 \\ 0.8083 \end{pmatrix} \quad p_3 = \begin{pmatrix} 0.4910 \\ 0.4910 \\ 0.2365 \\ 0.7455 \\ 0.6606 \end{pmatrix} \quad p_4 = \begin{pmatrix} 0.3185 \\ 0.3942 \\ 0.3942 \\ 0.7728 \\ 0.6214 \end{pmatrix}.$$

Table 1 shows the error committed by each of these approximations. The error has been computed taking the euclidean distance between the original fuzzy subset and the extensional approximation of it.

Table 1 Table comparing the error committed by the different methods proposed for μ_1, μ_2, μ_3 and μ_4

	μ_1	μ_2	μ_3	μ_4
ϕ_i	e=0.3	e=0.1	e=0.3162	0.8062
ψ_i	e=0.3	e=0.1414	e=0.3317	0.7071
m_i	e=0.2121	e=0.0816	e=0.2177	c=0.5316
p_i	e=0.2773	e=0.1243	e=0.2627	e=0.5973

Let us analyze now the results obtained for both methods.

A first consideration about the sets μ_1, μ_2, μ_3 and μ_4 is that they show different levels of variance and "'distance'" from extensionality. It can be observed for instance that the upper and lower approximation of μ_2 is very similar to the initial set while the approximations of μ_4 commit a much bigger error.

Another interesting previous consideration is that for μ_2 and μ_3, $\phi_E(\mu_i)$ commits a slightly lower error than $\psi(\mu_i)$ whereas the error committed in the approximation of μ_4 is lower in the case of $\psi_E(\mu_4)$. From here we can infer that there is no guarantee that the approximation provided by ϕ_E is better than the one given by ψ_E and viceversa

Comparing the results obtained with the two new methods proposed in the approximation of the given sets we observe that in all cases the approximation using the mean-based or the power-based method improves the approximation made either by the ϕ_E and ψ_E operators.

Finally it can be observed that in all cases the approximation made using the mean-based method is better than the one given by the power-based one.

According to the results obtained, the conclusion of this analysis is that there is evidence to consider that the two methods proposed improve the ones existing in the literature. However, it is a must to recall that the analysis made is not exhaustive and similar analysis should be done with other indistinguishabilities operators and extensional fuzzy subsets in order to assert this improvement formally.

6 Concluding Remarks

In this work we have recalled two of the methods proposed by the authors in [19] to approximate arbitrary fuzzy subsets by extensional ones for Ł-indistinguishability operators. Both methods are built upon two operators that provide the smallest extensional subset that contains μ ($\phi_E(\mu)$) and the biggest extensional one contained in μ ($\psi_E(\mu)$).

The first method consists in computing an adequate mean of the upper and the lower approximation of μ.

The second one is based on finding an adequate power (homotecy) for which $\psi(\mu)^r$ is a better extensional approximation of μ.

In both cases explicit formulas have been given to compute the best approximations following the methods.

Finally, both methods have been tested for a given indisitinguishability E and 4 extensional sets. Despite a deeper comparison should be done to extract conclusive assertions, the results obtained show relevant evidence that the two methods proposed improve the approximation given by $\phi_E(\mu)$ and $\psi_E(\mu)$. Among them, the results obtained suggest that the mean-based method is slightly better than the power-based one.

Acknowledgements. Research partially supported by project number TIN2009-07235. The first author is supported by the Universitat Politècnica de Catalunya (UPC).

References

[1] Aczél, J.: Lectures on functional equations and their applications. Academic Press, NY-London (1966)

[2] Bazaraa, M.S., Sherali, H.D., Shetty, C.M.: Nonlinear programming: theory and algorithms, 2nd edn. John Wiley & Sons, New York (1993)

[3] Bělohlávek, R.: Fuzzy Relational Systems: Foundations and Principles. Kluwer Academic Publishers, New York (2002)

[4] Boixader, D., Jacas, J., Recasens, J.: Fuzzy equivalence relations: advanced material. In: Dubois, D., Prade, H. (eds.) Fundamentals of Fuzzy Sets, pp. 261–290. Kluwer Academic Publishers, New York (2000)

[5] Boixader, D., Jacas, J., Recasens, J.: Upper and lower approximation of fuzzy sets. Int. J. of General Systems 29, 555–568 (2000)

[6] Bou, F., Esteva, F., Godo, L., Rodríguez, R.: On the Minimum Many-Valued Modal Logic over a Finite Residuated Lattice. Journal of Logic and Computation (2010), doi:10.1093/logcom/exp062

[7] Castro, J.L., Klawonn, F.: Similarity in Fuzzy Reasoning. Mathware & Soft Computing 2, 197–228 (1996)

[8] Elorza, J., Fuentes-González, R., Bragard, J., Burillo, P.: On the Relation Between Fuzzy Preorders, Fuzzy Closing Morphological Operators and Fuzzy Consequence Operators. In: Proceedings of ESTYLF 2010 Conference, Huelva, pp. 133–138 (2010)

[9] Flanders, H.: Differentiation under the integral sign. American Mathematical Monthly 80(6), 615–627 (1973)

[10] Garmendia, L., Recasens, J.: How to make T-transitive a proximity relation. IEEE Transactions on Fuzzy Systems 17(1), 200–207 (2009)

[11] Jacas, J., Recasens, J.: Fixed points and generators of fuzzy relations. J. Math. Anal. Appl. 186, 21–29 (1994)

[12] Jacas, J., Recasens, J.: Fuzzy T-transitive relations: eigenvectors and generators. Fuzzy Sets and Systems 72, 147–154 (1995)

[13] Jacas, J., Recasens, J.: Maps and Isometries between indistinguishability operators. Soft Computing 6, 14–20 (2002)

[14] Jacas, J., Recasens, J.: Aggregation of T-Transitive Relations. Int. J. of Intelligent Systems 18, 1193–1214 (2003)

[15] Klement, E.P., Mesiar, R., Pap, E.: Triangular norms. Kluwer Academic Publishers, Dordrecht (2000)

[16] Mattioli, G., Recasens, J.: Dualities and Isomorphisms between Indistinguishabilities and Related Concepts. In: FUZZ'IEEE, Taiwan, pp. 2369–2374 (2011)

[17] Mattioli, G., Recasens, J.: Natural Means of Indistinguishability Operators. In: Greco, S., Bouchon-Meunier, B., Coletti, G., Fedrizzi, M., Matarazzo, B., Yager, R.R. (eds.) IPMU 2012, Part III. CCIS, vol. 299, pp. 261–270. Springer, Heidelberg (2012)

[18] Mattioli, G., Recasens, J.: Structural Analysis of Indistinguishability Operators and Related Concepts. Informations Sciences (submitted)

[19] Mattioli, G., Recasens, J.: Approximating Arbitrary Fuzzy Subsets by Extensional Ones. IEEE Transactions on Fuzzy Systems (submitted)

[20] Morsi, N.N., Yakout, M.M.: Axiomatics for fuzzy rough sets. Fuzzy Sets and Systems 100, 327–342 (1998)

[21] Recasens, J.: Indistinguishability Operators. STUDFUZZ, vol. 260. Springer, Heidelberg (2010)

[22] Valverde, L.: On the Structure of F-indistinguishability Operators. Fuzzy Sets and Systems 17, 313–328 (1985)

[23] Williamson, J.H.: Lebesgue Integration. Holt, Rinehart and Winston (1962)

[24] Zadeh, L.A.: Similarity relations and fuzzy orderings. Inform. Sci. 3, 177–200 (1971)

[25] Zadeh, L.A.: Fuzzy sets and information granularity. In: Gupta, M.M., Ragade, R.K., Yager, R.R. (eds.) Advances in Fuzzy Set Theory and Applications, pp. 3–18. North-Holland, Amsterdam (1979)

[26] Zadeh, L.A.: Toward a theory of fuzzy information granulation and its centrality in human reasoning and fuzzy logic. Fuzzy Sets and Systems 90, 111–127 (1997)

On the Symmetrization of Quasi-metrics: An Aggregation Perspective

J. Martín, G. Mayor, and O. Valero

Abstract. In 1981, Borsík and Doboš studied the problem of how to combine, by means of a function, two metrics in order to obtain a single one as output. To this end, they introduced the notion of metric aggregation function and gave a characterization of such functions ([1]). Recently, in [14], Mayor and Valero have extended the original work of Borsík and Doboš to the context of quasi-metrics in such a way that a general description of how to merge through a function, called quasi-metric aggregation function, two quasi-metrics into a single one has been given. Since every quasi-metric induces, in a natural way, a metric the main purpose of this paper is to mastermind formally the problem of how to symmetrize a quasi-metric and to provide a solution to such a problem based on the main ideas that arise in the quasi-metric aggregation framework. To this end, the notion of metric generating function, those functions that allow to generate a metric from a quasi-metric, is introduced and a full description of such functions is given from an aggregation perspective. Moreover, a relationship between the quasi-metric aggregation problem and the symmetrization one is provided.

1 The Problem Statement

In 1981, Borsík and Doboš studied the problem of merging metrics in order to obtain a single new one ([1]). In order to concrete such a study let us recall a few pertinent concepts.

From now on, we shall use the letters $\mathbb{R}$ and $\mathbb{R}^+$ to denote the set of real numbers and the set of nonnegative real numbers, respectively.

J. Martín · G. Mayor · O. Valero
Dept. of Math. and Comput. Sci., University of Balearic Islands, Ctra. de Valldemossa km. 7.5, 07122 Palma de Mallorca, Spain
e-mail: {javier.martin,gmayor,o.valero}@uib.es

H. Bustince et al. (eds.), *Aggregation Functions in Theory and in Practise*,   319
Advances in Intelligent Systems and Computing 228,
DOI: 10.1007/978-3-642-39165-1_32, © Springer-Verlag Berlin Heidelberg 2013

Following [1], we will consider the set $\mathbb{R}_2^+ = \{(a,b) : a,b \in \mathbb{R}^+\}$ ordered by the pointwise order relation $\preceq$, i.e. $(a,b) \preceq (c,d) \Leftrightarrow a \leq c$ and $b \leq d$. Moreover, a function $\Phi : \mathbb{R}_2^+ \to \mathbb{R}^+$ will be said to be monotone provided that $\Phi(a,b) \leq \Phi(c,d)$ for all $(a,b),(c,d) \in \mathbb{R}_2^+$ with $(a,b) \preceq (c,d)$. Furthermore, a function $\Phi : \mathbb{R}_2^+ \to \mathbb{R}^+$ will be said to be subadditive if $\Phi((a,b)+(c,d)) \leq \Phi(a,b)+\Phi(c,d)$ for all $(a,b),(c,d) \in \mathbb{R}_2^+$.

According to [1], we will denote by $\mathscr{O}$ the set of all functions $\Phi : \mathbb{R}_2^+ \to \mathbb{R}^+$ satisfying: $\Phi(a,b) = 0 \Leftrightarrow (a,b) = (0,0)$.

With the aim of analyze in depth how to combine by means of a function two metrics in order to obtain a single one as a result, Borsík and Doboš introduced the notion of metric preserving function (metric aggregation function in [15] and [14]). Hence, a function $\Phi : \mathbb{R}_2^+ \to \mathbb{R}^+$ is a metric aggregation function provided that the function $d_\Phi : X \times X \to \mathbb{R}^+$ is a metric for every pair of metric spaces (X_1,d_1) and (X_2,d_2), where $X = X_1 \times X_2$ and $d_\Phi((x,y),(z,w)) = \Phi(d_1(x,z),d_2(y,w))$ for all $(x,y),(z,w) \in X$.

In [1], Borsík and Doboš gave a characterization, and thus a solution of the problem of merging two metrics into a single one, of metric aggregation functions in terms of the so-called triangle triplets. In order to present such a characterization let us recall that, given $a,b,c \in \mathbb{R}^+$, the triplet (a,b,c) forms a triangle triplet whenever $a \leq b+c$, $b \leq a+c$ and $c \leq b+a$. The aforementioned characterization can be enunciated as follows:

Theorem 1. *Let* $\Phi : \mathbb{R}_2^+ \to \mathbb{R}^+$. *Then the below assertions are equivalent:*

(1) Φ *is a metric aggregation function.*
(2) Φ *holds the following properties:*

　(2.1) $\Phi \in \mathscr{O}$.
　(2.2) Let $a,b,c,d,f,g \in \mathbb{R}^+$. *If* (a,b,c) *and* (d,f,g) *are triangle triplets, then so*
　　is $(\Phi(a,d),\Phi(b,f),\Phi(c,g))$.

Since Borsik and Doboš solved, by means of Theorem 1, the problem of merging two metrics, several authors have provided new advances in the study of the aggregation problem of a few kinds of generalized metrics. Specifically E. Castiñeira, A. Pradera and E. Trillas have provided a solution to the aggregation problem for C-generalized metrics, S-generalized distances and pseudometrics in [21], [20] and [19].

Inspired by the original work of Borsik and Doboš and motivated by the utility of quasi-metrics in several fields of Artificial Intelligence and Computer Science (see, for instance, [27], [26], [25], [4], [5], [6], [28], [16], [17],[2], [18], [7]), [29], [23], [24], Mayor and Valero studied the aggregation problem in the quasi-metric context. Thus they introduced the notion of quasi-metric aggregation function in [14] (asymmetric distance function in the quoted reference) and gave a characterization of such functions in the spirit of Theorem 1. With the aim to present such a characterization let us recall a few concepts about quasi-metrics.

Following [8], a quasi-metric on a (nonempty) set X is a function $d : X \times X \to \mathbb{R}^+$ such that for all $x,y,z \in X$:

(i) $d(x,y) = d(y,x) = 0 \Leftrightarrow x = y$.
(ii) $d(x,z) \le d(x,y) + d(y,z)$.

Note that a metric on a set X is a quasi-metric d on X satisfying, in addition, the following condition for all $x, y \in X$:

(iii) $d(x,y) = d(y,x)$.

A quasi-metric space is a pair (X,d) such that X is a (nonempty) set and d is a quasi-metric on X.

A well-known example of quasi-metric space is given by the pair $(\mathbb{R}^+, u)$ where u is defined by $u(x,y) = \max\{y - x, 0\}$ for all $x, y \in \mathbb{R}^+$.

According to [15] and [14], a function $\Phi : \mathbb{R}_2^+ \to \mathbb{R}^+$ is a quasi-metric aggregation function if the function $d_\Phi : X \times X \to \mathbb{R}^+$ is a quasi-metric for every pair of quasi-metric spaces (X_1, d_1) and (X_2, d_2), where $X = X_1 \times X_2$ and $d_\Phi((x,y),(z,w)) = \Phi(d_1(x,z), d_2(y,w))$ for all $(x,y),(z,w) \in X$.

The announced quasi-metric formulation of Theorem 1 given in [14] was stated in the following way:

Theorem 2. Let $\Phi : \mathbb{R}_2^+ \to \mathbb{R}^+$. Then the below assertions are equivalent:

(1) Φ is a quasi-metric aggregation function.
(2) Φ holds the following properties:

(2.1) $\Phi \in \mathcal{O}$.
(2.2) Let $a,b,c,d,f,g \in \mathbb{R}^+$. If $(a,d) \preceq (b,f) + (c,g)$, then $\Phi(a,d) \le \Phi(b,f) + \Phi(c,g)$.

(3) $\Phi \in \mathcal{O}$, and Φ is subadditive and monotone.

Several recent works on quasi-metric aggregation functions and the aggregation problem of generalized metric structures can be found in [11], [12], [13], [9] and [10].

According to [3], each quasi-metric d can be symmetrized. Indeed, on the one hand, it is clear that the function $d_{\max} : X \times X \to \mathbb{R}^+$ defined by $d_{\max}(x,y) = \max\{d(x,y), d(y,x)\}$ is a metric. On the other hand, it is a simple matter to see that the function $d_+ : X \times X \to \mathbb{R}^+$ given by $d_+(x,y) = d(x,y) + d(y,x)$ is also a metric. Of course it is obvious that the metrics $d_{\max}$ and d_+ are obtained by means of an appropriate aggregation function acting on the numerical values $d(x,y)$ and $d(y,x)$. In fact, $d_{\max}(x,y) = \Phi_{\max}(d(x,y), d(y,x))$ and $d_+(x,y) = \Phi_+(d(x,y), d(y,x))$, where $\Phi_{\max}(a,b) = \max\{a,b\}$ and $\Phi_+(a,b) = a + b$ for all $a,b \in \mathbb{R}^+$. Observer, in addition, that $\Phi_{\max}$ and Φ_+ are, by Theorem 2, quasi-metric aggregation functions.

Inspired, on the one hand, by the fact that the preceding processes of symmetrization of a quasi-metric can be formulated in the context of aggregation theory and, on the other hand, by the fact that we have not found a comprehensive study on how a quasi-metric can be symmetrized in the literature, in this paper we focus our study on the description of a general method of symmetrization of quasi-metrics from an aggregation perspective and, likewise, on the connection between this method and those given by the quasi-metric aggregation functions. Concretely our aim is to characterize, in the spirit of Theorem 2, those functions that provide the aforesaid symmetrization methods. To this end, we will introduce the notion of "metric generation function".

A function $\Phi : \mathbb{R}_2^+ \to \mathbb{R}^+$ will be called a metric generating function if the function $d_\Phi : X \times X \to \mathbb{R}^+$ is a metric on X for every quasi-metric space (X,d), where the function d_Φ is defined by

$$d_\Phi(x,y) = \Phi(d(x,y),d(y,x))$$

for all $x,y \in X$.

Now, in the light of the definition of metric generating function, it is evident that such functions represent mathematical methods to symmetrize quasi-metrics providing, thus, metrics as a result. Of course, our main purpose is to provide a characterization, in the spirit of Theorem 2, of the metric generating functions and, thus, to solve formally the symmetrization problem.

2   The Solution

In this section we provide a solution to the problem of how to symmetrize a quasi-metric such as we have posed in Section 1. With this aim, we prove the following result which will be useful in our subsequent work.

Lemma 1. *The following assertions hold:*

(1) The pair $(\mathbb{R}^2, p)$ *is a quasi-metric space, where*

$$p(x,y) = \max\{y_1 - x_1, 0\} + \max\{y_2 - x_2, 0\}$$

for all $x = (x_1, x_2), y = (y_1, y_2) \in \mathbb{R}^2$.
(2) The pair $(\mathbb{R}^3, q)$ *is a quasi-metric space, where*

$$q(x,y) = \max\{\max\{y_1 - x_1, 0\}, \max\{y_2 - x_2, 0\}, \max\{y_3 - x_3, 0\}\}$$

for all $x = (x_1, x_2, x_3), y = (y_1, y_2, y_3) \in \mathbb{R}^3$.

Proof. It is a routine to check that the functions $p : \mathbb{R}_+^2 \to \mathbb{R}_+$ and $q : \mathbb{R}_+^3 \to \mathbb{R}_+$ are quasi-metrics.   □

With the aim of discussing, as mentioned before, which properties a metric generating function must satisfy we need to introduce a new sort of triplets that we have called mixed triplets.

Let $a,b,c,d,f,g \in \mathbb{R}^+$. We will say that the triplets (a,b,c) and (d,f,g) are mixed provided that the following inequalities hold:

$$a \leq b+c, \; b \leq a+f, \; c \leq g+a,$$

$$d \leq f+g, \; f \leq d+b, \; g \leq c+d.$$

Observe that the notion of mixed triplets is somewhat related to the triangle triplet one. Concretely we have that a triplet (a,b,c) forms a triangle triplet if and only if the triplets (a,b,c) and (a,c,b) are mixed.

The next result gives a general description of those properties that a metric generating function $\Phi : \mathbb{R}_2^+ \to \mathbb{R}^+$ must hold in order to induce a metric d_Φ from every quasi-metric d in the terms exposed in Section 1.

Theorem 3. *Let* $\Phi : \mathbb{R}_2^+ \to \mathbb{R}^+$. *Then the below assertions are equivalent:*

(1) Φ *is a metric generating function.*
(2) Φ *holds the following properties:*

   (2.1) $\Phi \in \mathcal{O}$.
   (2.2) Φ *is symmetric, i.e.,* $\Phi(a,b) = \Phi(b,a)$ *for all* $(a,b) \in \mathbb{R}_2^+$.
   (2.3) $\Phi(a,d) \le \Phi(b,g) + \Phi(c,f)$ *for all* $a,b,c,d,f,g \in \mathbb{R}^+$ *such that* (a,b,c)
   and (d,f,g) *are mixed triplets.*

Proof. (1) $\implies$ (2). First of all we prove that $\Phi \in \mathcal{O}$. Indeed, let $a,b \in \mathbb{R}^+$ such that $\Phi(a,b) = 0$. Take $x,y \in \mathbb{R}^2$ given by $x = (\frac{-a}{2},b)$ and $y = (\frac{a}{2},0)$. Then a simpler computation shows that $p(x,y) = a$ and $p(y,x) = b$, where p is the quasi-metric on $\mathbb{R}^2$ introduced in assertion (1) in statement of Lemma 1. It follows that $p_\Phi(x,y) = \Phi(p(x,y),p(y,x)) = \Phi(a,b) = 0$. It follows that $\Phi(a,b) = 0 \Leftrightarrow a = b = 0$, since the function p_Φ is a metric on $\mathbb{R}^2$ and $x = y \Leftrightarrow a = b = 0$. Thus $\Phi \in \mathcal{O}$.

The symmetry of Φ follows from the fact that

$$\Phi(a,b) = p_\Phi(x,y) = p_\Phi(y,x) = \Phi(b,a)$$

for all $a,b \in \mathbb{R}^+$, where $x = (\frac{-a}{2},b)$ and $y = (\frac{a}{2},0)$ as before.

Next we prove that $\Phi(a,d) \le \Phi(b,g) + \Phi(c,f)$ for all $a,b,c,d,f,g \in \mathbb{R}^+$ such that (a,b,c) and (d,f,g) are mixed triplets. To this end, take $x,y,z \in \mathbb{R}^3$ given by $x = (-c+g,d,-b)$, $y = (0,0,a-b)$ and $z = (-c,f,0)$. It is a simpler matter to check that $q(x,y) = a$, $q(y,x) = d$, $q(x,z) = b$, $q(z,x) = g$, $q(z,y) = c$ and $q(y,z) = f$, where q is the quasi-metric introduced in assertion (2) in the statement of Lemma 1. It follows that

$$\begin{aligned}\Phi(a,d) &= \Phi(q(x,y),q(y,x)) = q_\Phi(x,y)\\ &\le q_\Phi(x,z) + q_\Phi(z,y)\\ &= \Phi(q(x,z),q(z,x)) + \Phi(q(z,y),q(y,z))\\ &= \Phi(b,g) + \Phi(c,f),\end{aligned}$$

since Φ is a metric generating function.

(2) $\implies$ (1). Let $x,y \in X$. Assume that $d_\Phi(x,y) = 0$. Then $\Phi(d(x,y),d(y,x)) = 0$. The fact that $\Phi \in \mathcal{O}$ guarantees that $\Phi(d(x,y),d(y,x)) = 0 \Leftrightarrow d(x,y) = d(y,x) = 0$. Whence we immediately obtain that $d_\Phi(x,y) = 0 \Leftrightarrow x = y$, since d is a quasi-metric on X.

Since Φ is symmetric we have that

$$d_\Phi(x,y) = \Phi(d(x,y),d(y,x)) = \Phi(d(y,x),d(x,y)) = d_\Phi(y,x)$$

for all $x,y \in X$.

Next consider $x, y, z \in X$. Then, by the fact that the quasi-metric d satisfies the triangle inequality, we obtain the following inequalities:

$$d(x,y) \le d(x,z) + d(z,y), \ d(x,z) \le d(x,y) + d(y,z), \ d(z,y) \le d(z,x) + d(x,y),$$

$$d(y,x) \le d(y,z) + d(z,x), \ d(y,z) \le d(y,x) + d(x,z), \ d(z,x) \le d(z,y) + d(y,x).$$

It follows that the triplets (a,b,c) and (d,f,g) are mixed, where $a = d(x,y)$, $d = d(y,x)$, $b = d(x,z)$, $c = d(z,y)$, $f = d(y,z)$ and $g = d(z,x)$.

Thus assertion (2.3) gives that

$$\begin{aligned} d_\Phi(x,y) &= \Phi(d(x,y), d(y,x)) \\ &\le \Phi(d(x,z), d(z,x)) + \Phi(d(z,y), d(y,z)) \\ &= d_\Phi(x,z) + d_\Phi(z,y). \end{aligned}$$

So we have shown that the function d_Φ is a metric on X.   □

Examples 3, 4 and 6, below, yield different instances of metric generating functions. Moreover, Examples 1 and 2 provide functions that are not metric generating functions.

Example 1. Consider the function $\Phi : \mathbb{R}_+^2 \to \mathbb{R}^+$ given by $\Phi(a,b) = a + \frac{1}{2}b$. It is clear that $1 = \Phi(1,0) \ne \Phi(0,1) = \frac{1}{2}$. So Φ is not symmetric and, thus, Theorem 3 gives that Φ is not a metric generating function.

An easy computation shows that a function $\Phi : \mathbb{R}_2^+ \to \mathbb{R}^+$ satisfying condition (2.3) in statement of Theorem 3 is subadditive. Consequently we have the next result.

Corollary 1. *Let* $\Phi : \mathbb{R}_2^+ \to \mathbb{R}^+$. *If* Φ *is a metric generating function, then* Φ *is subadditive.*

As a consequence of Corollary 1, one can wonder whether condition (2.3) in statement (2) in Theorem 3 can be replaced by subadditivity. However, the next example shows that Theorem 3 is not true if we replace the aforesaid condition by subadditivity.

Example 2. Let $\Phi : \mathbb{R}_2^+ \to \mathbb{R}^+$ be the function given by $\Phi(a,b) = \sqrt{a^2 + b^2 - ab}$. It is clear that $\Phi \in \mathcal{O}$ and that Φ is symmetric. Moreover, it is not hard to check that Φ is subadditive. Nevertheless, it is easily seen that the triplets $(2,1,1)$ and $(0, \frac{1}{2}, 0)$ are mixed triplets and that $\Phi(2,0) = 2$, $\Phi(1,0) = 1$ and $\Phi(1, \frac{1}{2}) = \frac{\sqrt{3}}{2}$. So, by Theorem 3, we conclude that Φ is not a metric generating function because of $\Phi(2,0) > \Phi(1,0) + \Phi(1, \frac{1}{2})$.

The next example provides instances of metric generating functions.

Example 3. The functions $\Phi_{\max} : \mathbb{R}_2^+ \to \mathbb{R}^+$, $\Phi_+ : \mathbb{R}_2^+ \to \mathbb{R}^+$ introduced in Section 1 and the function $\Phi_{med} : \mathbb{R}_2^+ \to \mathbb{R}^+$, given for all a, b by $\Phi_{med}(a,b) = \frac{a+b}{2}$, are, by statement (2) in Theorem 3, metric generating functions.

The next results, which are inspired by those given in [19], provide a little more information on the description of metric generating functions.

Let us recall that a function $\Phi : \mathbb{R}_2^+ \to \mathbb{R}^+$ has $e \in \mathbb{R}^+$ as an absorbent (neutral) element if $\Phi(e,a) = \Phi(a,e) = e$ ($\Phi(e,a) = \Phi(a,e) = a$) for all $a \in \mathbb{R}^+$. Moreovoer, a function $\Phi : \mathbb{R}_2^+ \to \mathbb{R}^+$ has $e \in \mathbb{R}^+$ as an idempotent element if $\Phi(e,e) = e$.

Corollary 2. *Let $\Phi : \mathbb{R}_2^+ \to \mathbb{R}^+$ be a metric generating functions. Then Φ has not 0 as an absorbent element.*

Proof. Assume that 0 is an absorbent element of Φ. It follows that $\Phi \notin \mathcal{O}$. So, by Theorem 3, we obtain that Φ is not a metric generating function which is a contradiction. $\quad\square$

Corollary 3. *Let $\Phi : \mathbb{R}_2^+ \to \mathbb{R}^+$ be a metric generating function. Then does not exist $e \in \mathbb{R}^+ \setminus \{0\}$ such that Φ has e as a neutral element.*

Proof. Assume for the purpose of contradiction that e is a neutral element of Φ with $e \neq 0$. Then we have that $\Phi(e,0) = 0$. It follows that $\Phi \notin \mathcal{O}$ and, by Theorem 3, that Φ is not a metric generating function, which contradicts our hypothesis. $\quad\square$

Corollary 4. *Let $\Phi : \mathbb{R}_2^+ \to \mathbb{R}^+$ be a metric generating function. Then do no exist $e,c \in \mathbb{R}^+$ with $0 < e < \frac{c}{2}$ such that Φ has $c \in \mathbb{R}^+$ as an idempotent element and e as an absorbent element.*

Proof. Suppose that Φ has $e \in \mathbb{R}^+$ as an absorbent element and c, with $0 < e < \frac{c}{2}$, as an idempotent element. Then we have that

$$\Phi(e + (c - e), e + (c - e)) = c > 2e = \Phi(e, c - e) + \Phi(c - e, e).$$

It follows that Φ is not subadditive and thus, by Corollary 1, we conclude that Φ is not a metric generating function. $\quad\square$

Corollary 5. *Let $\Phi : \mathbb{R}_2^+ \to \mathbb{R}^+$. If Φ is a metric generating function which has 0 as a neutral element, then $\Phi(a,b) \leq a + b$ for all $a,b \in \mathbb{R}^+$.*

Proof. Take $a,b \in \mathbb{R}^+$. Then, Corollary 1 guarantees that Φ is subadditive and, thus, that

$$\Phi(a,b) = \Phi((a,0) + (0,b)) \leq \Phi(a,0) + \Phi(0,b).$$

Since Φ has 0 as a neutral element we obtain that $\Phi(0,a) = a$ and that $\Phi(0,b) = b$. Hence we conclude that

$$\Phi(a,b) \leq a + b.$$

$\quad\square$

The functions $\Phi_{\max}$ and Φ_+, given in Example 3, have obviously 0 as neutral element.

Since every metric is a quasi-metric it seems natural to wonder whether a metric generating function Φ preserves metrics in the sense that $d_\Phi(x,y) = d(x,y)$ for every metric space (X,d) and for all $x,y \in X$. However Example 3 gives a negative answer to that question. Indeed, Φ_+ is a metric generating function and, however, $d_{\Phi_+}(x,y) = 2d(x,y)$ for all $x,y \in X$ and every metric space (X,d).

The next result provides a characterization of those metric generating functions which preserve metrics.

Theorem 4. *Let* $\Phi : \mathbb{R}_2^+ \to \mathbb{R}^+$ *be a metric generating function. Then the following assertions are equivalent:*

1) $d_\Phi = d$ *for every metric space* (X,d).
2) Φ *is idempotent, i.e.,* $\Phi(a,a) = a$ *for all* $a \in \mathbb{R}^+$.

Proof. (1) $\Longrightarrow$ (2). Let $a \in \mathbb{R}^+$. Consider the Euclidean metric space $(\mathbb{R}_2^+, d_E)$. Take $x,y \in \mathbb{R}_2^+$ given by $x = (a,0), y = (0,0)$. Then it is clear that $d_E(x,y) = a$. Since Φ is a metric generating function we have that $d_{E\Phi}$ is a metric on $\mathbb{R}_2^+$. Moreover, from (1) we obtain that $\Phi(a,a) = \Phi(d_E(x,y), d_E(x,y)) = d_{E\Phi}(x,y) = d_E(x,y) = a$.

(2) $\Longrightarrow$ (1). Since Φ is idempotent we have that $\Phi(a,a) = a$ for all $a \in \mathbb{R}^+$. So $d_\Phi(x,y) = \Phi(d(x,y), d(x,y)) = d(x,y)$ for all $x,y \in X$. $\qquad\square$

The functions $\Phi_{\max}, \Phi_{med}$ introduced in Example 3 are instances of idempotent metric generating functions.

3   A Connection between the Quasi-metric Aggregation Problem and the Symmetrization of Quasi-metrics

Since quasi-metric aggregation functions allow to merge two quasi-metrics into a single one and taking into account that a metric is a particular case of a quasi-metric, it seems natural to discuss when such functions are exactly metric generating functions. Next we provide information on the relationship between quasi-metric aggregation functions and metric generating functions.

Proposition 1. *Let* $\Phi : \mathbb{R}_2^+ \to \mathbb{R}^+$. *If* Φ *is a symmetric quasi-metric aggregation function, then* Φ *is a metric generating function.*

Proof. Theorem 2 guarantees that Φ is monotone and subadditive and that $\Phi \in \mathcal{O}$. Next we show that
$$\Phi(a,d) \leq \Phi(b,g) + \Phi(c,f)$$
for all $a,b,c,d,f,g \in \mathbb{R}^+$ such that (a,b,c) and (d,f,g) are mixed triplets. Indeed, if (a,b,c) and (d,f,g) are mixed triplets, then we have that $a \leq b+c$ and $d \leq f+g$ and, thus, the monotonicity of Φ provides that
$$\Phi(a,d) \leq \Phi(b+c, f+g) = \Phi((b,g) + (c,f)).$$
Whence we have, by the subadditivity of Φ, that
$$\Phi(a,d) \leq \Phi(b,g) + \Phi(c,f).$$
It follows, by Theorem 3, that Φ is a metric generating function. $\qquad\square$

The following example provides quasi-metric aggregation function which are metric generating functions.

Example 4. It is clear that the below functions $\Phi : \mathbb{R}_+^2 \longrightarrow \mathbb{R}_+$ satisfy statement (3) in Theorem 2 and, in addition, they are symmetric functions. Thus, by Proposition 1, all of them are metric generating functions:

(1) $\Phi(a,b) = \begin{cases} 0 \text{ if } a = b = 0 \\ 1 \text{ otherwise} \end{cases}$.

(2) $\Phi(a,b) = (w(a^p + b^p))^{\frac{1}{p}}$ for all $w \in \mathbb{R}_+ \setminus \{0\}$, where $p \in [1,\infty[$.

(3) $\Phi(a,b) = w\max\{a,b\}$ for all $w \in \mathbb{R}_+ \setminus \{0\}$.

(4) $\Phi(a,b) = w(a+b)$ for all $w \in \mathbb{R}_+ \setminus \{0\}$.

In the light of Proposition 1, it seems natural to wonder if every metric generating functions is a quasi-metric aggregation function. However, the next example yields a negative answer to the preceding question, i.e., it shows that there are metric generating functions that are not quasi-metric aggregation functions.

Example 5. Let $\Phi : \mathbb{R}_2^+ \to \mathbb{R}^+$ be the function defined by

$$\Phi(a,b) = \begin{cases} 0 \text{ if } \max\{a,b\} = 0 \\ 2 \text{ if } \max\{a,b\} \in]0,1[\\ 1 \text{ if } \max\{a,b\} \geq 1 \end{cases}.$$

It is a simple matter to check that Φ holds all conditions in assertion (2) in statement of Theorem 3 and, thus, that Φ is a metric generating function. However, Φ is not monotone. Indeed, $(\frac{1}{2},\frac{1}{2}) \leq (1,1)$ but $2 = \Phi(\frac{1}{2},\frac{1}{2}) \nleq \Phi(1,1) = 1$. Therefore, by Theorem 2, Φ is not a quasi-metric aggregation function.

The preceding example motivates that one can wonder which is the condition that a metric generating function must satisfy in order to be a quasi-metric aggregation function. The next result provides the answer to the preceding question and, concretely, it describes exactly the relationship between quasi-metric aggregation functions and metric generating functions.

Theorem 5. *Let* $\Phi : \mathbb{R}_2^+ \to \mathbb{R}^+$ *be a metric generating function. Then the following assertions are equivalent:*

(1) Φ *is a quasi-metric aggregation function.*
(2) Φ *is monotone.*

Proof. (1) $\Longrightarrow$ (2). Assume that Φ is a quasi-metric aggregation function. It follows, from Theorem 2, that Φ is monotone.

(2) $\Longrightarrow$ (1). Next suppose that Φ is a monotone metric generating function. Then, by Theorem 3, we obtain that $\Phi \in \mathcal{O}$. Moreover, by Proposition 1, we have that Φ is subadditive. Thus, by statement (3) in Theorem 2, we conclude that Φ is a quasi-metric aggregation function. $\square$

We end the paper giving a necessary condition, in the spirit of statement (2.2) in Theorem 2, that every monotone metric generating function must satisfy.

Corollary 6. *Let* $\Phi : \mathbb{R}_2^+ \to \mathbb{R}^+$ *be a monotone metric generating function. Then the following condition holds for all* $a, b \in \mathbb{R}^+$:

$$\Phi(a,b) \le \Phi(a,a) + \Phi(b,b).$$

Proof. Take $a, b \in \mathbb{R}^+$. Since Φ is monotone we have that

$$\Phi(a,b) \le \Phi(a+b, a+b)$$

for all $a, b \in \mathbb{R}^+$. By Corollary 1 we obtain that Φ is subadditive and, thus, that

$$\Phi(a+b, a+b) \le \Phi(a,a) + \Phi(b,b)$$

for all $a, b \in \mathbb{R}^+$. Therefore we conclude that

$$\Phi(a,b) \le \Phi(a,a) + \Phi(b,b)$$

for all $a, b \in \mathbb{R}^+$. $\square$

Next, Corollary 7 provides a kind of averaging behavior for those monotone metric generating functions that are also idempotent. Observe that Example 4 provides metric generating functions that are monotonic but they are not idempotent. Moreover, the functions $\Phi_{\max}, \Phi_{med}$ given in Example 3 are idempotent metric generating function which are, in addition, monotone.

Corollary 7. *Let* $\Phi : \mathbb{R}_2^+ \to \mathbb{R}^+$ *be a monotone metric generating function which is, in addition, idempotent. Then the following condition holds for all* $a, b \in \mathbb{R}^+$:

$$\min\{a,b\} \le \Phi(a,b) \le a+b \le 2\max\{a,b\}.$$

Proof. Take $a, b \in \mathbb{R}^+$. By Corollary 6 we have that

$$\Phi(a,b) \le \Phi(a,a) + \Phi(b,b).$$

Since Φ is idempotent we deduce that $\Phi(a,a) = a$ and $\Phi(b,b) = b$. So we have that

$$\Phi(a,b) \leq a+b.$$

Moreover, the fact that Φ is monotone provides that $\Phi(\min\{a,b\},\min\{a,b\}) \leq \Phi(a,b)$. The idempotency of Φ gives that $\Phi(\min\{a,b\},\min\{a,b\}) = \min\{a,b\}$. Hence we conclude that

$$\min\{a,b\} \leq \Phi(a,b) \leq a+b \leq 2\max\{a,b\}.$$

$\square$

4 Conclusion

Motivated, one the one hand, by the Borsík and Doboš work on the context of metric aggregation ([1]) and, on the other hand, by the fact that a wide number of distances that are useful in formal methods used in Artificial Intelligence and Computer Science are quasi-metrics that can be retrieved by means of appropriate aggregation functions, Mayor and Valero proposed and solved the problem of how to merge two quasi metrics in order to obtain a new single one ([14]). Inspired by the quasi-metric aggregation problem, in this paper we have masterminded formally the problem of how to symmetrize a quasi-metric. A solution to such a problem has been yielded in such a way that a general method to this end has been introduced from an aggregation perspective.

Acknowledgements. The authors acknowledge the support of the Spanish Ministry of Science and Innovation, grant MTM2009-10962.

References

[1] Borsík, J., Doboš, J.: On a product of metric spaces. Math. Slovaca 31, 193–205 (1981)
[2] Casanovas, J., Valero, O.: A connection between computer science and fuzzy theory: midpoints and and running time of computing. Mathware Soft. Comput. 15, 251–261 (2008)
[3] Deza, M.M., Deza, E.: Encyclopedia of Distances. Springer, Berlin (2009)
[4] García-Raffi, L.M., Romaguera, S., Sánchez-Pérez, E.A.: The supremum asymmetric norm on sequence algebras: a general framework to measure complexity spaces. Electronic Notes in Theoret. Comput. Sci. 74, 39–50 (2003)
[5] García-Raffi, L.M., Romaguera, S., Sánchez-Pérez, E.A.: Sequence spaces and asymmetric norms in the theory of computational complexity. Math. Comput. Model. 36, 1–11 (2002)
[6] García-Raffi, L.M., Romaguera, S., Sánchez-Pérez, E.A., Valero, O.: Normed semialgebras: a mathematical model for the complexity analysis of programs and algorithms. In: Atlam, E., et al. (eds.) Proc. 7th World Multiconference on Systemics, Cybernetics and Informatics, vol. II, pp. 55–58. IIIS, Orlando (2003)
[7] Hitzler, P., Seda, A.K.: Generalized Distance Functions in the Theory of Computation. Comput. J. 53, 443–464 (2010)

[8] Künzi, H.P.A.: Nonsymmetric distances and their associated topologies: About the origins of basic ideas in the area of asymmetric topology. In: Aull, C.E., Lowen, R. (eds.) Handbook of the History of General Topology, vol. 3, pp. 853–968. Kluwer Acad. Publ., Netherlands (2001)

[9] Martín, J., Mayor, G., Valero, O.: On quasi-metric aggregation functions and fixed point theorems. Fuzzy Sets Syst. (2012), doi:10.1016/j.fss.2012.08.009

[10] Martín, J., Mayor, G., Valero, O.: On aggregation of normed structures. Math. Comput. Model. 54, 815–827 (2011)

[11] Massanet, S., Valero, O.: On aggregation of metric structures: the extended quasi-metric case. Int. J. Comput. Int. Sys. 6, 115–126 (2013)

[12] Massanet, S., Valero, O.: New results on metric aggregation. In: Sainz-Palmero, G.I., et al. (eds.) Proc. 17th Spanish Conference on Fuzzy Technology and Fuzzy Logic, pp. 558–563. European Society for Fuzzy Logic and Technology, Valladolid (2012)

[13] Massanet, S., Valero, O.: On midpoints and aggregation of quasi-metrics. In: Macario, S., et al. (eds.) Proc. Workshop in Applied Topology WiAT 2012, pp. 115–124. Publicacions de la Universitat Jaume I, Castellón (2012)

[14] Mayor, G., Valero, O.: Aggregation of asymmetric distances in computer science. Inform. Sci. 180, 803–812 (2010)

[15] Mayor, G., Valero, O.: Aggregating asymmetric distances in Computer Sciences. In: Ruan, D., et al. (eds.) Proc. 8th FLINS Conference on Computational Intelligence in Decision and Control, vol. I, pp. 477–482. World Scientific, Singapore (2008)

[16] Pestov, V., Stojmirović, A.: Indexing schemes for similarity search: an illustrated paradigm. Fund. Inform. 70, 367–385 (2006)

[17] Pestov, V., Stojmirović, A.: Indexing schemes for similarity search in datasets of short protein fragments. Inform. Syst. 32, 1145–1165 (2007)

[18] Rodríguez-López, J., Romaguera, S., Valero, O.: Denotational semantics for programming languages, balanced quai-metrics and fixed points. Int. J. Comput. Math. 85, 623–630 (2008)

[19] Pradera, A., Trillas, E.: A note on pseudometrics aggregation. Int. J. Gen. Syst. 31, 41–51 (2002)

[20] Pradera, A., Trillas, E., Castiñeira, E.: On the aggregation of some classes of fuzzy relations. In: Bouchon-Meunier, B., Gutiérrez-Ríos, J., Magdalena, L., Yager, R.R. (eds.) Technologies for Constructing Intelligent Systems. STUDFUZZ, vol. 90, pp. 125–147. Springer, Heidelberg (2002)

[21] Pradera, A., Trillas, E., Castiñeira, E.: On distances Aggregation. In: Bouchon-Meunier, B., et al. (eds.) Proc. Information Processing and Management of Uncertainty in Knowledge-Based Systems International Conference, vol. II, pp. 693–700. Univ. Politécnica Madrid Press, Madrid (2000)

[22] Romaguera, S., Sánchez-Pérez, E.A., Valero, O.: Computing complexity distances between algorithms. Kybernetika 39, 569–582 (2003)

[23] Romaguera, S., Schellekens, M.P., Valero, O.: The complexity space of partial functions: A connection between complexity analysis and denotational semantics. Int. J. Comput. Math. 88, 1819–1829 (2011)

[24] Romaguera, S., Tirado, P., Valero, O.: New results on mathematical foundations of asymptotic complexity analysis of algorithms via complexity space. Int. J. Comput. Math. 89, 1728–1741 (2012)

[25] Seda, A.K.: Quasi-metrics and the semantics of logic programs. Fundamenta Informaticae 29, 97–117 (1997)

[26] Seda, A.K.: Some Issues Concerning Fixed Points in Computational Logic: Quasi-Metrics, Multivalued Mappings and the Knaster-Tarski Theorem. Topology Proc. 24, 223–250 (1999)

[27] Schellekens, M.P.: The Smyth completion: a common foundation for denonational semantics and complexity analysis. Electronic Notes in Theoret. Comput. Sci. 1, 211–232 (1995)

[28] Stojmirović, A.: Quasi-metric spaces with measure. Topology Proc. 28, 655–671 (2004)

[29] Tirado, P., Valero, O.: The average running time of an algorithm as a mifpoint between fuzzy sets. Math. Comput. Modelling 49, 1852–1868 (2009)

Aggregation Operators and Quadric Hypersurfaces

J. Recasens

Abstract. Aggregation operators that are quadric hypersurfaces are studied. The interest lays in the fact that the most popular aggregation operators are indeed quadric hypersurfaces.

Keywords: Aggregation Operator, Idempotent, Symmetric, Quadric Hypersurface.

1 Introduction

If some laymen were asked to aggregate two or more numerical values, they would probably suggest the use of the arithmetic mean. More sophisticated people would suggest the geometric, quadratic or harmonic mean while experts would also propose the use of the maximum t-conorm, the minimum t-norm, Łukasiewicz and product t-norms and t-conorms or OWA operators. These are indeed the most popular aggregation operators and they are (part of) ruled quadratic hyper surfaces. Apart form the Łukasiewicz and product t-norms and t-conorms, they are idempotent as well.

This paper studies other ruled quadric surfaces that correspond to aggregation operators in two variables. In this way new families of aggregation operators, some of them combinations of the previous ones, are obtained. The results are then generalized to aggregation operators in several variables that are (part of) quadric hypersurfaces.

Let us recall the definition of aggregation operator.

Definition 1. [2] An *aggregation operator* is a map $h : \bigcup_{n \in N}[0,1]^n \to [0,1]$ satisfying

J. Recasens
Sec Matemàtiques i Informàtica, ETS Arquitectura del Vallès,
Universitat Politècnica de Catalunya,
C. Pere Serra 1-15, 08190 Sant Cugat del Vallès, Spain
e-mail: j.recasens@upc.edu

H. Bustince et al. (eds.), *Aggregation Functions in Theory and in Practise*,
Advances in Intelligent Systems and Computing 228,
DOI: 10.1007/978-3-642-39165-1_33, © Springer-Verlag Berlin Heidelberg 2013

1. $h(0,...,0) = 0$ and $h(1,...,1) = 1$
2. $h(x) = x \ \forall x \in [0,1]$
3. $h(x_1,...,x_n) \leq h(y_1,...,y_n)$ if $x_1 \leq y_1,...,x_n \leq y_n$ (*monotonicity*).

$$\overbrace{}^{n \text{ times}}$$

h is idempotent if and only if $h(\overbrace{x,...,x}) = x$ for all $x \in [0,1]$ and for all $n \in N$.

h is symmetric if and only if $h(x_1,...,x_n) = h(x_{\pi(1)},...,x_{\pi(n)})$ for any permutation π of $\{1,2,...,n\}$.

Given $n \in N$, the restriction of an aggregation operator h to $[0,1]^n$ will be called an aggregation operator in n variables.

2 Ruled Quadric Surfaces

In this section the idempotent ans symmetric aggregation operators in two variables that are quadric surfaces will be studied.

Definition 2. A quadric surface is a surface defined in implicit form by a second degree polynomial

$$ax^2 + by^2 + cz^2 + dxy + exz + fyz + gx + hy + iz + j = 0. \tag{1}$$

In order to find the ruled quadric surfaces which are aggregation operators, we will consider separately the cases $c \neq 0$ and $c = 0$.

2.1 Case $c = 0$

If $c = 0$, then isolating z from (1) we obtain

$$z = -\frac{ax^2 + by^2 + dxy + gx + hy + j}{ex + fy + i}. \tag{2}$$

Replacing e by $-e$, f by $-f$ and i by $-i$, (2) is

$$z = \frac{ax^2 + by^2 + dxy + gx + hy + j}{ex + fy + i}.$$

If we want the last map to be symmetric, we must have $b = a$, $g = h$ and $e = f$, obtaining

$$z = \frac{ax^2 + ay^2 + dxy + gx + gy + j}{ex + ey + i}.$$

$z(0,0)$ must be 0. From this we have $j = 0$.

If z is idempotent, writing explicitly $z(x,x) = x$ we obtain

$$z(x,x) = \frac{2ax^2 + dx^2 + 2gx}{2ex + i} = x.$$

or

$$(2a - 2e + d)x^2 = (i - 2g)x.$$

This equation is satisfied for all $x \in [0, 1]$ if and only if

$$d = 2e - 2a.$$

and

$$i = 2g.$$

The formula of the quadric surface becomes then

$$z = \frac{ax^2 + ay^2 + (2e - 2a)xy + gx + gy}{ex + ey + 2g}. \tag{3}$$

Now we can consider two cases: $e \neq 0$ and $e = 0$.

2.1.1 Case $c = 0$ and $e \neq 0$

In this case we can divide the numerator and the denominator of (3) by e. Renaming $\frac{a}{e}$ by a, and $\frac{g}{e}$ by g, we get

$$z = \frac{a(x-y)^2 + 2xy + gx + gy}{x + y + 2g}.$$

The denominator must be different from 0 for all $x, y \in (0, 1)$. This means

$$g \geq 0 \text{ or } g \leq -1.$$

For $x = 0$ and $y = 1$, we obtain

$$z(0, 1) = \frac{a + g}{1 + 2g}$$

This value must be between 0 and 1. Imposing that it must be greater or equal than 0, we obtain the following conditions for a and g.

$$g \geq -\frac{1}{2} \text{ and } a \geq -g$$

or

$$g \leq -\frac{1}{2} \text{ and } a \leq -g$$

Imposing that it must be smaller or equal than 1, we obtain the following conditions for a and g.

$$g \geq -\frac{1}{2} \text{ and } a \leq g + 1$$

or

$$g \leq -\frac{1}{2} \text{ and } a \geq g + 1$$

The partial derivatives $\frac{\partial z}{\partial x}(1,0)$ and $\frac{\partial z}{\partial x}(0,1)$ must be greater or equal than 0.

$$\frac{\partial z}{\partial x}(1,0) = \frac{(2a+g)(1+2g)-a-g}{(1+2g)^2} \geq 0$$

is satisfied if and only if

$$g \geq -\frac{1}{4} \text{ and } a \geq \frac{-2g^2}{1+4g}$$

or

$$g \leq -\frac{1}{4} \text{ and } a \leq \frac{-2g^2}{1+4g}.$$

$$\frac{\partial z}{\partial x}(0,1) = \frac{(-2a+g+2)(1+2g)-a-g}{(1+2g)^2} \geq 0$$

is satisfied if and only if

$$g \geq -\frac{3}{4} \text{ and } a \leq \frac{2+2g^2+4g}{3+4g}$$

or

$$g \leq -\frac{3}{4} \text{ and } a \geq \frac{2+2g^2+4g}{3+4g}.$$

Summarizing, the conditions on g and a are

$$g \geq 0 \text{ and} \frac{-2g^2}{1+4g} \leq a \leq \frac{2+2g^2+4g}{3+4g}$$

or

$$g \leq -1 \text{ and} \frac{2+2g^2+4g}{3+4g} \leq a \leq \frac{-2g^2}{1+4g}.$$

2.1.2 Case $c = 0$ and $e = 0$

In this case, putting $\frac{a}{2g} = b$,

$$z = b(x-y)^2 + \frac{x+y}{2}.$$

$z(1,0)$ is then

$$b + \frac{1}{2}.$$

Imposing again that this value must be between 0 and 1, we get that

$$-\frac{1}{2} \leq b \leq \frac{1}{2}.$$

Imposing that the partial derivative $\frac{\partial z}{\partial x}(1,0)$ must be greater or equal than 0, we get

$$b \geq -\frac{1}{4}.$$

Imposing that the partial derivative $\frac{\partial z}{\partial x}(0,1)$ must be greater or equal than 0, we get

$$b \leq \frac{1}{4}.$$

Summarizing,

$$-\frac{1}{4} \leq b \leq \frac{1}{4}.$$

2.2 Case $c \neq 0$

If $c \neq 0$, we can divide (1) by c. Renaming $\frac{a}{c}$ by $-a$, $\frac{b}{c}$ by $-b$, etc, we obtain

$$z = \frac{1}{2}((ex + fy + i\pm$$

$$\sqrt{(ex + fy + i)^2 - 4ax^2 - 4by^2 - 4gx - 4dxy - 4hy - 4j}).$$

If we impose symmetry we get

$$z = \frac{1}{2}(ex + ey + i\pm \qquad (4)$$

$$\sqrt{(e(x+y) + i)^2 - 4ax^2 - 4ay^2 - 4gx - 4dxy - 4gy - 4j}.$$

We can distinguish the cases where the square root is added or subtracted.

2.2.1 Adding the Square Root

In this case, (4) becomes

$$z = \frac{1}{2}(ex + ey + i+$$

$$\sqrt{(e(x+y) + i)^2 - 4ax^2 - 4ay^2 - 4gx - 4dxy - 4gy - 4j}.$$

Imposing $z(0,0) = 0$, we get

$$i + \sqrt{i^2 - 4j} = 0$$

and therefore $i \leq 0$ and $j = 0$.

From $z(1,1) = 1$, we get

$$2 = 2e + i + \sqrt{(2e + i)^2 - 8a - 8g - 4d}$$

and form this, $1 - 2e - i + 2a + 2g + d = 0$.

From $z(\frac{1}{2}, \frac{1}{2}) = \frac{1}{2}$, we get

$$1 = e + i + \sqrt{(e+i)^2 - 2a - 4g - 2d}$$

and form this, $1 - 2e - 2i + 2a + 4g + d = 0$.
   So $i = 2g$ (and $g \leq 0$) and $d = -1 + 2e - 2a$.
   Now imposing $z(k,k) = k$ we get

$$k = \frac{1}{2}\left(2ek + i + \sqrt{(2ke+i)^2 - 8ak^2 - 8gk - 4dk^2}\right).$$

which is equivalent to

$$k = \frac{1}{2}\left(2ek + i + \sqrt{(2ke+i)^2 - 8ek^2 - 4ik + 4k^2}\right) =$$

$$\frac{1}{2}\left(2ek + i + \sqrt{(2ke+i-2k)^2}\right).$$

Then

$$2ke + i \leq 2k \text{ for all } k \in [0,1].$$

This is satisfied for all $k \in [0,1]$ if and only if

$$2e + i \leq 2. \tag{5}$$

Putting $b = \frac{e}{2}$, the equation of the quadric surface is then

$$z = b(x+y) + g +$$

$$\sqrt{(b(x+y)+g)^2 - a(x-y)^2 - g(x+y) + (1-4b)xy}.$$

and (5) becomes

$$2b + g \leq 1.$$

$$z(1,0) = b + g + \sqrt{b+g)^2 - a - g}$$

which implies

$$g + a \leq (b+g)^2.$$

$0 \leq z(1,0) \leq 1$ gives

$$b + g \leq 1, g + 2b - a \leq 1$$

and if $b + g \leq 0$, then

$$a + g \leq 0.$$

Now imposing that $\frac{\partial z}{\partial x}(1,0) \geq 0$ we obtain

$$b + \frac{1}{2} \frac{2b^2 + 2bg - 2a - g}{\sqrt{(b+g)^2 - a - g}} \geq 0$$

and $\frac{\partial z}{\partial x}(0,1) \geq 0$ gives

$$b + \frac{1}{2} \frac{2b^2 + 2bg + 2a - g + 1 - 4b}{\sqrt{(b+g)^2 - a - g}} \geq 0.$$

2.2.2 Subtracting the Square Root

Similar calculations as in the previous subsection leads to

$$z = b(x+y) + g -$$

$$\sqrt{(b(x+y) + g)^2 - a(x-y)^2 - g(x+y) + (1 - 4b)xy}$$

with

$g \geq \max(0, 1 - 2b, -b)$

$a \leq \min\left((b+g)^2 - g, 1 + g + 2b\right)$

$b - \frac{1}{2} \frac{2b^2 + 2bg - 2a - g}{\sqrt{(b+g)^2 - a - g}} \geq 0$

$b - \frac{1}{2} \frac{2b^2 + 2bg + 2a - g + 1 - 4b}{\sqrt{(b+g)^2 - a - g}} \geq 0$

If $b + g \geq 1$, then $g + 2b - a \geq 1$.

3 Summarizing the Results for Two Variables

Table 1 summarizes the results obtained in the previous section.

- If in the first equation $a = g = 0$, we recover the harmonic mean.
- If in the second equation $b = 0$, we recover the arithmetic mean.
- If in the third equation $b = g = a = 0$ we recover the geometric mean.
- If in the third equation $a = -\frac{1}{2}$ and $b = g = 0$, we recover the quadratic mean.
- If in the third equation $a = g = 0$ and $b = \frac{1}{2}$, we recover the Maximum aggregation operator.
- If in the fourth equation $a = g = 0$ and $b = \frac{1}{2}$, we recover the Minimum aggregation operator.
- If $\frac{1}{2} \leq p \leq 1$ and in the third equation $a = g = 0$ and $b = \frac{p}{2}$, we recover the OWA operator with weights p and $1 - p$.
- If $0 \leq p \leq \frac{1}{2}$ and in the fourth equation $a = g = 0$ and $b = \frac{p}{2}$, we recover the OWA operator with weights p and $1 - p$.

Table 2 shows the ruled quadric surfaces corresponding to Table 1.

Table 1 Ruled quadric surfaces that are idempotent and symmetric aggregation operators

1 $z = \dfrac{a(x-y)^2 + 2xy + gx + gy}{x+y+2g}$ $g \geq 0$ and $\dfrac{-2g^2}{1+4g} \leq a < \dfrac{2+2g^2+4g}{3+4g}$ or $g \leq -1$ and $\dfrac{2+2g^2+4g}{3+4g} \leq a \leq \dfrac{-2g^2}{1+4g}$
2 $z = b(x-y)^2 + \dfrac{x+y}{2}$ $-\frac{1}{4} \leq b \leq \frac{1}{4}$
3 $z = b(x+y) + g +$ $\sqrt{(b(x+y)+g)^2 - a(x-y)^2 - g(x+y) + (1-4b)xy}$ $g \leq \min(0, 1-2b, 1-b)$ $-1+g+2b \leq a \leq (b+g)^2 - g$ $b + \frac{1}{2}\dfrac{2b^2+2bg-2a-g}{\sqrt{(b+g)^2-a-g}} \geq 0$ $b + \frac{1}{2}\dfrac{2b^2+2bg+2a-g+1-4b}{\sqrt{(b+g)^2-a-g}} \geq 0$ If $b+g \leq 0$, then $a+g \leq 0$.
4 $z = b(x+y) + g -$ $\sqrt{(b(x+y)+g)^2 - a(x-y)^2 - g(x+y) + (1-4b)xy}$ $g \geq \max(0, 1-2b, -b)$ $a \leq \min\left((b+g)^2 - g, 1+g+2b\right)$ $b - \frac{1}{2}\dfrac{2b^2+2bg-2a-g}{\sqrt{(b+g)^2-a-g}} \geq 0$ $b - \frac{1}{2}\dfrac{2b^2+2bg+2a-g+1-4b}{\sqrt{(b+g)^2-a-g}} \geq 0$ If $b+g \geq 1$, then $g+2b-a \geq 1$.

4 Aggregation Operators in More Than Two Variables

Aggregation operators in n variables corresponding to quadric hypersurfaces can be studied in a similar way. In this case, the equation of the corresponding quadric hypersurface is

$$\sum_{i=0}^{n} a_i x_i^2 + \sum_{i,j=0, i>j}^{n} b_{ij} x_i x_j + \sum_{i=0}^{n} c_i x_i + d = 0$$

and we want to isolate a variable (say x_0, $x_0 = f(x_1, x_2, ..., x_n)$).

Table 2 Ruled quadric surfaces corresponding to Table 1

1	$z = \dfrac{a(x-y)^2 + 2xy + gx + gy}{x + y + 2g}$ If $a \neq \frac{1}{2}$, then it is a cone If $a = \frac{1}{2}$, then it is a couple of non-parallel planes
2	$z = b(x-y)^2 + \dfrac{x+y}{2}$ If $b \neq \frac{1}{4}$, then it is a hyperbolic paraboloid If $b = \frac{1}{4}$, then it is a parabolic cylinder
3 and 4	$z = b(x+y) + g \pm$ $\sqrt{(b(x+y)+g)^2 - a(x-y)^2 - g(x+y) + (1-4b)xy}$ If $b = \frac{1}{2}, a = \frac{1}{4}, g \neq 0$, then it is a couple of non-parallel planes If $b = \frac{1}{2}, a = \frac{1}{4}, g = 0$, then it is a double plane If $b = \frac{1}{2}, a > \frac{1}{4}, g \neq 0$, then it is an elliptical cylinder If $b = \frac{1}{2}, a < \frac{1}{4}, g = 0$, then it is a couple of non-parallel planes If $b = \frac{1}{2}, a < \frac{1}{4}, g \neq 0$, then it is a hyperbolic cylinder If $b \neq \frac{1}{2}, a = b - \frac{1}{4}$, then it is a couple of non-parallel planes If $b < \frac{1}{4}, a \neq b - \frac{1}{4}$, then it is a cone If $b > \frac{3}{4}, a \neq b - \frac{1}{4}$, then it is a cone

1. If $a_0 = 0$, then

$$x_0 = -\frac{\sum_{i=1}^n a_i x_i^2 + \sum_{i,j=1,i>j}^n b_{ij} x_i x_j + \sum_{i=1}^n c_i x_i + d}{\sum_{i=1}^n b_{0i} x_i + c_0}$$

Imposing symmetry and $f(0,0,...,0) = 0$ we get

$$x_0 = \frac{a \sum_{i=1}^n x_i^2 + b \sum_{i,j=1,i>j}^n x_i x_j + c \sum_{i=1}^n x_i}{d \sum_{i=1}^n x_i + e}$$

Idempotency leads to

$$b = \frac{2(d-a)}{n-1}$$

and
$$e = nc.$$

The quadric hypersurface is then

$$x_0 = \frac{a \sum_{i=1}^{n} x_i^2 + \frac{2(d-a)}{n-1} \sum_{i,j=1,i>j}^{n} x_i x_j + c \sum_{i=1}^{n} x_i}{d \sum_{i=1}^{n} x_i + nc}$$

If $d \neq 0$, then the equation has the form

$$x_0 = \frac{a \sum_{i=1}^{n} x_i^2 + \frac{2(1-a)}{n-1} \sum_{i,j=1,i>j}^{n} x_i x_j + c \sum_{i=1}^{n} x_i}{\sum_{i=1}^{n} x_i + nc}$$

If $d = 0$, then

$$x_0 = a \sum_{i=1}^{n} x_i^2 + \frac{2a}{1-n} \sum_{i,j=1,i>j}^{n} x_i x_j + \frac{\sum_{i=1}^{n} x_i}{n}.$$

2. If $a_0 \neq 0$, imposing symmetry and $0 = f(0,0,...,0)$, we get

$$x_0 = \frac{1}{2}(-b \sum_{i=1}^{n} x_i - c \pm$$

$$\sqrt{(b \sum_{i=1}^{n} x_i + c)^2 - 4(\sum_{i=1}^{n} x_i^2 + d \sum_{i,j=1,i>j}^{n} x_i x_j + c \sum_{i=1}^{n} x_i))}.$$

From idempotency we obtain
$$c = -en$$

and

$$1 + bn + b^2 n^2 = -n - \binom{n}{2} d.$$

Constraints for the parameters can be obtained in a similar way as for the two variables case.

5   Concluding Remarks

The quadric hypersurfaces that can be considered as idempotent and symmetric aggregation operators have been studied.

In forthcoming works, other aggregation operators such as t-norms, t-conorms or uninorms that are quadric hypersurfaces will be studied (see [2]),

References

[1] Alsina, C., Sklar, A.: A characterization of continuous associative operations whose graphs are ruled surfaces. Aequationes Mathematicae 33, 114–119 (1987)

[2] Beliakov, G., Pradera, A., Calvo, T.: Aggregation Functions: A Guide for Practitioners. STUDFUZZ, vol. 221. Springer, Heidelberg (2008)

[3] Calvo, T., Kolesárova, A., Komorníková, M., Mesiar, R.: Aggregation Operators: Properties, Classes and Construction Methods. In: Calvo, T., Mayor, G., Mesiar, R. (eds.) Aggregation Operators: New Trends and Applications. STUDFUZZ, vol. 97, pp. 3–104. Springer, Heidelberg (2002)

[4] Recasens, J.: Aggregation Operators and Ruled Surfaces. In: Alsinet, T., Puyol-Gruart, J., Torras, C. (eds.) Artificial Intelligence Research and Development, pp. 206–214. IOS Press, Amsterdam (2008)

An Analysis of Bilevel Linear Programming Solving Parameters Based on Factoraggregation Approach

Pavels Orlovs, Olga Montvida, and Svetlana Asmuss

Abstract. We introduce the notion of factoraggregation, which is a special construction of general aggregation operators, and apply it for an analysis of optimal solution parameters for bilevel linear programming problems. The aggregation observes lower level objective functions considering the classes of equivalence generated by an objective function on the upper level. The proposed method is illustrated with numerical and graphical examples.

1 Introduction

Aggregation of several input values into a single output value is an important tool of mathematics, physics, as well as of engineering, economical, social and other sciences. As the widely used examples of aggregation operators we can mention arithmetic and geometric means, minimum and maximum operators, t-norms and others (see e.g. [1], [2], [4]). In 2003 A. Takaci [9] introduced the notion of a general aggregation operator acting on fuzzy structures. In this paper we define a general aggregation operator, named a factoraggregation operator, the idea of which is based on factorization by some equivalence relation. We illustrate how this approach could be applied to analyse bilevel linear programming problems (BLPP). BLPP is a special class of multi-objective linear programming problems (MOLP), where some hierarchy between objective functions is involved. We consider a case, when there is only one objective on the upper level and multiple objectives on the lower level. In 1978 H.J. Zimmermann [10] described a fuzzy algorithm, based on membership functions of objectives, for solving multi-objective linear programming problems without any hierarchy between objectives. For bilevel linear programming problems M. Sakawa and I. Nishizaki [7], [8] proposed an interactive method of solution, involving some

Pavels Orlovs · Olga Montvida · Svetlana Asmuss
University of Latvia, Department of Mathematics, Zellu street 8, Riga, LV-1002, Latvia,
e-mail: `pavels.orlovs@gmail.com,olgamontvida@yahoo.com,`
`svetlana.asmuss@lu.lv`

H. Bustince et al. (eds.), *Aggregation Functions in Theory and in Practise*,
Advances in Intelligent Systems and Computing 228,
DOI: 10.1007/978-3-642-39165-1_34, © Springer-Verlag Berlin Heidelberg 2013

parameters for upper and lower level objectives. We describe how a factoraggregation operator specially designed for BLPP allows to analyse the optimal solution
depending on parameters of the method, and as a result to make a decision on the
choice of these parameters. According to this approach the set of variables is factor-
ized by the equivalence relation generated by the membership function of the upper
level objective, and the lower level objectives are aggregated taking into account this
factorization. This method is illustrated with one particular problem similar to the
mixed production planning problem described by J.C. Figueroa-Garcia et al. in [1].

2   BLPP Fuzzy Solution Approach

In this paper we observe bilevel linear programming problem with one objective on
the upper level P^U with higher priority in optimization than multiple objectives on
the lower level $P^L = (P_1^L, P_2^L, ..., P_n^L)$:

$$P^U: \quad y_0(x) = c_{01}x_1 + c_{02}x_2 + ... + c_{0k}x_k \longrightarrow \min$$
$$P_1^L: \quad y_1(x) = c_{11}x_1 + c_{12}x_2 + ... + c_{1k}x_k \longrightarrow \min$$
$$...$$
$$P_n^L: \quad y_n(x) = c_{n1}x_1 + c_{n2}x_2 + ... + c_{nk}x_k \longrightarrow \min$$

(1)

$$D: \begin{cases} a_{j1}x_1 + a_{j2}x_2 + ... + a_{jk}x_k \le b_j, \ j = \overline{1,m}, \\ x_l \ge 0, \ l = \overline{1,k}, \end{cases}$$

(2)

where $k, m, n \in \mathbb{N}$ and D is a non-empty bounded set.

As all objectives rarely reach their optimal values in one point, a compromise solution should be found. H.J. Zimmermann [10] proposed a fuzzy solution approach
by introducing membership functions of objectives. The membership function char-
acterises the degree of satisfaction for each objective, i.e. it shows how close is the
objective function to its optimal value. The construction of the membership function
of objective y_i could be based on the following function:

$$z_i(t) = \begin{cases} 1, & t < y_i^{min}, \\ \dfrac{t - y_i^{max}}{y_i^{min} - y_i^{max}}, & y_i^{min} \le t \le y_i^{max}, \\ 0, & t > y_i^{max}, \end{cases}$$

(3)

where y_i^{min} and y_i^{max} are the individual minimum and the individual maximum of
objective y_i respectively:

$$y_i^{min} = \min_{x \in D} y_i(x), \ y_i^{max} = \max_{x \in D} y_i(x), \ i = \overline{0,n}.$$

(4)

We obtain membership functions of the objectives by denoting

$$\mu_i(x) = z_i(y_i(x)), i = \overline{0, n}. \tag{5}$$

Here $\mu_0, \mu_1, ..., \mu_n : D \rightarrow [0, 1]$ are fuzzy subsets of D:

$$\mu_i \in [0, 1]^D, i = \overline{0, n}. \tag{6}$$

A solution x^* for MOLP (1) – (2) without any hierarchy could be found by solving the following linear programming problem:

$$\min(\mu_0(x), \mu_1(x), \ldots, \mu_n(x)) \longrightarrow \max_{x \in D} \tag{7}$$

or, in general,

$$A(\mu_0(x), \mu_1(x), \ldots, \mu_n(x)) \longrightarrow \max_{x \in D}, \tag{8}$$

where A is some aggregation operator. However, in case of objectives divided between levels of hierarchy, the present method does not reflect any priority of the upper level objective over the lower level.

3 Factoraggregation

Considering the case when there is one objective function on the upper level and multiple objectives on the lower level, a special aggregation has been constructed. The aggregation observes objective functions on the lower level considering the classes of equivalence generated by a function on the upper level:

$$A_{\mu_0}(\mu_1, \mu_2, ..., \mu_n)(x) = \max_{\mu_0(x) = \mu_0(u)} A(\mu_1(u), \mu_2(u), ..., \mu_n(u)), \tag{9}$$

where

$$x, u \in D, \quad \mu_0, \mu_1, ..., \mu_n \in [0, 1]^D.$$

Aggregation (9) is a specially designed construction of a general aggregation operator, based on an ordinary aggregation operator A. We start with the classical notion of an aggregation operator (see e.g. [1], [2], [4]).

Definition 1. A mapping $A : \bigcup_n [0, 1]^n \rightarrow [0, 1]$ is called an aggregation operator if the following conditions hold:

(A1) $A(0, 0, \ldots, 0) = 0$;
(A2) $A(1, 1, \ldots, 1) = 1$;
(A3) $\forall n \in \mathbb{N} \; \forall x_1, x_2, \ldots, x_n, y_1, y_2, \ldots, y_n \in [0, 1]$:

   if $x_1 \leq y_1, x_2 \leq y_2, \ldots, x_n \leq y_n$, then $A(x_1, x_2, \ldots, x_n) \leq A(y_1, y_2, \ldots, y_n)$.

Conditions (A1) and (A2) are called boundary conditions of A, but (A3) means the monotonicity of A.

The general aggregation operator $\tilde{A}$ acting on $[0, 1]^D$, where $[0, 1]^D$ is the set of all fuzzy subsets of D, was introduced in [9]. We denote the order on $[0, 1]^D$ by $\preceq$, the least and the greatest elements of this order are denoted by $\tilde{0}$ and $\tilde{1}$, which are indicators of $\varnothing$ and D respectively.

Definition 2. A mapping $\tilde{A}: \bigcup_n ([0,1]^D)^n \to [0,1]^D$ is called a general aggregation operator if the following conditions hold:

($\tilde{A}$1) $\tilde{A}(\tilde{0},\tilde{0},\ldots,\tilde{0}) = \tilde{0}$;
($\tilde{A}$2) $\tilde{A}(\tilde{1},\tilde{1},\ldots,\tilde{1}) = \tilde{1}$;
($\tilde{A}$3) $\forall n \in \mathbb{N} \; \forall \mu_1,\mu_2,\ldots,\mu_n, \eta_1,\eta_2,\ldots,\eta_n \in [0,1]^D$:
    if $\mu_1 \preceq \eta_1, \mu_2 \preceq \eta_2,\ldots,\mu_n \preceq \eta_n$, then $\tilde{A}(\mu_1,\mu_2,\ldots,\mu_n) \preceq \tilde{A}(\eta_1,\eta_2,\ldots,\eta_n)$.

There exist several approaches to construct the general aggregation operator $\tilde{A}$, based on an ordinary aggregation operator A. The most simplest one is the point-wise extension of an aggregation operator A:

$$\tilde{A}(\mu_1,\mu_2,\ldots,\mu_n)(x) = A(\mu_1(x)\mu_2(x),\ldots,\mu_n(x)), \qquad (10)$$

which, for example, was already used in expression (8).

Another method of constructing the general aggregation operator $\tilde{A}$ is the T - extension [9], the idea of which comes from the classical extension principle (see e.g. [6]):

$$\tilde{A}(\mu_1,\mu_2\ldots,\mu_n)(x) = \sup_{x=A(x_1,x_2,\ldots,x_n)} T(\mu_1(x_1),\mu_2(x_2)\ldots,\mu_n(x_n)), \qquad (11)$$

where A is a continuous aggregation operator and T is a t-norm.

We introduce a factoraggregation of fuzzy sets from $[0,1]^D$ by means of a given fuzzy set $\mu_0 \in [0,1]^D$ using the following construction:

$$\tilde{A}_{\mu_0}(\mu_1,\mu_2,\ldots,\mu_n)(x) = \sup_{\mu_0(u)=\mu_0(x)} A(\mu_1(u),\mu_2(u),\ldots,\mu_n(u)), \qquad (12)$$

where

$$\mu_1,\mu_2,\ldots,\mu_n \in [0,1]^D, x,u \in D.$$

In construction (11) for evaluation of general aggregation $\tilde{A}(\mu_1,\mu_2\ldots,\mu_n)$ at a point x we take the supremum of t-norm T of values $\mu_1(x_1),\mu_2(x_2)\ldots,\mu_n(x_n)$ on preimage $A^{-1}(x)$. In our construction (12) for evaluation $\tilde{A}_{\mu_0}(\mu_1,\mu_2,\ldots,\mu_n)(x)$ we take the supremum of aggregation A of values $\mu_1(u),\mu_2(u),\ldots,\mu_n(u)$ on the set of all points u, which are equivalent to x with respect to μ_0, i.e. we consider all elements $u \in D$ such that $\mu_0(u) = \mu_0(x)$. In the scope of this paper we are interested in the case, when $A = \min$. It is clear that in (11) one can also take the minimum t-norm. Even in the case, when $A = \min$ and $T = \min$, approaches (11) and (12) are quiet different. Let us note that, applying construction (12), instead of min operator one could consider different mean operators or max operator as well.

The motivation of choosing the name factoraggregation for (12) is that μ_0 generates the equivalence relation $\sim_{\mu_0}$:

$$u \sim_{\mu_0} v \iff \mu_0(u) = \mu_0(v), \qquad (13)$$

which factorizes D into the classes D^α of equivalence:

$$D^\alpha = \{x \in D | \mu_0(x) = \alpha\}. \tag{14}$$

Operator $\tilde{A}_{\mu_0}$ aggregates fuzzy sets in accordance with these classes of equivalence. It is easy to show that properties $(\tilde{A}1) - (\tilde{A}3)$ hold for operator $\tilde{A}_{\mu_0}$, therefore we can be sure that $\tilde{A}_{\mu_0}$ is a general aggregation operator.

4 Factoraggregation Applied for Analysis of BLPP Solving Parameters

By using membership functions $\mu_0, \mu_1, ..., \mu_n$ multi-objective linear programming problem (8) can be reduced to the classical linear programming (LPP):

$$\sigma \longrightarrow \max_{x,\sigma} \tag{15}$$

$$\begin{cases} \mu_i(x) \geq \sigma, & i = \overline{0,n}, \\ x \in D. \end{cases} \tag{16}$$

Let us denote by (x^*, σ^*) the solution of (15) – (16). In [3] is described how to verify if x^* is a Pareto optimal solution.

Definition 3. A vector $x^* \in D$ is said to be a Pareto optimal solution if and only if there does not exist another vector $x \in D$ such that $y_i(x) \leq y_i(x^*)$ for all $i = \overline{0,n}$ and $y_j(x) \neq y_j(x^*)$ for at least one j.

The algorithm proposed by M. Sakawa and I. Nishizaki (see e.g. [7]) specifies optimal solution x^{**} for BLPP (1) – (2) according to chosen values of parameters $\delta, \Delta_L, \Delta_U$, where

$$\mu_0(x^{**}) \geq \delta, \tag{17}$$

$$\Delta_L \leq \Delta = \frac{\min\{\mu_1(x^{**}), \mu_2(x^{**}), ..., \mu_n(x^{**})\}}{\mu_0(x^{**})} \leq \Delta_U. \tag{18}$$

By this method we solve LPP

$$\sigma \longrightarrow \max_{x,\sigma} \tag{19}$$

$$\begin{cases} \mu_0 \geq \delta, \\ \mu_i(x) \geq \sigma, & i = \overline{1,n}, \\ x \in D, \end{cases} \tag{20}$$

afterwards we check if $\Delta \in [\Delta_L, \Delta_U]$ and specify parameters again if it is necessary.

Parameter δ describes the minimal satisfactory level for membership function μ_0, but Δ characterizes the overall balance between the upper and lower levels. Taking into account that all three parameters are dependent one from another, the problem of the choice of parameters becomes important.

We specify the general construction of factoraggregation (12) by taking $A = $ min to apply it for analysis of parameters of BLPP solving algorithm. Fuzzy sets $\mu_1, \mu_2, ..., \mu_n$ in this case are the membership functions of the lower level objectives, but fuzzy set μ_0 is the membership function of the upper level objective:

$$\tilde{A}_{\mu_0}(\mu_1, \mu_2, ..., \mu_n)(x) = \max_{\mu_0(x)=\mu_0(u)} \min(\mu_1(u), \mu_2(u), ..., \mu_n(u)). \qquad (21)$$

Denoting $\tilde{A}_{\mu_0}(\mu_1, \mu_2, ..., \mu_n)(x) = \mu(x)$ we rewrite μ as $\mu(x) = z(y_0(x))$. Now introducing the notation $t = y_0(x)$ we consider two functions $\alpha = z_0(t)$ and $\alpha = z(t)$. The graphical analysis of these functions helps us to choose parameters Δ_L, Δ_U and δ correctly.

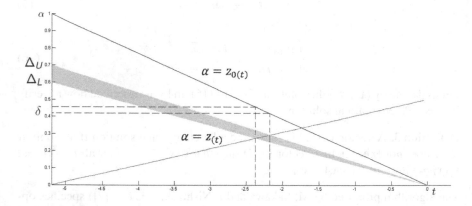

Fig. 1 Factoraggregation based analysis of BLPP solving parameters

Fig. 1 shows, that if we put $[\Delta_L, \Delta_U] = [0.6, 0.7]$, then the maximal possible satisfactory level for membership function μ_0 should be $\delta = 0.46$. The similar graphical analysis could be performed, when we choose the value of parameter δ first.

Let us consider the following example:

$$P^U : \quad y_0(x) = x_1 - x_2 \longrightarrow \min$$

$$P_1^L : \quad y_1(x) = -0.2x_1 - x_2 \longrightarrow \min \qquad (22)$$

$$P_2^L : \quad y_2(x) = x_2 \longrightarrow \min$$

$$D: \begin{cases} x_2 \le 6, \\ 5x_1 + x_2 \le 15, \\ x_l \ge 0, \ x_2 \ge 0. \end{cases} \tag{23}$$

The graphical analysis of the solving parameters could be performed by Fig. 2.

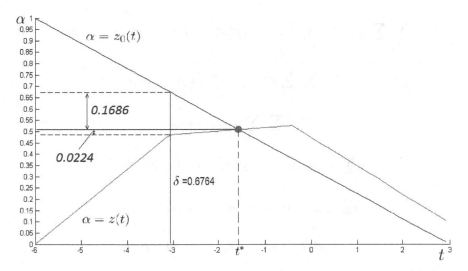

Fig. 2 Graphical analysis of solving parameters for BLPP (22) – (23)

The intersection of lines $\alpha = z_0(t)$ and $\alpha = z(t)$ on Fig. 2 points out an optimal solution x^* of MOLP problem without any hierarchy between objectives: $t^* = y_0(x^*)$. In our case we are dealing with BLPP, when objective function y_0 is minimized with the higher priority than objectives y_1 and y_2. The compromised solution x^* gives us the degree of satisfaction of the upper level objective $\delta = 0.51$. But the analysis of Fig. 2 allows us to see, that a minor decrease for 0.0224 in the degree of minimization on the lower level, which is characterized by the the result of factoraggregation $z(t)$, will give us the significant increase for 0.1686 in the degree of minimization δ on the upper level. It means, that we would rather choose point $t^{**} = -3.05$ to obtain the optimal solution x^{**} for the BLPP (22) – (23), than point $t^* = -1.6$, which gives us a solution without priority for the upper level objective. The similar graphical analysis could be performed, when we first choose the values of parameters Δ_L and Δ_U, which characterize the degree of minimization on the lower level, and then we can find out the possible values of the degree of minimization δ for the upper level objective.

5 Mixed Production Planning Problem

We consider the following modification of the mixed production planning problem described in [1]. The goal of the mixed production planning problem is to determine the most profitable manufacturing plan while minimizing ecologically dangerous products and dependence on outsource companies:

$$P^U : \sum_{j \in \mathbb{N}_J} \sum_{i \in \mathbb{N}_I} sp_{ji}(x^r_{ji} + x^o_{ji} + x^s_{ji}) - (cp^r_{ji}x^r_{ji} + cp^o_{ji}x^o_{ji} + cp^s_{ji}x^s_{ji}) \longrightarrow \max \qquad (24)$$

$$P^L_1 : \sum_{j \in \mathbb{N}_J} \sum_{i \in \mathbb{N}_I} ec_{ji}(x^r_{ji} + x^o_{ji} + x^s_{ji}) \longrightarrow \min$$

$$P^L_2 : \sum_{j \in \mathbb{N}_J} \sum_{i \in \mathbb{N}_I} w_{ji}x^s_{ji} \longrightarrow \min$$

$$\sum_{j \in \mathbb{N}_J} rm_{jir}(x^r_{ji} + x^o_{ji}) \leq am_{ir}, \quad i \in \mathbb{N}_I, \ r \in \mathbb{N}_R,$$

$$x^r_{ji}, x^o_{ji}, x^s_{ji} \geq 0, \quad x^s_{ji} \leq as_{ji}, \quad j \in \mathbb{N}_J, \ i \in \mathbb{N}_I,$$

$$d^{(-)}_{ji} \leq x^r_{ji} + x^o_{ji} + x^s_{ji} \leq d^{(+)}_{ji}, \quad j \in \mathbb{N}_J, \ i \in \mathbb{N}_I.$$

Index sets:
set $\mathbb{N}_R = \{1, 2, ..., R\}$ of all resources $r \in \mathbb{N}_R$,
set $\mathbb{N}_J = \{1, 2, ..., J\}$ of all products $j \in \mathbb{N}_J$,
set $\mathbb{N}_I = \{1, 2, ..., I\}$ of all periods $i \in \mathbb{N}_I$.
Decision variables:
x^r_{ji} – quantity of product j to be manufactured in regular time in the period i,
x^o_{ji} – quantity of product j to be manufactured in overtime in the period i,
x^s_{ji} – quantity of product j to be manufactured by outsourcing in the period i.
Parameters:
$SP = (sp_{ji} \mid j \in \mathbb{N}_J, i \in \mathbb{N}_I)$, where sp_{ji} is a sell price of product j in the period i,
$CP^r = (cp^r_{ji} \mid j \in \mathbb{N}_J, i \in \mathbb{N}_I)$, where cp^r_{ji} is a product j production cost in the period i for regular time,
$CP^o = (cp^o_{ji} \mid j \in \mathbb{N}_J, i \in \mathbb{N}_I)$, where cp^o_{ji} is a product j production cost in the period i for overtime,
$CP^s = (cp^s_{ji} \mid j \in \mathbb{N}_J, i \in \mathbb{N}_I)$, where cp^s_{ji} is a product j production cost in the period i for outsourcing,
$EC = (ec_{ji} \mid j \in \mathbb{N}_J, i \in \mathbb{N}_I)$, where ec_{ji} is an evaluation of harm to ecology caused by product j in the period i,
$W = (w_{ji} \mid j \in \mathbb{N}_J, i \in \mathbb{N}_I)$, where w_{ji} is a weight of outsource product j in the period i,
$RM = (rm_{jir} \mid j \in \mathbb{N}_J, i \in \mathbb{N}_I, r \in \mathbb{N}_R)$, where rm_{jir} is an amount of the r raw material units used to manufacture product j in the period i,

$AM = (am_{ir} \mid r \in \mathbb{N}_R, i \in \mathbb{N}_I)$, where am_{ir} is an availability of the raw material type r in the period i,

$AS = (as_{ji} \mid j \in \mathbb{N}_J, i \in \mathbb{N}_I)$, where as_{ji} is a number of available outsourced units of product j in the period i,

$D^- = (d_{ji}^{(-)} \mid j \in \mathbb{N}_J, i \in \mathbb{N}_I)$, where $d_{ji}^{(-)}$ is a minimum demand of product j in the period i,

$D^+ = (d_{ji}^{(+)} \mid j \in \mathbb{N}_J, i \in \mathbb{N}_I)$, where $d_{ji}^{(+)}$ is a maximum (potential) demand of product j in the period i.

Maximization problem (24) is reduced to the minimization problem by taking the negative profit function. We consider the case with no overtime production. The numerical example uses the following values of parameters:

$I = 1, J = 10, R = 5$,

$CP^r = (225, 165, 160, 105, 205, 175, 160, 225, 53, 74)$,

$CP^s = (260, 200, 185, 130, 240, 190, 210, 245, 105, 120)$,

$SP = (350, 300, 280, 210, 300, 305, 270, 315, 190, 220$,

$AM = (9000000, 4000000, 4500000, 3000000, 5500000)$,

$AS = (1237, 1107, 1519, 2636, 1979, 1617, 1442, 1527, 2266, 2500)$,

$D^+ = (8775, 7650, 6075, 7875, 6300, 7650, 5000, 6300, 4725, 8775)$,

$D^- = (3900, 3400, 2700, 1500, 3400, 4000, 1500, 4000, 1000, 2600)$,

$W = (0.07, 0.15, 0.07, 0.05, 0.05, 0.08, 0.05, 0.05, 0.18, 0.25)$,

$EC = (8, 9, 5, 4, 3, 7, 2, 1, 6, 10)$.

The values of RM are given by Table 1.

Table 1 Values rm_{jr}

j \ r	1	2	3	4	5
1	50.47	83.37	90.29	133.27	71.75
2	53.46	79.93	84.88	133.87	55.69
3	106.49	75.30	101.81	113.06	96.03
4	125.26	103.13	94.35	59.82	134.97
5	93.96	120.50	100.36	134.71	78.87
6	137.24	87.68	40.55	110.17	93.26
7	136.14	112.83	67.93	96.40	77.01
8	72.47	53.75	124.05	110.74	99.43
9	56.53	42.53	44.42	66.05	97.82
10	98.72	109.48	56.77	103.07	95.72

The graphical analysis of lines $\alpha = z_0(t)$ and $\alpha = z(t)$ is given by Fig. 3. The intersection of lines $\alpha = z_0(t)$ and $\alpha = z(t)$ points out an optimal solution x^* of MOLP problem without any hierarchy between objectives: $t^* = y_0(x^*)$. By setting $\Delta_L = 0.7$ and $\Delta_U = 0.8$ we can observe that δ should lie in the interval $[0.765, 0.799]$, otherwise there doesn't exist a solution x^{**} such that $(17) - (18)$ fulfil. The graphical analysis shows that as optimal solution of the mixed production planning problem it is rational to take x^{**} such that $t^{**} = y_0(x^{**})$.

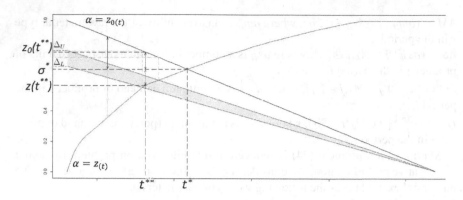

Fig. 3 Analysis of solving parameters for the mixed production planning problem

Acknowledgements. This work has been supported by the European Social Fund within the projects "Support for Doctoral Studies at University of Latvia" and "Support for Master Studies at University of Latvia".

References

[1] Calvo, T., Mayor, G., Mesiar, R.: Aggregation Operators. Physical-Verlag, Heidelberg (2002)
[2] Detyniecki, M.: Fundamentals on Aggregation Operators, Berkeley (2001)
[3] Figueroa-Garcia, J.C., Kalenatic, D., Lopez-Bello, C.A.: Multi-period mixed production planning with uncertain demands: fuzzy and interval fuzzy sets approach. Fuzzy Sets and Systems 206, 21–38 (2012)
[4] Grabisch, M., Marichal, J.-L., Mesiar, R., Pap, E.: Aggregation Functions. Cambridge University Press (2009)
[5] Jiménez, M., Bilbao, A.: Pareto-optimal solutions in fuzzy multi-objective linear programming. Fuzzy Sets and Systems 160, 2714–2721 (2009)
[6] Klement, E.P., Mesiar, R., Pap, E.: Triangular Norms. Kluwer Academic Publishers, Dordrecht (2000)
[7] Sakawa, M., Nishizaki, I.: Interactive fuzzy programming for decentralized two-level linear programming problems. Fuzzy Sets and Systems 125, 301–315 (2000)
[8] Sakawa, M., Nishizaki, I., Uemura, Y.: Interactive fuzzy programming for multi-level linear programming problems. Comput. Math. Appl. 36, 71–86 (1998)
[9] Takaci, A.: General aggregation operators acting on fuzzy numbers induced by ordinary aggregation operators. Novi. Sad. J. Math. 33(2), 67–76 (2003)
[10] Zimmermann, H.J.: Fuzzy programming and linear programming with several objective functions. Fuzzy Sets and Systems 1, 45–55 (1978)

Upper Bounding Overlaps by Groupings[*]

Nicolás Madrid, Edurne Barrenechea, Humberto Bustince, Javier Fernadez,
and Irina Perfilieva

Abstract. This paper focuses on determining conditions to ensure that certain overlaps are less than or equal to certain groupings. We also show that the meaning of such conditions is connected with the use of overlaps and groupings in image processing. Hence we present some results that are connected with the mentioned goal and formulated by using the notions of f-bound overlap and f-bound grouping.

1 Introduction

The notion of overlap operator was introduced originally in [3] with the aim of determining the overlapping of two different classes of fuzzy sets. Actually, the definition of such family of operators was motivated by the differentiation between an object and the background in an image. Hence, overlaps seem to be a family of aggregations potentially applicable to edge detection in image processing [8, 2, 5, 6, 9].

This paper presents a part of a more general study related to the substitution of t-norms a t-conorms by overlaps and groupings in [1], where an image preprocessing based on defining an interval-valued image from a grey-scale one was proposed. Unfortunately no every pair composed by an overlap operator and by a grouping operator can be considered for such intention. But in the definition of the interval

Nicolás Madrid · Irina Perfilieva
Institute for Research and Applications of Fuzzy Modeling, University of Ostrava. 30. dubna
22 701 03 Ostrava 1 Czech Republic
e-mail: {nicolas.madrid,irina.perfilieva}@osu.cz

Edurne Barrenechea · Humberto Bustince · Javier Fernández
Dept. Computer Science and Artificial Intelligence, Public University of Navarra,
C/ Arrosadía, s/n 31006 Pamplona, Spain
e-mail: {edurne.barrenechea,bustince,fcojavier.fernandez}@unavarra.es

* This work was supported by the European Regional Development Fund projects
  CZ.1.05/1.1.00/02.0070 and CZ.1.07/2.3.00/30.0010

H. Bustince et al. (eds.), *Aggregation Functions in Theory and in Practise*,
Advances in Intelligent Systems and Computing 228,
DOI: 10.1007/978-3-642-39165-1_35, © Springer-Verlag Berlin Heidelberg 2013

valued image in [1], it is crucial that the t-norm considered is less than or equal to the t-conorm considered. However, although this property always holds for all pairs of t-norms and t-conorms, not each overlap is less than or equal to all grouping. In this paper we present conditions to ensure that an overlap is upper bounded by a certain grouping.

The structure of the paper is described as follows. In Section 2 we recall the notions of overlap and grouping operator; moreover we recall the dual relationship between overlaps and groupings. Subsequently, in Section 3 we define the notions of f-bound overlap and f-bound grouping, which generalize somehow the existence of a neutral element in t-norms and t-conorms respectively. Moreover, we present also in such section some results guaranteeing that certain overlap is less than certain grouping and the relationship between the f-bound condition and dual operators. Finally, in Section 4 we describe conclusions and future work.

2 Preliminaries

Let us begin this section by recalling the definitions of overlap and grouping operators. For motivational aspects concerning with these definitions the reader is referred to [3, 4].

Definition 1. [4] An overlap operator is a mapping $G_O : [0,1]^2 \to [0,1]$ such that:

(G_O1) $G_O(x,y) = G_O(y,x)$ for all $x,y \in [0,1]$;
(G_O2) $G_O(x,y) = 0$ if and only if $x = 0$ or $y = 0$;
(G_O3) $G_O(x,y) = 1$ if and only if $x = y = 1$;
(G_O4) G_O is non-decreasing;
(G_O5) G_O is continuous.

Definition 2. [4] A *grouping* operator is a mapping $G_G : [0,1]^2 \to [0,1]$ such that:

(G_G1) $G_G(x,y) = G_G(y,x)$ for all $x,y \in [0,1]$;
(G_G2) $G_G(x,y) = 0$ if and only if $x = y = 0$;
(G_G3) $G_G(x,y) = 1$ if and only if $x = 1$ or $y = 1$;
(G_G4) G_G is non-decreasing;
(G_G5) G_G is continuous.

Both notions are evidently related because conditions (G_O1)-(G_O4)-(G_O5) and (G_G1)-(G_G4)-(G_G5) are identical and conditions (G_O2)-(G_O3) and (G_G2)-(G_G3) are dual with respect to each other. Actually, there is a dual relationship between overlaps and groupings which reminds the duality between t-norms and t-conorms. We recall that a negation operator n is any decreasing operator from $[0,1]$ to $[0,1]$ satisfying that $n(0) = 1$ and $n(1) = 0$. Moreover, a negation operator n is called *non-vanish* if $n(x) = 0$ if and only if $x = 1$, and *non-filling* if $n(x) = 1$ if and only if $x = 0$.

Definition 3. Let G be an overlap (resp. a grouping) and let n_1 and n_2 two non-filling and non-vanish negation operators. Then, the operator

$$\overline{G}^{n_1,n_2}(x,y) = n_1(G(n_2(x),n_2(y)))$$

is called the dual grouping (resp. overlap) of G with respect to n_1 and n_2.

The reader should note that the definition above needs a proof; i.e it is necessary to prove that effectively the operator $\overline{G}^{n_1,n_2}$ is a grouping (resp. an overlap). Such result can be found in [7].

To the end of this section we remind the fact that we have a structure of *complete lattice* on the set of operators of arity two defined on the unit interval $[0,1]$ (denoted by $(\mathcal{W}, \leq)$) by considering the point-wise ordering; i.e. given two operators $A, B: [0,1]^2 \to [0,1]$ we say that $A \leq B$ if and only if $A(x,y) \leq B(x,y)$ for all $x,y \in [0,1]$. Moreover, it is not difficult to check that the set of overlaps (resp. grouping) determines a sublattice of $(\mathcal{W}, \leq)$. However, such lattice structure on the set of overlaps (resp. groupings) is not complete (see [3]). Therefore, in this paper the operator supremum and infimum applied on an arbitrary set of overlaps (resp. groupings) is always considered on $(\mathcal{W}, \leq)$.

3 f-bound Groupings and Overlaps

As we have described in Introduction, t-norms and t-conorms are used to establish an interval of values in [1]. The crux of the matter in that approach is that a t-norm is less than a t-conorm. Hence t-norms and t-conorms are used to determine lower bounds and upper bounds of intervals, respectively. However, by considering an arbitrary overlap G_O and an arbitrary grouping G_G, we can not ensure that $G_O \leq G_G$, as the following example shows.

Example 1. Consider the overlap given by $G_O(x,y) = \min(\sqrt{x}, \sqrt{y})$ and the grouping given by $G_G(x,y) = \max(x^2, y^2)$. Then:

$$G_O(0.5, 0.25) = \min(\sqrt{0.5}, \sqrt{0.25}) \not\leq \max((0.5)^2, (0.25)^2) = G_G(0.5, 0.25)$$

So in this case $G_O \not\leq G_G$.

Therefore, as the aim of the research in course is to develop an approach similar to the one presented in [1] but by using overlaps and groupings instead of t-norms and t-conorms, a study on requirements to ensure that certain groupings are greater than or equal to certain overlaps is necessary.

We begin by defining the notions of f-bound overlaps and f-bound groupings, which generalize somehow a well known property of t-norms and t-conorms: namely, the role of 1 and 0 as neutral elements respectively; i.e any t-norm T and t-conorm S satisfy $T(x,1) = S(x,0) = x$.

Definition 4. Let f be a mapping from $[0,1]$ to $[0,1]$. An overlap G_O (resp. a grouping G_G) is called f-bound if the equality $G_O(x,1) = f(x)$ (resp. $G_G(x,0) = f(x)$) holds for all $x \in [0,1]$

Some remarks about this definition are needed. Firstly, note that every overlap G_O is $G_O(x,1)$-bound and all grouping G_G is $G_G(x,0)$-bound. Hence, the terminology of f-bound overlaps and f-bound groupings is included just for the sake of clarity. Secondly, it is easy to check that every t-norm (resp. t-conorm) is an x-bounded overlap (resp. grouping). Actually, an overlap (resp. a grouping) G is x-bounded if and only if 1 (resp. 0) is a neutral element of G. Finally, we will use sometimes the following abuse of notation: we consider in the rest of the paper f-bound overlaps without specifying explicitly that f is actually a mapping. Note that this consideration should not entail misunderstandings. Actually, we do not specify neither that such mappings f have to verify the following properties:

Proposition 1. *Let G be either an f-bound overlap or an f-bound grouping. Then:*

- *f is continuous,*
- *f is non-decreasing,*
- *f is surjective,*
- *$f(x) = 0$ if and only if $x = 0$,*
- *$f(x) = 1$ if and only if $x = 1$.*

Proof. Let us assume that G is an overlap (the proof by assuming that G is a grouping is similar). The continuity and non-decreasingness of f is a direct consequence of the continuity and non-decreasingness of G. Moreover, as $G(x,y) = 0$ if and only if $x = 0$ or $y = 0$, we have that $f(x) = G(x,1) = 0$ if and only if $x = 0$. Likewise, as $G(x,y) = 1$ if and only if $x = y = 1$, we have that $f(x) = G(1,x) = 1$ if and only if $x = 1$. Finally, as f is a continuous mapping from $[0,1]$ to $[0,1]$ and verifies $f(0) = 0$ and $f(1) = 1$, necessarily f has to be surjective. □

It is convenient to take into account that the set of mappings verifying the properties exposed in Proposition 1 has a structure of *non-complete* lattice. Actually, if we denote by Ω the set of mappings verifying the properties exposed in Proposition 1, one can prove that for all mapping $f \in \Omega$, there exist two mappings $f_1, f_2 \in \Omega$ such that $f_1 < f < f_2$. Thus, there is no maximal (minimal) element of Ω. On the other hand, it is also interesting to note that for all $f \in \Omega$ the families of f-bound overlaps (denoted hereafter by $\mathscr{O}_f$ and f-bound groupings (denoted hereafter by $\mathscr{G}_f$) are non-empty; just consider the overlap $\min(f(x), f(y))$ and the grouping $\max(f(x), f(y))$.

The name of "f-bound" has been chosen because the mapping f determines an upper bound in the case of overlaps and a lower bound in the case of groupings.

Proposition 2. *Let G_O and G_G be an overlap and a grouping, respectively. Then:*

- *if $G_O \in \mathscr{O}_f$, then $G_O(x,y) \leq \min\{f(x), f(y)\}$ for all $x,y \in [0,1]$.*
- *if $G_G \in \mathscr{G}_f$, then $G_G(x,y) \geq \max\{f(x), f(y)\}$ for all $x,y \in [0,1]$.*

Proof. Let G be an f-bound overlap. Then by monotonicity of G:

$$\left. \begin{array}{l} G(x,y) \leq G(x,1) = f(x) \\ G(x,y) \leq G(1,y) = f(y) \end{array} \right\} \Rightarrow G(x,y) \leq \min\{f(x), f(y)\}$$

for all $x,y \in [0,1]$. The case G is an f-bound grouping has a similar proof. □

The use of the notion of f-bound overlaps is motivated by the following result, which determines a sufficient condition to assert that an overlap is less than or equal to a grouping.

Proposition 3. *Let G_O and G_G be an overlap and a grouping respectively. If $G_O \in \mathscr{O}_{f_1}$ and $G_G \in \mathscr{G}_{f_2}$ with $f_1 \leq f_2$, then $G_O(x,y) \leq G_G(x,y)$ for all $x,y \in [0,1]$.*

Proof. By monotonicity of G_O and G_G we have the following chain of inequalities

$$G_O(x,y) \leq G_O(x,1) = f_1(x) \leq f_2(x) = G_G(x,0) \leq G_G(x,y)$$

for all $x,y \in [0,1]$. $\qquad\square$

Note that as a direct consequence of the above proposition we have the following result:

Corollary 1. *Let G_O be an overlap and let G_G be a grouping. If G_O and G_G are both f-bound then $G_O \leq G_G$.*

Note that Proposition 3 imposes a sufficient condition to guarantee that certain grouping is greater than or equal to certain overlap, but not a necessary condition. Actually, the following result states that for all f-bound overlap G_O there exists a family of infinite groupings $\{G_i\}_{i \in \mathbb{I}}$ such that each G_i is f_i-bound verifying that $f \geq f_i$ and $G_O \leq G_i$ for all $i \in \mathbb{I}$.

Proposition 4. *Let G_O be an overlap in $\mathscr{O}_f$ (resp. let G_G a grouping in $\mathscr{G}_f$), then for each mapping $\overline{f} \in \Omega$ there exists $G_G \in \mathscr{G}_{\overline{f}}$ (resp. $G_O \in \mathscr{O}_{\overline{f}}$) such that $G_O \leq G_G$.*

Proof. Let G_O be an f-bound overlap and let $\overline{f} \in \Omega$. Let us show that the operator

$$G_G(x,y) = \max(\overline{f}(x), \overline{f}(y), G_O(x,y))$$

is an $\overline{f}$-bound grouping such that $G_O \leq G_G$. Effectively, the operator G_G is a grouping since:

- G_G is symmetric (straightforward)
- $G_G(x,y) = 0$ if and only if $\overline{f}(x) = \overline{f}(y) = G_O(x,y) = 0$, if and only if $x = y = 0$.
- $G_G(x,y) = 1$ if and only if $\overline{f}(x) = 1$ or $\overline{f}(y) = 1$ or $G_O(x,y) = 1$, which is equivalent to $x = 1$ or $y = 1$.
- G_G is not decreasing and continuous, since f, G_O and max are non-decreasing and continuous operators.

Moreover, G_G is an $\overline{f}$-bound grouping since

$$G_G(0,x) = \max(\overline{f}(0), \overline{f}(x), G_O(0,x)) = \max(0, \overline{f}(x), 0) = \overline{f}(x).$$

Finally, the inequality $G_O \leq G_G$ holds by definition of G_G. $\qquad\square$

It is well known that among all t-norms, there is one which is the greatest one, namely the minimum operator. Dually, it is also well known that the operator maximum is the least t-conorm. As we say previously in Introduction, that feature does not hold when we consider the set of overlaps and the set of groupings (see [7]). However, a similar result is obtained if we consider just the set of f-bound overlaps and the set of f-bound groupings.

Proposition 5. *Let f be a mapping belonging to Ω. Then*

- *the operator $G_O(x,y) = f(\min(x,y))$ is the greatest f-bound overlap and*
- *the operator $G_G(x,y) = f(\max(x,y))$ is the least f-bound grouping.*

Proof. The proof of these two items are similar, so we give only the proof of the first one. Note previously that effectively $f(\min(x,y))$ defines an overlap thanks to the condition imposed on f. Actually, $f(\min(x,y))$ is an f-bound overlap, since $f(\min(x,1)) = f(x)$ for all $x \in [0,1]$. So, to end the proof we only have to show that $G_O(x,y) \leq f(\min(x,y))$ for all f-bound overlap G_O.

Let G_O be an f-bound overlap. Consider $x, y \in [0,1]$ and let us assume without lost of generality that $x \leq y$, then:

$$G_O(x,y) \leq G_O(x,1) = f(x) = f(\min(x,y))$$

Therefore, $G_O(x,y) \leq f(\min(x,y))$ for all $x, y \in [0,1]$. $\square$

The following result shows that neither the infimum of $\mathscr{O}_f$ is an overlap nor the supremum of $\mathscr{G}_f$ is a grouping for all $f \in \Omega$.

Proposition 6. *Let f be a mapping belonging to Ω. Then the infimum of $\mathscr{O}_f$ is:*

$$\mathscr{O}_f^{\wedge}(x,y) = \begin{cases} f(x) & \text{if } y = 1 \\ f(y) & \text{if } x = 1 \\ 0 & \text{Otherwise} \end{cases}$$

and the supremum of $\mathscr{G}_f$ is:

$$\mathscr{G}_f^{\vee}(x,y) = \begin{cases} f(x) & \text{if } y = 0 \\ f(y) & \text{if } x = 0 \\ 1 & \text{Otherwise} \end{cases}$$

Proof. Let us begin by proving that $\mathscr{O}_f^{\wedge}$ is the infimum of the subset $\{G_O^k\}$ of $\mathscr{O}_f$ defined as follows:

$$G_O^k(x,y) = \max(f(x^k) \cdot f(y), f(x) \cdot f(y^k))$$

for each $k \in \mathbb{N} \setminus \{0\}$. Let us show that each $G_O^k(x,y)$ is really an f-bound overlap for all $k \in \mathbb{N} \setminus \{0\}$; it is not difficult to prove the symmetry, non-decreasingness and continuity. Now, $G_O^k(x,y) = 0$ if and only if $f(x^k) \cdot f(y) = 0$ and $f(x) \cdot f(y^k) = 0$, if and only if (by properties of f) $x = 0$ or $y = 0$. Moreover, $G_O^k(x,y) = 1$ if and only if

$f(x^k) \cdot f(y) = 1$ or $f(x) \cdot f(y^k) = 1$, and both cases, by using the properties of f, are equivalent to $x = 1$ and $y = 1$. Finally, by using that f is non-decreasing, $f(1) = 1$ and that $x^k \leq x$ for all $x \in [0,1]$ and all $k \in \mathbb{N} \setminus \{0\}$, we obtain

$$G_O^k(x,1) = \max(f(x^k) \cdot f(1), f(x) \cdot f(1^k)) = \max(f(x^k), f(x)) = f(x)$$

So, effectively, each $G_O^k(x,y)$ is really an f-bound overlap for all $k \in \mathbb{N} \setminus \{0\}$.

Now let us show that $\inf_{k \in \mathbb{N}}(G_O^k)$ coincides with $\mathcal{O}_f^\wedge$. Let us distinguish two cases. Firstly, if $y = 1$ (or $x = 1$) we have that:

$$G_O^k(x,1) = f(x) \text{ for all } k \in \mathbb{N} \setminus \{0\}$$

So, $\inf_{k \in \mathbb{N}}(G_O^k)(x,1) = \inf_{k \in \mathbb{N}}(G_O^k(x,1)) = \inf_{k \in \mathbb{N}}(f(x)) = f(x) = \mathcal{O}_f^\wedge(x,1)$.

Secondly let us assume that $x, y \neq 1$. So note that by properties of f, we have that $f(x) \neq 1 \neq f(y)$ and then $\inf_{k \in \mathbb{N}}(f(x^k) \cdot f(y)) = \inf_{k \in \mathbb{N}}(f(y^k) \cdot f(x)) = 0$ for all $x, y \in [0,1)$. So we obtain that:

$$\inf_{k \in \mathbb{N}}(G_O^k)(x,y) = \inf_{k \in \mathbb{N}}(\max(f(x^k) \cdot f(y), f(x) \cdot f(y^k))) = 0 = \mathcal{O}_f^\wedge(x,y)$$

for all $x, y \in [0,1)$.

Finally, to end the proof, note that obviously $\mathcal{O}_f^\wedge$ is a lower bound of $\mathcal{O}_f$. Hence, since $\mathcal{O}_f^\wedge$ is in fact an infimum of a subset of $\mathcal{O}_f$, then $\mathcal{O}_f^\wedge$ has to be necessarily the infimum of $\mathcal{O}_f$.

To prove that $\mathcal{G}_f^\vee$ is the supremum of $\mathcal{G}_f$ we proceed similarly than above but considering the subset of groupings defined by $G_G^k(x,y) = 1 - G_O^k(1 - x, 1 - y)$ for all $k \in \mathbb{N} \setminus \{0\}$. $\square$

Corollary 2. *Let f be a mapping belonging to Ω. Then the set $\mathcal{O}_f$ (resp. $\mathcal{G}_f$) does not have the least element (resp. greatest element).*

Proof. Simply note that the infimum of the set of f-bound overlaps (resp. groupings) given by Proposition 6 is not an overlap (resp. a grouping) since is not continuous. $\square$

The following result shows that we can consider infinite strictly descending (resp. ascending) chains of f-bounds overlaps (resp. f-bound groupings).

Proposition 7. *If $G \in \mathcal{O}_f$ (resp. $G \in \mathcal{G}_f$), then there exists $\overline{G} \in \mathcal{O}_f$ (resp. $\overline{G} \in \mathcal{G}_f$) such that $\overline{G} < G$ (resp. $G < \overline{G}$)*

Proof. Just note that in the opposite case G would be minimal. Then, since $\mathcal{O}_f$ is a sublattice of $(\mathcal{W}, \leq)$, G has to be necessarily the least element $\mathcal{O}_f$. But that contradicts Proposition 6. $\square$

The result below shows that, independently of the $f_1, f_2 \in \Omega$ considered, we can find always pairs of overlaps $G_O \in \mathcal{O}_{f_1}$ and groupings $G_G \in \mathcal{G}_{f_2}$ such that $G_O \leq G_G$.

Corollary 3

- *Let G_O be an overlap in $\mathcal{O}_f$, then for all $\overline{f} \leq f$ there exists an infinite set $\overline{\mathcal{G}} \subset \mathcal{G}_{\overline{f}}$ such that $G_O \leq G_G$ for all $G_G \in \overline{\mathcal{G}}$.*
- *Respectively, let G_G be a grouping in $\mathcal{G}_f$, then for all $\overline{f} \geq f$ there exists an infinite set $\overline{\mathcal{O}} \subset \mathcal{O}_{\overline{f}}$ such that $G_O \leq G_G$ for all $G_O \in \overline{\mathcal{O}}$.*

Proof. It is a consequence of Proposition 4 and Proposition 7.   □

The following result deals with arbitrary overlaps and groupings instead of only with f-bound ones.

Proposition 8. *Let G be an overlap (resp. grouping), then there exist two overlaps (resp. groupings) G^1 and G^2 such that $G^1 < G < G^2$*

Proof. Let us assume that G is an overlap (the proof by assuming that G is a grouping is similar). We begin by recalling that G is in fact a $G(x, 1)$-boundary overlap. So the existence of G^1 is guaranteed by Proposition 7. The existence of G^2 is proven as follows. It is straightforward to check that Ω with the ordering induced by the natural ordering of $[0, 1]$ has not a maximum element. So there exists a mapping f in Ω and such that $G(x, 1) < f(x)$. Then, the overlap defined by $G^2(x, y) = \min(f(x), f(y))$ is the searched one, since by using Proposition 5 we have that:
$$G(x, y) \leq \min(G(x, 1), G(y, 1)) < \min(f(x), f(y)) = G^2(x, y)$$
for all $x, y \in [0, 1]$.   □

As a consequence of Proposition 8 we obtain an already known result about overlaps and groupings (see [3]).

Corollary 4. *The set of overlaps (resp. groupings) does not have neither maximal nor minimal elements.*

Another interesting property of the set of f-bound overlaps and the set of f-bound groupings is that they are dense[2].

Proposition 9. *Let G_1 and G_2 be two overlaps in $\mathcal{O}_f$ (resp. two groupings in $\mathcal{G}_f$) such that $G_1 < G_2$. Then there exists a $G \in \mathcal{O}_f$ (resp. $G \in \mathcal{G}_f$) such that $G_1 < G < G_2$.*

Proof. Just consider the f-bound overlap (resp. f-bound grouping) defined by $G(x, y) = \frac{G_1(x, y) + G_2(x, y)}{2}$.   □

At the end of this section we consider the notions of duality and f-boundary. First of all, it is convenient to mention that the dual construction is not enough to ensure that an overlap is less than or equal to one of its respective dual groupings; as the example below shows.

---

[2] Dense as poset

Example 2. Let us define a dual grouping G_G from an overlap G_O such that the inequality $G_O \leq G_G$ does not hold. Consider the overlap $G_O(x,y) = (x \cdot y)^{1/3}$ and its dual grouping $G_{G_O}^{1-x}(x,y) = 1 - ((1-x) \cdot (1-y))^{1/3}$. Then, $G_O(0.5, 0.5) = (0.5 \cdot 0.5)^{1/3} \approx 0.63 \nleq 0.37 \approx 1 - (0.5 \cdot 0.5)^{1/3} = G_G(0.5, 0.5)$.

The following definition and the subsequent result establish a restriction on the negations chosen in the dual construction in order to ensure that the dual operator maintains the f-bound condition.

Definition 5. Let G be an overlap (resp. grouping), let f be a bijective mapping belonging to Ω, and let n be a bijective, non-filling and non-vanish negation. Then, the operator defined by

$$\overline{G}^{n,f}(x,y) = f \circ n(G((n \circ f)^{-1}(x), (n \circ f)^{-1}(y)))$$

is called the f-dual grouping (resp. f-dual overlap) of G with respect to n.

Note that the f-dual grouping of an overlap G with respect to n is actually a dual grouping of G (according to Definition 3), since both $f \circ n$ and $(n \circ f)^{-1}$ are non-filling and non-vanish negation operators. So the f-dual construction simply determines dual overlaps and dual groupings by establishing a restriction on the negations chosen.

The definition above is motivated by the following result.

Proposition 10. *Let G be an f-bound overlap (resp. grouping), then all f-dual grouping (resp. overlap) of G is also f-bound.*

Proof. The proof only shows the case of overlaps; the proof for groupings is similar. If f is not bijective, the result holds trivially, since the set of f-dual groupings is empty.

So let us assume that f is bijective and let us show that $\overline{G}^{n,f}(x,0) = f(x)$ for all bijective, non-filling and non-vanish negation n. Let n be a negation operator, then by properties of f, $(n \circ f)^{-1}(0) = 1$. So:

$$\overline{G}^{n,f}(x,0) = f \circ n(G((n \circ f)^{-1}(x), (n \circ f)^{-1}(0))) = f \circ n(G((n \circ f)^{-1}(x), 1))$$

for all $x \in [0,1]$. By using now that $G(x,1) = f(x)$ we obtain finally:

$$\overline{G}^{n,f}(x,0) = f \circ n \circ f((n \circ f)^{-1}(x)) = f \circ n \circ f \circ f^{-1} \circ n^{-1}(x) = f(x)$$

for all $x \in [0,1]$. $\qquad\square$

Just two remarks about underlying aspects concerning the result above and its proof. The result is applicable to arbitrary f-bound overlaps and to arbitrary f-bound groupings. However when the mapping f is not bijective the f-dual operator is not defined. So the result provides useful information only for the case of f-bound overlaps and f-bound groupings with f bijective. Moreover, the reader can check that the conditions imposed on the negation (i.e. the negation has to be non-filling and non-vanish) are not used in the proof. These conditions are imposed to ensure that $f \circ n$ and $(n \circ f)^{-1}$ are non-filling and non-vanish negation operators. Otherwise we cannot ensure that $\overline{G}^{n,f}$ is an overlap or a grouping. That feature has not been included in the proof since, as we said above, the definition of f-dual construction is a specific case of duality presented in Definition 3.

Finally, as a consequence of Proposition 10, we can ensure that an f-dual overlap is less than all its f-dual groupings.

Corollary 5. *Let G_O be an f-bound overlap and let $\overline{G_O}^{n,f}$ be an f-dual grouping of G_O. Then the inequality $G_O \leq \overline{G_O}^{n,f}$ holds.*

And reciprocally:

Corollary 6. *Let G_G be an f-bound grouping and let $\overline{G_G}^{n,f}$ be an f-dual overlap of G_G. Then the inequality $\overline{G_G}^{n,f} \leq G_G$ holds.*

Example 3. Let us reconsider the overlap given in Example 2; i.e $G_O(x,y) = (x \cdot y)^{1/3}$. As $G_O(x,1) = x^{1/3}$, then G_O is an $x^{1/3}$-boundary overlap. So, the $x^{1/3}$-dual grouping of G_O with respect to the standard negation $n(x) = 1 - x$ is:

$$\overline{G_O}^{1-x,x^{1/3}}(x,y) = \left(1 - ((1-x)^3 \cdot (1-y)^3)^{1/3}\right)^{1/3} = (1 - (1-x) \cdot (1-y))^{1/3}$$

Moreover, thanks to Corollary 5 we have that

$$G_O(x,y) = (x \cdot y)^{1/3} \leq (1 - (1-x) \cdot (1-y))^{1/3} = \overline{G_O}^{1-x,x^{1/3}}(x,y)$$

for all $x,y \in [0,1]$.

4   Conclusion and Future Work

In this paper we have presented results that guarantee that certain overlap operator is less than or equal to certain grouping. Specifically, the notions of f-bound grouping and f-bound overlap have been introduced to achieve the goal. We have shown that given an f-bound overlap G_O, all $\overline{f}$-bound grouping with $f \leq \overline{f}$ is greater than or equal to G_O. Additionally, we have proven that for the case $f > \overline{f}$, although there exists some $\overline{f}$-bound grouping G_G such that $G_O \not\leq G_G$, we can find always an $\overline{f}$-bound grouping greater than or equal to G_O.

Moreover, some results concerning the structure of the set of f-bound groupings (f-bound overlaps) have been presented. Finally we have related the construction of dual overlaps and groupings to the notion of f-bound. Specifically, we have shown that the dual construction is not enough to guarantee that an overlap is less than or equal to one of its dual groupings; subsequently we have defined a new construction of dual operators, included by the original definition of duality but which preserves, somehow, the property of being f-bound; and finally we shown that, as a consequence of this feature, we can construct dual groupings G_G from certain overlaps G_O verifying that $G_O \leq G_G$ and vice-versa.

The future work will be mainly related to the use of overlaps and groupings operations to image processing. In particular, one of our goals is to substitute efficiently t-norms and t-conorms by overlaps and groupings in the approach described in [1]. In order to obtain a consistent extension of the approach, the overlap G_O and the grouping G_G chosen have to verify the inequality $G_O \leq G_G$. However, the presence of idempotence seems to be necessary in order to achieve a valuable approach; or at least that both operators coincide on the diagonal. Therefore, before to implement the algorithm allowing the use of groupings and overlaps, it is still necessary to develop theoretical results concerning with the idempotence of overlaps G_O and groupings G_G and conserving the ordering $G_O \leq G_G$.

References

[1] Barrenechea, E., Bustince, H., De Baets, B., Lopez-Molina, C.: Construction of interval-valued fuzzy relations with application to the generation of fuzzy edge images. IEEE Transactions on Fuzzy Systems 19(5), 819–830 (2011)

[2] Bowyer, K., Kranenburg, C., Dougherty, S.: Edge detector evaluation using empirical roc curves. In: IEEE Computer Society Conference on Computer Vision and Pattern Recognition, vol. 1, (xxiii+637+663), p. 2 (1999)

[3] Bustince, H., Fernandez, J., Mesiar, R., Montero, J., Orduna, R.: Overlap functions. Nonlinear Analysis: Theory, Methods & Applications 72(3-4), 1488–1499 (2010)

[4] Bustince, H., Pagola, M., Mesiar, R., Hullermeier, E., Herrera, F.: Grouping, overlap, and generalized bientropic functions for fuzzy modeling of pairwise comparisons. IEEE Transactions on Fuzzy Systems 20(3), 405–415 (2012)

[5] Canny, J.: A computational approach to edge detection. IEEE Transactions on Pattern Analysis and Machine Intelligence PAMI-8(6), 679–698 (1986)

[6] Dankova, M., Hodakova, P., Perfilieva, I., Vajgl, M.: Edge detection using f-transform. In: 2011 11th International Conference on Intelligent Systems Design and Applications (ISDA), pp. 672–677 (November 2011)

[7] Jurio, A., Bustince, H., Pagola, M., Pradera, A., Yager, R.: Some properties of overlap and grouping functions and their application to image thresholding. Fuzzy Sets and Systems (2013)

[8] Melin, P., Mendoza, O., Castillo, O.: An improved method for edge detection based on interval type-2 fuzzy logic. Expert Systems with Applications 37(12), 8527–8535 (2010)

[9] Rosenfeld, A., Thurston, M.: Edge and curve detection for visual scene analysis. IEEE Transactions on Computers C-20(5), 562–569 (1971)

A Preliminary Study of the Usage of Similarity Measures to Detect Singular Points in Fingerprint Images

Juan Cerrón, Mikel Galar, Carlos Lopez-Molina, Edurne Barrenechea, and Humberto Bustince

Abstract. One of the most prominent features of a fingerprint is the presence of singular points, which are locations of the fingertip at which unusual ridge patterns take place. They allow the classification of fingerprint images into different subclasses, accelerating the posterior matching process, but they can also be used to improve matching accuracy. In this work, we put forward a new method for singular point detection based on similarity measures. These measures are used to compare the orientations in a fingerprint orientation map to those in pre-established templates representing the canonical form of different singular points. This method provides a simple, yet effective, way to detect singular points in fingerprint images. Moreover, it is more flexible than the commonly used Poincarè method and also significantly simpler than other approaches, such as those based on complex filters. Preliminary experiments on two datasets show promising results.

1 Introduction

Among biometric systems, fingerprint-based identification systems are the most extended ones. These systems analyze the pattern of ridges and valleys located in the fingertips surface, and match it to the stored information of the corresponding individual. The individuality of a fingerprint is determined by the local ridge features and relationships [9]. One of the most used ridge features are the so-called minutiae, which are geometric disturbances of the ridges. Their distribution in the fingertip is frequently used to decide whether two fingerprints belongs to the same individual or not [10, 7, 5]. However, the process of minutiae extraction and matching is not trivial, and demands a certain computational effort. Considering the increasing number of the size of the biometric databases, it is necessary to reduce the individuals a new fingerprint has to be compared with. Hence, indexing schemes appear as a

Juan Cerrón · Mikel Galar · Carlos Lopez-Molina · Edurne Barrenechea · Humberto Bustince
Departamento de Automática y Computación, Universidad Pública de Navarra
e-mail: juan.cerron@unavarra.es

H. Bustince et al. (eds.), *Aggregation Functions in Theory and in Practise*,
Advances in Intelligent Systems and Computing 228,
DOI: 10.1007/978-3-642-39165-1_36, © Springer-Verlag Berlin Heidelberg 2013

good solution to avoid the comparison with a large portion of such databases. The most usual indexing schemes are based on the classification of the fingerprints intro five major classes [12], namely: *left loop*, *right loop*, *whorl*, *arch* and *tented arch*. Each of these classes corresponds to a characteristic general pattern of the ridge flow in the fingerprint. The classification of a candidate fingerprint into one of these categories can be carried out by analyzing the presence of Singular Points (SPs), which are the locations in the fingerprint at which very large orientation variance occurs. Apart from being of major importance for classifying the fingerprint into classes, SPs can also be employed in matching algorithms to both register a fingerprint, and to improve the accuracy of the matching by of hybrid methods [4]. As further detailed in this work, the main SPs are *cores* and *deltas*.

Singular points are identified as uncommon local patters in the ridge flow. Consequently, SP detection methods must rely on some kind of representation of such flow, which we refer to as *orientation map*. A major concern in the detection of singular resides in how the orientation map is obtained from the original fingerprint image. The most commonly employed method to compute the orientation map of a fingerprint is the gradient method [1], which uses the pixel gradients to compute the local orientation in adjacent blocks of the image. It is well known that imperfections in the fingerprint such as noise, spots, cuts, scars, etc. can induce errors in the local gradients, consequently producing wrong values in the orientation map, which might hinder the correct detection of the SPs. Given an orientation map, the most common method to detect singular points is the Poincarè method [12], which evaluates the curve formed by the neighbors of each orientation block to label it as a core, a delta or nothing. That is, it analyzes the variability of the orientations in adjacent blocks to discover unusual orientation patterns. Other methods in the literature make use of complex filters [11] or zero-pole models [8] to detect SPs.

In this work, we aim to overcome the problems of Poincarè, which highly relies on the orientation map, and to present a simpler and more intuitive way of computing SPs. We establish an orientation template for each kind of singular point, and then seek for the areas in the orientation map having the highest similarity to the templates. Then, we decide whether the similarity is enough as though label it as a SP. This method offers more flexibility than the existing ones due to the wide variety of similarity measures available in the literature. In order to properly assess the usefulness of the similarity-based singular points detection, we have carried out a preliminary experimental study. In our study we evaluate the accuracy of the singular points extracted with our method in the well-known NIST DB4 database, as well as in a dataset created with SFinGe software tool[1] [6, 12], which allows us to create synthetic fingerprints.

The rest of the paper is organized as follows: In Section 2 we recall several preliminary concepts needed to develop the work. Our proposal for SP detection is presented in Section 3. Section 4 shows the results obtained with this new method, while Section 5 recalls the contents of the work and outlines some future work.

---

[1] Synthetic Fingerprint Generator: `http://biolab.csr.unibo.it/`
`research.asp?organize=Activities&select=&selObj=12`

2 Preliminaries

This section recalls the concepts needed to develop the rest of the work starting with the basic concepts of Restricted Equivalence Function (REF) and Similarity Measure (SM), as well as some fingerprint recognition concepts required for the proposed method.

2.1 Restricted Equivalence Functions and Similarity Measures

Before explaining the similarity measures considered in our algorithm, we have to define some concepts and operations such as negation, which models the concept of *opposite*:

Definition 1. A mapping $n : [0,1] \rightarrow [0,1]$ with $n(0) = 1$, $n(1) = 0$, strictly decreasing, and continuous is called *strict negation*. Moreover, if n is involutive, i.e., if $n(n(x)) = x$ for all $x \in [0,1]$, then n is called a *strong negation*.

In this paper, we use the concept of Restricted Equivalence Function (REF), which relies on the concept of negation, to build the similarity measures. Such similarity measures are used to compare the angles in the orientation map and those in the singular point templates. Initially, REFs were introduced in [2, 3] to measure the degree of equivalence (*closeness*) between two points.

Definition 2. [2, 3] A function $r : [0,1]^2 \rightarrow [0,1]$ is called restricted equivalence function associated to the strong negation n, if it satisfies the following conditions

(1) $r(x,y) = r(y,x)$ for all $x,y \in [0,1]$;
(2) $r(x,y) = 1$ if and only if $x = y$;
(3) $r(x,y) = 0$ if and only if $x = 1$ and $y = 0$ or $x = 0$ and $y = 1$;
(4) $r(x,y) = r(n(x),n(y))$ for all $x,y \in [0,1]$;
(5) For all $x,y,z \in [0,1]$, if $x \leq y \leq z$, then $r(x,y) \geq r(x,z)$ and $r(y,z) \geq r(x,z)$.

A REF can be constructed using a pair of automorphisms of the unit interval, as proposed in [2]. We start recalling the concept of automorphism.

Definition 3. A continuous, strictly increasing function $\varphi : [a,b] \rightarrow [a,b]$ such that $\varphi(a) = a$ and $\varphi(b) = b$ is called automorphism of the interval $[a,b] \subset \mathbb{R}$.

Proposition 1. *[2] Let φ_1, φ_2 be two automorphisms of the interval $[0,1]$. Then*

$$r(x,y) = \varphi_1^{-1}(1 - |\varphi_2(x) - \varphi_2(y)|)$$

is a restricted equivalence function associated with the strong negation $n(x) = \varphi_2^{-1}(1 - \varphi_2(x))$.

Example 1. Let $\varphi_1(x) = x$ and $\varphi_2(x) = \sqrt{x}$, then

$$r(x,y) = 1 - |\sqrt{x} - \sqrt{y}| \qquad (1)$$

is a REF associated with $n(x) = (1 - \sqrt{x})^2$.

We now introduce an extension of the concept of *closeness* to compare tuples $(\mathbf{a} = (a_1, ..., a_N) \in [0,1]^N)$ instead of scalar values. This extension gives rise to the Similarity Measures (SMs). In our work, the elements of a tuple will be the components of an angle, i.e., its sine or cosine, respectively.

Definition 4. [2] A function $s : [0,1]^N \times [0,1]^N \to [0,1]$ is called a similarity measure with respect to the strong negation n if it satisfies the following properties:

(1) $s(\mathbf{a}, \mathbf{b}) = s(\mathbf{b}, \mathbf{a})$;
(2) $s(\mathbf{a}, n(\mathbf{a})) = 0$ if and only if $a_i = 0$ or $a_i = 1$ for all $i \in \{1, ..., N\}$, where $n(\mathbf{a}) = (n(a_1), ..., n(a_N))$;
(3) $s(\mathbf{a}, \mathbf{b}) = 1$ if and only if $a_i = b_i$ for all $i \in \{1, ..., N\}$;
(4) If $a_i \le b_i \le c_i$, for all $i \in \{1, ..., N\}$, then $s(\mathbf{a}, \mathbf{b}) \ge s(\mathbf{a}, \mathbf{c})$ and $s(\mathbf{c}, \mathbf{b}) \ge s(\mathbf{c}, \mathbf{a})$;
(5) $s(n(\mathbf{a}), n(\mathbf{b})) = s(\mathbf{a}, \mathbf{b})$.

Aggregation functions are useful in subsequent developments, since they provide provide a rule to combine different input values into a single output value.

Definition 5. An aggregation function is a mapping $M : [0,1]^N \to [0,1]$ such that:

(1) $M(0,0,...,0) = 0$,
(2) $M(1,1,...,1) = 1$, and
(3) M is nondecreasing.

Proposition 2. [2] Let $M : [0,1]^N \to [0,1]$ be an aggregation function such that : $M(x_1, ..., x_N) = 0$ if and only if $x_1 = ... = x_N = 0$; and $M(x_1, ..., x_N) = 1$ if and only if $x_1 = ... = x_N = 1$. Let $r : [0,1]^2 \to [0,1]$ be a restricted equivalence function associated with the strong negation n. Under these condition s, given by

$$s : [0,1]^N \times [0,1]^N \to [0,1]$$
$$s(\mathbf{a}, \mathbf{b}) = \overset{N}{\underset{i=1}{M}} r(a_i, b_i) \tag{2}$$

is a similarity measure associated with the strong negation n.

Example 2. Let M be the arithmetic mean and let $r(x, y) = \sqrt{1 - |x - y|}$ be the RDF associated to the strong negation $n(x) = 1 - x$, then

$$s(\mathbf{a}, \mathbf{b}) = \frac{1}{N} \sum_{i=1}^{N} \sqrt{1 - |a_i - b_i|} \tag{3}$$

is a similarity measure associated with $n(x) = 1 - x$.

2.2 Extraction of the Orientation Map

An orientation map is the representation of the ridge flow of the fingerprint image in terms of angles, usually in the range $\left(-\frac{\pi}{2}, \frac{\pi}{2}\right]$. The most commonly used method to obtain an orientation map is the gradient method [1], which we also consider in this work. Each pixel gradient is estimated with Sobel masks [13] to calculate the vertical (G_x) and horizontal (G_y) components. Note that gradients are computed for each pixel and the result of a single pixel may not be reliable enough to compute the orientation. That is why the gradients are averaged in $W_1 \times W_1$ blocks to obtain the estimation of the orientations according to:

$$G_{xx} = \sum_{i=1}^{W_1} \sum_{j=1}^{W_1} G_x^2(i,j), \quad G_{yy} = \sum_{i=1}^{W_1} \sum_{j=1}^{W_1} G_y^2(i,j) \tag{4}$$

and

$$G_{xy} = \sum_{i=1}^{W_1} \sum_{j=1}^{W_1} G_x(i,j) G_y(i,j) . \tag{5}$$

Now, the averaged gradient direction Φ, with $-\frac{1}{2}\pi < \Phi \leq \frac{1}{2}\pi$, is given by:

$$\Phi = \frac{1}{2} \angle (G_{xx} - G_{yy}, 2G_{xy}), \tag{6}$$

where $\angle(x,y)$ is defined as:

$$\angle(x,y) = \begin{cases} \tan^{-1}(x/y) & \text{if } x \geq 0, \\ \tan^{-1}(y/x) + \pi & \text{if } x < 0 \wedge y \geq 0, \\ \tan^{-1}(y/x) - \pi & \text{otherwise} \end{cases} \tag{7}$$

and the final ridge and valley orientation θ, with $-\frac{1}{2}\pi < \theta \leq \frac{1}{2}\pi$ is perpendicular to Φ:

$$\theta = \begin{cases} \Phi + \frac{1}{2}\pi & \text{if } \Phi \leq 0 \\ \Phi - \frac{1}{2}\pi & \text{otherwise} \end{cases} \tag{8}$$

After obtaining the first estimation of the orientation map, a smoothing phase is commonly carried out to obtain a more robust orientation map. Such smoothing is performed using a uniform window of size $W_2 \times W_2$. An example of the estimation of an orientation map using this method can be seen in Fig. 1.

3 Singular Point Detection Algorithm

In this section, we present our new algorithm for SP detection. First, we show how SPs manifest in an orientation image in order to clarify the construction of the templates and then, we explain our proposal based on SMs.

The algorithm is broken down into the following phases:

1. Establish the templates of the cores and deltas (2 for cores, 1 for deltas).
2. Obtain the orientation map of the fingerprint image.
3. Compute the similarity map, which represents how similar are the orientation map and each template.

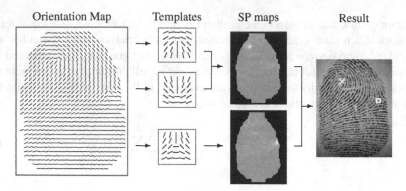

Fig. 1 Schematic representation of the of the proposed method

4. Identify the local maxima in each similarity map.
5. Confirm or reject the existence of SPs in such maxima.

An schematic representation of the algorithm can be seen in Fig. 1. Hereafter, we provide details of each one of the phases of the algorithm.

3.1 Definition of the Templates

This method uses two different templates of size $W_3 \times W_3$ to represent cores and deltas. Initially, the templates are defined with orientations, like the orientation map. However, both the templates and the orientation map are squared (every value θ is transformed to $2 \cdot \theta$) to ease the process of detecting SPs because of its new symmetry characteristic (see Figs. 2 and 3). The previous orientations, which were in the range $\left(-\frac{\pi}{2}, \frac{\pi}{2}\right]$, are now directions in the range $(-\pi, \pi]$. This fact lead us to consider two different templates to represent core singular points since convex core and concave core have opposite directions (Fig. 3). Notice also that working with directions (2θ) instead of orientations (θ) makes the information rotation invariant, which is a desired characteristic. The three templates centered in the coordinate axis are modeled as follows:

$$F_1(x,y) = \tan^{-1}(x, -y) \tag{9}$$

$$F_2(x,y) = \tan^{-1}(-x, y) \tag{10}$$

$$F_3(x,y) = \tan^{-1}(-y, -x) \tag{11}$$

where F_1 and F_2 represent the two possible core templates, and F_3 represents the delta one (Fig. 3).

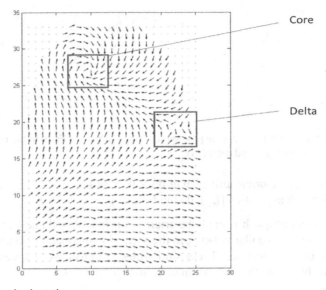

Fig. 2 Squared orientation map

3.2 Similarity Proposal

Intuitively, the similarity between the singular points areas of the squared orientation map and the templates proposed can be appreciated. In these conditions the SM becomes a valid tool to detect detect singular points. As indicated in Definition 4, a SM compares two tuples by means of REFs and aggregation functions. The tuples to be compared are: (a) each $W_3 \times W_3$ squared orientation map block, transformed into a row ordered tuple **b** and (b) the templates F_k, $k = 1, 2, 3$, also transformed into a row ordered tuple $\mathbf{f}_k$. However, as **b** and $\mathbf{f}_k$ are formed by angles, we decompose each one into two components, vertical (v) and horizontal (h), and compare these components independently. The horizontal and vertical components of a tuple **t** of N elements are computed as:

$$v(\mathbf{t}) = (v(t_1), ..., v(t_N)) \quad \text{and} \quad h(\mathbf{t}) = (h(t_1), ..., h(t_N)) \tag{12}$$

where

$$v(t_i) = \frac{\sin(t_i) + 1}{2} \quad \text{and} \quad h(t_i) = \frac{\cos(t_i) + 1}{2} . \tag{13}$$

Note that the original values of $\cos(t_i)$ and $\sin(t_i)$, for all $i \in \{1, ..., N\}$, are in the range $[-1, 1]$ but $v(t_i)$ and $h(t_i)$ re-scale these values to $[0, 1]$, as it is the range used for similarity measures.

The detailed explanation of the whole SP detection method follows. Note that the tuples $\mathbf{f}_k$, with $k \in \{1, ..., k\}$, are the vectorial representations of the orientations in the templates F_k, as in Eqs. (9)–(11).

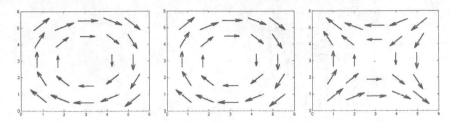

Fig. 3 Templates used to represent the prototypical singular points. From left to right: convex core (F_1), concave core (F_2) and delta (F_3).

1. Scroll the squared orientation map in $W_3 \times W_3$ blocks.
2. For each block at position (i, j), do the following:

    a. Construct a tuple **b** with all the orientations in a $W_3 \times W_3$ neighbourhood.
    b. Compute the similarity between the vertical components of **b** and those of each of the templates. These similarities are given as matrices Ω_k^v of as many blocks as the orientation map, so that

    $$\Omega_k^v(i, j) = s(v(\mathbf{f}_k), v(\mathbf{b})), \tag{14}$$

    where s is a similarity function.
    c. Compute the similarity between the horizontal components of **b** and those of each of the templates:

    $$\Omega_k^h(i, j) = s(h(\mathbf{f}_k), h(\mathbf{b})). \tag{15}$$

    d. Aggregate, for each template, the vertical and horizontal similarities with the geometric mean. We recall that any aggregation function could be used instead of such mean:

    $$\Omega_k(i, j) = \sqrt{\Omega_k^v(i, j) \cdot \Omega_k^h(i, j)}. \tag{16}$$

3. For each position (i, j) compute the *total core similarity* (Ω_C) as the maxima of Ω_1 and Ω_2:

    $$\Omega_C(i, j) = \max(\Omega_1(i, j), \Omega_2(i, j)). \tag{17}$$

4. Find the local maxima in each similarity map (Ω_C and Ω_3). The two maxima of each similarity map are taken as SPs if they exceed a given threshold, which we refer to as T. In any case, a fingerprint must contain, at most, four SPs: two cores and two deltas.

4   Experiments

We have done a preliminary study testing our method in two different datasets. The first one is the commonly used NIST DB4, produced by the National Institute of

Table 1 Settings used in the generation of the SFinGe dataset for the experimental test

Scanner parameters	Acquisition area: 0.58" x 0.77" (14.6mm x 19.6mm); Resolution: 500 dpi, Image size: 288 x 384; Background type: Optical, Background noise: Default; Crop borders: 0 x 0.
Generation parameters	Seed: 1; Impression per finger: 1. Class distribution: Natural; Set all distributions as: *Varying quality and perturbations*; Generate pores: enabled, Save ISO templates: enabled.
Output settings	Output file type: WSQ.

Table 2 Parameter setting for the SP detection

Setting	NIST experiment	SFinGe experiment
Gradient block size (W_1)	8	4
Smoothing size (W_2)	8	4
Template size (W_3)	5	5
SP threshold (T)	0.8	0.8

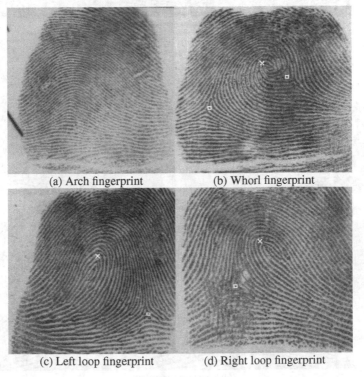

(a) Arch fingerprint   (b) Whorl fingerprint

(c) Left loop fingerprint   (d) Right loop fingerprint

Fig. 4 Singular points detected by our method in NIST DB4 images

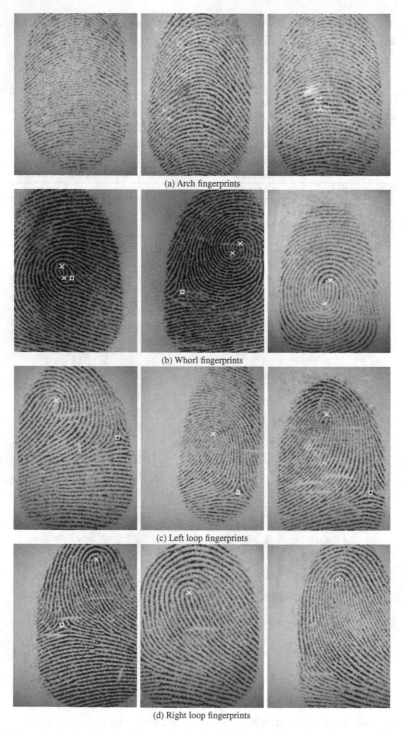

(a) Arch fingerprints

(b) Whorl fingerprints

(c) Left loop fingerprints

(d) Right loop fingerprints

Fig. 5 Singular points detected by our method in SFinGe images

Standards and Technology. The second dataset has been created with SFinGe, a software tool [12, 6], which allows us to create realistic fingerprints from a certain number of settings. We provide the detailed settings for the generation of the SFinGe fingerprints in Tables1.

The parameters used in the experiments for the computation of the orientation map have been selected as the standard values for fingerprint images in the specialized literature. The values of the parameters of our method (template size and threshold for the SP acceptance) have been set from the expertise obtained in preliminary experiments, as listed in Table 2. Apart from these parameters, a similarity measure must be chosen to the orientation map and the templates. For this experiment we have selected the one constructed from the REF based on $\varphi_1(x) = x$ and $\varphi_2(x) = x$, as in Proposition 2.

The results obtained for a series of fingerprint images from both datasets used can be observed in Figs. 4 and 5. In these images, detected cores are point out with crosses, whereas delta points are represented by square boxes. Note that the expected behaviour depends on the type of fingerprint. For example, in the case of *arch*, the desired behaviour is to detect no singular point (which is achieved in these fingerprints). It is interesting to observe that our method is able to detect singular points even in relatively noisy or healed areas, such as those in Fig. 4(d). Hence, the proposed methodology seems to be promising to detect SPs in varying quality fingerprints, what constitutes one of the major challenges in the area. Nevertheless, these a preliminary results, which should be further backed up by more intensive experimental studies.

5 Conclusions and Future Lines

In this contribution, we have proposed a method for singular point detection based on the comparison of orientation maps using similarity measures. The key advantages of our method with respect to the existing ones are the simplicity and flexibility. A preliminary experiment has shown promising accuracy in different conditions. Despite not having carried out experiments in significantly large datasets, we are confident in the potential of our method, which needs to be validated in exhaustive experimental studies. As future work, we aim to test different similarity measures, apart from comparing our algorithm with other proposals in the literature such as the Poincarè method, or those methods based complex filters and zero-pole models.

References

[1] Bazen, A.M., Gerez, S.H.: Systematic methods for the computation of the directional fields and singular points of fingerprints. IEEE Trans. on Pattern Analysis and Machine Intelligence 24(7), 905–919 (2002)
[2] Bustince, H., Barrenechea, E., Pagola, M.: Restricted equivalence functions. Fuzzy Sets and Systems 157(17), 2333–2346 (2006)

[3] Bustince, H., Barrenechea, E., Pagola, M.: Image thresholding using restricted equivalence functions and maximizing the measures of similarity. Fuzzy Sets Syst. 158(5), 496–516 (2007)

[4] Cappelli, R., Ferrara, M.: A fingerprint retrieval system based on level-1 and level-2 features. Expert Systems with Applications 39(12), 10,465–10,478 (2012)

[5] Cappelli, R., Ferrara, M., Maltoni, D.: Minutia cylinder-code: A new representation and matching technique for fingerprint recognition. IEEE Trans. on Pattern Analysis and Machine Intelligence 32(12), 2128–2141 (2010)

[6] Cappelli, R., Maio, D., Maltoni, D.: Sfinge: an approach to synthetic fingerprint generation. In: Eighth International Conference on Control, Automation, Robotics and Vision (ICARCV 2004), Kunming, China (2004)

[7] Chen, X., Tian, J., Yang, X.: A new algorithm for distorted fingerprints matching based on normalized fuzzy similarity measure. IEEE Trans. on Image Processing 15(3), 767–776 (2006)

[8] Fan, L., Wang, S., Wang, H., Guo, T.: Singular points detection based on zero-pole model in fingerprint images. IEEE Trans. on Pattern Analysis and Machine Intelligence 30(6), 929–940 (2008)

[9] Hong, L., Wan, Y., Jain, A.: Fingerprint image enhancement: algorithm and performance evaluation. IEEE Trans. on Pattern Analysis and Machine Intelligence 20(8), 777–789 (1998)

[10] Jiang, X., Yau, W.: Fingerprint minutiae matching based on the local and global structures. In: Proc. of the International Conference on Pattern Recognition, vol. 2, pp. 1038–1041 (2000)

[11] Liu, M.: Fingerprint classification based on adaboost learning from singularity features. Pattern Recognition 43(3), 1062–1070 (2010)

[12] Maltoni, D., Maio, D., Jain, A., Prabhakar, S.: Handbook of fingerprint recognition. Springer-Verlag New York Inc. (2009)

[13] Sobel, I., Feldman, G.: A 3x3 isotropic gradient operator for image processing. Presented at a talk at the Stanford Artificial Intelligence Project (1968)

Part VII
Fuzzy Transforms

F-transform in View of Aggregation Functions

Irina Perfilieva and Vladik Kreinovich

Abstract. A relationship between the discrete F-transform and aggregation functions is analyzed. We show that the discrete F-transform (direct or inverse) can be associated with a set of linear aggregation functions that respect a fuzzy partition of a universe. On the other side, we discover conditions that should be added to a set of linear aggregation functions in order to obtain the discrete F-transform. Last but not least, the relationship between two analyzed notions is based on a new (generalized) definition of a fuzzy partition without the Ruspini condition.

1   Introduction

In the last ten years, the theory of F-transforms has been intensively developed in many directions and especially in connection with image processing. The following topics have been newly elaborated on the *F-transform* platform: image compression and reconstruction [1, 2, 3], image reduction and sharpening [4], edge detection [5, 16], etc. On the other side, similar applications can be produced with the help of *aggregation functions*, see e.g., [7, 8]. The goal of this contribution is to discover a relationship between both notions.

Irina Perfilieva
Centre of Excellence IT4Innovations,
Division of the University of Ostrava
Institute for Research and Applications of Fuzzy Modelling,
30. dubna 22, 701 03 Ostrava 1, Czech Republic
e-mail: Irina.Perfilieva@osu.cz

Vladik Kreinovich
Department of Computer Science
University of Texas at El Paso
500 W. University, El Paso, TX 79968, USA
e-mail: vladik@utep.edu

H. Bustince et al. (eds.), *Aggregation Functions in Theory and in Practise*,
Advances in Intelligent Systems and Computing 228,
DOI: 10.1007/978-3-642-39165-1_37, © Springer-Verlag Berlin Heidelberg 2013

To comment the goal, we notice that it is not difficult to show that the discrete F-transform (direct or inverse) can be associated with a set of linear aggregation functions. However, the opposite characterization is not so obvious. In this contribution, we found conditions that should be added to a set of linear aggregation functions of the same number of variables in order to obtain the discrete F-transform. Last but not least, the proposed relationship between two notions is based on a new (generalized) definition of a fuzzy partition without the Ruspini condition.

We believe that this investigation contributes to a mutual success of both theories.

2   F-transform on a Generalized Fuzzy Partition

The F-transform technique was introduced in [9]. Below we remind its main principles for the so called *discrete* functions. The latter means that an original function f is known (may be computed) on a finite set $P = \{p_1, \ldots, p_l\} \subseteq [a, b]$. The interval $[a, b]$ will be considered as a universe of discourse that is partitioned into $n \geq 3$ fuzzy sets $A_1, \ldots, A_n$. We identify fuzzy sets $A_1, \ldots, A_n$ with their membership functions that map $[a, b]$ onto $[0, 1]$ and call them *basic functions*.

2.1   Generalized Fuzzy Partition

The following is a new definition of a *generalized fuzzy partition* which differs from that in [9] by using a smaller number of axioms.

Definition 1. Let $[a, b]$ be an interval on the real line $\mathbb{R}$, $n > 2$, and let $x_1, \ldots, x_n$ be nodes such that $a \leq x_1 < \ldots < x_n \leq b$. Let $[a, b]$ be covered by the intervals $[x_k - h'_k, x_k + h''_k] \subseteq [a, b]$, $k = 1, \ldots, n$, such that their left and right margins $h'_k, h''_k \geq 0$ fulfill $h'_k + h''_k > 0$.

We say that fuzzy sets $A_1, \ldots, A_n : [a, b] \to [0, 1]$ constitute a *generalized fuzzy partition* of $[a, b]$ (with nodes $x_1, \ldots, x_n$ and margins h'_k, h''_k, $k = 1, \ldots, n$), if for every $k = 1, \ldots, n$, the following three conditions are fulfilled:

1. (*locality*) — $A_k(x) > 0$ if $x \in (x_k - h'_k, x_k + h''_k)$, and $A_k(x) = 0$ if $x \in [a, b] \setminus (x_k - h'_k, x_k + h''_k)$;
2. (*continuity*) — A_k is continuous on $[x_k - h'_k, x_k + h''_k]$;
3. (*covering*) — for $x \in [a, b]$, $\sum_{k=1}^n A_k(x) > 0$.

We say that fuzzy sets $I_1, \ldots, I_n : [a, b] \to \{0, 1\}$ constitute a *(0-1)-generalized partition* of $[a, b]$ with nodes and margins as above, if for every $k = 1, \ldots, n$, I_k fulfills (*locality*) as above, (*continuity*) on $(x_k - h'_k, x_k + h''_k)$ and (*covering*) as above.

If nodes and margins are the same for generalized fuzzy and *(0-1)*-partitions $A_1, \ldots, A_n$ and $I_1, \ldots, I_n$, respectively, then we say that the latter is a "mask" of the former.

It is worth to remark that given nodes $x_1, \ldots, x_n$ and margins h'_k, h''_k, $k = 1, \ldots, n$, within $[a, b]$, a *(0-1)*-generalized partition $I_1, \ldots, I_n$ of $[a, b]$ is uniquely determined.

We say that a generalized fuzzy partition $A_1, \ldots, A_n$ of $[a, b]$ with nodes $x_1, \ldots, x_n$ and margins h'_k, h''_k, $k = 1, \ldots, n$, is *centered at nodes* if basic functions are bell-shaped, i.e. for each $k = 1, \ldots, n$, A_k is monotonically increasing on $[x_k - h'_k, x_k]$ and monotonically decreasing on $[x_k, x_k + h''_k]$.

Further on, the word "generalized" in characterization of fuzzy partitions will be omitted and left only when this fact is essential.

2.2 Discrete F-transform

We assume that a discrete function $f : P \to [0,1]^1$ on a finite domain $P = \{p_1, \ldots, p_l\}$, $P \subseteq [a,b]$, is given and that P is *sufficiently dense with respect to a fixed partition* $A_1, \ldots, A_n$, of $[a,b]$, i.e.,

$$(\forall k)(\exists j) A_k(p_j) > 0.$$

Then, the (discrete) F-transform of f and its inverse are defined as follows.

Definition 2. Let $A_1, \ldots, A_n$, for $n > 2$, be basic functions that form a generalized fuzzy partition of $[a,b]$, and let function f be defined on the set $P = \{p_1, \ldots, p_l\} \subseteq [a,b]$, which is sufficiently dense with respect to the partition. We assume that $n \le l$. The n-tuple of real numbers $F_n[f] = (F_1, \ldots, F_n)$ is the *discrete F-transform* of f with respect to $A_1, \ldots, A_n$ if

$$F_k = \frac{\sum_{j=1}^{l} f(p_j) A_k(p_j)}{\sum_{j=1}^{l} A_k(p_j)}, k = 1, \ldots, n. \tag{1}$$

The *inverse F-transform* $\hat{f}$ of f is a function that is defined on the same set P as above and represented by the following inversion formula:

$$\hat{f}(p_j) = \frac{\sum_{k=1}^{n} F_k A_k(p_j)}{\sum_{k=1}^{n} A_k(p_j)}, \quad j = 1, \ldots, l. \tag{2}$$

Assume that the elements of P are numbered in accordance with their order, i.e., $p_1 < \cdots < p_l$. Denote $P_k = \{p_j | A_k(p_j) > 0\}$, $k = 1, \ldots, n$. Because P is sufficiently dense with respect to $A_1, \ldots, A_n$, each set P_k, $k = 1, \ldots, n$ is not empty. Moreover, from the property *locality* it follows that for all $k = 1, \ldots, n$, there exist integers k_1, k_2 such that $1 \le k_1 \le k_2 \le l$ and $P_k = \{p_j \mid k_1 \le j \le k_2\}$. We say that P_k is covered by A_k or A_k covers P_k.

Let us identify the function f on P with the l-dimensional vector $(f_1, \ldots, f_l) \in [0,1]^l$ of its values such that $f_j = f(p_j)$, $j = 1, \ldots, l$. Because $A_1, \ldots, A_n$ is a fixed partition on $[a,b]$ and f is an arbitrary function on P, the F-transform $F_n[f]$ of f can be considered as a result of a linear map $F_n[f] : [0,1]^l \to [0,1]^n$ between linear vector spaces $[0,1]^l$ and $[0,1]^n$. We split this map into n separate maps $F_k : [0,1]^l \to [0,1]$ where $F_k(f_1, \ldots, f_l) = F_k$, $k = 1, \ldots, n$, and consider each map F_k as a real function of l arguments. In the sequel, we will be keeping at this viewpoint.

Let us list basic properties of the map $F_k : [0,1]^l \to [0,1]$, $k = 1, \ldots n$:

---

[1] The restriction of the range of f to $[0,1]$ is not principal and was assigned due to further correspondence with aggregation functions.

P1. *(linearity)* - for all $\mathbf{x}, \mathbf{y} \in [0,1]^l$ and $\alpha, \beta \in [0,1]$ such that $\alpha \mathbf{x} + \beta \mathbf{y} \in [0,1]^l$,
   $F_k(\alpha \mathbf{x} + \beta \mathbf{y}) = \alpha F_k(\mathbf{x}) + \beta F_k(\mathbf{y})$;

P2. *(idempotency)* - for all $c \in [0,1]$, $F_k(c,\ldots,c) = c$;

P3. *(non-decreasing)* - if $\mathbf{x}, \mathbf{y} \in [0,1]^l$ and $\mathbf{x} \leq \mathbf{y}$, then $F_k(\mathbf{x}) \leq F_k(\mathbf{y})$;

P4. *(redundancy)* - if basic function A_k covers the set $P_k = \{p_j | k_1 \leq j \leq k_2\}$, then
   only those arguments x_j among $x_1,\ldots,x_l$, whose indices are within the interval $k_1 \leq j \leq k_2$, are essential, i.e. for all $x_1,\ldots,x_l \in [0,1]$, $F_k(x_1,\ldots,x_l) =$
   $F_k(0,\ldots,x_{k_1},\ldots,x_{k_2},\ldots,0)$.

It easily follows from properties P1 and P3 that the map F_k, $k = 1,\ldots n$, is
monotonously non-decreasing. This fact together with the property P2 proves that
the map F_k is an additive and idempotent *aggregation function*[2] (see [10]). Moreover, from property P4 we deduce that the following derived function $F_k' : [0,1]^{l_k} \to$
$[0,1]$ where $l_k = (k_2 - k_1)$ and $F_k'(x_{k_1},\ldots,x_{k_2}) = F_k(0,\ldots,x_{k_1},\ldots,x_{k_2},\ldots,0)$ is an
aggregation function as well.

   In the following section, we will analyze the inverse problem, i.e., under which
conditions n aggregation functions determine the F-transform.

3   Discrete F-transform and Aggregation Functions

The goal of this section is to find conditions that characterize aggregation functions
as the F-transform components.

3.1   Aggregation Functions and Generic Fuzzy Partition

In this section, we will see that two kinds of properties: functional (additivity, etc.)
and spacial (correspondence with a certain partition), should be demanded from a set
of aggregation functions if we want them to represent the F-transform components.

Theorem 1. *Let $I_1,\ldots,I_n$, $n > 2$, be a (0-1)-generalized partition of $[a,b]$ with nodes
$x_1,\ldots,x_n$ and margins h_k', h_k'', $k = 1,\ldots,n$, and let finite set $P = \{p_1,\ldots,p_l \subseteq [a,b]\}$
where $l \geq n$ be sufficiently dense with respect to it. Then for any additive,
non-decreasing, idempotent aggregation functions $F_1,\ldots,F_n : [0,1]^l \to [0,1]$, that
fulfill the property P4 (with respect to $I_1,\ldots,I_n$) there exists a fuzzy partition
$A_1,\ldots,A_n$ of $[a,b]$ with the mask $I_1,\ldots,I_n$, such that for each $k = 1,\ldots,n$, the
k-th F-transform component F_k of a discrete function $f : P \to [0,1]$, identified with
$(f_1,\ldots,f_l)$, is equal to the value of the corresponding aggregation function F_k at
point $(f_1,\ldots,f_l)$.*

Proof. Let us fix k, $1 \leq k \leq n$, and prove the assertion for the aggregation function
$F_k : [0,1]^l \to [0,1]$. By the assumption, F_k fulfills the properties in the formulation.
From the first three, namely additivity, non-decreasing and idempotency, it follows

---

[2] An aggregation function of l variables in $[0,1]$ is a function which is non-decreasing in
   each argument and idempotent at boundaries $(0,\ldots,0)$ and $(1,\ldots,1)$.

(see, e.g , Proposition 4.21 from [10]) that there exist "weights" $w_{k1}, \ldots, w_{kl} \in [0, 1]$ such that $\sum_{j=1}^{l} w_{kj} = 1$ and

$$F_k(f_1, \ldots, f_l) = \sum_{j=1}^{l} w_{kj} f_j, \text{ where } (f_1, \ldots, f_l) \in [0, 1]^l. \tag{3}$$

Let $P_k = \{p_j \mid k_1 \leq j \leq k_2\}$ be covered by I_k. By the assumption, $P_k \neq \emptyset$. By the property P4, for all $f_1, \ldots, f_l \in \mathbb{R}$, $F_k(f_1, \ldots, f_l) = F_k(0, \ldots, f_{k_1}, \ldots, f_{k_2}, \ldots, 0)$. Therefore,

$$F_k(f_1, \ldots, f_l) = \sum_{j=1}^{l} w_{kj} f_j = \sum_{j=k_1}^{k_2} w_{kj} f_j.$$

In the above given equality, $(f_1, \ldots, f_l)$ is an arbitrary vector in $[0, 1]^l$, and this fact implies that coefficients $w_{kj} = 0$, if $j \in \{1, \ldots, l\} \setminus \{k_1, \ldots, k_2\}$. Let us define the basic function A_k on P as

$$A_k(p_j) = \begin{cases} w_{kj}, & \text{if } k_1 \leq j \leq k_2, \\ 0, & \text{otherwise,} \end{cases} \tag{4}$$

and prove that the k-th F-transform component F_k of $f : P \rightarrow [0, 1]$ with respect to A_k in (4) is equal to the aggregation $F_k(f_1, \ldots, f_l)$ where $f_j = f(p_j)$, $j = 1, \ldots, l$. Indeed by (1),

$$F_k = \frac{\sum_{j=1}^{l} f(p_j) A_k(p_j)}{\sum_{j=1}^{l} A_k(p_j)} = \frac{\sum_{j=1}^{l} f_j w_{kj}}{\sum_{j=1}^{l} w_{kj}} = F_k(f_1, \ldots, f_l).$$

To complete the proof it is sufficient to show that A_k can be continuously extended to the whole interval $[a, b]$ with the mask I_k.

By the *locality* of a generalized fuzzy partition, $I_k(x) > 0$ if and only if $x \in (x_k - h'_k, x_k + h''_k)$. By (4), $A_k(p_j) > 0$ if and only if $p_j \in P_k$. Because P_k is covered by I_k, $P_k \subset (x_k - h'_k, x_k + h''_k)$. Therefore, on the first step we construct a continuous extension of A_k to $[x_k - h'_k, x_k + h''_k]$. It can be obtained if we continuously connect the following points on the real plane: $(x_k - h'_k, 0), (p_{k_1}, w_{k,k_1}), \ldots, (p_{k_2}, w_{k,k_2}), (x_k + h''_k, 0)$. On the second step we put $A_k(x) = 0$ for all $x \in [a, b] \setminus [x_k - h'_k, x_k + h''_k]$, which is a continuous extension of A_k to $[a, b] \setminus [x_k - h'_k, x_k + h''_k]$. It is easy to see that thus extended A_k fulfills all requirements from Definition 1.

In the following corollary, we compose a matrix W so that the vector of F-transform components of f is the product of W by the vector of f.

Corollary 1. *Let the assumptions of Theorem 1 be fulfilled. Then for any additive, non-decreasing, idempotent aggregation functions* $F_1, \ldots, F_n : [0, 1]^l \rightarrow [0, 1]$, *that fulfill the property P4, there exists a $n \times l$ matrix W such that the F-transform $F_n[f] = (F_1, \ldots, F_n)$ of any discrete function $f : P \rightarrow [0, 1]$ such that $f(p_j) = f_j, j = 1, \ldots, l$, can be computed by the product $W\mathbf{f}$ where $\mathbf{f} = (f_1, \ldots, f_l)$, i.e.*

$$F_n[f] = \mathbf{Wf}. \tag{5}$$

Proof. Under the denotation of Theorem 1 and its proof, elements w_{kj} of the matrix W are weights that determine aggregation functions in accordance with (3).

We say that W is an *aggregation matrix that corresponds to the F-transform.*

3.2   Aggregation Functions and Centered Fuzzy Partition

This section is focused on fuzzy partitions that are centered at nodes. Our goal is to analyze under which conditions aggregating functions represent the F-transform with respect to this type of partition.

Let us consider aggregation functions of l variables, each one runs over $[0,1]$. We say that the point $\mathbf{y} \in [0,1]^l$ is a result of a *point-spread noise* applied to a point $\mathbf{x} \in [0,1]^l$ if both points differ exactly in one coordinate.

Definition 3. Let $F : [0,1]^l \to [0,1]$ be an aggregation function, $1 \le s \le l$ and $\mathbf{0}_q \in [0,1]^l$ be a point whose coordinates are 0s, except for the q-th one which is equal to 1. We say that aggregation F works as a *"noise damper"* centered at s, if it fulfills the following condition:

$$\text{if } (s \le q_2 < q_1 \le l) \text{ or } (1 \le q_1 < q_2 \le s) \text{ then } F(\mathbf{0}_{q_1}) \le F(\mathbf{0}_{q_2}). \tag{6}$$

Let us explain the above given notions. The value "1" at the q-coordinate in $\mathbf{0}_q$ represents a noise. The "noise damper" centered at s property of F means that the farther is the position of noise "1" from the s-th coordinate, the less is the value of aggregation performed by F.

The following theorem shows that aggregating functions that fulfill conditions of Theorem 1 and work as noise dampers centered at certain nodes represent the F-transform components with respect to a fuzzy partition that is centered at these nodes.

Theorem 2. Let $I_1, \ldots, I_n$, $n > 2$, be a $(0-1)$-generalized partition of $[a,b]$ with nodes $x_1, \ldots, x_n$ and margins h'_k, h''_k, $k = 1, \ldots, n$, and let finite set $P = \{p_1, \ldots, p_l \subseteq [a,b]\}$ where $l \ge n$ be sufficiently dense with respect to it. Assume that $x_1, \ldots, x_n \in P$, i.e. for all $1 \le k \le n$, there exists $1 \le j_k \le l$ such that $x_k = p_{j_k}$. Let $F_1, \ldots, F_n : [0,1]^l \to [0,1]$ be additive, non-decreasing, idempotent aggregation functions that fulfill the property P4 (with respect to $I_1, \ldots, I_n$) and work as noise dampers centered at respective positions $j_1, \ldots, j_n$. Then there exists a fuzzy partition $A_1, \ldots, A_n$ of $[a,b]$ with the mask $I_1, \ldots, I_n$, such that it is centered at nodes $x_1, \ldots, x_n$, and for each $k = 1, \ldots, n$, the k-th F-transform component F_k of any discrete function $f : P \to [0,1]$ is equal to $F_k(f_1, \ldots, f_l)$ where $f_j = f(p_j)$, $j = 1, \ldots, l$.

Proof. Assume that assumptions above are fulfilled. Let us fix k, $1 \le k \le n$, and prove the claim for the aggregation function $F_k : [0,1]^l \to [0,1]$. By Theorem 1, there exist coefficients $w_1, \ldots, w_l \in [0,1]$ such that $\sum_{j=1}^l w_j = 1$ and

$$F_k(f_1,\ldots,f_l) = \sum_{j=1}^{l} w_j f_j, \text{ where } (f_1,\ldots,f_l) \in [0,1]^l. \tag{7}$$

Let $P_k = \{p_j \mid k_1 \leq j \leq k_2\}$ be covered by I_k. By the assumption, $x_k \in P_k$ so that $x_k = p_{j_k}$ for some $k_1 \leq j_k \leq k_2$. Let us prove that the sequence of coefficients $w_1,\ldots,w_l$ non-strictly increases for $i \leq j_k$ and non-strictly decreases for $i \geq j_k$, i.e.,

$$w_1 \leq \ldots \leq w_{j_k} \geq w_{j_k+1} \geq \ldots \geq w_l. \tag{8}$$

By (6), the aggregation function F_k works as a "noise damper" centered at j_k. Let $1 \leq q \leq l$, and $\mathbf{0}_q$ be the l-tuple whose elements are 0s, except for the q-th one which is equal to 1. By (7), $F_k(\bar{0}_q) = w_q$. Therefore, by (6),

$$\text{if } (k \leq q_2 < q_1 \leq l) \text{ or } (1 \leq q_1 < q_2 \leq l) \text{ then } w_{q_1} \leq w_{q_2}.$$

This proves (8). The rest of the proof coincides with the proof of Theorem 1.

4 Inverse F-transform and Aggregation Functions

If we compare expressions (1) and (2) for the direct and inverse F-transform, then we see that they have similar structures. Therefore, the inverse F-transform is expected to be represented by aggregation functions too. The aim of this section is to find a relationship between a set of aggregation functions which determine the direct F-transform and another set of aggregation functions which determine the inverse F-transform.

Assume that the direct F-transform of a discrete function $f : P \to [0,1]$, where the set $P = \{p_1,\ldots,p_l\} \subseteq [a,b]$ is sufficiently dense with respect to a certain fuzzy partition $A_1,\ldots,A_n$ of $[a,b]$, is determined by a corresponding set of aggregation functions $F_1,\ldots,F_n : [0,1]^l \to [0,1]$ such that for every $(f_1,\ldots,f_l) \in [0,1]^l$,

$$F_k(f_1,\ldots,f_l) = \frac{\sum_{j=1}^{l} f_j A_k(p_j)}{\sum_{j=1}^{l} A_k(p_j)}, k = 1,\ldots,n. \tag{9}$$

By this we mean that the k-th F-transform component F_k of the function f is equal to $F_k(f_1,\ldots,f_l)$, provided that $f_j = f(p_j)$, $j = 1,\ldots,l$.

The inverse F-transform $\hat{f}$ of f with respect to the same partition $A_1,\ldots,A_n$ is a function on P that is determined by another set of functions $\hat{f}_j : [0,1]^n \to [0,1]$ such that $\hat{f}(p_j) = \hat{f}_j(F_1,\ldots,F_n)$, $j = 1,\ldots,l$, where $F_1,\ldots,F_n$ are the F-transform components of f and

$$\hat{f}_j(F_1,\ldots,F_n) = \frac{\sum_{k=1}^{n} F_k A_k(p_j)}{\sum_{k=1}^{n} A_k(p_j)}, \quad j = 1,\ldots,l. \tag{10}$$

The following reasoning (similar to that in Subsection 2.2) aims at proving that the functions $\hat{f}_j$, $j = 1,\ldots,l$, are aggregations. Indeed, the inverse F-transform (10) can be considered as a result of a linear map $\hat{f} : [0,1]^n \to [0,1]^l$ between linear vector spaces $[0,1]^n$ and $[0,1]^l$. We split this map into l separate maps $\hat{f}_j : [0,1]^n \to [0,1]$ so that each one is a real function of n arguments.

The basic properties of $\hat{f}_j : [0,1]^n \to [0,1]$, $j = 1, \ldots l$ are the same as they are for the maps $F_k : [0,1]^l \to [0,1]$, $k = 1, \ldots n$: linearity, idempotency, non-decreasing and redundancy. The latter differs from the above formulated P4 in interchanging j and k. Let us give the precise formulation.

P5. (*redundancy*) - if a point p_j, $j = 1, \ldots, l$, is covered by several basic functions A_k, i.e. $A_k(p_j) > 0$, where $j_1 \leq k \leq j_2$, then only those arguments x_k among $x_1, \ldots, x_n$, whose indices are within the interval $j_1 \leq k \leq j_2$, are essential, i.e. for all $x_1, \ldots, x_n \in [0,1]$, $\hat{f}_j(x_1, \ldots, x_n) = \hat{f}_j(0, \ldots, x_{j_1}, \ldots, x_{j_2}, \ldots, 0)$.

Therefore, the maps $\hat{f}_j : [0,1]^n \to [0,1]$, $j = 1, \ldots, l$ are linear aggregation functions on $[0,1]^n$ that fulfill the property P5. Conversely, similarly to Theorem 1, any l additive, non-decreasing, idempotent aggregation functions $\hat{f}_j$ on $[0,1]^n$ that fulfill the property P5 can be combined into one function $\hat{f} : P \to [0,1]$ such that $\hat{f}(p_j) = \hat{f}_j(F_1, \ldots, F_n)$, $j = 1, \ldots, l$.

Our goal is to find conditions on aggregation functions $F_1, \ldots, F_n : [0,1]^l \to [0,1]$ and aggregation functions $\hat{f}_j : [0,1]^n \to [0,1]$, $j = 1, \ldots, l$, such that they determine the direct and inverse F-transforms with respect to the same partition $A_1, \ldots, A_n$. The following theorem gives the solution.

Theorem 3. *Let $I_1, \ldots, I_n$, $n > 2$, be a (0-1)-generalized partition of $[a,b]$ with nodes $x_1, \ldots, x_n$ and margins h'_k, h''_k, $k = 1, \ldots, n$, and let finite set $P = \{p_1, \ldots, p_l \subseteq [a,b]\}$ where $l \geq n$ be sufficiently dense with respect to it. Then for any additive, non-decreasing, idempotent aggregation functions $F_1, \ldots, F_n : [0,1]^l \to [0,1]$, that fulfill the property P4 there exist additive, non-decreasing, idempotent aggregation functions $\hat{f}_1, \ldots, \hat{f}_l : [0,1]^n \to [0,1]$, that fulfill the property P5, both with respect to $I_1, \ldots, I_n$, and a fuzzy partition $A_1, \ldots, A_n$ of $[a,b]$ with the mask $I_1, \ldots, I_n$ such that for any discrete function $f : P \to [0,1]$ such that $f(p_j) = f_j$, $j = 1, \ldots, l$,*

(i) *the F-transform component F_k, $k = 1, \ldots, n$, of f is the value of the corresponding aggregation function F_k at point $(f_1, \ldots, f_l)$,*
(ii) *the inverse F-transform $\hat{f}(p_j)$, $j = 1, \ldots, l$, is equal to the corresponding aggregation function $\hat{f}_j$ at point $(F_1, \ldots, F_n)$.*

In Corollary 1, the aggregation matrix W that corresponds to the F-transform was introduced. A similar result will be established for the inverse F-transform.

Corollary 2. *Let the assumptions of Theorem 1 be fulfilled and $W = (w_{kj})$ be a $n \times l$ matrix that corresponds to the F-transform so that for a function f, (5) holds. Then the related inverse F-transform $\hat{f}$ of f is characterized by the $l \times n$ matrix $\tilde{W} = (\tilde{w}_{jk})$ so that*

$$\hat{f} = \tilde{W} F_n[f]$$

where

$$\tilde{w}_{jk} = \frac{w_{kj}}{\sum_{k=1}^n w_{kj}}, \, j = 1, \ldots, l, \, k = 1, \ldots, n.$$

Conclusion

In this contribution, we focused on a relationship between the F-transform and aggregation functions. We showed that the F-transform components can be obtained by linear aggregation functions that respect a fuzzy partition of a universe. On the other side, we discovered conditions that should be added to a set of linear aggregation functions in order to obtain the F-transform components. Similarly, the inverse F-transform can be associated with another set of linear aggregation functions that respect a fuzzy partition of a co-universe. Two sets of linear aggregation functions that are associated with the direct and inverse F-transforms are connected via the so called aggregation matrix. The relationship between two analyzed notions is based on a new (generalized) definition of a fuzzy partition without the Ruspini condition.

Acknowledgements. This work relates to Department of the Navy Grant N62909-12-1-7039 issued by Office of Naval Research Global. The United States Government has a royalty-free license throughout the world in all copyrightable material contained herein. Additional support was given also by the European Regional Development Fund in the IT4Innovations Centre of Excellence project. (CZ.1.05/1.1.00/02.0070).

References

[1] Perfilieva, I.: Fuzzy transforms and their applications to image compression. In: Bloch, I., Petrosino, A., Tettamanzi, A.G.B. (eds.) WILF 2005. LNCS (LNAI), vol. 3849, pp. 19–31. Springer, Heidelberg (2006)

[2] Di Martino, F., Loia, V., Perfilieva, I., Sessa, S.: Int. Journ. of Appr. Reasoning 48, 110 (2008)

[3] Hurtik, P., Perfilieva, I.: Image compression methodology based on fuzzy transform. In: Herrero, Á., et al. (eds.) Int. Joint Conf. CISIS'12-ICEUTE'12-SOCO'12. AISC, vol. 189, pp. 525–532. Springer, Heidelberg (2013)

[4] Hurtik, P., Perfilieva, I.: Proc. of the Intern. MIBISOC 2013 Conference, Brussels (to appear, 2013)

[5] Daňková, M., Hodáková, P., Perfilieva, I., Vajgl, M.: Proc. Intelligent Systems Design and Applications (ISDA 2011), Cordoba, Spain, pp. 672–677 (2011)

[6] Perfilieva, I., Hodáková, P., Hurtík, P.: f^1-transform edge detector inspired by canny's algorithm. In: Greco, S., Bouchon-Meunier, B., Coletti, G., Fedrizzi, M., Matarazzo, B., Yager, R.R. (eds.) IPMU 2012, Part I. CCIS, vol. 297, pp. 230–239. Springer, Heidelberg (2012)

[7] Barrenechea, E., Bustince, H., De Baets, B., Lopez-Molina, C.: IEEE Transactions on Fuzzy Systems 19(5), 819 (2011)

[8] Beliakov, G., Bustince, H., Paternain, D.: IEEE Transactions on Image Processing 21(3), 1070 (2012)

[9] Perfilieva, I.: Fuzzy Sets and Systems 157, 993 (2006)

[10] Grabisch, M., Marichal, J.L., Mesiar, R., Pap, E.: Aggregation Functions. Cambridge Univ. Press, Cambridge (2009)

Fuzzy Hit-or-Miss Transform Using the Fuzzy Mathematical Morphology Based on T-norms

M. González-Hidalgo, S. Massanet, A. Mir, and D. Ruiz-Aguilera

Abstract. The extension of the Hit-or-Miss transform (HMT) to grey-level images is difficult due to the problem of defining the complement of an image in this context. Thus, several extensions have been proposed in the literature avoiding the use of the complement. However, in the fuzzy framework, the complement is well-established by means of a fuzzy negation and the binary HMT can be extended preserving its geometrical interpretation. In this paper, we extend the binary HMT to a fuzzy HMT (FHMT) using the mathematical morphology based on t-norms. Some properties of this operator are studied and some initial experimental results are presented proving the potential of the FHMT in shape recognition and pattern matching.

Keywords: Fuzzy hit-or-miss transform, fuzzy mathematical morphology, t-norm, fuzzy implication.

1 Introduction

Mathematical Morphology (MM) was introduced in the early sixties by Matheron [23] and Serra [35]. Originally, it was developed for binary images where this theory provided an extremely interesting set of tools for the analysis and shape recognition in this class of images. Soon thereafter, it was extended to grey-level images (GL) following different approaches (see [15, 35, 37]). Among the tools provided by the MM, there is the hit-or-miss transform (HMT) [15, 35], which is capable of identifying in a binary image groups of connected pixels satisfying certain geometric restrictions or forming a certain configuration. For the processing of binary images the HMT is widely used and well defined [35, 37] and involves the search and location in an image of a predefined shape (called structuring element (SE)). The SE are

M. González-Hidalgo · S. Massanet · A. Mir · D. Ruiz-Aguilera
Department of Mathematics and Computer Science,
University of the Balearic Islands, E-07122 Palma de Mallorca, Spain
e-mail: {manuel.gonzalez,s.massanet,arnau.mir,daniel.ruiz}@uib.es

H. Bustince et al. (eds.), *Aggregation Functions in Theory and in Practise*,
Advances in Intelligent Systems and Computing 228,
DOI: 10.1007/978-3-642-39165-1_38, © Springer-Verlag Berlin Heidelberg 2013

designed to match the geometry of objects of interest in the foreground and background of the image. Despite this, there are few authors who have considered its possible extension to grey-level images. The main difficulty, as we shall see, resides in that this operator uses in its definition the image and its complement, and this last concept in the Grey-Level Mathematical Morphology (GLMM) is not clearly established. In an effort to avoid its use, several researchers have proposed definitions and methods to extend the HMT to grey-level images, and recently an unified framework for calculating a grey-level HMT (GLHMT) has been presented in [26].

The fuzzy mathematical morphology (FMM) is a generalization of the binary MM using concepts and techniques from the fuzzy sets theory ([6, 25]). This theory allows a better treatment and a representation with greater flexibility of the uncertainty and ambiguity present in any level of an image. In this framework, the analysis of the fuzzy Hit-or-Miss transform (FHMT) is much narrower. Sinha et al. in [36] discussed a FHMT based on his FMM to achieve word recognition. In such a paper, no properties are analysed but they indicated how to choose the structuring elements. Later, in an unpublished work, Deng [10] introduced a FHMT based on fuzzy logic and with a similar approach to the one we will introduce, but the theory remains to be developed. Popov [30] introduced a FHMT using intuitionistic fuzzy sets and after that in [11] they applied it to find face features without analysing the properties or explaining how to choose the SE. Finally, in [16] a FHMT is used to perform an automate boundary extraction by a fuzzy gradient.

The hit-or-miss transform is often used to detect specific configurations of pixels and it has interesting applications. We can highlight applications to document analysis [7]; template and pattern matching [4, 19, 33]; boundary and edge extraction [16, 22]; face detection and localization [11, 32]; medical image analysis [27, 8]; building and vehicle detection [18, 21, 38]; satellite and astronomical image analysis [1, 29, 17] and analysis of geographic and topographic data [37, 40]. But in many of these applications the hit-or-miss transform is used after preprocessing the image and performing a threshold to it. In addition, no analysis of the proposed transformation is realized and there is no explanation how the structuring elements are selected. This is an indication of how difficult is to apply the GLHMT to image processing and therefore, a detailed analysis is necessary.

In this paper we define a general fuzzy hit-or-miss transform for grey-level image following the FMM introduced by De Baets [9] and further developed in [12, 14, 13]. We analyse their properties, the selection of the structuring element and how the properties have applications in the experimental results. This paper is organised as follows. In the next section, we recall the definitions of fuzzy logic operators which are needed in subsequent sections. Next, we review briefly in this work, for the sake of clarity, the binary HMT and the most prevalent extensions of the HMT for grey-level images. We show how our definition is related with the classical binary definition and we generalise it to the fuzzy environment. The fuzzy HMT generalization and its properties, when we use t-norms is the aim of Section 3. In Section 4 we show how this operator can be used to achieve object detection and illustrate its abilities by various experiments in different situations. Finally, we conclude with some conclusions and future work.

2 Preliminaries

Fuzzy morphological operators are defined using fuzzy operators such as fuzzy conjunctions and fuzzy implications. More details on these logical connectives can be found in [5] and [3], respectively.

Definition 1. A non-decreasing binary operator $C : [0,1]^2 \to [0,1]$ is called a *fuzzy conjunction* if it satisfies

$$C(0,1) = C(1,0) = 0 \text{ and } C(1,1) = 1.$$

The most well-known kind of conjunctions is the class of t-norms [20].

Definition 2. A conjunction T on $[0,1]$ is called a *t-norm* when it is commutative, associative and it satisfies $T(1,x) = x$ for all $x \in [0,1]$.

Next we recall the definitions of strong fuzzy negations and fuzzy implications.

Definition 3. A non-increasing function $N : [0,1] \to [0,1]$ is called a *strong fuzzy negation* if it is an involution, i.e., if $N(N(x)) = x$ for all $x \in [0,1]$.

Definition 4. A binary operator $I : [0,1]^2 \to [0,1]$ is a *fuzzy implication* if it is non-increasing in the first variable, non-decreasing in the second one and it satisfies $I(0,0) = I(1,1) = 1$ and $I(1,0) = 0$.

A well-known way to obtain fuzzy implications is the residuation method. Given a conjunction C such that $C(1,x) > 0$ for all $x > 0$ the binary operator

$$I_C(x,y) = \sup\{z \in [0,1] \mid C(x,z) \le y\}$$

is a fuzzy implication called the *residual implication* or *R-implication* of C (see [28]). In Table 1, the most important t-norms and their corresponding R-implications are collected.

Using the previous operators, we can define the basic fuzzy morphological operators such as dilation and erosion. We will use the following notation: C denotes a

Table 1 Some t-norms and their corresponding R-implications

t-norm	Expression	R-implication
Łukasiewicz	$T_{LK}(x,y) = \max\{x+y-1,0\}$	$I_{LK}(x,y) = \min\{1, 1-x+y\}$
Minimum	$T_M(x,y) = \min\{x,y\}$	$I_{GD}(x,y) = \begin{cases} 1 & \text{if } x \le y \\ y & \text{if } x > y \end{cases}$
Product	$T_P(x,y) = xy$	$I_{GG}(x,y) = \begin{cases} 1 & \text{if } x \le y \\ \frac{y}{x} & \text{if } x > y \end{cases}$
Nilp. Minimum	$T_{nM}(x,y) = \begin{cases} 0 & \text{if } x+y \le 1, \\ \min\{x,y\} & \text{otherwise.} \end{cases}$	$I_{FD}(x,y) = \begin{cases} 1 & \text{if } x \le y \\ \max\{1-x,y\} & \text{if } x > y \end{cases}$

fuzzy conjunction, I a fuzzy implication, A a grey-level image, and B a grey-level structuring element (see [35] and [37] for formal definitions). In addition, d_A denotes the set of points where A is defined and $T_v(A)$ is the translation of a fuzzy set A by $v \in \mathbb{R}^n$ defined by $T_v(A)(x) = A(x-v)$.

Definition 5. ([25]) The *fuzzy dilation* $D_C(A,B)$ and the *fuzzy erosion* $E_I(A,B)$ of A by B are the grey-level images defined by

$$
D_C(A,B)(y) = \sup_{x \in d_A \cap T_y(d_B)} C(B(x-y), A(x))
$$
$$
E_I(A,B)(y) = \inf_{x \in d_A \cap T_y(d_B)} I(B(x-y), A(x)).
$$

Now we recall the binary and grey-level approaches for Hit-or-Miss transform. First of all, the hit-or-miss transform (HMT) of a binary image is a classic morphology operator [35, 37], that uses two structuring elements B_{FG} and B_{BG}. The basic idea consists on extracting all pixels in a binary image that are surrounded by areas on the image where both foreground (represented by B_{FG}) and background (represented by B_{BG}) match predefined patterns. The pattern that should match the foreground is defined by B_{FG}, while B_{BG} defines the pattern that should match the background. By definition B_{FG} and B_{BG} share the same origin and $B_{FG} \cap B_{BG} = \emptyset$. We use $B = (B_{FG}, B_{BG})$ to denote the composite structuring element (SE).

The HMT of a binary image A by the composite SE B is the set of points x such that when the origin of B coincides with x, B_{FG} fits A while B_{BG} fits A^c (the complement set of A):

$$
A \circledast B = \{x : (B_{FG})_x \subseteq A, \ (B_{BG})_x \subseteq A^c\} = (A \ominus B_{FG}) \cap (A^c \ominus B_{BG}),
$$

where $(\cdot)_x$ denotes the translation by x and $\ominus$ is the binary erosion operator

$$
A \ominus B = \{x : B_x \subseteq A\}.
$$

The binary HMT operator is not easily extended to grey-level images. The several definitions of grey-level HMT (GLHMT) were unified by Naegel et al. in [26]. The common denominator in these extensions is to avoid the use of A^c in the definition; this is due to the difficulty in formalising this concept for grey-level images. First, Khosravi and Schafer [19] developed the so-called "template matching" based on a generalization of the HMT in grey-level images with noise. Furthermore, Schaefer and Casasent [34] presented a version of HMT for object detection, while Raducanu and Grana [31] proposed a greyscale HMT based on level sets (LSHMT). On the other hand, Soille [37] introduced two extensions for Hit-or-Miss: the so-called unconstrained HMT and the constrained HMT and he applied them to the analysis of topographic maps. A detailed account of these extensions can be found in [26, 29]. Finally, some extensions of HMT to multivariate images have been recently proposed in [2, 39, 41] and see also [24] for a new conceptual view of the HMT.

3   Fuzzy Hit-or-Miss Transform and Its Properties

In the fuzzy set theory, the complement operation is modelled using a fuzzy negation and the intersection of fuzzy sets is modelled using a fuzzy conjunction, so we can transform directly the HMT definition to the fuzzy set theory using concepts and techniques of the FMM based on fuzzy conjunctions and fuzzy implications (see [25] and in particular, [12, 14, 13] when the fuzzy conjunction is a conjunctive uninorm or a discrete t-norm). Since, B_{FG} and B_{BG} are fuzzy sets, the rigid condition $B_{FG} \cap B_{BG} = \emptyset$ in the binary case is not mandatory in our extension since in the fuzzy case, when this condition holds, the fuzzy hit-or-miss transform may be a non-empty set.

Definition 6. Let N be a strong fuzzy negation, I a fuzzy implication and C a fuzzy conjunction. The *fuzzy Hit-or-Miss transform* (FHMT) of the grey-level fuzzy image A with respect to the grey-scale structuring element $B = (B_1, B_2)$ is defined by

$$FHMT_B(A) = C\left(E_I(A, B_1), E_I(co_N A, B_2)\right), \qquad (1)$$

where $co_N A(x) = N(A(x))$ for all $x \in \mathbb{R}^n$.

Recall that under the necessary conditions (see [25]), the fuzzy erosion and the fuzzy dilation based on a fuzzy conjunction, are N-duals, that is $E_I(co_N A, B_2) = co_N D_C(A, B_2)$. In this case we have that

$$FHMT_B(A) = C\left(E_I(A, B_1), co_N D_C(A, B_2)\right).$$

In the rest of this section, some interesting properties of FHMT are presented. We will use the following notation: C denotes a fuzzy conjunction, I a fuzzy implication and N a strong fuzzy negation.

The first result shows that we retrieve the binary hit-or-miss transform when we apply the FHMT to a binary image A and a binary structuring element B. This is a direct consequence of the coincidence of the fuzzy conjunction and the fuzzy implication with their binary counterparts when we restrict the values to $\{0, 1\}^2$.

Theorem 1. *Let A be a binary image and $B = (B_1, B_2)$ a binary structuring element. Then the FHMT coincides with the binary Hit-or-Miss transform, that is,*

$$FHMT_B(A) = A \circledast B.$$

From now on, we will say that B is an empty structuring element if $B(x) = 0$ for all x. In this case we put $B = \emptyset$.

Theorem 2. *Let B_1 be a grey-scale structuring element. Then, taking $B = (B_1, \emptyset)$, we obtain that the fuzzy erosion is a particular case of the FHMT transform*

$$E_I(A, B_1) = FHMT_B(A).$$

Moreover, if the fuzzy erosion and the fuzzy dilation are N-duals with respect to the fuzzy negation used in the FHMT, we have that

$$D_C(A, B_1) = co_N(FHMT_B(co_N A)).$$

The FHMT transform is invariant under translations.

Proposition 1. *Let A be a grey-scale image, let $B = (B_1, B_2)$ be a grey-scale structuring element and let $v \in \mathbb{R}^n$. Then it holds:*

$$FHMT_B(T_v(A)) = T_v(FHMT_B(A)) \quad and \quad FHMT_{T_v(B)}(A) = T_{-v}(FHMT_B(A)),$$

where $T_v(B) = (T_v(B_1), T_v(B_2))$.

Let $B = (B_1, B_2)$ and $\overline{B} = (B_2, B_1)$ be two grey-scale structuring elements. The next result establishes the consequence of the choice of $\overline{B}$ instead of B.

Proposition 2. *Let A be a grey-scale image, let $B = (B_1, B_2)$ be a grey-scale structuring element. Then, we have that*

$$FHMT_B(A) = FHMT_{\overline{B}}(co_N A).$$

As we have already commented, FHMT can be used to find patterns with a given shape and size in an image. In contrast to the binary hit-or-miss operator (which either finds fully coincidence or not), the fuzzy hit-or-miss operator always finds the searched pattern, if it exists, giving a degree of truth as a result. The advantage of using FHMT over the other methods for template matching is that FHMT finds the searched pattern (no matter if the coincidence is full or partial) and the degree of matching is measured by a degree of truth, that corresponds with the similarity of the searched pattern with a part of the image, through a membership function. Specifically, given a point $x \in \mathbb{R}^n$ we can understand the fuzzy hit-or-miss transform

$$FHMT_B(A)(x) = T(u, v),$$

as a "degree of similarity" related with the value of the aggregation operator C at the point (u, v), where u and v are the degrees of truth indicating how B_1 is included in the grey-level image A, and B_2 included in $co_N(A)$ (that is, the degree of truth indicating how B_2 is excluded from A), respectively. Therefore, the "similarity degree" given by the FHMT is the grey-level value which the closer to 1 the measure is, more certainly we have that the shape of B_1 is contained in the image A. The next results deal with the study of the similarity degree given by the FHMT when we consider a fuzzy mathematical morphology based on t-norms as conjunctions and their R-implications as fuzzy implications.

Definition 7. Let B_1 be a grey-level image (a structuring element) and let A be a grey-level image. We say that B_1 is a part of A if there exists a point y such that if we translate B_1 to y, we have $B_1(x - y) = A(x)$ for all $x \in T_y(B_1)$. In this case we say that B_1 is *a part of A at the point y*.

Theorem 3. *Let A be a grey-level image, $B = (B_1, B_2)$ a grey-scale structuring element satisfying that $B_2 = co_N(B_1)$, T a t-norm and I_T its R-implication and $y \in \mathbb{R}^n$. Then B_1 is a part of A at the point y if, and only if, $FHMT_B(A)(y) = 1$.*

Now we study the case when B_1 and B_2 are constant grey-level images, A is constant in a square of the same dimensions of B and the grey level of A is greater than the one of B_1. In the next result we obtain the expression of the FHMT considering the t-norms and their corresponding R-implications given in Table 1.

Theorem 4. *Let A be a grey-level image, $B = (B_1, B_2)$ a grey-scale structuring element such that $B_1(x) = m$ for all x, $B_2 = co_N(B_1)$, T a t-norm, I_T its corresponding R-implication and $y \in \mathbb{R}^n$. Suppose that $A(x) = k$ for all $x \in T_y(B_1)$ with $m < k$. Then we have that*

$$FHMT_B(A)(y) = \begin{cases} N(k) & \text{if } T = T_M, \\ \frac{N(k)}{N(m)} & \text{if } T = T_P, \\ 1 - N(m) + N(k) & \text{if } T = T_{LK}, \\ \max\{m, N(k)\} & \text{if } T = T_{nM}. \end{cases}$$

The previous result suggests that the most interesting t-norms to use in the FHMT are T_{LK} and T_P. This is because in this case, we obtain values of the FHMT that depend on k and m and get closer to 1 in a continuous way as m approaches to k. On the other hand, the value of FHMT when T_M is used only depends on k and when T_{nM} is applied, the resulting function is not continuous.

4 Experimental Results

In the first example we check the functionality of the fuzzy Hit-or-Miss transformation that we have proposed. It consists of a synthetic image with geometric figures (see Figure 1-(a)). The aim is to detect a specific combination of grey level and square size. In the figure we can see the structuring element that we have used to calculate the FHMT. In Figure 1-(f), all the squares with a size equal to or greater than the size of the structuring element have been detected. The grey level of the pixels belonging to the square that exactly matches the structuring element (in shape and grey level) is 1, the maximum value, which is predicted by Theorem 3. Furthermore, we have observed in experiments with similar images that the FHMT value decreases when the difference between the grey level of the desired shape into the image with respect to the grey level of the structuring element increases. For example, the square of the Figure 1-(a) with grey level 195 has been detected in the FHMT image with grey level 187 (0.7333), and the square with grey level 197 has been detected with grey level 185 (0.7255). This is reasonable since the square with gray level 195 is more similar to the chosen structuring element than the square with gray level 197. In addition, as we can see, the squares with a size greater than or equal to the structuring element have been detected with a grey level higher than the background in the FHMT image. In Figure 1-(g) we show the thresholded version of FHMT to see better how all squares with a size greater than or equal to the structuring element have been found. Obviously one could argue that if an ellipse or another figure has a size larger than the square with the same grey level, it would

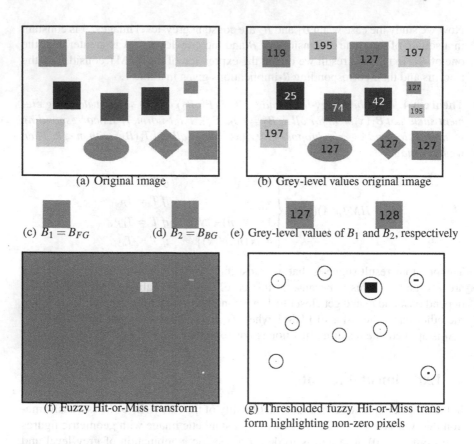

(a) Original image   (b) Grey-level values original image

(c) $B_1 = B_{FG}$ (d) $B_2 = B_{BG}$ (e) Grey-level values of B_1 and B_2, respectively

(f) Fuzzy Hit-or-Miss transform   (g) Thresholded fuzzy Hit-or-Miss transform highlighting non-zero pixels

Fig. 1 Fuzzy Hit-or-miss transform using T_{LK}, I_{LK} and $N_C(x) = 1 - x$ of the original image displayed in (a) using $B = (B_1, B_2)$

be detected as a square in the FHMT. This is true, but this fact can be easily avoided by surrounding the structuring element with a white border.

In addition, we can observe that there appear brighter regions larger than one pixel in the FHMT image (see Figure 1-(f)). This is because the structuring element is included in a geometric figure of the image when the structuring element is centred in all the pixels of this region.

Next, in Figure 2, we can see that all the smaller blobs than the structuring element have been detected with a grey level higher than the background. Note that in this case there is a lighter border in the structuring element and we do not detect the larger blobs.

Finally we show a different experiment from the previous ones, where we try to find the typefaces E in a dollar image. As shown in Figure 3, the brightest areas correspond to the E's which we want to identify, and if we threshold the image, we can see how they have been fully identified.

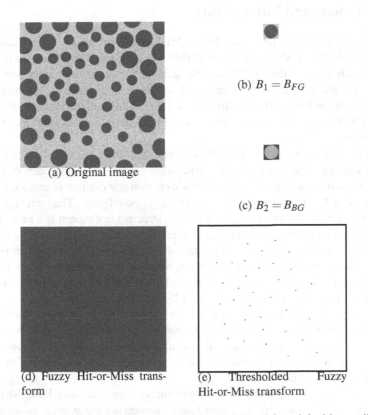

(a) Original image

(b) $B_1 = B_{FG}$

(c) $B_2 = B_{BG}$

(d) Fuzzy Hit-or-Miss transform

(e) Thresholded Fuzzy Hit-or-Miss transform

Fig. 2 Fuzzy Hit-or-miss transform using T_P, I_{GG} and N_C of the original image displayed in (a) using $B = (B_1, B_2)$

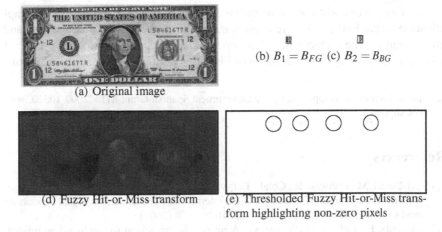

(a) Original image

(b) $B_1 = B_{FG}$ (c) $B_2 = B_{BG}$

(d) Fuzzy Hit-or-Miss transform

(e) Thresholded Fuzzy Hit-or-Miss transform highlighting non-zero pixels

Fig. 3 Hit or miss transform using T_{LK}, I_{LK} and N_C of the original image displayed in (a) using $B = (B_1, B_2)$

5 Conclusions and Future Work

In this work, a new extension of the Hit-or-Miss transform for grey-level images, called the Fuzzy Hit-or-Miss transform (FHMT), is introduced using the fuzzy sets theory. The definition of the FHMT uses the generalizations of the binary conjunction and implication to the fuzzy framework in order to preserve the geometrical interpretation of the binary Hit-or-Miss transform. In addition, the concept of complement in the binary approach is well-defined in our definition due to the use of a strong negation.

First of all, after proving that FHMT coincides with the binary HMT when it is applied to a binary image and a binary structuring element, we have seen that this transform is a generalization of the well-known erosion and dilation operators. Next, the invariance of this transform under translations is established. The next step was to study the behaviour of this operator when the structuring element is a part of the fuzzy image. In this case it was found that the transform detects all parts of the image which includes the structuring element with value 1. However, the FHMT does not only detect the parts of the image which are equal to the structuring element. Its value is the aggregation of a similarity degree between the structuring element B_1 and the image A, and the structuring element B_2 and the complement image $co_N A$.

The result to apply this transform using known t-norms to an image satisfying certain conditions with a particular structuring element (see Theorem 4) has also been studied. Only the expression of the result obtained by the FHMT operator applying the product t-norm or the Łukasiewicz t-norm is a continuous function depending on the structuring element.

In the section of experimental results, some images were analysed. First, we have seen that the FHMT was able to detect some geometric figures such as squares, rhombus, circles, ellipses, etc. and second, the typeface E was also detected in a dollar image.

The future work will consist on the study of more theoretical properties, the application of the FHMT to more complicated images, the use of more t-norms and how to improve the performance of this operator. Finally, we want to study the use of other classes of fuzzy conjunctions into FHMT, in particular conjunctive uninorms.

Acknowledgements. Supported by the Government Spanish Grant MTM2009-10320, with FEDER support.

References

[1] Al-Omari, M., Qahwaji, R., Colak, T., Ipson, S.: Morphological-based filtering of noise: Practical study on solar images. In: IEEE International Conference on Signal Processing and Communications, ICSPC 2007, pp. 1075–1078 (2007)

[2] Aptoula, E., Lefèvre, S., Ronse, C.: A hit-or-miss transform for multivariate images. Pattern Recogn. Lett. 30(8), 760–764 (2009), http://dx.doi.org/10.1016/j.patrec.2009.02.007, doi:10.1016/j.patrec.2009.02.007

[3] Baczyński, M., Jayaram, B. (eds.): Fuzzy Implications. STUDFUZZ, vol. 231. Springer, Heidelberg (2008)

[4] Barat, C., Ducottet, C., Jourlin, M.: Pattern matching using morphological probing. In: Proc. of Int. Conf. on Image Processing, ICIP 2003, vol. 1, pp. 369–372 (2003), doi:10.1109/ICIP.2003.1246975

[5] Beliakov, G., Pradera, A., Calvo, T. (eds.): Aggregation Functions: A Guide for Practitioners. STUDFUZZ, vol. 221. Springer, Heidelberg (2007)

[6] Bloch, I., Maître, H.: Fuzzy mathematical morphologies: a comparative study. Pattern Recognition 28, 1341–1387 (1995)

[7] Bloomberg, D.S., Maragos, P.: Generalized hit-miss operators. Proc SPIE 1350, Image Algebra and Morphological Image Processing 116, 116–128 (1990), http://dx.doi.org/10.1117/12.23580, doi:10.1117/12.23580

[8] Bouraoui, B., Ronse, C., Baruthio, J., Passat, N., Germain, P.: 3D segmentation of coronary arteries based on advanced mathematical morphology techniques. Computerized Medical Imaging and Graphics 34(5), 377–387 (2010), http://www.sciencedirect.com/science/article/pii/S0895611110000121, doi:10.1016/j.compmedimag.2010.01.001

[9] De Baets, B.: Fuzzy morphology: A logical approach. In: Ayyub, B.M., Gupta, M.M. (eds.) Uncertainty Analysis in Engineering and Science: Fuzzy Logic, Statistics, and Neural Network Approach, pp. 53–68. Kluwer Academic Publishers, Norwell (1997)

[10] Deng, T.Q.: Fuzzy logic and mathematical morphology. Tech. rep., Centrum voor Wiskunde en Informatica, Amsterdam, The Netherlands (2000)

[11] Dimitrova, D., Popov, A.: Finding face features in color images using fuzzy hit-or-miss transform. In: Proceedings of the 9th WSEAS International Conference on Fuzzy Systems, FS 2008, pp. 79–84. World Scientific and Engineering Academy and Society (WSEAS), Stevens Point (2008), http://dl.acm.org/citation.cfm?id=1416056.1416069

[12] González, M., Ruiz-Aguilera, D., Torrens, J.: Algebraic properties of fuzzy morphological operators based on uninorms. In: Artificial Intelligence Research and Development. Frontiers in Artificial Intelligence and Applications, vol. 100, pp. 27–38. IOS Press, Amsterdam (2003)

[13] González-Hidalgo, M., Mir-Torres, A., Ruiz-Aguilera, D., Torrens, J.: Fuzzy morphology based on uninorms: Image edge-detection. opening and closing. In: Tavares, J., Jorge, N. (eds.) Computational Vision and Medical Image Processing, pp. 127–133. Taylor & Francis Group, London (2008)

[14] González-Hidalgo, M., Massanet, S., Torrens, J.: Discrete t-norms in a fuzzy mathematical morphology: Algebraic properties and experimental results. In: Proceedings of WCCI-FUZZ-IEEE, Barcelona, Spain, pp. 1194–1201 (2010)

[15] Heijmaans, H.: Morphological Image Operators. Academic Press, Boston (1994)

[16] Intajag, S., Paithoonwatanakij, K.: Automated boundary extraction by gradient fuzzy gray-scale hit-or-miss transformation. In: The 2000 IEEE Asia-Pacific Conference on Circuits and Systems, IEEE APCCAS 2000, pp. 465–468 (2000), doi:10.1109/APCCAS.2000.913537

[17] Intajag, S., Paithoonwatanakij, K., Cracknell, A.: Iterative satellite image segmentation by fuzzy hit-or-miss and homogeneity index. IEEE Proceedings on Vision, Image and Signal Processing 153(2), 206–214 (2006), doi:10.1049/ip-vis:20045211

[18] Jin, X., Davis, C.: Vector-guided vehicle detection from high-resolution satellite imagery. In: Proceedings of IEEE International Geoscience and Remote Sensing Symposium, IGARSS 2004, vol. 2, pp. 1095–1098 (2004), doi:10.1109/IGARSS.2004.1368603

[19] Khosravi, M., Schafer, R.W.: Template matching based on a grayscale hit-or-miss transform. IEEE Trans. Image Processing 5(6), 1060–1066 (1996)

[20] Klement, E., Mesiar, R., Pap, E.: Triangular norms. Kluwer Academic Publishers, Dordrecht (2000)

[21] Lefèvre, S., Weber, J.: Automatic building extraction in VHR images using advanced morphological operators. In: Urban Remote Sensing Joint Event, pp. 1–5 (2007), 10.1109/URS.2007.371825

[22] Mansoor, A.B., Mian, A.S., Khan, A., Khan, S.A.: Fuzzy morphology for edge detection and segmentation. In: Bebis, G., Boyle, R., Parvin, B., Koracin, D., Paragios, N., Tanveer, S.-M., Ju, T., Liu, Z., Coquillart, S., Cruz-Neira, C., Müller, T., Malzbender, T. (eds.) ISVC 2007, Part II. LNCS, vol. 4842, pp. 811–821. Springer, Heidelberg (2007), http://dl.acm.org/citation.cfm?id=1779090.1779177

[23] Matheron, G.: Random Sets and Integral Geometry. Wiley (1975)

[24] Murray, P., Marshall, S.: A new design tool for feature extraction in noisy images based on grayscale hit-or-miss transforms. IEEE T. Image Process. 20(7), 1938–1948 (2011), doi:10.1109/TIP.2010.2103952

[25] Nachtegael, M., Kerre, E.E.: Classical and fuzzy approaches towards mathematical morphology. In: Kerre, E.E., Nachtegael, M. (eds.) Fuzzy Techniques in Image Processing. STUDFUZZ, vol. 52, ch. 1, Physica-Verlag, New York (2000)

[26] Naegel, B., Passat, N., Ronse, C.:) Grey-level hit-or-miss transforms-Part I: Unified theory. Pattern Recognition 40, 635–647 (2007)

[27] Naegel, B., Passat, N., Ronse, C.:) Grey-level hit-or-miss transforms-Part II: Application to angiographic image processing. Pattern Recognition 40, 648–658 (2007)

[28] Ouyang, Y.: On fuzzy implications determined by aggregation operators. Inform. Sci. 193, 153–162 (2012)

[29] Perret, B., Lefèvre, S., Collet, C.: A robust hit-or-miss transform for template matching applied to very noisy astronomical images. Pattern Recognition 42, 2470–2480 (2009)

[30] Popov, A.T.: General approach for fuzzy mathematical morphology. In: Proc. of 8th International Symposium on Mathematical Morphology (ISMM), pp. 39–48 (2007)

[31] Raducanu, B., Graña, M.: A grayscale hit-or-miss transform based on level sets. In: Proc. IEEE Intl. Conf. Image Processing, Vancouver, BC, Canada, pp. 931–933 (2000)

[32] Raducanu, B., Graña, M., Albizuri, X., d'Anjou, A.: A probabilistic hit-and-miss transform for face localization. Pattern Anal. Appl. 7(2), 117–127 (2004), http://dx.doi.org/10.1007/s10044-004-0207-4, doi:10.1007/s10044-004-0207-4

[33] Ronse, C.: A lattice-theoretical morphological view on template extraction in images. Journal of Visual Communication and Image Representation 7(3), 273–295 (1996), http://www.sciencedirect.com/science/article/pii/S1047320396900243, doi:10.1006/jvci.1996.0024

[34] Schaefer, R., Casasent, D.: Nonlinear optical hit—miss transform for detection. Appl. Opt. 34(20), 3869–3882 (1995), http://ao.osa.org/abstract.cfm?URI=ao-34-20-3869, doi:10.1364/AO.34.003869

[35] Serra, J.: Image analysis and mathematical morphology, vol. 1. Academic Press, London (1982)

[36] Sinha, D., Sinha, P., Douherty, E., Batman, S.: Design and analysis of fuzzy morphological algorithms for image processing. IEEE Trans on Fuzzy Systems 5(4) (1997)

[37] Soille, P.: Morphological Image Analysis, 2nd edn. Springer, Heidelberg (2003)

[38] Stankov, K., He, D.C.: Building detection in very high spatial resolution multispectral images using the hit-or-miss transform. IEEE Geosci. Remote S. 10(1), 86–90 (2013)

[39] Velasco-Forero, S., Angulo, J.: Hit-or-miss transform in multivariate images. In: Blanc-Talon, J., Bone, D., Philips, W., Popescu, D., Scheunders, P. (eds.) ACIVS 2010, Part I. LNCS, vol. 6474, pp. 452–463. Springer, Heidelberg (2010), http://dx.doi.org/10.1007/978-3-642-17688-342

[40] Wang, M., Leung, Y., Zhou, C., Pei, T., Luo, J.: A mathematical morphology based scale space method for the mining of linear features in geographic data. Data Min. Knowl. Disc. 12, 97–118 (2006), http://dx.doi.org/10.1007/s10618-005-0021-7, doi:10.1007/s10618-005-0021-7

[41] Weber, J., Lefèvre, S.: Spatial and spectral morphological template matching. Image and Vision Computing 30, 934–945 (2012)

Image Reduction Operators as Aggregation Functions: Fuzzy Transform and Undersampling

D. Paternain, A. Jurio, R. Mesiar, G. Beliakov, and H. Bustince

Abstract. After studying several reduction algorithms that can be found in the literature, we notice that there is not an axiomatic definition of this concept. In this work we propose the definition of weak reduction operators and we propose the properties of the original image that reduced images must keep. From this definition, we study whether two methods of image reduction, undersampling and fuzzy transform, satisfy the conditions of weak reduction operators.

1 Introduction

In image processing and computer vision there exist several fields in which very large images are used (high spatial resolution). In remote sensing, for instance, images taken from planes or even satellites are handled, and they may contain hundreds of millions of pixels [5]. The time cost of enhancing, filtering or even more complex algorithms such as those for segmentation is so high that it can be unacceptable.

Image reduction consists of diminishing the spatial resolution of an image, that is, obtaining an image with less pixels, in such a way that the largest possible amount of information is kept [2, 9]. In this way, we get images of great quality but with less pixels. This can be very useful because, instead of processing the original image, we can apply algorithms to the reduced images diminishing the required running time.

D. Paternain · A. Jurio · H. Bustince
Departamento de Automatica y Computacion,
Universidad Publica de Navarra, Pamplona, Spain
e-mail: {daniel.paternain,aranzazu.jurio,bustince}@unavarra.es

R. Mesiar
Faculty of Civil Engineering, Slovak University of Technology, Bratislava, Slovakia
e-mail: mesiar@math.sk

G. Beliakov
School of Information Technology, Deakin University, Burwood, Australia
e-mail: gleb@deakin.edu.au

H. Bustince et al. (eds.), *Aggregation Functions in Theory and in Practise*,   405
Advances in Intelligent Systems and Computing 228,
DOI: 10.1007/978-3-642-39165-1_39, © Springer-Verlag Berlin Heidelberg 2013

In previous works [10, 11], the concept of local reduction operator has been studied. These functions take small parts of an image and reduce them into a single pixel. However, in this work we focus on the concept of reduction operator that, in comparison with local reduction operator, acts globally over the image.

The goal of this work is to define axiomatically the concept of reduction operator as a mapping that takes an image and gives back a smaller (reduced) image keeping some properties of the original image. In the definition we specify the set of minimal properties that the images must kept. We note that these properties are similar to those of idempotent aggregation operators, so we study the relation with them.

From the definition, we analyze two well-known image reduction algorithms, fuzzy transform and undersampling. We study if they are reduction operators in our sense.

Finally we have compared both methods using a set of images, analyzing in a visual way the differences and, so, the advantages and disadvantages of their possible use in real problems.

2   Preliminaries

In this work we relate reduction operators with the concept of aggregation function [3, 6, 8, 3].

Definition 1. An aggregation function of dimension n is an increasing mapping $M :$ $[0,1]^n \to [0,1]$ such that $M(0,\ldots,0) = 0$ and $M(1,\ldots,1) = 1$.

Definition 2. An aggregation function M is said to be idempotent if $M(x,\ldots,x) = x$ for all $x \in [0,1]$.

For aggregation functions, idempotency is equivalent to

$$\min(x_1,\ldots,x_n) \le M(x_1,\ldots,x_n) \le \max(x_1,\ldots,x_n)$$

for all $x_1,\ldots,x_n \in [0,1]$.

Definition 3. Let $M : [0,1]^n \to [0,1]$ be an aggregation function.

- M is said to be homogeneous if $M(\lambda x_1,\ldots,\lambda x_n) = \lambda M(x_1,\ldots,x_n)$ for any $\lambda \in$ $[0,1]$ and for any $x_1,\ldots,x_n \in [0,1]$.
- M is said to be shift-invariant if $M(x_1 + r,\ldots,x_n + r) = M(x_1,\ldots,x_n) + r$ for all $r > 0$ such that $0 \le x_i + r \le 1$ for any $i = 1,\ldots,n$.

3   Image Reduction Operators

In this work we understand an image of $n \times m$ pixels as a set of $n \times m$ elements arranged in rows and columns. That is, we consider an image as a $n \times m$ matrix. Each element of the matrix is a value in $[0,1]$. We get this values by normalizing the intensity of each pixel in the image (dividing by $L-1$ being L the number of possible gray levels). We denote by $\mathcal{M}_{n \times m}$ the set of all possible matrices of dimension $n \times m$ over $[0,1]$.

As it has been stated in the introduction of this work, for us a reduction operator is a function that takes an image and obtains a smaller image (lower spatial resolution) containing as much information as possible from the original. But we believe that the new image cannot be built in any way, since it must keep some properties. Clearly, these properties may change depending of the application. In this work we fix a minimal set of properties that every reduced image should keep from the original image, regardless which the application is.

In this work we consider that reduction operators must satisfy, at least, the following conditions:

- (Monotonicity) Given two images such that one is darker than the other, the reduced image of the former must be darker than the reduced of the latter.
- (Idempotence) Given a flat image such that every pixel has the same value, the reduced image must be also flat with every pixel having that value.

Depending on the application, we can increase the set of properties, as for example:

- (Homogeneity) If we multiply each pixel of an image by a constant factor, the reduced image corresponds to the reduced image of the original image multiplied by the same factor.
- (Shift invariance) If we add to each pixel the same value, the reduced image corresponds to the reduced image of the original image added by the same value.

All these considerations have led us to propose the following definition:

Definition 4. A mapping $O_{WR} : \mathcal{M}_{n \times m} \to \mathcal{M}_{n' \times m'}$ with $n' \leq n$ and $m' \leq m'$ is a weak reduction operator if the following properties hold:

- (OWR1) For all $A, B \in \mathcal{M}_{n \times m}$, if $A \leq B$, then $O_{WR}(A) \leq O_{WR}(B)$.
- (OWR2) For each $A \in \mathcal{M}_{n \times m}$ if $A = c$, then $O_{WR}(A) = c$.

Definition 5. We say that a weak reduction operator O_{WR} is

- (OWR3) homogeneous if $O_{WR}(\lambda A) = \lambda O_{WR}(A)$ for all $A \in \mathcal{M}_{n \times m}$ and for all $\lambda \in [0,1]$.
- (OWR4) stable under translation if $O_{WR}(A + r) = O_{WR}(A) + r$ holds for all $A \in \mathcal{M}_{n \times m}$ and for all $r > 0$ such that $a_{ij} + r \leq 1$ for any $i \in \{1, \ldots, n\}, j \in \{1, \ldots, m\}$.

In Theorem 1 we establish the relationship between reduction operators and idempotent aggregation functions.

Theorem 1. Let $O_{WR} : \mathcal{M}_{n \times m} \to \mathcal{M}_{n' \times m'}$ be a weak reduction operator. Then, the function $M : [0,1]^{n \times m} \to [0,1]$ given by

$$M(A_{11}, \ldots, A_{1m}, \ldots, A_{n1}, \ldots, A_{nm}) = P_{i,j}(O_{WR}(A))$$

where $P_{i,j}$ denotes the projection of the i-th row j-th column (the element of i-th row and j-th column) is an idempotent aggregation function for all $A \in \mathcal{M}_{n \times m}$.

Proof. Monotonicity of M is given by $(OWR1)$ of weak reduction operators. Boundary conditions and idempotency are given by $(OWR2)$.

Theorem 1 shows that each pixel in a reduced image obtained by a weak reduction operator can be obtained from applying an idempotent aggregation function to every pixel of the original image.

4 Image Reduction Algorithms as Reduction Operators

In this section we study two examples of image reduction algorithms in the literature. Our objective is to study whether these algorithms are reduction operators in the sense of Definition 4.

4.1 Image Undersampling

One of the simplest and fastest methods to reduce images consists in removing a given number of pixels in order to reduce the spatial dimension of the image. For example, removing one of every two rows and columns. This procedure gets a reduced image with only 25% of pixels of the original image.

Proposition 1. *Let $A \in \mathcal{M}_{n \times m}$, and let $A' \in \mathcal{M}_{n' \times m'}$ be obtained by suppressing from A $n - n'$ rows and $m - m'$ columns. If we fix the rows and columns to be suppressed, the operator defined as*

$$O_{WR}(A) = A'$$

is a reduction operator with respect to any strong negation N that satisfies $(OW3)$ and $(OW4)$ for all $A \in \mathcal{M}_{n \times m}$.

Proof. Fixing the pixels to be eliminated is essential to satisfy monotonicity of reduction operators $(OWR1)$. $(OWR2)$, $(OWR3)$ and $(OWR4)$ are direct.

Corollary 1. *Under conditions of Proposition 1, the function $M : [0,1]^{n \times m} \rightarrow [0,1]$ given by*

$$M(A_{11}, \ldots, A_{nm}) = A_{ij}$$

is an idempotent aggregation function for all $i \in \{1, \ldots, n\}, j \in \{1, \ldots, m\}$.

Proof. Direct

In Figure 1 we show an example of reduction operator based on undersampling. From the original image Cameraman ($n = m = 256$) we reduce it to obtain two new images with $n' = m' = 128$ by eliminating one of each two rows/columns and with $n' = m' = 85$ by eliminating two of each three rows/columns.

(a) (b) (c)

Fig. 1 Original image cameraman and two reduced images obtained by reduction operator given by undersampling algorithm with $n' = m' = 128$ (image (b)) eliminating rows/columns=$2,4,\ldots$ and $n' = m' = 85$ (image (c)) eliminating rows/columns=$2,3,5,6,\ldots$

4.2 Fuzzy Transform

The fuzzy transform [12] is a widely studied technique during last years, due to its applicability to image processing [6, 7, 13]. In this work we study the fuzzy transform as an image reduction operator..

In previous section we have established a relation between images and matrices. Now, in order to follow the original notation of [12], we consider an image A as a bivariante function $f_A : \{1,\ldots,n\} \times \{1,\ldots,m\} \to [0,1]$. Evidently, $f_A \in \mathcal{M}_{n\times m}$. The 2-dimensional discrete fuzzy transform is a mapping that associates each f_A with a $f_{A'}$ such that $f_{A'} \in \mathcal{M}_{n'\times m'}$ with $n' \leq n$ and $m' \leq m$.

Definition 6. Let $n' \geq 2$ and $x_1 < \ldots < x_{n'}$ be fixed points in the interval $[a,b]$ such that $x_1 = a$ and $x_{n'} = b$. The fuzzy sets $P_1,\ldots,P_{n'} : [a,b] \to [0,1]$ form a fuzzy partition of $[a,b]$ if the following conditions hold for each $k = 1,\ldots,n'$

1. $P_k(x_k) = 1$;
2. $P_k(x) = 0$ if $x \notin (x_{k-1},x_{k+1})$ (being $x_0 = a$ and $x_{n'+1} = b$);
3. $P_k(x)$ is continuous;
4. $P_k(x), k = 2,\ldots,n'$ is strictly increasing in $[x_{k-1},x_k]$ and $P_k(x), k = 1,\ldots,n'-1$ is strictly decreasing in $[x_k,x_{k+1}]$.
5. For all $x \in [a,b]$, $\sum_{k=1}^{n'} P_k(x) = 1$

Definition 7. Let $f_A \in \mathcal{M}_{n\times m}$. Let $P_1,\ldots,P_{n'}$ and $Q_1,\ldots,Q_{m'}$ be two fuzzy partitions of $[1,n]$ and $[1,m]$, respectively. We say that the $n' \times m'$ dimensional matrix $f_{A'}$ is the discrete fuzzy transform of f_A with respect to $P_1,\ldots,P_{n'}$ and $Q_1,\ldots,Q_{m'}$ if

$$f_{A'}(k,l) = \frac{\sum_{j=1}^{m} \sum_{i=1}^{n} f_A(i,j) P_k(i) Q_l(j)}{\sum_{j=1}^{m} \sum_{i=1}^{n} P_k(i) Q_l(j)} \tag{1}$$

Theorem 2. For any fixed partitions $P_1,\ldots,P_{n'}$ and $Q_1,\ldots,Q_{m'}$, the discrete fuzzy transform is a weak reduction operator in the sense of Definition 4.

Proof. (OWR1) Let $f_A, f_B \in \mathcal{M}_{n \times m}$ such that $f_A \leq f_B$. It is clear that $f_{A'}(k,l) \leq f_{B'}(k,l)$ for all $k \in \{1, \ldots, n'\}, l \in \{1, \ldots, m'\}$.

(OWR2) If $f_A = c$, then

$$f_{A'}(k,l) = \frac{\sum_{j=1}^{m} \sum_{i=1}^{n} f_A(i,j) P_k(i) Q_l(j)}{\sum_{j=1}^{m} \sum_{i=1}^{n} P_k(i) Q_l(j)} = \frac{\sum_{j=1}^{m} \sum_{i=1}^{n} c P_k(i) Q_l(j)}{\sum_{j=1}^{m} \sum_{i=1}^{n} P_k(i) Q_l(j)} =$$

$$= \frac{c \sum_{j=1}^{m} \sum_{i=1}^{n} P_k(i) Q_l(j)}{\sum_{j=1}^{m} \sum_{i=1}^{n} P_k(i) Q_l(j)} = c$$

Corollary 2. *Under conditions of Theorem 2, the function* $M : [0,1]^{n \times m} \to [0,1]$ *given by*

$$M(f_A(1,1), \ldots, f_A(n,m)) = \frac{\sum_{j=1}^{m} \sum_{i=1}^{n} f_A(i,j) P_k(i) Q_l(j)}{\sum_{j=1}^{m} \sum_{i=1}^{n} P_k(i) Q_l(j)}$$

is an indempotent aggregation function for all $k \in \{1, \ldots, n'\}, l \in \{1, \ldots, m'\}$.

Proof. Direct

Theorem 3. *Under conditions of Theorem 2, the discrete fuzzy transform is a weak reduction operator that satisfies* (OWR3) *and* (OWR4).

Proof. Let $f_{B'} = \lambda f_A + r$ with $\lambda, r \in [0,1]$ such that $f_B \leq 1$ for all $i = 1, \ldots, n$ and $j = 1, \ldots, m$. Then

$$f_{B'}(k,l) = \frac{\sum_{j=1}^{m} \sum_{i=1}^{n} (\lambda f_A(i,j) + r) P_k(i) Q_l(j)}{\sum_{j=1}^{m} \sum_{i=1}^{n} P_k(i) Q_l(j)} =$$

$$\frac{\sum_{j=1}^{m} \sum_{i=1}^{n} \lambda f_A(i,j) P_k(i) Q_l(j) + r P_k(i) Q_j(l)}{\sum_{j=1}^{m} \sum_{i=1}^{n} P_k(i) Q_l(j)} =$$

$$\frac{\sum_{j=1}^{m} \sum_{i=1}^{n} \lambda f_A(i,j) P_k(i) Q_l(j)}{\sum_{j=1}^{m} \sum_{i=1}^{n} P_k(i) Q_l(j)} + \frac{\sum_{j=1}^{m} \sum_{i=1}^{n} r P_k(i) Q_j(l)}{\sum_{j=1}^{m} \sum_{i=1}^{n} P_k(i) Q_l(j)} =$$

$$\lambda f_{A'}(k,l) + r$$

From the original image Lena (256×256 pixels), we apply fuzzy transform to obtain reduced versions of this image: we first apply fuzzy transform with $n' = m' = 128$ (image (b) of Figure 2) and with $n' = m' = 85$ (image (c) of Figure 2). In both experiments, fuzzy partitions are constructed by means of triangular fuzzy sets forming a uniform fuzzy partition.

In Figure 3 we visually test homogeneity of fuzzy transform by multiplying each original image by a factor $\lambda = 0.5$. Observe that the original image becomes darker as well as the reduced images obtained by the fuzzy transform with $n' = m' = 128$ and $n' = m' = 85$.

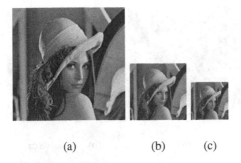

(a)   (b)   (c)

Fig. 2 Original image (a) and reduced images obtained by reduction operator given by fuzzy transform with $n' = m' = 128$ (image (b)) and with $n' = m' = 85$ (image (b))

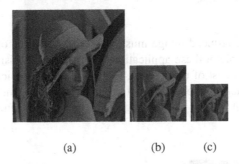

(a)   (b)   (c)

Fig. 3 Image (a) obtained by multiplying each pixel of Lena by $\lambda = 0.5$. Reduced images obtained by fuzzy transform with $n' = m' = 128$ (image(b)) and with $n' = m' = 85$ (image (c))

4.3 Visual Comparison of Reduction Algorithms

In this subsection we visually compare each one of the reduction methods studied in this work. It is important to recall that in the fuzzy transform, we always take a uniform partition of intervals $[1,n], [1,m]$. For this reason, each pixel of the reduced image is obtained taking into account all the pixels of the original image (all the pixels of the original image are used to calculate each pixel of the reduced). However, in undersampling algorithm, we take only into account one pixel of the original image. Therefore, with this method we can obtain reduced images where we keep the details of certain area of the image while we loose other areas. This can be seen in Figure 4. Image (a) is original Lena image with $n = m = 256$. Both images (b) and (c) are reduced images taking $n' = m' = 128$. The first image has been obtained by eliminating certain rows/columns (therefore it is a reduction operator). The second image has been obtained by the fuzzy transform. While both images have been obtained by a reduction operator, the images are completely different.

From the images obtained in Figure 4 we notice that, if in a certain application we are interested in keeping the details of an specific area, then we should apply a reduction operator based on undersampling. However, there exist many other ap-

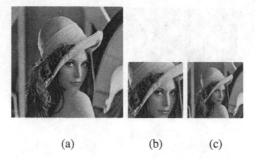

(a)   (b)   (c)

Fig. 4 Original image Lena (a) and two reduced images obtained by subsampling keeping information of an specific are of the image (image (b)) and by fuzzy transform with uniform partition taking $n' = m' = 128$.

plications where the reduced image must represent the whole original image. The most common example of these applications is image compression, that consists in reducing the storage cost of the images. For this kind of application we have two options: 1) use reduction operator based on undersampling where the pixels eliminated are situated along the whole image or 2) to apply the fuzzy transform with uniform partitions.

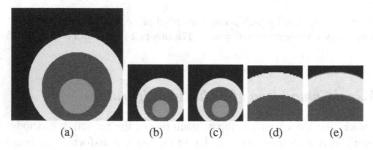

(a)   (b)   (c)   (d)   (e)

Fig. 5 Original image Circles (a) and reduced images obtained by subsampling and fuzzy transform with $n' = m' = 128$

In Figure 5 we compare both techinques 1) and 2) of previous paragraph in the original image Circles. In images (b) and (c) we show the two images obtained by both reduction operators. In order to see the differences, in images (d) and (e) we make a zoom of the same area of both images. Notice that the differences between the two images are located in the edges of objects, that is, in areas where the variation of intensities of pixels are very large. Observe that while the image obtained by fuzzy transform presents more blur, the shape of the object is better preserved. In contrast, when we use undersampling (we eliminate one of each row/column) edges are more sharpened.

Another interesting difference (related to the images of Figure 5) is the fact that the value of a certain pixel in the reduced image obtained by the fuzzy transform may not appear in the original image. That is, a pixel of a reduced image can be different to every pixel of the original image when we use fuzzy transform. However, this behavior do not apper when we use undersampling as reduction operator.

5 Conclusions

In this work we have presented and studied the properties that image reduction algorithms must satisfy. We have formalized these properties in the definition of weak reduction operator. We have also studied the close relation between weak reduction operators and idempotent aggregation functions.

Finally, given two image reduction algorithms of the literature, undersampling and fuzzy transform, we have studied whether these algorithms are weak reduction operators in our sense and we have analyzed simmilarities and differences of these techniques.

Acknowledgements. This work has been partially supported by the Government of Spain of Spain under grants TIN2011-29520 and TIN2010-15055, the Research Services of the Public University of Navarra and the grant VEGA 1/0143/11.

References

[1] Beliakov, G., Pradera, A., Calvo, T.: Aggregation Functions: A Guide for Practitioners. STUDFUZZ, vol. 221. Springer, Heidelberg (2007)

[2] Beliakov, G., Bustince, H., Paternain, D.: Image Reduction Using Means on Discrete Product Lattices. IEEE Transactions on Image Processing 21, 1070–1083 (2012)

[3] Bustince, H., Calvo, T., De Baets, B., Fodor, J., Mesiar, R., Montero, J., Paternain, D., Pradera, A.: A class of aggregation functions encompassing two-dimensional OWA operators. Information Sciences 180, 1977–1989 (2010)

[4] Calvo, T., Mayor, G., Mesiar, R.: Aggregation Operators. New Trends and Applications. STUDFUZZ, vol. 97. Springer, Heidelberg (2002)

[5] Cui, L., Cheng, S.S., Stankovic, L.: Onboard Low-Complexity Compression of Solar Stereo Images. IEEE Transactions on Image Processing 21, 3114–3118 (2012)

[6] Di Martino, F., Sessa, S.: Compression and decompression of images with discrete fuzzy transforms. Information Sciences 177, 2349–2362 (2007)

[7] Di Martino, F., Sessa, S.: Fragile watermarking tamper detection with images compressed by fuzzy transform. Information Sciences 195, 62–90 (2012)

[8] Grabisch, M., Marichal, J.-L., Mesiar, R., Pap, E.: Aggregation Functions. Cambridge University Press, Cambridge (2009)

[9] Karabassis, E., Spetsakis, M.E.: An Analysis of Image Interpolation, Differentiation, and Reduction Using Local Polynomial Fits. Graphical Models and Image Processing 57, 183–196 (1995)

[10] Paternain, D., Bustince, H., Fernandez, J., Beliakov, G., Mesiar, R.: Image reduction with local reduction operators. In: Proceedings of the IEEE International Conference of Fuzzy Systems (FUZZ-IEEE 2010), Barcelona (2010)

[11] Paternain, D., Lopez-Molina, C., Bustince, H., Mesiar, R., Beliakov, G.: Image reduction using fuzzy quantifiers. In: Melo-Pinto, P., Couto, P., Serôdio, C., Fodor, J., De Baets, B. (eds.) Eurofuse 2011. AISC, vol. 107, pp. 351–362. Springer, Heidelberg (2011)

[12] Perfilieva, I.: Fuzzy transforms: Theory and applications. Fuzzy Sets and Systems 157, 993–1023 (2006)

[13] Perfilieva, I., De Baets, B.: Fuzzy transforms of monotone functions with application to image compression. Information Sciences 180, 3304–3315 (2010)

Part VIII
Implications

Implications Satisfying the Law of Importation with a Given T-norm

S. Massanet and J. Torrens

Abstract. The main goal of this paper is to characterize all fuzzy implications with continuous natural negation that satisfy the law of importation with a given continuous t-norm T. Particular cases when the fixed t-norm T is the minimum, the product and the Łukasiewicz t-norm are deduced from the general result and the corresponding characterizations are presented separately.

Keywords: Implication function, Law of importation, t-norm, t-conorm.

1 Introduction

One of the most important connectives used in fuzzy control and approximate reasoning are fuzzy implications. This is because they are the generalization of binary implications in classical logic to the framework of fuzzy logic and consequently they are used to perform fuzzy conditionals [15, 18, 24]. In addition of modelling fuzzy conditionals, they are also used to perform backward and forward inferences in any fuzzy rules based system through the inference rules of modus ponens and modus tollens [17, 24, 33].

Moreover, fuzzy implications have proved to be useful in many other fields like fuzzy relational equations [24], fuzzy DI-subsethood measures and image processing [11, 1], fuzzy morphological operators [19], computing with words [24], data mining [41] and rough sets [32], among others. Thus, it is not surprising that fuzzy implications have attracted the efforts of many researchers not only from the point of view of their applications, but also from the purely theoretical perspective. See for instance the surveys [6] and [24] and the book [5], entirely devoted to fuzzy implications.

S. Massanet · J. Torrens
University of the Balearic Islands, Palma de Mallorca 07122, Spain
e-mail: {s.massanet,jts224}@uib.es

H. Bustince et al. (eds.), *Aggregation Functions in Theory and in Practise*,   417
Advances in Intelligent Systems and Computing 228,
DOI: 10.1007/978-3-642-39165-1_40, © Springer-Verlag Berlin Heidelberg 2013

From this theoretical point of view, there are several lines of research that have been specially developed. Among them we can highlight the following ones:

1. The study of the different classes of fuzzy implications and their axiomatic characterization (see [5] and the references therein, but also the recent works [1, 8]).
2. The relationship among these classes and the intersections between them (see again [5] and the references therein, as well as the recent works [9] and [26]).
3. The study of new construction methods of fuzzy implications (see [5, 27, 28, 31, 30, 36]).
4. The analysis of additional properties of fuzzy implications.

In the last line, there are a lot of properties that have been studied in detail by many authors. In almost all the cases the interest of each property comes from its specific applications and its theoretical study usually reduces to the solution of a functional equation. Some of the most studied properties are:

a) *The modus ponens*, because it becomes crucial in the inference process through the compositional rule of inference (CRI). Some works on this property are [21, 38, 39, 40].
b) *The distributivity properties* over conjunctions and disjunctions. In this case, these distributivities allow to avoid the combinatorial rule explosion in fuzzy systems ([13]). They have been extensively studied again by many authors, see [2, 3, 7, 10, 34, 35, 37].
c) *The law of importation*. This property is extremely related to the exchange principle (see [25]) and it has proved to be useful in simplifying the process of applying the CRI in many cases, see [16] and [5]. It can be written as

$$I(T(x,y),z) = I(x,I(y,z)) \quad \text{for all} \quad x,y,z \in [0,1],$$

where T is a t-norm (or a more general conjunction) and I is a fuzzy implication. The law of importation has been studied in [5, 16, 22, 23, 25]. Moreover, in this last article the law of importation has also been used in new characterizations of some classes of implications like (S,N)-implications and R-implications. Finally, it is a crucial property to characterize Yager's implications (see [29]).

Although all these works devoted to the law of importation, there are still some open problems involving this property. In particular, given any t-norm T, it is an open problem to find all fuzzy implications I such that they satisfy the law of importation with respect to this fixed t-norm T. That is, find all fuzzy implications such that

$$I(T(x,y),z) = I(x,I(y,z)) \quad \text{for all} \quad x,y,z \in [0,1], \tag{LI}$$

being T any fixed (continuous) t-norm.

In this paper we want to deal with this problem and we will give some partial solutions (in the sense that we will find all solutions involving fuzzy implications with an additional property). Specifically, we will characterize all fuzzy implications with continuous natural negation that satisfy the law of importation with any given continuous t-norm T. Particular cases when the fixed t-norm T is the minimum, the product and the Łukasiewicz t-norm are deduced from the general result and the corresponding characterizations are presented separately.

2 Preliminaries

We will suppose the reader to be familiar with the theory of t-norms. For more details in this particular topic, we refer the reader to [20]. To make this work self-contained, we recall here some of the concepts and results used in the rest of the paper. First of all, the definition of fuzzy negation is given.

Definition 1. (Definition 1.1 in [14]) A decreasing function $N : [0,1] \to [0,1]$ is called a *fuzzy negation*, if $N(0) = 1, N(1) = 0$. A fuzzy negation N is called

(i) *strict*, if it is strictly decreasing and continuous,
(ii) *strong*, if it is an involution, i.e., $N(N(x)) = x$ for all $x \in [0,1]$.

Next lemma plays an important role in the results presented in this paper. Essentially, given a fuzzy negation, it defines a new fuzzy negation which in some sense can perform the role of the inverse of the original negation.

Lemma 1. *(Lemma 1.4.10 in [5]) If N is a continuous fuzzy negation, then the function $\mathfrak{R}_N : [0,1] \to [0,1]$ defined by*

$$\mathfrak{R}_N(x) = \begin{cases} N^{(-1)}(x) & \text{if } x \in (0,1], \\ 1 & \text{if } x = 0, \end{cases}$$

where $N^{(-1)}$ stands for the pseudo-inverse of N given by

$$N^{(-1)}(x) = \sup\{z \in [0,1] \mid N(z) > x\}$$

for all $x \in [0,1]$, is a strictly decreasing fuzzy negation. Moreover,

$$\mathfrak{R}_N^{(-1)} = N,$$

$$N \circ \mathfrak{R}_N = id_{[0,1]},$$

$$\mathfrak{R}_N \circ N|_{Ran(\mathfrak{R}_N)} = id|_{Ran(\mathfrak{R}_N)},$$

where $Ran(\mathfrak{R}_N)$ stands for the range of function $\mathfrak{R}_N$.

Next, we introduce the concept of automorphism and conjugate function.

Definition 2. A function $\varphi : [0,1] \to [0,1]$ is an automorphism if it is continuous and strictly increasing and satisfies the boundary conditions $\varphi(0) = 0$ and $\varphi(1) = 1$, i.e., if it is an increasing bijection from $[0,1]$ to $[0,1]$.

Definition 3. Let $\varphi : [0,1] \to [0,1]$ be an automorphism. Two functions $f, g : [0,1]^n \to [0,1]$ are φ-conjugate if $g = f_\varphi$, where

$$f_\varphi(x_1, \ldots, x_n) = \varphi^{-1}(f(\varphi(x_1), \ldots, \varphi(x_n))), \quad x_1, \ldots, x_n \in [0,1].$$

Note that given an automorphism $\varphi : [0,1] \to [0,1]$, the φ-conjugate of a t-norm T, that is T_φ, and the φ-conjugate of an implication I (see Definition 4), that is I_φ, are again a t-norm and an implication, respectively.

Now, we recall the definition of fuzzy implications.

Definition 4. (Definition 1.15 in [14]) A binary operator $I : [0,1]^2 \to [0,1]$ is said to be a *fuzzy implication* if it satisfies:

(I1) $I(x,z) \geq I(y,z)$ when $x \leq y$, for all $z \in [0,1]$.
(I2) $I(x,y) \leq I(x,z)$ when $y \leq z$, for all $x \in [0,1]$.
(I3) $I(0,0) = I(1,1) = 1$ and $I(1,0) = 0$.

Note that, from the definition, it follows that $I(0,x) = 1$ and $I(x,1) = 1$ for all $x \in [0,1]$ whereas the symmetrical values $I(x,0)$ and $I(1,x)$ are not derived from the definition. Fuzzy implications can satisfy additional properties coming from tautologies in crisp logic. In this paper, we are going to deal with the law of importation, already presented in the introduction.

The natural negation of a fuzzy implication will be also useful in our study.

Definition 5. (Definition 1.4.15 in [5]) Let I be a fuzzy implication. The function N_I defined by $N_I(x) = I(x,0)$ for all $x \in [0,1]$, is called the *natural negation* of I.

Remark 1
(i) If I is a fuzzy implication, N_I is always a fuzzy negation.
(ii) Given a binary function $F : [0,1]^2 \to [0,1]$, we will denote by $N_F(x) = F(x,0)$ for all $x \in [0,1]$ its 0-horizontal section. In general, N_F is not a fuzzy negation. In fact, it is trivial to check that N_F is a fuzzy negation if, and only if, $F(x,0)$ is a decreasing function satisfying $F(0,0) = 1$ and $F(1,0) = 0$.

3   On the Satisfaction of (LI) with a Given T-norm T

In this section, we want to characterize all fuzzy implications with a continuous natural negation which satisfy the Law of Importation (LI) with a fixed t-norm T. Until now, all the previous studies on the law of importation have focused on the satisfaction of (LI) by concrete classes of fuzzy implications. Thus, some results involving this property and (S,N)-implications are presented in [16] and [25]; R-implications in [16]; QL-implications in [5] and [22] and Yager's implications in [4]. On the other hand, fixed a concrete t-norm T, it is still an open problem to know which fuzzy implications satisfy (LI) with this T.

First of all, it is worth to study if fixed a concrete t-norm T, any fuzzy negation can be the natural negation of a fuzzy implication satisfying (LI) with T. In fact, there exists some dependence between the t-norm T and the natural negation of the fuzzy implication I. Thus, not all fuzzy negations can be natural negations of a fuzzy implication satisfying (LI) with a concrete t-norm. To characterize which fuzzy negations are compatible with a t-norm T in this sense, the following property will be considered:

$$\text{if } N(y) = N(y') \text{ for some } y, y' \in [0,1], \text{ then } N(T(x,y)) = N(T(x,y')) \quad \forall x \in [0,1].$$
$$(1)$$

Proposition 1. *Let $I : [0,1]^2 \to [0,1]$ be a binary function such that N_I is a fuzzy negation. If I satisfies (LI) with a t-norm T, then N_I and T satisfy Property (1).*

The following example illustrates the previous result.

Example 1. Let N be the continuous (non-strict) fuzzy negation given by

$$N_1(x) = \begin{cases} -2x+1 & \text{if } 0 \le x < 0.25, \\ 0.5 & \text{if } 0.25 \le x \le 0.75, \\ 2-2x & \text{otherwise}, \end{cases}$$

and $T = T_P$, the product t-norm. Consider now a fuzzy implication I with $N_I = N_1$. Then it can not satisfy (LI) with T_P since in this case Property (1) does not hold: $N_1(0.25) = 0.5 = N_1(0.75)$ but

$$N_1(T_P(0.1, 0.25)) = N_1(0.025) = 0.95 \neq 0.85 = N_1(0.075) = N_1(T_P(0.1, 0.75)).$$

This implies that on the one hand,

$$I(0.1, I(0.25, 0)) = I(0.1, N_1(0.25)) = I(0.1, N_1(0.75)) = I(0.1, I(0.75, 0)),$$

but on the other hand,

$$I(T_P(0.1, 0.25), 0) = N_1(T_P(0.1, 0.25)) \neq N_1(T_P(0.1, 0.75)) = I(T_P(0.1, 0.75), 0),$$

and (LI) does not hold.

Next result gives the expression of any binary function with N_I a continuous fuzzy negation satisfying (LI) with a t-norm T. The binary function only depends on the t-norm T and its natural negation.

Proposition 2. *Let $I : [0,1]^2 \to [0,1]$ be a binary function with N_I a continuous fuzzy negation satisfying (LI) with a t-norm T. Then*

$$I(x,y) = N_I(T(x, \mathfrak{R}_{N_I}(y))).$$

From now on, we will denote these implications generated from a t-norm T and a fuzzy negation N by $I_{N,T}(x,y) = N(T(x, \mathfrak{R}_N(y)))$.

Remark 2. Instead of $\mathfrak{R}_{N_I}$, we can consider any function N_1 such that $N_1^{(-1)} = N_I$ and $N_I \circ N_1 = \mathrm{id}_{[0,1]}$. This is a straightforward consequence of the satisfaction of Property (1) in this case. Since $N_I(\mathfrak{R}_{N_I}(y)) = N_I(N_1(y))$, then using the aforementioned property, $N_I(T(x, \mathfrak{R}_{N_I}(y))) = N_I(T(x, N_1(y)))$ and therefore, $I_{N_I,T}$ can be computed using either $\mathfrak{R}_{N_I}$ or N_1.

This class of implications is contained into the class of (S,N)-implications generated from a continuous negation N. This fact is coherent with the characterization of (S,N)-implications, where (LI) is involved, given in Theorem 22 in [25].

Theorem 1. *Let N be a continuous negation and T a t-norm satisfying Property (1). Then $I_{N,T}$ is an (S,N)-implication generated from $S(x,y) = N(T(\mathfrak{R}_N(x), \mathfrak{R}_N(y)))$ and N.*

Moreover, this class of implications satisfies (LI) with the same t-norm T from which they are generated.

Proposition 3. *Let N be a continuous fuzzy negation and T a t-norm satisfying Property (1). Then $I_{N,T}$ satisfies (LI) with T.*

Now, we are in condition to fully characterize the binary functions I with N_I a continuous fuzzy negation satisfying (LI) with a t-norm T.

Theorem 2. *Let $I : [0,1]^2 \to [0,1]$ be a binary function with N_I a continuous fuzzy negation and T a t-norm. Then*

$$I \text{ satisfies (LI) with } T \Leftrightarrow N_I \text{ and } T \text{ satisfy Property (1) and } I = I_{N_I,T}.$$

Note that it remains to know when N_I and T satisfy Property (1). From now on, we will try given a concrete continuous t-norm T, to determine which fuzzy negations satisfy the property with T.

4 Characterization of Fuzzy Implications Satisfying (LI) with a Continuous T-norm

In the previous section, Example 1 shows that Property (1) does not hold for any t-norm and fuzzy negation. Consequently, given a fixed t-norm T, in order to characterize all fuzzy implications with a continuous natural negation satisfying (LI) with T, we need to know which fuzzy negations are compatible with the t-norm T. In this section, we will answer this question for some continuous t-norms presenting for each one, which fuzzy negations can be considered and which fuzzy implications satisfying (LI) with T are generated in that case.

First of all, some negations satisfy Property (1) with any t-norm T (not necessarily continuous).

Proposition 4. *Let N be a continuous fuzzy negation. If there exists $x_0 \in [0,1)$ such that $N(x_0) = 1$ and N is strictly decreasing in $(x_0, 1)$ then Property (1) holds for any t-norm T.*

Remark 3. Note that the previous result includes strict fuzzy negations which are compatible with any t-norm T.

4.1 Minimum T-norm

The first t-norm we are going to study is the minimum t-norm $T_M(x,y) = \min\{x,y\}$. This t-norm performs well with any continuous negation not restricting the choice of the fuzzy negation.

Proposition 5. *If $T = T_M$, then Property (1) holds for any continuous negation N.*

At this point, we can characterize all fuzzy implications with continuous natural negation satisfying (LI) with T_M.

Theorem 3. *Let $I : [0,1]^2 \to [0,1]$ be a binary function with N_I a continuous fuzzy negation. Then the following statements are equivalent:*

(i) I satisfies (LI) with T_M.
(ii) I is given by $I(x,y) = \max\{N_I(x),y\}$.

Remark 4. The fuzzy implications satisfying (LI) with T_M arc, in fact, the so-called generalized Kleene-Dienes implications. In particular, if $N_I(x) = N_C(x) = 1 - x$, we retrieve the Kleene-Dienes implication $I_{KD}(x,y) = \max\{1-x,y\}$. On the other hand, note that there are other implications satisfying (LI) with T_M than those given in Theorem 3. Of course, they must have non-continuous natural negation like the Gödel implication given by

$$I_{GD}(x,y) = \begin{cases} 1 \text{ if } x \leq y, \\ y \text{ if } x > y, \end{cases}$$

which satisfies (LI) with T_M using Theorem 7.3.5 in [5].

4.2 Continuous Archimedean T-norms

In contrast with the minimum t-norm, not all continuous fuzzy negations are compatible with Archimedean t-norms.

Proposition 6. *If T is an Archimedean t-norm, Property (1) holds if, and only if, N is a continuous fuzzy negation being strictly decreasing for all $x \in (0,1)$ such that $N(x) < 1$.*

Remark 5. Note that if we consider Archimedean t-norms, compatible fuzzy negations are strict ones and those given by the expression:

$$N(x) = \begin{cases} 1 & \text{if } x \in [0,x_0], \\ N'\left(\frac{x-x_0}{1-x_0}\right) & \text{otherwise,} \end{cases} \tag{2}$$

where $x_0 \in (0,1)$ and N' is a strict negation. Thus, if the fuzzy negation is a continuous (non-strict) negation, it can only have a unique constant region and it must value 1 there.

Recall that continuous Archimedean t-norms are divided in two subsets: nilpotent t-norms and strict t-norms. So, from now on, we will study these two cases separately.

4.2.1   Nilpotent T-norms

Nilpotent t-norms are φ-conjugated with the Łukasiewicz t-norm $T_{LK}(x,y) = \max\{x+y-1,0\}$, i.e., $T = (T_{LK})_\varphi$ for some automorphism φ. The following result characterizes completely fuzzy implications satisfying (LI) with these t-norms.

Theorem 4. *Let $I : [0,1]^2 \to [0,1]$ be a binary function with N_I a continuous fuzzy negation and $\varphi : [0,1] \to [0,1]$ an automorphism. Then the following statements are equivalent:*

(i) I satisfies (LI) with $(T_{LK})_\varphi$.
(ii) One of the following two cases hold:

    (a) If N_I is strict, then I is given by

$$I(x,y) = \begin{cases} 1 & \text{if } y > (N_I \circ (N_C)_\varphi)(x), \\ f^{-1}(f(N_I(x)) + f(y) - 1) & \text{if } y \leq (N_I \circ (N_C)_\varphi)(x). \end{cases}$$

    where $f = \varphi \circ N_I^{-1}$, and $N_C(x) = 1 - x$ denotes the classical negation.
    (b) If N_I is given by Equation (2) with $x_0 \in (0,1)$ and N' a strict negation, then I is given by $I(x,y) =$

$$= \begin{cases} 1 & \text{if } y > N'\left(\frac{\varphi^{-1}(1-\varphi(x)+\varphi(x_0))-x_0}{1-x_0}\right), \\ N'\left(\frac{\varphi^{-1}(\varphi(x)+\varphi(x_0+(1-x_0)N'^{-1}(y))-1)-x_0}{1-x_0}\right) & \text{if } y \leq N'\left(\frac{\varphi^{-1}(1-\varphi(x)+\varphi(x_0))-x_0}{1-x_0}\right). \end{cases}$$

Taking $\varphi(x) = x$, a particular result for the Łukasiewicz t-norm can be deduced.

Corollary 1. *Let $I : [0,1]^2 \to [0,1]$ be a binary function with N_I a continuous fuzzy negation. Then the following statements are equivalent:*

(i) I satisfies (LI) with T_{LK}.
(ii) One of the two following cases hold:

    (a) If N_I is strict, then I is given by

$$I(x,y) = \begin{cases} 1 & \text{if } y > (N_I \circ N_C)(x), \\ N_I(x + N_I^{-1}(y) - 1) & \text{if } y \leq (N_I \circ N_C)(x). \end{cases}$$

    (b) If N_I is given by Equation (2) with $x_0 \in (0,1)$ and N' a strict negation, then

$$I(x,y) = \begin{cases} 1 & \text{if } y > N'\left(\frac{1-x}{1-x_0}\right), \\ N'\left(N'^{-1}(y) - \frac{1-x}{1-x_0}\right) & \text{if } y \leq N'\left(\frac{1-x}{1-x_0}\right). \end{cases}$$

Remark 6. Taking $N_I = N_C$, the Łukasiewicz implication $I_{LK}(x,y) = \min\{1, 1-x+y\}$ is obtained from case (a) in the previous corollary. Again note that there are other implications satisfying (LI) with T_{LK} than those given in Corollary 1. Of course, they must have non-continuous natural negation like the Weber implication, which satisfies (LI) with T_{LK} by Example 7.3.3-(ii) in [5] and is given by

$$I_{WB}(x,y) = \begin{cases} 1 & \text{if } x < 1, \\ y & \text{if } x = 1. \end{cases}$$

4.2.2 Strict T-norms

Strict t-norms are those t-norms T which are φ-conjugated with the product t-norm $T_P(x,y) = xy$, i.e., $T = (T_P)_\varphi$ for some automorphism φ. Example 1 shows that strict t-norms do not satisfy Property (1) with every fuzzy negation, as it is stated in Proposition 6.

The following result allows us to characterize fuzzy implications satisfying (LI) with a fixed strict t-norm.

Theorem 5. *Let* $I : [0,1]^2 \to [0,1]$ *be a binary function with* N_I *a continuous fuzzy negation and* $\varphi : [0,1] \to [0,1]$ *an automorphism. Then the following statements are equivalent:*

(i) I satisfies (LI) with $(T_P)_\varphi$.
(ii) One of the two following cases hold:

(a) If N_I *is strict, then I is given by*

$$I(x,y) = g^{-1}(g(N_I(x)) \cdot g(y))$$

where $g = \varphi \circ N_I^{-1}$.
(b) If N_I *is given by Equation (2) with* $x_0 \in (0,1)$ *and* N' *a strict negation, then*

$$I(x,y) = \begin{cases} 1 & \text{if } y \geq N'\left(\frac{\varphi^{-1}\left(\frac{\varphi(x_0)}{\varphi(x)}\right) - x_0}{1-x_0}\right), \\ N'\left(\frac{\varphi^{-1}(\varphi(x) \cdot \varphi(x_0 + (1-x_0) \cdot N'^{-1}(y))) - x_0}{1-x_0}\right) & \text{if } y < N'\left(\frac{\varphi^{-1}\left(\frac{\varphi(x_0)}{\varphi(x)}\right) - x_0}{1-x_0}\right). \end{cases}$$

Remark 7. The fuzzy implications obtained in case (a) of the previous result are in fact φ-conjugates of Yager's f-generated implications with $f(0) < \infty$ such that $f = g \circ \varphi^{-1}$, since

$$(I_f)_\varphi(x,y) = \varphi^{-1}(f^{-1}(\varphi(x) \cdot f(\varphi(y)))) = g^{-1}(g(N_I(x)) \cdot g(y)).$$

Recall that φ-conjugates of Yager's f-generated implications with $f(0) < \infty$ are characterized as the only binary operations satisfying (LI) with $(T_P)_\varphi$ and with N_I a strict fuzzy negation (see Theorem 8 in [29]).

Taking $\varphi(x) = x$, a particular result for the product t-norm can be deduced.

Corollary 2. *Let $I : [0,1]^2 \to [0,1]$ be a binary function with N_I a continuous fuzzy negation. Then the following statements are equivalent:*

(i) I satisfies (LI) with T_P.
(ii) One of the two following cases hold:

   (a) If N_I is strict, then I is given by

$$I(x,y) = N_I(x \cdot N_I^{-1}(y)).$$

   (b) If N_I is given by Equation (2) with $x_0 \in (0,1)$ and N' a strict negation, then

$$I(x,y) = \begin{cases} 1 & \text{if } y \geq N'\left(\frac{x_0(1-x)}{x(1-x_0)}\right), \\ N'\left(\frac{x \cdot (x_0 + (1-x_0) \cdot N'^{-1}(y)) - x_0}{1-x_0}\right) & \text{if } y < N'\left(\frac{x_0(1-x)}{x(1-x_0)}\right). \end{cases}$$

Remark 8. The fuzzy implications obtained in case (a) of the previous result are in fact Yager's f-generated implications with $f(0) < \infty$ such that $f = N_I^{-1}$. Recall that Yager's f-generated implications with $f(0) < \infty$ are characterized as the only binary operations satisfying (LI) with T_P and with N_I a strict fuzzy negation (see Theorem 6 in [29]).

Remark 9. Note that there are other implications satisfying (LI) with T_P than those given in Corollary 2. Of course, they must have non-continuous natural negation like the Yager implication given by

$$I_{YG}(x,y) = \begin{cases} 1 & \text{if } x = 0 \text{ and } y = 0, \\ y^x & \text{if } x > 0 \text{ or } y > 0, \end{cases}$$

which satisfies (LI) with T_P using Theorem 7.3.4 in [5].

5 Conclusions and Future Work

In this paper, we have characterized all fuzzy implications satisfying (LI) with a t-norm T when the natural negation of the implication is continuous. Moreover, we have determined in particular the expression of these implications when some continuous t-norms are considered: the minimum t-norm and the Archimedean continuous ones. The fuzzy implications obtained in these cases are always (S,N)-implications but often, they belong also to other well-known classes as Yager's f-generated implications with $f(0) < \infty$ and their conjugate implications.

As a future work, we want to study the case when an ordinal sum t-norm is considered in order to cover all continuous t-norms. In addition, some non-continuous t-norms as the drastic t-norm and the nilpotent minimum worth to be studied.

Acknowledgements. This paper has been partially supported by the Spanish Grant MTM2009-10320 with FEDER support.

References

[1] Aguiló, I., Suñer, J., Torrens, J.: A characterization of residual implications derived from left-continuous uninorms. Information Sciences 180(20), 3992–4005 (2010)

[2] Baczyński, M.: On the distributivity of fuzzy implications over continuous and archimedean triangular conorms. Fuzzy Sets and Systems 161(10), 1406–1419 (2010)

[3] Baczyński, M.: On the distributivity of fuzzy implications over representable uninorms. Fuzzy Sets and Systems 161(17), 2256–2275 (2010)

[4] Baczyński, M., Jayaram, B.: Yager's classes of fuzzy implications: some properties and intersections. Kybernetika 43, 157–182 (2007)

[5] Baczyński, M., Jayaram, B.: Fuzzy Implications. Springer, Heidelberg (2008)

[6] Baczyński, M., Jayaram, B.: (S,N)- and R-implications: A state-of-the-art survey. Fuzzy Sets and Systems 159, 1836–1859 (2008)

[7] Baczyński, M., Jayaram, B.: On the distributivity of fuzzy implications over nilpotent or strict triangular conorms. Transactions on Fuzzy Systems 17(3), 590–603 (2009)

[8] Baczyński, M., Jayaram, B.: (U,N)-implications and their characterizations. Fuzzy Sets and Systems 160, 2049–2062 (2009)

[9] Baczyński, M., Jayaram, B.: Intersections between some families of (U,N)- and RU-implications. Fuzzy Sets and Systems 167(1), 30–44 (2011); Special Issue: Aggregation techniques and applications Selected papers from AGOP 2009 dedicated to Jaume Casasnovas

[10] Balasubramaniam, J., Rao, C.: On the distributivity of implication operators over T and S norms. IEEE Transactions on Fuzzy Systems 12, 194–198 (2004)

[11] Bustince, H., Mohedano, V., Barrenechea, E., Pagola, M.: Definition and construction of fuzzy DI-subsethood measures. Information Sciences 176, 3190–3231 (2006)

[12] Bustince, H., Pagola, M., Barrenechea, E.: Construction of fuzzy indices from fuzzy DI-subsethood measures: application to the global comparison of images. Information Sciences 177, 906–929 (2007)

[13] Combs, W., Andrews, J.: Combinatorial rule explosion eliminated by a fuzzy rule configuration. IEEE Transactions on Fuzzy Systems 6(1), 1–11 (1998)

[14] Fodor, J.C., Roubens, M.: Fuzzy Preference Modelling and Multicriteria Decision Support. Kluwer Academic Publishers, Dordrecht (1994)

[15] Gottwald, S.: Treatise on Many-Valued Logic. Research Studies Press, Baldock (2001)

[16] Jayaram, B.: On the law of importation $(x \wedge y) \rightarrow z \equiv (x \rightarrow (y \rightarrow z))$ in fuzzy logic. IEEE Transactions on Fuzzy Systems 16, 130–144 (2008)

[17] Jayaram, B.: Rule reduction for efficient inferencing in similarity based reasoning. International Journal of Approximate Reasoning 48(1), 156–173 (2008)

[18] Kerre, E., Huang, C., Ruan, D.: Fuzzy Set Theory and Approximate Reasoning. Wu Han University Press, Wu Chang (2004)

[19] Kerre, E., Nachtegael, M.: Fuzzy techniques in image processing. STUDFUZZ, vol. 52. Springer, New York (2000)

[20] Klement, E., Mesiar, R., Pap, E.: Triangular norms. Kluwer Academic Publishers, Dordrecht (2000)

[21] Mas, M., Monserrat, M., Torrens, J.: Modus ponens and modus tollens in discrete implications. Int. J. Approx. Reasoning 49(2), 422–435 (2008)

[22] Mas, M., Monserrat, M., Torrens, J.: The law of importation for discrete implications. Information Sciences 179, 4208–4218 (2009)

[23] Mas, M., Monserrat, M., Torrens, J.: A characterization of (U,N), RU, QL and D-implications derived from uninorms satisfying the law of importation. Fuzzy Sets and Systems 161, 1369–1387 (2010)

[24] Mas, M., Monserrat, M., Torrens, J., Trillas, E.: A survey on fuzzy implication functions. IEEE Transactions on Fuzzy Systems 15(6), 1107–1121 (2007)

[25] Massanet, S., Torrens, J.: The law of importation versus the exchange principle on fuzzy implications. Fuzzy Sets and Systems 168(1), 47–69 (2011)

[26] Massanet, S., Torrens, J.: Intersection of Yager's implications with QL and D-implications. International Journal of Approximate Reasoning 53, 467–479 (2012)

[27] Massanet, S., Torrens, J.: On a generalization of yager's implications. In: Greco, S., Bouchon-Meunier, B., Coletti, G., Fedrizzi, M., Matarazzo, B., Yager, R.R. (eds.) IPMU 2012, Part II. CCIS, vol. 298, pp. 315–324. Springer, Heidelberg (2012)

[28] Massanet, S., Torrens, J.: On some properties of threshold generated implications. Fuzzy Sets and Systems 205, 30–49 (2012)

[29] Massanet, S., Torrens, J.: On the characterization of Yager's implications. Information Sciences 201, 1–18 (2012)

[30] Massanet, S., Torrens, J.: Threshold generation method of construction of a new implication from two given ones. Fuzzy Sets and Systems 205, 50–75 (2012)

[31] Massanet, S., Torrens, J.: On the vertical threshold generation method of fuzzy implication and its properties. Fuzzy Sets and Systems (2013), doi:10.1016/j.fss.2013.03.003

[32] Radzikowska, A.M., Kerre, E.E.: A comparative study of fuzzy rough sets. Fuzzy Sets and Systems 126(2), 137–155 (2002)

[33] Ruan, D., Kerre, E.E. (eds.): Fuzzy IF-THEN Rules in Computational Intelligence: Theory and Applications. Kluwer Academic Publishers, Norwell (2000)

[34] Ruiz-Aguilera, D., Torrens, J.: Distributivity and conditional distributivity of a uninorm and a continuous t-conorm. IEEE Transactions on Fuzzy Systems 14(2), 180–190 (2006)

[35] Ruiz-Aguilera, D., Torrens, J.: Distributivity of residual implications over conjunctive and disjunctive uninorms. Fuzzy Sets and Systems 158, 23–37 (2007)

[36] Shi, Y., Van Gasse, B., Ruan, D., Kerre, E.: On a new class of implications in fuzzy logic. In: Hüllermeier, E., Kruse, R., Hoffmann, F. (eds.) IPMU 2010. CCIS, vol. 80, pp. 525–534. Springer, Heidelberg (2010)

[37] Trillas, E., Alsina, C.: On the law $[(p \wedge q) \to r] = [(p \to r) \vee (q \to r)]$ in fuzzy logic. IEEE Transactions on Fuzzy Systems 10(1), 84–88 (2002)

[38] Trillas, E., Alsina, C., Pradera, A.: On MPT-implication functions for fuzzy logic. Rev. R. Acad. Cien. Serie A. Mat. 98, 259–271 (2004)

[39] Trillas, E., Alsina, C., Renedo, E., Pradera, A.: On contra-symmetry and MPT condicionality in fuzzy logic. International Journal of Intelligent Systems 20, 313–326 (2005)

[40] Trillas, E., del Campo, C., Cubillo, S.: When QM-operators are implication functions and conditional fuzzy relations. International Journal of Intelligent Systems 15, 647–655 (2000)

[41] Yan, P., Chen, G.: Discovering a cover set of ARsi with hierarchy from quantitative databases. Information Sciences 173, 319–336 (2005)

Fuzzy Implication Classes Satisfying a Boolean-Like Law

Anderson Cruz, Benjamín Bedregal, and Regivan Santiago

Abstract. Properties that are always true in the classical theory (Boolean laws) have been extended to fuzzy theory and so-called Boolean-like laws. The fact that they do not remain valid in any standard fuzzy set theory has induced a broad investigation. In this paper we show the sufficient and necessary conditions that a fundamental Boolean-like law — $y \leq I(x,y)$ — holds in fuzzy logics. We focus the investigation on the following classes of fuzzy implications: (S,N)-, R-, QL-, D-, (N,T)-, f-, g- and h-implications.

1 Introduction

The well-known material implication was the first one to be studied and disseminated. This fact induces us to believe that material implication is the correct notion (common sense) of what actually is a logical implication. However, other Boolean implications, such as intuitionistic and quantum implications, do give acceptable implication models and they must be regarded to understand the meaning of a logical implication.

In fuzzy logics, the lack of a consensus on Boolean implication meaning entails non-equivalent acceptable definitions of fuzzy implications (see [15, 16, 27, 28, 34] as examples) and classes of fuzzy implications, such as [4, 5, 10, 12, 20, 22].

Therefore, for a better understanding of what is a logical implication (in Boolean and fuzzy contexts) and also to characterize the approximate reasoning in accordance to the implication definition of a given logic, several Boolean laws have been generalized and studied as (in)equations in fuzzy logics in which t-norms,

Anderson Cruz
Federal Rural University of Semi-Árido - UFERSA, 59515-000, Angicos, RN, Brazil
e-mail: anderson@ufersa.edu.br

Benjamín Bedregal · Regivan Santiago
Federal University of Rio Grande do Norte - UFRN, 59072-970, Natal, RN, Brazil
e-mail: {bedregal,regivan}@dimap.ufrn.br

H. Bustince et al. (eds.), *Aggregation Functions in Theory and in Practise*,
Advances in Intelligent Systems and Computing 228,
DOI: 10.1007/978-3-642-39165-1_41, © Springer-Verlag Berlin Heidelberg 2013

t-conorms, fuzzy negations and implications are used — see [1], [2], [7], [8], [28], [32] and [33]. Although in classic-like fuzzy semantics [11] Boolean-like laws are valid, if, and only if,[1] their original Boolean laws are tautologies. A lot of Boolean laws do not remain valid when are generalized to the fuzzy setting.

In this scenario, this paper deals with the functional inequation (1), where I is a fuzzy implication. (1) generalizes the relation $v(q) \leq (v(p) \Rightarrow v(q))$, in which $\Rightarrow$ is the implication operator and $v(p), v(q) \in \{0, 1\}$ are the truth values of p and q, respectively. Such relation is a well-known Boolean law that entails other fundamental Boolean rules, e.g. $(v(p) \Rightarrow 1) = 1$ which can be translated as *If the consequent of an implication is true, then the implication is also true*[2].

$$y \leq I(x, y), \text{ for all } x, y \in [0, 1] \tag{1}$$

This paper extends [14]. This one provided sufficient and necessary conditions under which (1), holds for (S,N)-, R- and QL-implications. In this current paper, the main concern is extending such investigation for D-, (N,T)-, and h-implications.

The paper is organized as follows. Section 2 recalls some basic definitions and results about fuzzy operators. Section 3 recall a fuzzy implication definition I, some of their classes and some useful equivalences between those classes. In such section we also show some results about interrelationship between (1) and other fuzzy implication properties. Section 4 shows up, for those classes, under which conditions (1) holds. Finally, the final remarks of the paper is exposed in section 5.

2   Basic Definitions

In this section we mention the preliminaries definitions and previous results about t-norms, t-conorms and fuzzy negations.

Definition 1. A function $T : [0, 1]^2 \to [0, 1]$ is a t-norm if T satisfies commutativity (T1), associativity (T2), monoticity (T3) and 1-identity (T4).

Remark 1. Considering the partial order on the family of all t-norms induced from the order on [0,1], $T_M(x, y) = min(x, y)$ is the greatest t-norm. Therefore for any t-norm T, $T(x, y) \leq T_M(x, y) \leq y$, for all $x, y \in [0, 1]$.

Definition 2. A function $S : [0, 1]^2 \to [0, 1]$ is a t-conorm if S satisfies commutativity (S1), associativity (S2), monoticity (S3) and 0-identity (S4).

Remark 2. Considering the partial order on the family of all t-conorms induced from the order on [0,1] on the family of all t-conorms, $S_M(x, y) = max(x, y)$ is the least t-conorm. So for any t-conorm S, $S(x, y) \geq S_M(x, y) \geq y$. Moreover, $S_M(x, 1) = S_M(1, x) = 1$, so $S(x, 1) = S(1, x) = 1$ for any t-conorm S.

---

[1] iff, for short.

[2] $(v(p) \Rightarrow 1) = 1$ can be generalized to fuzzy logic as $I(x, 1) = 1$ (denoted in this paper by (I8)).

Definition 3. A function $N : [0,1] \to [0,1]$ is a fuzzy negation, if $N(0) = 1, N(1) = 0$ (N1), and N is decreasing (N2). Beyond this, a fuzzy negation is called **strong**, if it is involutive, i.e., $N(N(x)) = x$, for all $x \in [0,1]$.

As examples of fuzzy negations, we cite the greatest fuzzy negation $N_\top$:

$$N_\top(x) = \begin{cases} 0 \,, \text{if } x = 1 \\ 1 \,, \text{if } x \in [0,1[. \end{cases}$$

2.1 Properties Involving Fuzzy Operators

In this subsection we address three properties: Distributivity of t-conorms over t-norms; Law of excluded middle; and N-duality.

- In classical logic, the distributivity of disjunction over conjunction is a well-known property, its extension to fuzzy logic takes into account t-norms and t-conorms: A t-conorm S is distributive over a t-norm T if

$$S(x, T(y,z)) = T(S(x,y), S(x,z)). \tag{2}$$

An important result about such property is the following:

Proposition 1. *[17, Proposition 2.22] Let T be a t-norm and S a t-conorm, then S is distributive over T iff $T = T_M$.*

- One of the fundamental Boolean laws of classical theory is the Law of Excluded Middle (LEM). As LEM in classical logic states that $\neg p \lor p$ is always true, we have the following extension to fuzzy logic.

Definition 4. Let S be a t-conorm and N a fuzzy negation, the pair (S,N) satisfies the LEM if

$$S(N(x), x) = 1, \text{ for all } x \in [0,1]. \tag{LEM}$$

Remark 3. [3] *[3, Remark 2.3.10]* $S(N_\top(x), x) = 1$, for any t-conorm S, that is, $(S, N_\top)$ satisfies (LEM), for any S. Moreover, if S is positive[4], (S,N) satisfies (LEM) only if $N = N_\top$.

- For any t-conorm S there exists a t-norm T such that, $S(x,y) = 1 - T(1-x, 1-y)$. Moreover, let T be a t-norm, S a t-conorm and N a fuzzy negation then S is said the N-dual of T, if

$$S(x,y) = N(T(N(x), N(y))). \tag{3}$$

---

[3] Some proofs will refer this remark.
[4] S is positive iff, if $S(x,y) = 1$ then $x = 1$ or $y = 1$

3 Fuzzy Implications

We adopt a well acceptable definition of fuzzy implication [13, 25, 26].

Definition 5. A function $I : [0,1]^2 \to [0,1]$ is called a fuzzy implication if it satisfies the following boundary conditions.

I1.: $I(0,0) = 1$. I2.: $I(0,1) = 1$. I3.: $I(1,0) = 0$. I4.: $I(1,1) = 1$.

Some other potential properties for fuzzy implications are given next:

I5. Left antitonicity: if $x_1 \leq x_2$ then $I(x_1,y) \geq I(x_2,y)$, for all $x_1,x_2,y \in [0,1]$;
I6. Right isotonicity: if $y_1 \leq y_2$ then $I(x,y_1) \leq I(x,y_2)$, for all $x,y_1,y_2 \in [0,1]$;
I7. Left boundary condition: $I(0,y) = 1$, for all $y \in [0,1]$;
I8. Right boundary condition: $I(x,1) = 1$, for all $x \in [0,1]$;
I9. Left neutrality: $I(1,y) = y$, for all $y \in [0,1]$;
I10. Identity property: $I(x,x) = 1$, for all $x \in [0,1]$;
I11. Exchange principle: $I(x,I(y,z)) = I(y,I(x,z))$, for all $x,y,z \in [0,1]$;
I12. Ordering property[5]: $x \leq y$ iff $I(x,y) = 1$, for all $x,y, \in [0,1]$;

I12a. Left ordering property: if $x \leq y$ then $I(x,y) = 1$, for all $x,y, \in [0,1]$;
I12b. Right ordering property: if $I(x,y) = 1$ then $x \leq y$, for all $x,y, \in [0,1]$;

I13. Gen. of the first classical axiom: $I(y,I(x,y)) = 1$, for all $x,y, \in [0,1]$;
I14. Contraposition: Let N be a fuzzy negation, $I(x,y) = I(N(y),N(x))$, for all $x,y, \in [0,1]$;
I15. Continuity.

There are some relations between above properties as exposed in [13], [3], [29] and [30]. We highlight the previous study about (1) by [13] and [29] where Bustince, Shi et al. investigated the interrelationship between some fuzzy implications properties. In this scenario we also expose some relations between (1) and above properties. See the following propositions and lemmas.

Proposition 2. *[13] If a fuzzy implication I satisfies (1) then I satisfies (I8).*

Lemma 1. *[13, Lemma 1 viii] If a fuzzy implication I satisfies (I5) and (I9), then I satisfies (1).*

Adapting the results of [29, Remark 7.5] we have the next proposition.

Proposition 3. *Let I be a fuzzy implication:*
If I satisfies (I11) and (I12) then I satisfies (1);
If I satisfies (I5), (I11) and $I(x,0) = N_I(x)$ then I satisfies (1);
If I satisfies (I5), (I11) and (I15) then I satisfies (1).

Proposition 4. *If a fuzzy implication I satisfies (1) and (I12a) then I satisfies (I13).*

Proof. Straightforward.

---

[5] Also called confinement property.

Proposition 5. *If a fuzzy implication I satisfies (I13) and (I12b) then I satisfies (1).*

Proof. Straightforward.

Corollary 1. *If a fuzzy implication I satisfies (I12), then I satisfies (1) iff I satisfies (I13).*

Proof. Straightforward from Propositions 4 and 5.

3.1   Classes of Fuzzy Implications

There are three main classes of fuzzy implications, namely: (S,N)-, R- and QL-implications. Other classes can be generated from those ones, namely: D- and (N,T)-implications are generated from QL- and (S,N)-implications, respectively. The h-implications were defined by Massanet et al. in [22] in a similar way done by Yager in [11] — defined by a generator function. Each one of these fuzzy implication classes has a specific motivation. In the following definition we are going to recall them.

Definition 6. Let T be a t-norm, S a t-conorm and N a fuzzy negation, then:

- A function $I : [0,1]^2 \to [0,1]$ is called an (S,N)-implication (denoted by $I_{S,N}$) if

$$I(x,y) = S(N(x),y). \tag{4}$$

- A function $I : [0,1]^2 \to [0,1]$ is called an R-implication (denoted by I_T) if

$$I(x,y) = sup\{t \in [0,1] \,|\, T(x,t) \leq y\}. \tag{5}$$

- A function $I : [0,1]^2 \to [0,1]$ is called a QL-implication[6] (denoted by $I_{S,N,T}$) if

$$I(x,y) = S(N(x),T(x,y)). \tag{6}$$

- A function $I : [0,1]^2 \to [0,1]$ is called a D-implication[7] (denoted by $I_{S,T,N}$) if

$$I(x,y) = S(T(N(x),N(y)),y). \tag{7}$$

- A function $I : [0,1]^2 \to [0,1]$ is called a (N,T)-implication (denoted by $I_{N,T}$) if

$$I(x,y) = N(T(x,N(y))) \tag{8}$$

- A function $I : [0,1]^2 \to [0,1]$ is called an f-generated implication (denoted by I^f) if there exists a strictly decreasing and continuous function $f : [0,1] \to [0,\infty]$ with $f(1) = 0$ and with the understanding $0 \cdot \infty = 0$, such that

---

[6] In this paper we assume the QL-implication definition given in [3].

[7] D-implications are generally defined from a strong negation [20, 24, 25]. However, we maintain the standard of QL-implication definition and then we relax the D-implication definition and define it from any (strong or non-strong) fuzzy negation.

$$I^f(x,y) = f^{-1}(x \cdot f(y)). \tag{9}$$

- A function $I : [0,1]^2 \to [0,1]$ is called an f-generated implication (denoted by I^f) if there exists a strictly increasing and continuous function $g : [0,1] \to [0,\infty]$ with $g(0) = 0$ and with the understanding $0 \cdot \infty = \infty$ and $\frac{1}{0} = \infty$, such that

$$I^g(x,y) = g^{(-1)}\left(\frac{1}{x} \cdot g(y)\right), \tag{10}$$

where $g^{(-1)}$ is the pseudo-inverse of g given by

$$g^{(-1)}(x) = \begin{cases} g^{-1}(x) & , \text{if } x \in [0, g(1)] \\ 1 & , \text{if } x \in [g(1), \infty]. \end{cases} \tag{11}$$

- A function $I : [0,1]^2 \to [0,1]$ is called an h-implication (denoted by I^h) if there exist an $e \in\,]0,1[$ and, a strictly increasing and continuous function $h : [0,1] \to [-\infty, +\infty]$ in which $h(0) = -\infty$, $h(e) = 0$ and $h(1) = +\infty$, such that

$$I(x,y) = \begin{cases} 1 & , \text{if } x = 0 \\ h^{-1}(x \cdot h(y)) & , \text{if } x > 0 \text{ and } y \leq e \\ h^{-1}(\frac{1}{x} \cdot h(y)) & , \text{if } x > 0 \text{ and } y > e. \end{cases}$$

The function h is called an h-generator (with respect to e) of I^h.[8]

Lemma 2. *[3, Theorem 2.5.4] and [11, pp.359] Every R-implication satisfies (I1)-(I10) and (I12a).*

Lemma 3. *[3, Theorem 2.6.2] Every QL-implication satisfies (I1)-(I4), (I6), (I7) and (I9).*

Lemma 4. [9] *If (S,N) satisfies (LEM) and $T = T_M$, then a QL-implication $I_{S,N,T}$ satisfies (I10).*

Proof. Straightforward from Lemma 11.

Proposition 6. *Given a QL-implication $I_{S,N,T}$ and an (S,N)-implication $I_{S,N}$, $I_{S,N,T} \leq I_{S,N}$.*

Proof. By Eq. 6, $I_{S,N,T}(x,y) = S(N(x), T(x,y))$ and $T(x,y) \leq y$ (by Remark 1), so $S(N(x), T(x,y)) \leq S(N(x), y) = I_{S,N}(x,y)$.

Lemma 5. *Every D-implication satisfies (I1)-(I4), (I5) and (I9).*

Proof. Let $I_{S,T,N}$ be a D-implication and $I_{S,N,T}$ a QL-implication, so $I_{S,T,N}$ satisfies (I1)-(I4) by [3, Theorem 1.6.2], since $I_{S,T,N}$ is N-reciprocal of $I_{S,N,T}$ (that is, $I_{S,N,T}(N(y), N(x)) = I_{S,T,N}(x,y)$).

---

[8] The functions f and g are also known as f- and g-generators of I^f and I^g, respectively.
[9] A similar result is found in [3, Proposition 2.6.21].

Now, assume that $x_1, x_2, y \in [0,1]$ and $x_1 \leq x_2$. Then, by (N2), $N(x_1) \geq N(x_2)$. By (T3), $T(N(x_1), N(y)) \geq T(N(x_2), N(y))$, and by (S3) we have $S(T(N(x_1), N(y)), y) \geq S(T(N(x_2), N(y)), y)$. Hence $I_{S,T,N}(x_1, y) \geq I_{S,T,N}(x_2, y)$. Hence $I_{S,T,N}$ satisfies (I5).

For any $y \in [0,1]$, $I_{S,T,N}(1, y) = S(T(N(1), N(y)), y) = S(T(0, N(y)), y)$ and $T(0, N(y))$
$= 0$ (by Remark 1). Since $S(0, y) = y$, so $I_{S,T,N}(1, y) = y$. Hence $I_{S,T,N}$ satisfies (I9).

Lemma 6. *[10, Prop. 2.6] Every (N,T)-implication satisfies (I1)-(I6).*

Lemma 7. *An (N,T)-implication $I_{N,T}$ satisfies (I9), if N is a strong negation.*

Proof. Since N is a strong negation, then $I_{N,T}(1, y) = N(T(1, N(y))) = N(N(y))$
$= y$.

Lemma 8. *[3, Prop. 3.1.2 and Theo. 3.1.7] Every f-implication satisfies (I1)-(I6), (I9) and (I11).*

Lemma 9. *[3, Prop. 3.2.2 and Theo. 3.2.8] Every g-implication satisfies (I1)-(I6) and (I9)-(I11).*

Lemma 10. *[22, Prop. 1 and Theo. 5(i)] Let h be an h-generator w.r.t. a fixed $e \in$ $]0,1[$, then I^h satisfies (I1)-(I6) and (I9).[10] So I^h is a fuzzy implication.*

3.2 Equivalences Among the Fuzzy Implications Classes

Proposition 7. *[20], [16] Let N be a strong negation and, given a D-implication $I_{S,T,N}$ and a QL-implication $I_{S,N,T}$. If $I_{S,T,N}$ or $I_{S,N,T}$ satisfies the contraposition (I14), then $I_{S,T,N} = I_{S,N,T}$.*

Lemma 11. *[3, Proposition 4.2.2] Given an (S,N)-implication $I_{S,N}$ and a QL-implication $I_{S,N,T}$. If $T = T_M$ and (S,N) satisfies (LEM), then $I_{S,N,T} = I_{S,N}$.*

Lemma 12. *Given a D-implication $I_{S,T,N}$ and an (S,N)-implication $I_{S,N}$. If $T = T_M$ and (S,N) satisfies (LEM), then $I_{S,T,N} = I_{S,N}$.*

Proof. By Proposition 1, $T = T_M$ iff (S,T) satisfies (2). So, for all $x, y \in [0,1]$:
$$
\begin{aligned}
I_{S,T,N}(x,y) &= S(T(N(x), N(y)), y) & &\text{by} (7) \\
&= S(y, T(N(x), N(y))) & &\text{by} (S1) \\
&= T(S(y, N(x)), S(y, N(y))) & &\text{by} (2) \\
&= T(S(N(x), y), S(N(y), y)) & &\text{by} (S1) \\
&= T(S(N(x), y), 1) & &\text{by} (LEM) \\
&= S(N(x), y) & &\text{by} (T4) \\
&= I_{S,N}(x, y) & &\text{by} (4).
\end{aligned}
$$

---

[10] Truly, in [22, Prop. 1 and Theo. 5(i)] is demonstrated that I^h satisfies (I2) is trivially deduced from (I4) and (I5). Beyond that, we also can deduce straightforward (I7) from (I1) and (I6), and (I8) from (I4) and (I5).

Theorem 1. *Given a QL-implication $I_{S,N,T}$, a D-implication $I_{S,T,N}$ and an (S,N)-implication $I_{S,N}$. If $T = T_M$ and (S,N) satisfies (LEM), then $I_{S,T,N} = I_{S,N} = I_{S,N,T}$.*

Proof. Straightforward from Lemmas 11 and 12.

If we regard that N is a strong negation, then we got another result relating (S,N)-, QL- and D-implications.

Proposition 8. *[20, Proposition 6] Let N be a strong negation and given a QL-implication $I_{S,N,T}$, a D-implication $I_{S,T,N}$ and an (S,N)-implication $I_{S,N}$. If $I_{S,T,N}$ and $I_{S,N}$ satisfy (I1)-(I6), then the corresponding QL- and D-implication coincide and are given by:*

$$I_{S,N,T}(x,y) = I_{S,T,N}(x,y) = \begin{cases} 1 & , \text{if } x \leq y \\ I_{S,N}(x,y) & , \text{otherwise.} \end{cases}$$

Lemma 13. *Let N be a strong negation. $I_{S,N} = I_{N,T}$ iff S is N-dual of T.*

Proof. Straightforward.

Theorem 2. *Given a D-implication $I_{S,T,N}$, an (S,N)-implication $I_{S,N}$ and a QL-implication $I_{S,N,T}$. If $T = T_M$, S is N-dual of T and (S,N) satisfies (LEM). Then $I_{S,T,N} = I_{S,N} = I_{S,N,T} = I_{N,T}$.*

Proof. Straightforward from Lemma 13 and Theorem 1.

4   On the $y \leq I(x,y)$ of Fuzzy Logic

In the sequel we show some results about (1) and, (S,N)-, R-, QL-, D-, (N,T), f-, g- and h-implications.

Theorem 3. *Every (S,N)-implication satisfies (1).*

Proof. Straightforward, since S_M is the least t-conorm.

Theorem 4. *Every R-implication satisfies (1).*

Proof. Straightforward from Lemmas 1 and 2.

Theorem 5. *Every D-implication satisfies (1).*

Proof. Straightforward from Lemmas 1 and 5.

Theorem 6. *Every (N,T)-implication satisfies (1).*

Proof. Trivial, since $N(N(x)) \geq x$ implies $I_{N,T}(x,y) \geq y$.

Theorem 7. *Every f-implication satisfies (1).*

Proof. Straightforward from Lemmas 1 and 8.

Theorem 8. *Every g-implication satisfies (1).*

Proof. Straightforward from Lemmas 1 and 9.

Theorem 9. *Let h be an h-generator w.r.t. a fixed e $\in]0,1[$, then I^h satisfies (1).*

Proof. Straightforward from Lemmas 1 and 10.

We demonstrated that (S,N)-implications satisfy (1). Since there is an intersection between the (S,N)- and QL-implications classes, we verify the sufficient and necessary conditions in which the elements of such intersection satisfy (1).[11]

Theorem 10. *If (S,N) satisfies (LEM) and $T = T_M$ then a QL-implication $I_{S,N,T}$ satisfies (1).*

Proof. Straightforward from Theorem 3 and Lemma 11, we deduce the following theorem.

The reader will note that Lemma 11 and Theorem 10 give the sufficient conditions for $I_{S,N,T}$ to satisfy (1). In the sequel we present results that give the necessary conditions.

Lemma 14. *If a QL-implication $I_{S,N,T}$ satisfies (1) then (S,N) satisfies (LEM).*

Proof. By (1), $1 \leq I_{S,N,T}(y,1)$, then $I_{S,N,T}(y,1) = 1$ (i.e. $I_{S,N,T}$ satisfies (I8)). So $S(N(y),T(y,1)) = 1$, and by (T4) $S(N(y),y) = 1$. Hence (S,N) satisfies (LEM).

The reciprocal of Theorem 10 is not true (see [14, Example 5.1]).

Theorem 11. *Let S be a strictly increasing in [0,1[t-conorm. If a QL-implication $I_{S,N,T}$ satisfies (1) and (I10), then (S,N) satisfies (LEM) and $T = T_M$.*

Proof. By Lemma 14, if $I_{S,N,T}$ satisfies (1), then (S,N) satisfies (LEM). Now, by (I10), $S(N(x),T(x,x)) = 1$, and since (S,N) satisfies (LEM), then for any $x \in [0,1]$, $S(N(x),T(x,x)) = 1 = S(N(x),x)$. Case $x = 1$ so, trivially, $T(x,x) = x$. Case $x < 1$, since S is strictly increasing in [0,1[, then $S(N(x),T(x,x)) = S(N(x),x)$ implies $T(x,x) = x$. Therefore $T = T_M$, since T_M is the only idempotent t-norm [19, Theorem 3.9].

Corollary 2. *Let S be a strictly increasing in [0,1[t-conorm. Then the following statements are equivalent:*

1. *A QL-implication $I_{S,N,T}$ satisfies (1) and (I10);*
2. *(S,N) satisfies (LEM) and $T = T_M$.*

Proof. Straightforward from Lemma 4 and Theorems 10 and 11.

---

[11] Following results, about QL-implications, were rewritten from [14] for a better reading.

Note that, if $x \leq y$ iff $I(x,y) = 1$, then $I(x,x) = x$. In other words, if I satisfies (I12), then I satisfies (I10). Therefore, by Theorem 11, we deduce the following Corollary.

Corollary 3. *Let S be a strictly increasing in $[0,1[$ t-conorm. If a QL-implication $I_{S,N,T}$ satisfies (1) and (I12), then (S,N) satisfies (LEM) and $T = T_M$.*

The reciprocal of Corollary 3 is not true. Its counter-example is $I_{S',T_M,N_{\top}}$[12] (given below), since $(S',N_{\top})$ satisfies (LEM) and $I_{S',T_M,N_{\top}}$ satisfies (1), but $I_{S',T_M,N_{\top}}$ does not satisfy (I12).

$$I_{S',T_M,N_{\top}}(x,y) = \begin{cases} 1 \, , \text{if } x < 1 \\ y \, , \text{if } x = 1. \end{cases}$$

5 Final Remarks

This paper provided sufficient and necessary conditions under which the Boolean-like law $x \leq I(y,x)$, refereed by (1), holds for (S,N)-, R-, QL-, D-, (N,T)-, f-, g- and h-implications; beyond of analyzing the relations among fuzzy implication properties and (1). The property (1) was firstly studied in [13] where Bustince et. al. demonstrated, among other results, that, let I' be a fuzzy implication which satisfies (I1)-(I6), if I' satisfies (I9) then I' satisfies (1). The main results of this paper are stated by Theorems 3 to 9, and Corollary 2. From those results, we can conclude that, like in the Boolean context, (1) is also a fundamental property in the fuzzy context (in a wide quantity of fuzzy implication classes): Any (S,N)-, R-, D-, (N,T)-, f-, g- and h-implication satisfies (1).

We also note a close relation between (I10) and (1): Every R-implication satisfies both; every (S,N)-implication where (S,N) satisfies (LEM), also satisfies both; and only QL-implications which satisfy (I10) guarantee the reciprocal of Theorem 10.

A particular result was obtained to QL-implications ($I_{T,S,N}$): we prove that, regarding that S is strictly increasing in $[0,1[$, (S,N) satisfies (LEM) and $T = T_M$ iff $I_{T,S,N}$ satisfies (1) and (I10). We also proved that: if $I_{T,S,N}$ satisfies (1) and (I12), then (S,N) satisfies (LEM) and $T = T_M$ (Corollary 3); but the reciprocal of Corollary 3 is not true.

References

[1] Alsina, C., Trillas, E.: On iterative Boolean-like laws of fuzzy sets. In: EUSFLAT Conf., pp. 389–394 (2005)
[2] Alsina, C., Trillas, E.: On the law S(S(x,y),T((x,y))) = S(x,y) of fuzzy logic. Fuzzy Optim. Decis. Mak. 6(2), 99–107 (2007)
[3] Baczyński, M., Jayaram, M.: Fuzzy Implications. STUDFUZZ, vol. 231. Springer, Heidelberg (2008)
[4] Baczyński, M., Jayaram, M.: (S,N)- and R-implications: a state-of-art survey. Fuzzy Sets and Systems 159(14), 1836–1859 (2008)

---

[12] In which S' is any strictly increasing in $[0,1[$ t-conorm.

[5] Baczyński, M., Jayaram, B.: QL-implications: some properties and intersections. Fuzzy Sets and Systems 161(2), 158–188 (2010)

[6] Baldwin, J., Pilsworth, B.: Axiomatic approach to implication for approximate reasoning with fuzzy logic. Fuzzy Sets and Systems 3(2), 193–219 (1980)

[7] Balasubramaniam, J., Rao, C.J.M.: On the distributivity of implication operators over T and S norms. IEEE Trans. Fuzzy Systems 12(2), 194–198 (2004)

[8] Balasubramaniam, J.: On the law of importation $(x \wedge y) \longrightarrow z \equiv (x \longrightarrow (y \longrightarrow z))$ in fuzzy logic. IEEE Trans. Fuzzy Systems 16(1), 130–144 (2008)

[9] Bandler, W., Kohout, L.: Fuzzy power sets and fuzzy implication operators. Fuzzy Sets and Systems 4(1), 13–30 (1980)

[10] Bedregal, B.: A normal form which preserves tautologies and contradictions in a class of fuzzy logics. J. Algorithms 62(3-4), 135–147 (2007)

[11] Bedregal, B., Cruz, A.: A characterization of classic-like fuzzy semantics. Logic Journal of the IGPL 16(4), 357–370 (2008)

[12] Bedregal, B., Reiser, R.H.S., Dimuro, G.P.: Xor-implications and E-implications: Classes of fuzzy implications based on fuzzy xor. Electr. Notes Theor. Comput. Sci. 247, 5–18 (2009)

[13] Bustince, H., Burillo, P., Soria, F.: Automorphism, negations and implication operators. Fuzzy Sets and Systems 134(1), 209–229 (2003)

[14] Cruz, A., Bedregal, B., Santiago, R.H.N.: The law x≤ I(y,x) and the three main classes of fuzzy implications. In: IEEE International Conference on Fuzzy Systems (FUZZ-IEEE), pp. 1–5 (2012)

[15] Drewniak, J.: Invariant fuzzy implications. Soft Comput. 10(6), 506–513 (2006)

[16] Fodor, J., Roubens, M.: Fuzzy preference modelling and multicriteria decision support. Kluwer Academic Publishers (1994)

[17] Klement, E., Mesiar, R., Pap, E.: Triangular Norms. Triangular Norms. Kluwer Academic Publishers, Dordrecht (2000)

[18] Klir, G.J., Yuan, B.: Fuzzy Sets and Fuzzy Logic Theory and Applications. Prentice Hall PTR, New Jersey (1995)

[19] Mamdani, E.H.: Application of fuzzy logic to approximate reasoning using linguistic synthesis. IEEE Trans. Computers 26(12), 1182–1191 (1977)

[20] Mas, M., Monserrat, M., Torrens, J.: Ql-implications versus D-implications. Kybernetika 42(3), 315–366 (2006)

[21] Mas, M., Monserrat, M., Torrens, J., Trillas, E.: A survey of fuzzy implication functions. IEEE Trans. on Fuzzy Systems 15(6), 1107–1121 (2007)

[22] Massanet, S., Torrens, J.: On a new class of fuzzy implications: h-implications and generalizations. Information Sciences 181(11), 2111–2127 (2011)

[23] Rasiowa, H., Sikorski, R.: The Mathematics of Metamathematics. Monografie matematyczne. Polska Akademia Nauk. (1963)

[24] Reiser, R.H.S., Bedregal, B.C., Dimuro, G.P.: Interval-valued D-implications. Tend. Mat. Apl. Comput. (TEMA) 10(1), 63–74 (2009)

[25] Reiser, R.H.S., Bedregal, B.C., Santiago, R.H.N., Dimuro, G.P.: Analysing the relationship between interval-valued D-implications and interval-valued QL-implications. Tend. Mat. Apl. Comput. (TEMA) 11(1), 89–100 (2010)

[26] Ruan, D., Kerre, E.E.: Fuzzy implication operators and generalized fuzzy method of cases. Fuzzy Sets and Systems 54(1), 23–37 (1993)

[27] Sainio, E., Turunen, R.M.E.: A characterization of fuzzy implications generated by generalized quantifiers. Fuzzy Sets and Systems 159(4), 491–499 (2008)

[28] Shi, Y., Ruan, D., Kerre, E.E.: On the characterizations of fuzzy implications satisfying I(x, y)=I(x, I(x, y)). Information Sciences 177(14), 2954–2970 (2007)

[29] Shi, Y., Gasse, B.V., Kerre, E.E.: On dependencies and independencies of fuzzy implication axioms. Fuzzy Sets and Systems 161(10), 1388–1405 (2010)

[30] Shi, Y., Van Gasse, B., Ruan, D., Kerre, E.: On a new class of implications in fuzzy logic. In: Hüllermeier, E., Kruse, R., Hoffmann, F. (eds.) IPMU 2010, Part I. CCIS, vol. 80, pp. 525–534. Springer, Heidelberg (2010)

[31] Trillas, E., Valverde, L.: On some functionally expressable implications for fuzzy set theory. In: Proc. 3rd Inter. Seminar on Fuzzy Set Theory, Linz, Austria, pp. 173–190 (1981)

[32] Trillas, E., Alsina, C.: Standard theories of fuzzy sets with the law $(\mu \wedge \sigma')' = \sigma \vee (\mu' \wedge \sigma')$. Int. J. Approx. Reasoning 37(2), 87–92 (2004)

[33] Trillas, E., Alsina, C.: On the law $[(p \vee q) \rightarrow r] = [(p \rightarrow r) \wedge (q \rightarrow r)]$ in fuzzy logic. IEEE Trans. Fuzzy System 10(1), 84–98 (2002)

[34] Yager, R.: On the implication operator in fuzzy logic. Information Sciences 31(2), 141–164 (1983)

[35] Yager, R.: On some new classes of implication operators and their role in approximate reasoning. Information Sciences 167(1-4), 193–216 (2004)

[36] Zadeh, L.A.: Outline of a new approach to the analysis of complex systems and decision processes. IEEE Trans. on System, Man and Cybernetics 3(1), 28–44 (1973)

Implications Generated by Triples of Monotone Functions

Dana Hliněná, Martin Kalina, and Pavol Král'

Abstract. In this paper we deal with fuzzy implications generated via triples of monotone functions f, g, h. This idea has been presented for the first time at the IPMU 2012 conference, where we have introduced the generating formula and studied some special cases of these fuzzy implications. In our contribution we further develop this concept and study properties of generated fuzzy implications. More precisely, we study how some specific properties of generators f, g, h influence properties of the corresponding fuzzy implications.

We give also some examples of such generated fuzzy implications and examples illustrating the intersection of the system of fuzzy implications generated by this method with known types of generated fuzzy implications.

1 Introduction and Basic Notions

A fuzzy implication is a mapping $I : [0,1]^2 \to [0,1]$ that generalizes the classical implication to fuzzy logic case in a similar way as t-norms (t-conorms) generalize

Dana Hliněná
Dept. of Mathematics FEEC Brno Uni. of Technology
Technická 8, Cz-616 00 Brno, Czech Republic
e-mail: hlinena@feec.vutbr.cz

Martin Kalina
Slovak University of Technology in Bratislava
Faculty of Civil Engineering, Department of Mathematics
RadlinskÃ©ho 11, Sk-813 68 Bratislava, Slovakia
e-mail: kalina@math.sk

Pavol Král'
Dept. of Quantitative Methods and Information Systems,
Faculty of Economics, Matej Bel University
Tajovského 10, Sk-975 90 Banská Bystrica, Slovakia
e-mail: pavol.kral@umb.sk

H. Bustince et al. (eds.), *Aggregation Functions in Theory and in Practise*,
Advances in Intelligent Systems and Computing 228,
DOI: 10.1007/978-3-642-39165-1_42, © Springer-Verlag Berlin Heidelberg 2013

the classical conjunction (disjunction). It is well known that there exist many ways how to construct them (see e.g., [2, 3, 4, 5, 13, 16, 20, 22, 23]). Fuzzy implications, in some cases, can be represented in the form $h(f(x) * g(y))$, where $*$ is one of the usual arithmetic operations and f, g, h are monotone functions. A similar approach is used also for generating of aggregation functions, see e.g. [15]. We have started our research in this direction in 2012 at IPMU conference [11], where we have introduced the generating formula for the first time. Now we continue our research investigating properties of such generated implications. We will consider only the case when f is a decreasing function and g, h are increasing functions. We define an operator

$$I_{(f,g,h)}(x,y) = h(f(x) + g(y)). \tag{1}$$

For appropriately chosen triple (f,g,h) we get a fuzzy implication. The precise definition will be given in the next section. Formula (1) enables us to get a general view at fuzzy implications generated by means of a triple of monotone functions, regardless whether additively or multiplicatively, since we get one case from the other by exponential or logarithmic transformations of f and g, respectively.

In order to avoid confusion let us note that in the rest of the paper we use the following notions. If $x_1 \leq x_2$ implies $f(x_1) \leq (\geq, <, >)f(x_2)$, then f is an increasing (decreasing, strictly increasing, strictly decreasing) function, respectively.

Definition 1. (see, e.g., [8]) A decreasing function $N : [0,1] \to [0,1]$ is called a fuzzy negation if $N(0) = 1, N(1) = 0$. A fuzzy negation N is called

1. strict if it is strictly decreasing and continuous in $[0,1]$,
2. strong if it is an involution, i.e., if $N(N(x)) = x$ for all $x \in [0,1]$.

Fuzzy conjunction is typically modelled by triangular norms.

Definition 2. (see, e.g., [21]) A triangular norm T (t-norm for short) is a commutative, associative, monotone binary operator on the unit interval $[0,1]$, fulfilling the boundary condition $T(x,1) = x$, for all $x \in [0,1]$.

Remark 1. Note that, for a strict negation N, the N-dual operation to a t-norm T defined by $S(x,y) = N^{-1}(T(N(x),N(y)))$ is called t-conorm. For more information, see, e.g., [14].

Uninorms were introduced by Yager and Rybalov in 1996 [24] as a generalization of triangular norms and conorms.

Definition 3. An associative, commutative and increasing operation $U : [0,1]^2 \to [0,1]$ is called a uninorm, if there exists $e \in [0,1]$, called the neutral element of U, such that

$$U(e,x) = U(x,e) = x \quad \text{for all } x \in [0,1].$$

Fodor et al. [9] characterized the so-called representable uninorms.

Proposition 1. ([9]) *A uninorm U with neutral element $e \in]0,1[$ is representable if and only if there exists a strictly increasing and continuous function $h : [0,1] \to$*

$[-\infty, \infty]$ *such that* $h(0) = -\infty$, $h(e) = 0$, $h(1) = \infty$. *The uninorm U is in that case given by*

$$U(x,y) = h^{-1}(h(x) + h(y)) \quad \text{for } x, y \in [0,1],$$

where the value $\infty - \infty$ *can be defined as* ∞ *or* $-\infty$.

For each function h we have two possibilities how to define the corresponding representable uninorm. These two possibilities are the disjunctive (in case $\infty - \infty = \infty$) and conjunctive (in case $\infty - \infty = -\infty$) ones. The generator h of a uninorm can be also strictly decreasing. In this case $\infty - \infty = \infty$ gives a conjunctive uninorm.

In literature, we can find several definitions of fuzzy implications. In this paper we will use the following one equivalent to the definition introduced by Fodor and Roubens in [8]. For more details one can consult [1] or [16].

Definition 4. A function $I : [0,1]^2 \to [0,1]$ is called a fuzzy implication if it satisfies the following conditions:

(I1) I is decreasing in its first variable,
(I2) I is increasing in its second variable,
(I3) $I(1,0) = 0, I(0,0) = I(1,1) = 1$.

Next we list some important properties of fuzzy implications. For more information one can consult [10], [17] and [19].

Definition 5. A fuzzy implication $I : [0,1]^2 \to [0,1]$ satisfies:

(NP)   the left neutrality property (or is called left neutral) if
$$I(1,y) = y \text{ for all } y \in [0,1],$$
(EP)   the exchange principle if
$$I(x,I(y,z)) = I(y,I(x,z)) \text{ for all } x,y,z \in [0,1],$$
(IP)   the identity principle if
$$I(x,x) = 1 \text{ for all } x \in [0,1],$$
(OP)   the ordering property if
$$x \le y \quad \Leftrightarrow \quad I(x,y) = 1; \; x,y \in [0,1],$$
(CP)   the contrapositive symmetry with respect to a given fuzzy negation N if
$$I(x,y) = I(N(y),N(x)); \; x,y \in [0,1],$$
(LI)   the law of importation with respect to a t-norm T if
$$I(T(x,y),z) = I(x,I(y,z)); \; x,y,z \in [0,1],$$
(WLI)   the weak law of importation with respect to a commutative and increasing function $F : [0,1]^2 \to [0,1]$ if
$$I(F(x,y),z) = I(x,I(y,z)); \; x,y,z \in [0,1].$$

In papers [6, 7] some applications of fuzzy implications were studied and in this connection also some other properties of fuzzy implications were considered.

An important technical notion is that of pseudo-inverse.

Definition 6. (see e.g., [14]) Let $f : [0,1] \to [-\infty, \infty]$ be a decreasing function. Then $f^{(-1)} : [-\infty, \infty] \to [0,1]$ defined by

$$f^{(-1)}(x) = \sup\{z \in [0,1]; f(z) > x\},$$

is called the pseudo-inverse of f, with the convention $\sup \emptyset = 0$.

Definition 7. (see e.g., [8]) Let $g : [0,1] \to [-\infty, \infty]$ be an increasing function. The function $g^{(-1)}$ which is defined by

$$g^{(-1)}(x) = \sup\{z \in [0,1]; g(z) < x\},$$

is called the pseudo-inverse of g, with the convention $\sup \emptyset = 0$.

2 Generated Implications

Now we give conditions under which formula $I_{(f,g,h)}(x,y) = h(f(x) + g(y))$ leads to a fuzzy implication.

Theorem 1. *Let* $f : [0,1] \to [-\infty, \infty]$, $g : [0,1] \to [-\infty, \infty]$ *and* $h : [-\infty, \infty] \to [0,1]$ *be monotone functions such that* f *and* g *are real-valued in* $]0,1[$. *Let the following properties be satisfied*

(G1) $f(0) > f(1)$, $g(0) < g(1)$,
(G2) $h(f(1) + g(0)) = 0$,
(G3) $h(\min\{f(0) + g(0), f(1) + g(1)\}) = 1$ *with the convention* $\infty - \infty = \infty$.

Then $I_{(f,g,h)} : [0,1]^2 \to [0,1]$ *defined by*

$$I_{(f,g,h)}(x,y) = h(f(x) + g(y)), \tag{2}$$

is a fuzzy implication.

If a triple of monotone functions (f,g,h) fulfils all assumptions of Theorem 1, then we say that (f,g,h) is an *admissible triple of functions*.

Definition 8. Let (f,g,h) be an admissible triple of functions. Then we say that $I_{(f,g,h)}$ is an (f,g,h)-implication.

2.1 *Illustrative Examples and Relation to Known Types of Generated Fuzzy Implications*

First we show illustrative examples of (f,g,h)-implication.

Example 1. We will consider admissible triples of functions (f_i, g_i, h_i).

1. Assume continuous functions $f_1(x) = 1 - x$, $g_1(x) = 2x$ for $x \in [0,1]$ and

$$h_1(z) = \begin{cases} 1, & \text{if } z \geq 1, \\ 0, & \text{if } z \leq \frac{1}{2}, \\ 2z - 1, & \text{otherwise.} \end{cases}$$

for $z \in [-\infty, \infty]$. Then we get that (see Fig. 1)

$$I_{(f_1,g_1,h_1)}(x,y) = \begin{cases} 1, & \text{if } y \geq \frac{1}{2}x, \\ 0, & \text{if } y \leq \frac{1}{2}x - \frac{1}{4}, \\ 4y - 2x + 1, & \text{otherwise.} \end{cases}$$

2. Consider the following triple (f_2, g_2, h_2):

$$f_2(x) = \begin{cases} 1, & \text{if } x \in [0, \frac{1}{4}], \\ \frac{3}{4} - x, & \text{if } x \in]\frac{1}{4}, \frac{3}{4}[, \\ 0, & \text{if } x \in [\frac{3}{4}, 1], \end{cases} \qquad g_2(x) = \begin{cases} 2x - \frac{1}{2}, & \text{if } x \in [0, \frac{1}{4}], \\ 0, & \text{if } x \in]\frac{1}{4}, \frac{1}{2}[, \\ 2x - 1, & \text{if } x \in [\frac{1}{2}, 1], \end{cases}$$

$$h_2(z) = \begin{cases} 0, & \text{if } z < -\frac{1}{2}, \\ \frac{1}{2}z + \frac{1}{4}, & \text{if } z \in [-\frac{1}{2}, 0], \\ \frac{1}{2}z + \frac{3}{4}, & \text{if } z \in]0, \frac{1}{2}], \\ 1, & \text{if } z > \frac{1}{2}. \end{cases}$$

Then we get the following (f_2, g_2, h_2)-implication (see Fig. 2)

$$I_{(f_2,g_2,h_2)}(x,y) = \begin{cases} y, & \text{if } x \in [\frac{3}{4}, 1] \text{ and } y \in [0, \frac{1}{4}], \\ \frac{1}{4}, & \text{if } x \in [\frac{3}{4}, 1] \text{ and } y \in]\frac{1}{4}, \frac{1}{2}[, \\ y + \frac{1}{4}, & \text{if } x \in [\frac{3}{4}, 1] \text{ and } y \in [\frac{1}{2}, \frac{3}{4}[, \\ y - \frac{1}{2}x + \frac{3}{8}, & \text{if } x \in]\frac{1}{4}, \frac{3}{4}[\text{ and } y \leq \frac{1}{2}x - \frac{1}{8}, \\ y - \frac{1}{2}x + \frac{7}{8}, & \text{if } x \in]\frac{1}{4}, \frac{3}{4}[\text{ and } y \in]\frac{1}{2}x - \frac{1}{8}, \frac{1}{4}], \\ \frac{9}{8} - \frac{1}{2}x, & \text{if } x \in]\frac{1}{4}, \frac{3}{4}[\text{ and } y \in]\frac{1}{4}, \frac{1}{2}[, \\ y - \frac{1}{2}x + \frac{5}{8}, & \text{if } x \in]\frac{1}{4}, \frac{3}{4}[\text{ and } y \in [\frac{1}{2}, \frac{1}{2}x + \frac{3}{8}], \\ 1, & \text{otherwise.} \end{cases}$$

In the next example we illustrate relationships between (f, g, h)-implications and other known types of generated fuzzy implications. For an overview of generated fuzzy implications one can consult [12].

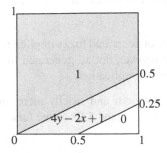

Fig. 1 (f_1, g_1, h_1)-implication

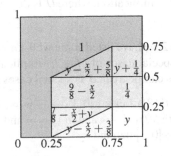

Fig. 2 (f_2, g_2, h_2)-implication

Example 2. In this example we will assume that functions f_i, g_i have domain equal to $[0,1]$ and h_i has range equal to $[0,1]$.

1. Let $f : [0,1] \to [0,-\infty]$ be a continuous strictly decreasing function with $f(1) = 0$. Set $g_1(y) = -f(y)$ and $h_1(z) = f^{(-1)}(-z)$. Then $I_{(f,g_1,h_1)}$ is the R_T-implication ([8]) where T is the generated t-norm with additive generator f ([14]).

2. Let $g : [0,1] \to [0,-\infty]$ be a continuous strictly increasing function with $g(0) = 0$. Further, let $N : [0,1] \to [0,1]$ be a fuzzy negation. Set $f_2(x) = g(N(x))$ and $h_2(x) = g^{(-1)}(x)$. Then $I_{(f_2,g,h_2)}$ is the (S,N)-implication ([1, 2]) where S is the generated t-conorm with additive generator g ([14]).

3. Let $f : [0,1] \to [0,\infty]$ be a continuous and strictly decreasing function with $f(1) = 0$. Set $f_3(x) = -\ln x$, $g_3(y) = -\ln(f(y))$, $h_3(z) = f^{(-1)}(e^{-z})$. Then

$$I_{(f_3,g_3,h_3)}(x,y) = f^{(-1)}(x \cdot f(y)),$$

what is Yager's f-implication ([23]).

4. Let $g : [0,1] \to [0,\infty]$ with $g(0) = 0$. Set $f_4(x) = -\ln x$, $g_4(y) = \ln(g(y))$, $h_4(z) = g^{(-1)}(e^z)$. Then

$$I_{(f_4,g_4,h_4)}(x,y) = g^{(-1)}\left(\frac{1}{x} \cdot g(y)\right),$$

what is Yager's g-implication ([23]).

5. Let $h : [0,1] \to [0,1]$ be a strictly decreasing and continuous function with $h(0) = 1$. Set $f_5(x) = -\ln x$, $g_5(y) = -\ln(h(y))$, $h_5(z) = h^{(-1)}(e^{-z})$. Then

$$I_{(f_5,g_5,h_5)}(x,y) = h^{(-1)}(x \cdot h(y)),$$

what is the h-generated implication introduced by Jayaram [13].

6. Let $h : [0,1] \to [-\infty,\infty]$ be a continuous and strictly increasing function with $h(0) = -\infty$ and $h(1) = \infty$. Let $N : [0,1] \to [0,1]$ be a fuzzy negation. Assume the convention $-\infty + \infty = \infty$. Set $f_6(x) = h(N(x))$. Then $I_{(f_6,h,h^{-1})}$ is the (U,N)-implication, where U is the representable uninorm whose generator is h ([4]).

7. Let $h : [0,1] \to [-\infty,\infty]$ be a continuous and strictly increasing function with $h(0) = -\infty$ and $h(1) = \infty$. Assume the convention $-\infty + \infty = \infty$. Then $I_{(-h,h,h^{-1})}$ is the R_U-implication, where U is the representable uninorm whose generator is h ([4]).

In Example 2 we have listed some well-known types of generated fuzzy implications which are special cases of (f,g,h)-implications. However, not all generated fuzzy implications can be treated as special cases of (f,g,h)-implications.

Example 3. ([18]) Let $h : [0,1] \to [-\infty,\infty]$ be a continuous and strictly increasing function with $h(0) = -\infty$ and $h(1) = \infty$ and $h(\tilde{e}) = 0$ for a fixed $\tilde{e}$. Then the function $I^h : [0,1]^2 \to [0,1]$ defined by

$$I^h(x,y) = \begin{cases} 1, & \text{if } x = 0, \\ h^{-1}(x \cdot h(y)), & \text{if } x > 0 \text{ and } y \le \tilde{e}, \\ h^{-1}\left(\frac{1}{x} \cdot h(y)\right), & \text{if } x > 0 \text{ and } y > \tilde{e}, \end{cases}$$

is a fuzzy implication and h is called h-generator.

This fuzzy implication cannot be constructed via a triple of monotone functions (f,g,h).

2.2 Properties of (f,g,h)-Implications

In this part we show some sufficient conditions in case of the Exchange Principle, the Contrapositive Symmetry and the (Weak) Law of Importation, and necessary and sufficient conditions for other properties presented in Definition 5 in order to hold for (f,g,h)-implications.

If a fuzzy implication is somehow derived from a uninorm with its neutral element $\tilde{e} \in\,]0,1[$, instead of properties (NP), (IP) and (OP) their modifications (NP$\tilde{e}$), (IP$\tilde{e}$) and (OP$\tilde{e}$) are studied. These modifications are defined as follows:

Definition 9. Let $I : [0,1]^2 \to [0,1]$ be a fuzzy implication and $\tilde{e} \in\,]0,1[$ be fixed. Then we say that I satisfies property

(NP$\tilde{e}$)   if for all $y \in [0,1]$ we have $I(\tilde{e},y) = y$,
(IP$\tilde{e}$)   if for all $x \in [0,1]$ we have $I(x,x) \ge \tilde{e}$,
(OP$\tilde{e}$)   if for all $x,y \in [0,1]$ we have $x \le y$ if and only if $I(x,y) \ge \tilde{e}$.

In admissible triple of functions we consider real-valued functions in $]0,1[$ (see Theorem 1). Since we assume $\tilde{e} \in\,]0,1[$, $f(\tilde{e}) > -\infty$. This property will be important in Theorem 3.

In some cases, e.g., in fuzzy control theory the value $\tilde{e} \in\,]0,1[$ may represent a level which makes working (fires) a criterion. This means that properties from Definition 9 can be interesting also without the fact whether an implication is derived from a uninorm or not. This is why we will study under which conditions (f,g,h)-implications fulfil also these properties.

Theorem 2. *Let (f,g,h) be an admissible triple of functions. Then the (f,g,h)-implication is left-neutral if and only if the following is fulfilled*

1. $f(1) > -\infty$,
2. g is strictly increasing,
3. $h(z) = g^{(-1)}(z - f(1))$.

Particularly, if $f(1) = 0$ then condition 3 in Theorem 2 says that $h = g^{(-1)}$.

Theorem 3. *Let (f,g,h) be an admissible triple of functions. Let $\tilde{e} \in\,]0,1[$ be fixed. Then the (f,g,h)-implication fulfils (NP$\tilde{e}$) if and only if the following holds*

1. g is strictly increasing,
2. $h(z) = g^{(-1)}(z - f(\tilde{e}))$.

For the Exchange Principle we have the following sufficient conditions.

Theorem 4. *Let* (f,g,h) *be an admissible triple of functions. Then the* (f,g,h)-*implication fulfils* (EP) *if one of the following conditions is fulfilled*

1. g *is continuous in* $[0,1]$, $g(0) \geq 0$, $f(1) \geq 0$ *and* $h = g^{(-1)}$.
2. g *is continuous in* $[0,1]$, $g(0) = -\infty$, $g(1) = \infty$, $h = g^{(-1)}$ *and* $f(0) = \infty$.
3. g *is bounded from below, continuous in* $]0,1]$, *discontinuous at* 0, $h = g^{(-1)}$ *and*

$$\lim_{x \to 0_+} f(x) \leq \lim_{y \to 0_+} g(y) - g(0).$$

4. $h\left(\lim_{x \to 1_-} g(x)\right) = 1$, g *is bounded from below and*

$$\lim_{x \to 1_-} f(x) \geq \lim_{y \to 1_-} g(y) - g(0).$$

5. $\lim_{x \to 0_+} g(x) > -\infty$, $\lim_{x \to 1_-} g(x) < \infty$ *and further*

$$h\left(\lim_{x \to 1_-} g(x)\right) = 1, \qquad h\left(\lim_{x \to 0_+} g(x)\right) = 0,$$
$$\lim_{x \to 0_+} f(x) \leq \lim_{y \to 0_+} g(y) - g(0), \quad \lim_{x \to 0_+} (g(1-x) - g(x)) \leq \lim_{x \to 1_-} f(x).$$

Remark 2. Condition 1 in Theorem 4 covers all generated (S,N)-implications from continuous t-conorms S and condition 2 covers all (U,N)- as well as R_U-implications from representable uninorms U. But both cases cover also other types of generated implications since g may have intervals of constantness.

Example 4. **(a)** Let us assume the following function $g_1 : [0,1] \to \mathbb{R}$:

$$g_1(x) = \begin{cases} -2+x, & \text{if } x \neq 1, \\ 0, & \text{if } x = 1. \end{cases}$$

Then the $(-g_1, g_1, g_1^{(-1)})$-implication fulfils condition 4 of Theorem 4, i.e., it satisfies (EP). This fuzzy implication, the well known Weber implication I_{WB}, is of the following form:

$$I_{\left(-g_1, g_1, g_1^{(-1)}\right)}(x,y) = \begin{cases} y, & \text{if } x = 1, \\ 1, & \text{otherwise.} \end{cases}$$

(b) Assume function $g_2 : [0,1] \to \mathbb{R}$:

$$g_2(x) = \begin{cases} -4, & \text{if } x = 0, \\ -2+x, & \text{if } x \in]0,1[, \\ 0, & \text{if } x = 1. \end{cases}$$

Then the $(-g_2, g_2, g_2^{(-1)})$-implication fulfils condition 5 of Theorem 4, i.e., it satisfies (EP). This fuzzy implication is of the following form:

$$I_{\left(-g_2, g_2, g_2^{(-1)}\right)}(x, y) = \begin{cases} y, & \text{if } x = 1, \\ 0, & \text{if } x \neq 0 \text{ and } y = 0, \\ 1, & \text{otherwise.} \end{cases}$$

Theorem 5. *Let (f, g, h) be an admissible triple of functions. Then the (f, g, h)-implication fulfils* (IP) *if and only if the following conditions are satisfied*

1. $\sup\{z \in \mathbb{R}; h(z) < 1\} \in \mathbb{R}$,
2. *for all $x \in {]0, 1[}$ we have $f(x) + g(x) \geq \sup\{z \in \mathbb{R}; h(z) < 1\}$.*

Remark 3. If we consider in Theorem 5 instead of a general function h the pseudo-inverse $g^{(-1)}$ then condition 1 turns into

$$\lim_{y \to 1_-} g(y) < \infty,$$

and condition 2 gives the following formula for all $x \in {]0, 1[}$:

$$f(x) \geq \lim_{y \to 1_-} g(y) - g(x). \tag{3}$$

This further implies that if $g(0) = 0$ and g is continuous in $[0, 1[$, then formula (3) is equivalent with the existence of a fuzzy negation N such that for all $x \in {]0, 1[}$ we have that $f(x) \geq g(N(x))$.

Theorem 6. *Let (f, g, h) be an admissible triple of functions. Let $\tilde{e} \in {]0, 1[}$ be fixed. Then the (f, g, h)-implication fulfils* (IP$\tilde{e}$) *if and only if for all $x \in {]0, 1[}$ there exists $z \in \mathbb{R}$ such that $h(z) \geq \tilde{e}$ and the following holds*

$$f(x) \geq z - g(x). \tag{4}$$

Remark 4. If (f, g, h) is an admissible triple of functions and $h(0) = \tilde{e} \in \mathbb{R}$ then formula (4) turns into

$$f(x) \geq -g(x).$$

Particularly, for $h(0) = \tilde{e}$ the $(-g, g, h)$-implication fulfils (IP$\tilde{e}$).

Theorem 7. *Let (f, g, h) be an admissible triple of functions. Then the (f, g, h)-implication fulfils* (OP) *if and only if the following holds*

1. *functions f and g are strictly monotone and bounded in $]0, 1[$,*
2. *functions f and g have the same continuity points in $]0, 1[$,*
3. $\sup\{z \in \mathbb{R}; h(z) < 1\} \in \mathbb{R}$,
4. *for all $x \in {]0, 1[}$ we have $f(x) + g(x) \geq \sup\{z \in \mathbb{R}; h(z) < 1\}$,*
5. *for all continuity points of f and g in $]0, 1[$ we have*

$$f(x) + g(x) = \sup\{z \in \mathbb{R}; h(z) < 1\}.$$

The last condition in Theorem 7 says that f is given uniquely at all continuity points by g and the value $\sup\{z \in \mathbb{R}; h(z) < 1\}$.

Theorem 8. *Let (f,g,h) be an admissible triple of functions. Let $\tilde{e} \in]0,1[$ be fixed. Then the (f,g,h)-implication fulfils $(OP\tilde{e})$ if and only if the following holds*

1. functions f and g are strictly monotone,
2. functions f and g have the same continuity points in $]0,1[$,
3. for all $x \in]0,1[$ we have $f(x) + g(x) \geq \sup\{z \in \mathbb{R}; h(z) < \tilde{e}\}$,
4. for all continuity points of f and g in $]0,1[$ we have

$$f(x) + g(x) = \sup\{z \in \mathbb{R}; h(z) < \tilde{e}\}. \tag{5}$$

Remark 5. If (f,g,h) is an admissible triple of functions and $h(0) = \tilde{e} \in \mathbb{R}$ then formula (5) turns into

$$f(x) = -g(x).$$

for all continuity points of g. Particularly, for $h(0) = \tilde{e}$ the $(-g,g,h)$-implication fulfils $(OP\tilde{e})$.

Now, we proceed to the Contrapositive Symmetry of (f,g,h)-implications. Concerning (CP) we have just the following sufficient conditions.

Theorem 9. *Let (f,g,h) be an admissible triple of functions and $N : [0,1] \to [0,1]$ be a fuzzy negation. Then the (f,g,h)-implication fulfils (CP) with respect to N if one of the following conditions is satisfied:*

- *N is strong and for all $x \in [0,1]$ we have $f(x) = g(N(x))$.*
- *There exists a set of disjoint sub-intervals of $[0,1]$ denoted by*

$$\mathscr{J} = \{J_1, J_2, J_3, \ldots, J_n, \ldots\}$$

such that each $J_i \in \mathscr{J}$ is an interval of constantness of g. Further, for all $x \in [0,1]$ we have $f(x) = g(N(x))$, and for N the following holds

1. $N(N(x)) = x$ for all $x \notin \bigcup_n J_n$,
2. arbitrary $J_n \in \mathscr{J}$ and $x \in J_n$ yield $N(N(x)) \in J_n$.

Finally, we finish our overview of properties of (f,g,h)-implications by the Law of Importation and its weak version. Also in this case we have just a sufficient condition.

Theorem 10. *Let $f : [0,1] \to [0,\infty]$ be an additive generator of a continuous t-norm T, $g : [0,1] \to [0,\infty]$ be a continuous increasing function and $h : [-\infty,\infty] \to [0,1]$ be equal to $h = g^{(-1)}$. Moreover, assume that (f,g,h) is an admissible triple of functions. Then the (f,g,h)-implication fulfils (LI) with respect to T.*

Theorem 11. *Let $U : [0,1]^2 \to [0,1[$ be a representable conjunctive uninorm and $f : [0,1] \to [-\infty,\infty]$ be its decreasing generator. Further, let one of the following conditions is fulfilled for functions $g : [0,1] \to [-\infty,\infty]$ and $h : [-\infty,\infty] \to [0,1]$:*

1. g is a continuous and strictly increasing function with $g(0) = -\infty$ and $g(1) = \infty$, and $h = g^{-1}$,
2. g is an increasing function with $g(0) = -\infty$ and $g(1) = \infty$, and $h(-\infty) = 0$, $h(\infty) = 1$ and $h(x) = c$ for $x \in \mathbb{R}$, where $c \in [0,1]$ is a constant.

Then (f,g,h) is an admissible triple of functions and the (f,g,h)-implication fulfils (LI) with respect to U.

In fact, the (f,g,h)-implication in Theorem 11, where g,h fulfil condition 1, is a $(\tilde{U},N)$-implication. $\tilde{U}$ is a representable uninorm whose generator is g and N is a strict fuzzy negation such that $f(N(x)) = g(x)$. Since the generator of U from Theorem 11 is f, we may have $U \neq \tilde{U}$.

As a generalisation of Theorems 10 and 11 we get the following.

Theorem 12. Let $f : [0,1] \to [-\infty,\infty]$ be a decreasing function, $g : [0,1] \to [-\infty,\infty]$ be an increasing function. Let they fulfil the following conditions:

- f is continuous in $]0,1[$ and g is continuous in $[0,1[$,
- $\lim_{x \to 0_+} f(x) + g(0) \geq \lim_{y \to 1_-} g(y)$,
- either $f(1) \geq 0$ and $g(0) \geq 0$, or

$$\lim_{x \to 1_-} (f(x) + g(x)) \leq \lim_{y \to 0_+} g(y) \quad and \quad f(1) + g(1) \geq \lim_{y \to 1_-} g(y).$$

Further, let $F : [0,1]^2 \to [0,1]$ be a commutative operation given by $F(x,y) = f^{(-1)}(f(x) + f(y))$ and $h : [-\infty,\infty] \to [0,1]$ be equal to $h = g^{(-1)}$. Assume that (f,g,h) is an admissible triple of functions. Then the (f,g,h)-implication fulfils (WLI) with respect to F.

3 Conclusion

In this paper we have studied properties of fuzzy implications generated via a triple of monotone functions (f,g,h). By means of Examples 2 and 3 we have shown relationships between (f,g,h)-implications and other known types of generated fuzzy implications.

Acknowledgements. Martin Kalina has been supported from the Science and Technology Assistance Agency under contract No. APVV-0073-10, and from the VEGA grant agency, grant numbers 1/0143/11 and 1/0297/11. Pavol Král' has been supported by the Agency of the Ministry of Education, Science, Research and Sport of the Slovak Republic for the Structural Funds of EU, under project ITMS: 26110230082.

References

[1] Baczyński, M., Jayaram, B.: Fuzzy implications. STUDFUZZ, vol. 231. Springer, Berlin (2008)
[2] Baczyński, M., Jayaram, B. (S,N)- and R-implications: A state-of-the-art survey. Fuzzy Sets and Systems 159(14), 1836–1859 (2008)

[3] Baczyński, M., Jayaram, B.: QL-implications: Some properties and intersections. Fuzzy Sets and Systems 161(2), 158–188 (2010)

[4] De Baets, B., Fodor, J.: Residual operators of uninorms. Soft Computing 3, 89–100 (1999)

[5] Biba, V., Hliněná, D.: Generated fuzzy implications and known classes of implications. Acta Univ. M. Belii, Ser. Math. 16, 25–34 (2010)

[6] Bustince, H., Burillo, P., Soria, F.: Automorphisms, negations and implication operators. Fuzzy Sets and Systems 134(2), 209–229 (2003)

[7] Bustince, H., Fernandez, J., Sanz, J., Baczyński, M., Mesiar, R.: Construction of strong equality index from implication operators. Fuzzy Sets and Systems 211, 15–33 (2013)

[8] Fodor, J., Roubens, M.: Fuzzy preference modelling and multicriteria decision support. Kluwer Academic Publishers, Dordrecht (1994)

[9] Fodor, J., Yager, R.R., Rybalov, A.: Structure of uninorms. International Journal of Uncertainty, Fuzziness and Knowledge-based Systems 5, 411–422 (1997)

[10] Hájek, P.: Mathematics of Fuzzy Logic. Kluwer, Dordrecht (1998)

[11] Hliněná, D., Kalina, M., Král', P.: Generated implications revisited. In: Greco, S., Bouchon-Meunier, B., Coletti, G., Fedrizzi, M., Matarazzo, B., Yager, R.R., et al. (eds.) IPMU 2012, Part II. CCIS, vol. 298, pp. 345–354. Springer, Heidelberg (2012)

[12] Hliněná, D., Kalina, M., Král', P.: Implication functions generated using functions of one variable. In: Baczynski, M., Beliakov, G., Bustince, H., Pradera, A. (eds.) Adv. in Fuzzy Implication Functions. STUDFUZZ, vol. 300, pp. 125–153. Springer, Heidelberg (2013)

[13] Jayaram, B.: Contrapositive symmetrisation of fuzzy implications-revisited. Fuzzy Sets and Systems 157, 2291–2310 (2006)

[14] Klement, E.P., Mesiar, R., Pap, E.: Triangular Norms, 1st edn. Springer (2000)

[15] Komorníková, M.: Aggregation operators and additive generators. Int. J. Fuzziness, Uncertainty, Fuzziness and Knowledge-Based Systems 0.2, 205–215 (2001)

[16] Mas, M., Monserrat, J., Torrens, M., Trillas, E.: A survey on fuzzy implication functions. IEEE T. Fuzzy Systems 15(6), 1107–1121 (2007)

[17] Massanet, S., Torrens, J.: The law of importation versus the exchange principle on fuzzy implications. Fuzzy Sets and Systems 168(1), 47–69 (2011)

[18] Massanet, S., Torrens, J.: On a new class of fuzzy implications: h-implications and generalizations. Information Sciences 181, 2111–2127 (2011)

[19] Novák, V., Perfilieva, I., Močkoř, J.: Mathematical Principles of Fuzzy Logic. Kluwer, Boston (1999)

[20] Ouyang, Y.: On fuzzy implications determined by aggregation operators. Information Sciences 193, 153–162 (2012)

[21] Schweizer, B., Sklar, A.: Probabilistic Metric Spaces. North Holland, New York (1983)

[22] Smutná, D.: On many valued conjunctions and implications. Journal of Electrical Engineering 50, 8–10 (1999)

[23] Yager, R.R.: On some new classes of implication operators and their role in approximate reasoning. Information Sciences 167(1-4), 193–216 (2004)

[24] Yager, R.R., Rybalov, A.: Uninorm aggregation operators. Fuzzy Sets and Systems 80, 111–120 (1996)

A Generalization of a Characterization Theorem of Restricted Equivalence Functions

Eduardo Palmeira, Benjamín Bedregal, and Humberto Bustince

Abstract. Fodor and Roubens' equivalence functions (for short EF) are mapping normally used for making a comparison between images by means it can be used for measuring the similarity of images. So, having a suitable way to construct these functions is very important. In these sense, we present in this work a characterization theorem for restricted equivalence functions (a particular case of EF) using aggregation functions which is able to describe them from implications and vice-versa. We also present similar results for restricted dissimilarity functions and normal $E_{e,N}$-functions.

1 Introdution

On image processing, a very studied problem is providing a suitable measure for making a global comparison of images [13, 14, 20, 23]. In this framework, Bustince et al. introduced in [7] the notion of restricted equivalence functions on $[0, 1]$ (for short REF) as a particular case of equivalence functions defined by J. Fodor and M. Roubens in [15]. In [7] it is also presented a theorem that characterizes REF via implications, i.e. a way to construct these kind of functions using fuzzy implications. Also, Bustince et al. in [8] provided the notions of restricted dissimilarity functions and normal E_N-functions other ways to make comparison between images.

Eduardo Palmeira
Universidade Estadual de Santa Cruz, Ilhéus, Brazil
e-mail: espalmeira@uesc.br

Benjamín Bedregal
Universidade Federal do Rio Grande do Norte, Natal, Brazil
e-mail: bedregal@dimap.ufrn.br

Humberto Bustince
Universidad Publica de Navarra, Pamplona, Spain
e-mail: bustince@unavarra.es

H. Bustince et al. (eds.), *Aggregation Functions in Theory and in Practise*,
Advances in Intelligent Systems and Computing 228,
DOI: 10.1007/978-3-642-39165-1_43, © Springer-Verlag Berlin Heidelberg 2013

The main objectives of this work are: (1) provide a generalization of Theorem 7 of [7] for restricted equivalence functions using aggregation functions, (2) restricted dissimilarity functions and normal $E_{e,N}$-functions.

This paper is organized as follows: In Section 2, we discuss about some concepts related to fuzzy sets. In Subsection 2.1 we present the notion of fuzzy negations, fuzzy implications in Subsection 2.2 and aggregation functions in Subsection 2.3. Also in this framework, we present the notion of restricted equivalence functions on $[0, 1]$ in Subsection 2.4, restricted dissimilarity functions in Subsection 2.5 and normal E_N-functions in Subsection 2.6. Section 3 is divided in three subsections (3.1, 3.2 and 33) where we present the characterization theorem for restricted equivalence functions, restricted dissimilarity functions and normal $E_{e,N}$-functions using aggregation functions and some other related results.

2 Preliminaries

In this section we recall some known definitions and results related to fuzzy logic which are important for us as a theoretical base. In the following two subsections we discuss about fuzzy negations and implications valued on $[0, 1]$. Also in the framework of fuzzy sets, we present the concept of aggregation functions in Subsection 2.3. In the following subsections we recall the notion of restricted equivalence functions, restricted dissimilarity functions and normal $E_{e,N}$-functions as proposed in [7, 8]. Some elementary concepts are used here without being defined but for a suitable formalization of them we recommend [11, 17, 18]

2.1 Fuzzy Negations

There are several ways to fuzzy negations as a generalization of classical concept [17]. Here, we present that we believe be the most general possible definition of it.

Definition 1. A mapping $N : [0, 1] \to [0, 1]$ is a *fuzzy negation* on $[0, 1]$, if the following properties are satisfied for each $x, y \in [0, 1]$:
(N1) $N(0) = 1$ and $N(1) = 0$ and
(N2) If $x \leqslant y$ then $N(y) \leqslant N(x)$.
Moreover, negation N is called *strong* if it also satisfies the involution property, i.e.
(N3) $N(N(x)) = x$ for each $x \in [0, 1]$.
N is *strict* if it is continuous and satisfies the property:
(N4) $N(x) < N(y)$ whenever $y < x$,
and N is called *frontier* if satisfies the property:
(N5) $N(x) \in \{0, 1\}$ if and only if $x = 0$ or $x = 1$.

Some important and very known facts from the literature that we would like to point out here are the following:

- Every strong negation is strict but the converse is not true in general. A contra-example for this is negation $N(x) = 1 - x^2$;

- $N = N^{-1}$ whereas N is a strong fuzzy negation;
- Every point $e \in [0,1]$ such that $N(e) = e$ is called an equilibrium point of negation N.

Remark 1. Klir and Yuan have proved in [19] that all fuzzy negation has at most an equilibrium point and hence if a fuzzy negation N has an equilibrium point then it is unique. Also, it is important to point out that every continuous negation has an equilibrium point. In [4] one can see an example showing that there is a negation with no equilibrium point, namely

$$N_\perp(x) = \begin{cases} 0 & if \ x > 0 \\ 1 & if \ x = 0 \end{cases}$$

Proposition 1. *[4] Let N_1 and N_2 be fuzzy negations such that $N_1 \leqslant N_2$. If e_1 and e_2 are equilibrium points of N_1 and N_2 respectively then $e_1 \leqslant e_2$.*

2.2 Fuzzy Implications

There are several ways to define fuzzy implication as one can see in the literature [1, 2, 5, 9, 15, 21, 24]. We consider here the notion of implication proposed in [15] by Fodor and Roubens.

Definition 2. A function $I : [0,1] \times [0,1] \longrightarrow [0,1]$ is a fuzzy implication if for each $x, y, z \in [0,1]$ the following properties hold:

1. *First place antitonicity (FPA): if $x \leqslant y$ then $I(y,z) \leqslant I(x,z)$;*
2. *Second place isotonicity (SPI): if $y \leqslant z$ then $I(x,y) \leqslant I(x,z)$;*
3. *Left boundary condition (LB): $I(0,y) = 1$;*
4. *Right boundary condition (RB) $I(x,1) = 1$;*
5. *Corner condition 3 (CC3): $I(1,0) = 0$.*

Example 1. Functions $I_\perp, I_\top : [0,1] \times [0,1] \to [0,1]$ given by

$$I_\perp(x,y) = \begin{cases} 1, & if \ x = 0 \ or \ y = 1; \\ 0, & otherwise. \end{cases}$$

and

$$I_\top(x,y) = \begin{cases} 0, & if \ x = 1 \ and \ y = 0; \\ 1, & otherwise. \end{cases}$$

for all $x, y \in [0,1]$ are fuzzy implications.

Consider also the following properties of an implication I:
(CC1): $I(0,0) = 1$ (corner condition 1);
(CC2): $I(1,1) = 1$ (corner condition 2);
(CC4) $I(0,1) = 1$ (corner condition 4);
(NP) $I(1,y) = y$ for each $y \in [0,1]$ (left neutrality principle);
(EP) $I(x,I(y,z)) = I(y,I(x,z))$ for all $x, y, z \in [0,1]$ (exchange principle);

(IP) $I(x,x) = 1$ for each $x \in [0,1]$ (identity principle);
(OP) $I(x,y) = 1$ if and only if $x \leqslant y$ (ordering property);
(IBL) $I(x,I(x,y)) = I(x,y)$ for all $x,y,z \in [0,1]$ (iterative Boolean law);
(CP) $I(x,y) = I(N(y),N(x))$ for each $x,y \in [0,1]$ with N a fuzzy negation (contraposition law);
(P) $I(x,y) = 0$ if and only if $x = 1$ and $y = 0$ (Positivety);
(SN) $I(x,0) = N(x)$ is a strong negation;
(SI) $I(x,y) \geqslant y$ for all $x,y \in [0,1]$;
(C) I is a continuous function (continuity).

Lemma 1. *[9] Let $I : [0,1] \times [0,1] \to [0,1]$ be a function. Then*

(i) if I satisfies (SPI) and (CC1) then I satisfies (LB);
(ii) if I satisfies (FPA) and (CC2) then I satisfies (RB);
(iii) if I satisfies (FPA) and (CP) then I satisfies (SPI);
(iv) if I satisfies (SPI) and (CP) then I satisfies (FPA);
(v) if I satisfies (LB) and (CP) then I satisfies (RB);
(vi) if I satisfies (RB) and (CP) then I satisfies (LB);
(vii) if I satisfies (NP) and (CP) then I satisfies (SN);
(viii) if I satisfies (SN) and (CP) then I satisfies (NP);
(ix) if I satisfies (SPI) and (SN) then I satisfies (LB);
(x) if I satisfies (FPA) and (NP) then I satisfies (SI);
(xi) if I satisfies (EP) and (SN) then I satisfies (CP);
(xii) if I satisfies (FPA), (NP) and (CP) then I satisfies (SPI), (LB), (RB), (CC3), (SN) and (SI);
(xiii) if I satisfies (SPI), (EP) and (OP) then I satisfies (FPA), (LB), (RB), (CC3), (NP), (SI) and (IP).

Lemma 2. *[9] Let $I : [0,1] \times [0,1] \to [0,1]$ be any function satisfying at least one of following items:*

(i) If I satisfies (SPI), (NP) and (SN), or
(ii) If I satisfies (FPA), (NP) and (SN), or
(iii) If I satisfies (RB), (EP), (SN) and (IP), or
(iv) If I satisfies (NP), (EP), (SN) and $I(x,x) = I(0,x)$ for all $x \in [0,1]$,

then I satisfies (P).

2.3 Aggregation Functions

As fuzzy implications, aggregation functions have more than one different way to be defined due to there is no consensus about the axioms that must be demanded from such operators [6, 12, 16]. We consider here the definition that does not go against the arithmetic mean.

Definition 3. A function $M : \bigcup_{n \in \mathbb{N}} [0,1]^n \to [0,1]$ for some $n \geqslant 2$ is called an aggregation function if it satisfies the following properties:

(A1) $M(x_1,\ldots,x_n) = 0$ if and only if $x_i = 0$ for all $i \in \{1,2,\ldots,n\}$;

(A2) $M(x_1,\ldots,x_n) = 1$ if and only if $x_i = 1$ for all $i \in \{1,2,\ldots,n\}$;

(A3) For any pairs $(x_1,\ldots,x_n)$ and $(y_1,\ldots,y_n)$ of elements of $[0,1]^n$, if $x_i \leqslant y_i$ for all $i \in \{1,2,\ldots,n\}$ then $M(x_1,\ldots,x_n) \leqslant M(y_1,\ldots,y_n)$;

(A4) $M(x_1,\ldots,x_n) \leqslant M(x_{p(1)},\ldots,x_{p(n)})$ for any permutation p on $\{1,2,\ldots,n\}$.

Moreover, if M satisfies

(A5) $M(x_1,\ldots,x_n) < M(y_1,\ldots,y_n)$ whenever $x_i < y_i$ for all $i \in \{1,2,\ldots,n\}$.

For $n = 2$, if M satisfies

(A6) $M(x,y) = M(y,x)$ it is called commutative.

An element $e \in [0,1]$ is a neutral element of M if it satisfies

(A7) For all $x \in [0,1]$

$$M(x,e) = M(e,x) = x.$$

Example 2. If $P = [0,1]^2$ then functions $R,L : P^n \to P$ defined by

$$R((x_1,y_1),\ldots,(x_n,y_n)) = (\min(x_1,\ldots,x_n),\min(y_1,\ldots,y_n))$$

and

$$L((x_1,y_1),\ldots,(x_n,y_n)) = (\max(x_1,\ldots,x_n),\max(y_1,\ldots,y_n))$$

are n-ary aggregation functions on P.

2.4 Restricted Equivalence Functions

In image processing equivalences are usually considered to make a comparison between two images. One of the most known equivalences was proposed by Fodor and Roubens in [15].

Definition 4. A function $EF : [0,1]^2 \to [0,1]$ is called an equivalence if it satisfies the following conditions:

1. $EF(x,y) = EF(y,x)$ for all $x,y \in [0,1]$;
2. $EF(0,1) = EF(1,0) = 0$;
3. $EF(x,x) = 1$ for all $x \in [0,1]$;
4. If $x \leqslant y \leqslant z \leqslant t$ then $EF(x,t) \leqslant EF(y,z)$.

But, equivalence functions as in Definition 4 do not allow us to ensure that only $(0,1)$ and $(1,0)$ are assigned to 0, i.e. could exist a pair $x,y \in [0,1] - \{0,1\}$ such that $EF(x,y) = 0$. This is a disadvantage for comparing images since it can not ensured that just an image (in black and white) and its negative must be measured as zero. In order to eliminate this problem and others, Bustince et al. in [7] redefined (on $[0,1]$) Fodor and Roubens' equivalence function adding some constraints as follows.

Definition 5. Let N be a strong negation on $[0,1]$. A function $REF : [0,1]^2 \to [0,1]$ is called a restricted equivalence function with respect to N if it satisfies, for all $x,y,z \in [0,1]$, the following conditions:

(L1) $REF(x,y) = REF(y,x)$;
(L2) $REF(x,y) = 1$ if and only if $x = y$;
(L3) $REF(x,y) = 0$ if and only if either $x = 1$ and $y = 0$ or $x = 0$ and $y = 1$;
(L4) $REF(x,y) = REF(N(x),N(y))$;
(L5) if $x \leqslant y \leqslant z$ then $REF(x,z) \leqslant REF(x,y)$.

Example 3. [7] Considering negation $N : [0,1] \to [0,1]$ given by $N(x) = 1 - x$, then function $REF(x,y) = 1 - |x - y|$ is a restricted equivalence function.

Theorem 1. *[7] Every restricted equivalence function is an equivalence function in the sense of Definition 4.*

The following theorem describe a characterization of restricted equivalence functions by implications proved by H. Bustince et al. in [7].

Theorem 2. *A function $REF : [0,1]^2 \to [0,1]$ is a restricted equivalence function (with respect to a strong negation N) if and only if there exists a function $I : [0,1]^2 \to [0,1]$ satisfying (FPA), (OP), (CP) and (P) such that*

$$REF(x,y) = I(x,y) \wedge I(y,x) \tag{1}$$

2.5 Restricted Dissimilarity Functions

In literature, there are some ways to define functions to measure similarity or dissimilarity between two images. Bustince et al. in [10] provided the notion of restricted dissimilarity functions based on error metric for images proposed by [3].

Definition 6. A function $d : [0,1]^2 \to [0,1]$ is called a restricted dissimilarity function (for short RDF) if it satisfies, for all $x,y,z \in [0,1]$, the following conditions:
(D1) $d(x,y) = d(y,x)$;
(D2) $d(x,y) = 1$ if and only if either $x = 1$ and $y = 0$ or $x = 0$ and $y = 1$;
(D3) $d(x,y) = 0$ if and only if $x = y$;
(D4) if $x \leqslant y \leqslant z$ then $d(x,y) \leqslant d(x,z)$ and $d(y,z) \leqslant d(x,z)$.

Theorem 3. *[8] Let $REF : [0,1]^2 \to [0,1]$ be a restricted equivalence function with respect to negation N. If N' a strong negation (not necessarily equal to N) then, the function defined by*

$$d(x,y) = N'(REF(x,y)) \quad for \ all \ \ x,y \in [0,1] \tag{2}$$

is a restricted dissimilarity function.

Corollary 1. *Under the same conditions of Theorem 3, it holds that*

$$d(x,y) = d(N(x),N(y)) \quad for \ all \ \ x,y \in [0,1]. \tag{3}$$

Proof. Straightforward from Theorem 3 and (L4). $\qquad\square$

2.6   Normal $E_{e,N}$-functions

Definition 7. [8] Let e be an equilibrium point of a strong negation N. A function $E_{e,N} : [0,1] \to [0,1]$ is called a normal $E_{e,N}$-function associated to N if it satisfies the following conditions:

1. $E_{e,N}(x) = 1$ if and only if $x = e$;
2. $E_{e,N}(x) = 0$ if and only if $x = 0$ or $x = 1$;
3. For all $x, y \in [0,1]$ such that either $e \leqslant x \leqslant y$ or $y \leqslant x \leqslant e$ it follows $E_{e,N}(y) \leqslant E_{e,N}(x)$;
4. $E_{e,N}(x) = E_{e,N}(N(x))$ for all $x \in [0,1]$.

Theorem 4. *[7] Let N be a strong negation and e be an equilibrium point of N. If $REF : [0,1]^2 \to [0,1]$ is a restricted equivalence function then the function given by*

$$E_{e,N}(x) = REF(x, N(x)) \tag{4}$$

for all $x \in [0,1]$ is a normal $E_{e,N}$-function.

Corollary 2. *If d is a restricted dissimilarity function then the function given by $E_{e,N}(x) = N(d(x, N(x)))$, for all $x, y \in [0,1]$, is a normal $E_{e,N}$-function.*

3   Characterization Theorem

In the framework of images defining a precise way to compare two images is a very studied issue. As we have seen in Section 2.4 restricted equivalence functions play an important role in this issue making a suitable measure on similarity of images. But it is not so easy to define these functions and hence having a general way to construct them from other more known function constitute a very important tool for providing them.

In this sense, Bustince et al. in [7] proposed a characterization theorem where they describe a way to construct restricted equivalence functions from implications and vice-versa using operator $\wedge$ of $[0,1]$ (see Theorem 2). Here, we present a generalization of this theorem using aggregation functions instead of $\wedge$. Also, we present generalizations of characterization theorems for restricted dissimilarity functions and normal $E_{e,N}$-functions.

3.1   For Restricted Equivalence Functions

Proposition 2. *Let $M : [0,1]^2 \to [0,1]$ be a function satisfying (A1), (A2), (A6) and (A7). Then, a function $REF : [0,1]^2 \to [0,1]$ is a restricted equivalence function (with respect to a strong negation N) if and only if there exists a function $I : [0,1]^2 \to [0,1]$ satisfying (FPA), (OP), (CP) and (P) such that*

$$REF(x,y) = M(I(x,y), I(y,x)). \tag{5}$$

Proof. (Necessity)

Suppose that *REF* is a restricted equivalence function and define the function I :
$[0,1]^2 \rightarrow [0,1]$ by

$$I(x,y) = \begin{cases} 1 \, , \, x \leq y \\ REF(x,y) \, , \, x > y \end{cases}$$

First, we will prove that $REF(x,y) = M(I(x,y), I(y,x))$ for all $x,y \in [0,1]$. In this
case, we have three possibilities:

1. If $x = y$ then $I(x,y) = I(y,x) = 1$ by (OP). Thus $REF(x,y) = 1 = M(1,1)$
   $= M(I(x,y), I(y,x))$.

2. Now, suppose that $x < y$. Then by (OP) $I(x,y) = 1$ and by $(A7)$ $M(I(x,y),$
   $I(y,x)) = I(y,x)$. Hence $REF(x,y) = REF(y,x) = I(y,x) = M(I(x,y), I(y,x))$.

3. Finally, if $x > y$ then $I(y,x) = 1$. Again by $(A7)$ we have $M(I(x,y), I(y,x)) =$
   $I(x,y)$ that implies $REF(x,y) = M(I(x,y), I(y,x))$.

It remains to prove that properties (FPA), (OP), (CP) and (P) hold.

- (FPA)
  Let $x,y,z \in [0,1]$ such that $x \leq z$. Again, we have three possibilities:
  (i) $y < x \leq z$
  In this case, $REF(y,x) \geq REF(y,z)$ by $(L5)$ and hence $REF(x,y) \geq REF(z,y)$
  by (1). Since $I(x,y) = REF(x,y)$ and $I(z,y) = REF(z,y)$ it follows that
  $I(x,y) \geq I(z,y)$.
  (ii) $x \leq y < z$
  According to definition of I we have that $I(x,y) = 1$ and $I(z,y) = REF(z,y)$.
  Hence $I(x,y) \geq I(z,y)$.
  (iii) $x \leq z \leq y$
  Again by definition of I it follows that $I(x,y) = 1$ and $I(z,y) = 1$. Thus
  $I(x,y) = I(z,y)$.
  Therefore, by (i), (ii) and (iii) it can be concluded that $I(x,y) \geq I(z,y)$ with
  $x \leq z$.

- (OP)
  If $x \leq y$ then $I(x,y) = 1$ by definition of I.
  Reciprocally, suppose that $I(x,y) = 1$. Thus either $x \leq y$ or $x > y$ and $REF(x,y)$
  $= 1$. But, in the second case we have a contradiction since $REF(x,y) = 1$ if
  and only if $x = y$. Hence $x \leq y$.

- (CP)
  By definition of I we have

$$I(N(y), N(x)) = \begin{cases} 1, \, N(y) \leq N(x) \\ REF(N(y), N(x)), \, N(y) > N(x) \end{cases}$$

$$= \begin{cases} 1, & x \le y \\ REF(x,y), & x > y \end{cases}$$

$$= I(x,y)$$

- (P)

  Suppose that $x = 1$ and $y = 0$. Thus $x > y$ and hence $I(x,y) = REF(x,y) = 0$. Reciprocally if $I(x,y) = 0$ then $x > y$ and $REF(x,y) = 0$ that implies $x = 1$ and $y = 0$.

(Sufficiency) Straightforward from Theorem 7 in [7]   □

3.2 For Restricted Dissimilarity Functions

Lemma 3. *Let* $G : [0,1]^2 \to [0,1]$ *be a function satisfying for all* $x,y \in [0,1]$
(G1) $G(x,y) = G(y,x)$;
(G2) $G(x,0) = x$;
(G3) $G(x,1) = 1$.
Thus, if N *is a strong negation then the function* $M : [0,1]^2 \to [0,1]$ *given by* $M(x,y) = N(G(N(x),N(y)))$ *satisfies* (A2), (A6) *and* (A7). *Moreover, the following equation holds:*

$$N(M(x,y)) = M(N(x),N(y)). \tag{6}$$

Proof. It is easy to see that property (A2) and equation (6) hold (since N is involutive). It remains to prove (A6) and (A7). In other words, we shall prove that 1 is a neutral element and 0 is a annihilator of M. Thus, for all $x \in [0,1]$, we have

$$M(x,1) = N(G(N(x),N(1))) = N(G(N(x),0)) = N(N(x)) = x$$

and

$$M(x,0) = N(G(N(x),N(0))) = N(G(N(x),1)) = N(1) = 0.$$

□

The following theorem is a generalization of characterization of restricted dissimilarity proposed in [8] (see Theorem 6).

Theorem 5. *Let* $G : [0,1]^2 \to [0,1]$ *be a function satisfying* (G1), (G2) *and* (G3). *Given a function* $d : [0,1]^2 \to [0,1]$, *if there exists a function* $I : [0,1]^2 \to [0,1]$ *for which the properties* (FPA), (OP), (CP) *and* (P) *hold and such that* $d(x,y) = G(N(I(x,y)),N(I(y,x)))$ *for all* $x,y \in [0,1]$, *then* d *is a restricted dissimilarity function satisfying equation* (3).

Proof. Given G satisfying properties (G1), (G2) and (G3) by Lemma 3 the function $M : [0,1]^2 \to [0,1]$ defined by $M(x,y) = N(G(N(x),N(y)))$ for all $x,y \in [0,1]$ is such that (A2), (A6) and (A7) hold. Thus, since I is a function satisfying (FPA), (OP), (CP) and (P) then by Theorem 2 a function $REF : [0,1]^2 \to [0,1]$ given by $REF(x,y) = M(I(x,y),I(y,x))$ for all $x,y \in [0,1]$ is a restricted equivalence function. Moreover,

$$REF(x,y) = M(I(x,y),I(y,x))$$
$$= N(G(N(I(x,y)),N(I(y,x))))$$
$$= N(d(x,y)).$$

Since N is involutive then $d(x,y) = N(REF(x,y))$ and hence by Theorem 3 we can affirm that d is a restricted dissimilarity function. Also, it is clear that d satisfies equation (3). □

3.3 For Normal $E_{e,N}$-Functions

Theorem 6. *Let $M : [0,1]^2 \to [0,1]$ a function satisfying (A2), (A6) and (A7). If $I : [0,1]^2 \to [0,1]$ satisfies (FPA), (OP), (CP) and (P) then*

$$E_{e,N}(x) = M(I(x,N(x)),I(N(x),x))$$

for all $x \in [0,1]$ is a normal $E_{e,N}$-function.

Proof. By Theorem 2 we know that $REF(x,y) = M(I(x,y),I(y,x))$ for all $x,y \in [0,1]$ is a restricted equivalence function. Thus

$$E_{e,N}(x) = REF(x,N(x)) = M(I(x,N(x)),I(N(x),x))$$

is a normal $E_{e,N}$-function by Theorem 4. □

Corollary 3. *Let e be an equilibrium point of the strong negation N. Under the same conditions of Theorem 6, we have $E_{e,N}(x) = I(x,N(x))$ for all $x \in [0,1]$ such that $e \leqslant x$.*

Proof. If $e \leqslant x$ then $N(x) \leqslant N(e)$ and hence $N(x) \leqslant e \leqslant x$ since $N(e) = e$. Thus by (OP) we have that $I(N(x),x) = 1$. Therefore $E_{e,N}(x) = M(I(x,N(x)),I(N(x),x)) = M(I(x,N(x)),1) = I(x,N(x))$ by Theorem 6 and (A6). □

4 Final Remarks

In this paper we have presented a generalization of characterization theorem (Theorem 7 of [7]) for restricted equivalence functions using aggregation functions and some related results for restricted dissimilarity functions and normal $E_{e,N}$-functions. The main advantage of this is that we can built REF's in a much more general way since different aggregation functions can be considered.

As a consequence of results presented here we have worked on a version of Theorem 2 for restricted equivalence functions on bounded lattices (also for Theorems 5 and 6) and the resulting paper will be submitted to an important journal in this area soon (see [22]).

For further works we have the interest of studying generalized versions of results presented in [10] about restricted dissimilarity functions and penalty functions.

References

[1] Baczyński, M.: Residual implications revised. notes on the Smets-Magrez theorem. Fuzzy Sets and Systems 145(2), 267–277 (2004)

[2] Baczyński, M., Jayaram, B.: Fuzzy Implications. STUDFUZZ, vol. 231. Springer, Heidelberg (2008)

[3] Baddeley, A.J.: An error metric for images. Robust Computer Vision, pp. 59–78. Wichmann, Karlsruhe (1992)

[4] Bedregal, B.: On interval fuzzy negations. Fuzzy Sets and Systems 161, 2290–2313 (2010)

[5] Bedregal, B., Beliakov, G., Bustince, H., Fernández, J., Pradera, A., Reiser, R.: (S,N)-Implications on Bounded Lattices. In: Baczynski, M., Beliakov, G., Bustince, H., Pradera, A. (eds.) Adv. in Fuzzy Implication Functions. STUDFUZZ, vol. 300, pp. 101–124. Springer, Heidelberg (2013)

[6] Beliakov, G.: Construction of aggregation functions from data using linear programming. Fuzzy Sets and Systems 160(1), 65–75 (2009)

[7] Bustince, H., Barrenechea, E., Pagola, M.: Restricted equivalence functions. Fuzzy Sets and Systems 157, 2333–2346 (2006)

[8] Bustince, H., Barrenechea, E., Pagola, M.: Relationship between restricted dissimilarity functions, restricted equivalence functions and normal E_N-functions: Image thresholding invariant. Pattern Recognition Letters 29, 525–536 (2008)

[9] Bustince, H., Burillo, P., Soria, F.: Automorphisms, negations and implication operators. Fuzzy Sets and Systems 134(2), 209–229 (2003)

[10] Bustince, H., Fernadez, J., Mesiar, R., Pradera, A., Beliakov, G.: Restricted dissimilarity functions and penalty functions. In: EUSFLAT-LFA 2011, Aix-les-Bains, France (2011)

[11] Burris, S., Sankappanavar, H.P.: A Course in Universal Algebra. The Millennium Edition, New York (2005)

[12] Calvo, T., Mayor, G., Mesiar, R.: Aggregation Operators: News trends and applications. STUDFUZZ, vol. 97. Physica-Verlag, Heidelberg (2002)

[13] Chaira, T., Ray, A.K.: Region extraction using fuzzy similarity measures. J. Fuzzy Math. 11(3), 601–607 (2003)

[14] Chaira, T., Ray, A.K.: Fuzzy measures for color image retrieval. Fuzzy Sets and Systems 150, 545–560 (2005)

[15] Fodor, J., Roubens, M.: Fuzzy Preference Modelling and Multicriteria Decision Support. Kluwer Academic Publisher, Dordrecht (1994)

[16] Grabisch, M., Marichal, J., Mesiar, R., Pap, E.: Aggregation functions: Means. Information Sciences 181(1), 1–22 (2011)

[17] Klement, E.P., Mesiar, R.: Logical, Algebraic, Analytic, and Probabilistic Aspects of Triangular Norms. Elsevier B.V., The Netherlands (2005)

[18] Klement, E.P., Mesiar, R., Pap, E.: Triangular Norms. Kluwer Academic Publishers, Dordrecht (2000)

[19] Klir, G.J., Yuan, B.: Fuzzy Sets and Fuzzy Logics: Theory and Applications. Prentice-Halls PTR, Upper Saddle River (2005)

[20] Lopez-Molina, C., De Baets, B., Bustince, H.: Generating fuzzy edge images from gradient magnitudes. Computer Vision and Image Understanding 115(11), 1571–1580 (2011)

[21] Mas, M., Monserrat, M., Torrens, J., Trillas, E.: A survey on fuzzy implication functions. IEEE Trans. on Fuzzy Systems 15(6), 1107–1121 (2007)

[22] Palmeira, E.S., Bedregal, B., Bustince, H., De Baets, B., Jurio, A.: Restricted equiva-
     lence functions on bounded lattices. Information Sciences (submitted, 2013)
[23] Van der Weken, D., Nachtegael, M., Kerre, E.E.: Using similarity measures and homo-
     geneity for the comparison of images. Image and Vision Comput. 22, 695–702 (2004)
[24] Yager, R.R.: On the Implication Operator in Fuzzy Logic. Information Sciences 31(2),
     141–164 (1983)

Part IX
Integrals

Some New Definitions of Indicators for the Choquet Integral

Jaume Belles-Sampera, José M. Merigó, and Miguel Santolino

Abstract. Aggregation operators are broadly used in decision making problems. These operators are often characterized by indicators. Numerous of these aggregation operators may be represented by means of the Choquet integral. In this article four different indicators usually associated to the ordered weighted averaging (OWA) operator are extended to the Choquet integral. In particular, we propose the extensions of the degree of balance, the divergence, the variance indicator and Rényi entropies. Indicators for the weighted ordered weighted averaging (WOWA) operator are derived to illustrate the application of results. Finally, an example is provided to show main contributions.

1 Introduction

Aggregation operators are broadly used to summarize information in decision making [1, 6, 18]. The ordered weighted averaging (OWA) operator [19] is one of the most extensively analyzed aggregation operator [23]. The OWA operator is commonly assessed by means of indicators that characterize the weighting vector of the operator. Initially, Yager [19] introduced the orness/andness indicators and the entropy of dispersion for this purpose. After that, he suggested additional indicators

Jaume Belles-Sampera · Miguel Santolino
Department of Econometrics, Riskcenter - IREA. University of Barcelona. Av. Diagonal 690, 08034-Barcelona, Spain
e-mail: jbellesa8@alumnes.ub.edu, msantolino@ub.edu

José M. Merigó
Riskcenter - IREA. University of Barcelona. Av. Diagonal 690, 08034-Barcelona, Spain.
Decision and Cognitive Science Research Centre, Manchester Business School. Booth Street West, M15 6PB, Manchester, United Kingdom
e-mail: jose.merigolindahl@mbs.ac.uk

H. Bustince et al. (eds.), *Aggregation Functions in Theory and in Practise*,
Advances in Intelligent Systems and Computing 228,
DOI: 10.1007/978-3-642-39165-1_44, © Springer-Verlag Berlin Heidelberg 2013

in order to cover exceptional situations that cannot be correctly characterized with
the degree of orness and the entropy of dispersion. In particular, he suggested the
balance indicator [20] and the divergence [22]. Other authors have also considered
this issue such as Fullér and Majlender [4] that suggested the use of a variance indicator and Majlender [9] that presented the Rényi entropy [13] for OWA operators
as a generalization of their Shannon entropy [14].

The OWA operator may be represented as a Choquet integral [5, 7]. The Choquet integral [3] is linked to non-additive measures. A wide range of aggregation
operators may be represented by means of this type of integral. Some of the indicators traditionally associated to OWA operators have been extended to the Choquet
integral. For example, Marichal [11] and Grabisch et al [6] presented several types
of degree of orness indicators: specifically for the Choquet integral the former, for
general aggregation functions the latter. As well, Yager [21], Marichal [10] and Kojadinovic et al [8] studied the entropy of dispersion in the framework of the Choquet
integral. However, up to the best of our knowledge, some of those indicators are not
yet defined at the Choquet aggregation level.

The aim of this article is to extend four indicators for the characterization of the
Choquet Integral. In particular, we develop extensions of the degree of balance, the
divergence, the variance indicator and Rényi entropies to characterize the Choquet
integral. The weighted ordered weighted averaging (WOWA) operator introduced
by Torra [15] may be understood as a particular case of the Choquet integral (see
Theorem 4, page 3 in Torra [16]). Indicators for characterizing the WOWA operator are derived to illustrate the application of results. Finally, a fictitious example
is developed to show how these extended indicators may be computed and interpreted. The additional information provided by these indicators to decision makers
is highlighted.

This paper is organized as follows. In section 2 the Choquet integral is briefly
reviewed and the four extended indicators for characterizing the Choquet integral
are shown. The indicators inherited by the WOWA operator understood as a Choquet
integral are provided in section 3. In section 4 an illustrative example is provided.
Finally, main conclusions are summarized in section 5.

2 Indicators for the Choquet Integral

A brief explanation of the Choquet integral is provided in this section and different
indicators associated to OWA operators are extended to the Choquet integral. In
particular, we suggest generalizations of the degree of balance, the divergence, the
variance indicator and Rényi entropies to characterize the Choquet integral.

2.1 The Choquet Integral

A definition of capacity is needed as a first step. Let $N = \{1, ..., n\}$ be a finite set
and $2^N = \wp(N)$ be the set of all subsets of N. A capacity on N is a mapping from

2^N to $[0,1]$ which satisfies that $\mu(\emptyset) = 0$ and that if $A \subseteq B$ then $\mu(A) \leq \mu(B)$, for any $A, B \in 2^N$ (monotonicity).

If μ is a capacity such that $\mu(N) = 1$, then we say that μ satisfies normalization. A capacity μ is additive[1] if $\mu(A \cup B) + \mu(A \cap B) = \mu(A) + \mu(B)$ for any $A, B \subseteq N$. A capacity μ is symmetric if $\mu(A) = \mu(B)$ for all A, B with the same cardinality (i.e., $|A| = |B|$).

Let μ be a capacity on N, and $f : N \to [0, +\infty)$ be a function. Let σ be a permutation of $(1, ..., n)$, such that $f(\sigma(1)) \leq f(\sigma(2)) \leq ... \leq f(\sigma(n))$ and $A_{\sigma,i} = \{\sigma(i), ..., \sigma(n)\}$, with $A_{\sigma,n+1} = \emptyset$. The Choquet integral of f with respect to μ is defined as

$$\mathscr{C}_\mu(f) := \sum_{i=1}^{n} f(\sigma(i))(\mu(A_{\sigma,i}) - \mu(A_{\sigma,i+1})). \tag{1}$$

Extensions to real functions may be found in the literature [12]. We follow the asymmetric extension which is formulated as in expression (1) but taking into account that the domain of f is $(-\infty, +\infty)$.

2.2 Extended Indicators

Extensions of the degree of balance, the divergence, the variance indicator and Rényi entropies to the Choquet integral are suggested in this section. These extensions satisfy that when the capacity μ linked to the Choquet integral $\mathscr{C}_\mu$ is symmetric and normalized, then the indicators for $\mathscr{C}_\mu$ coincide with the respective indicators for OWA (see, for more details, Belles-Sampera et al [2]).

The following notation will be used. If μ is a capacity on N, then

$$\mathbb{S}_i^\mu := \binom{n}{i}^{-1} \sum_{\substack{A \subseteq N \\ |A|=i}} \mu(A) \quad .$$

Degree of Balance

The concept of degree of balance for an OWA operator was introduced by Yager [20]. As the degree of orness introduced by Yager [19], the degree of balance measures the grade of favoring the lower-valued elements or the higher-valued ones when the weighting vector is applied. We suggest the expression (2) for the degree of balance indicator associated to the Choquet integral[2]. Note that the degree of balance introduced by Yager [20] was ranged in $[-1, 1]$. Here, the degree of balance is

---

[1] It is usual to present the additivity property as follows: $\mu(A \cup B) = \mu(A) + \mu(B)$ for any pair of disjoint subsets A and B of N. It has to be noted that the way in which we present the additivity property contains the preceding one as a particular case.

[2] This definition requires of the extension of the degree of orness to the Choquet integral due to Marichal [11], which has the following expression: $\omega(\mathscr{C}_\mu) = \dfrac{1}{n-1} \sum_{i=1}^{n-1} [\mathbb{S}_i^\mu]$.

defined for any interval $[a,b] \subseteq \mathbb{R}$ where $b > a$.

$$Bal_{[a,b]} (\mathscr{C}_\mu) := (b-a)\boldsymbol{\omega} (\mathscr{C}_\mu) + \mu (N) a, \tag{2}$$

Note that the definition (2) is a linear transformation of the degree of orness $\boldsymbol{\omega} (\mathscr{C}_\mu)$. In particular, $Bal_{[0,1]} (\mathscr{C}_\mu) = \boldsymbol{\omega} (\mathscr{C}_\mu)$.

Divergence Indicator

The divergence indicator of an OWA operator was introduced by Yager [22]. This indicator complements the characterization of the $\text{OWA}_{\vec{w}}$ operator, especially in situations where the degree of orness and the dispersion indicator would not be enough for characterizing a weighting vector $\vec{w}$, as stated by Yager [22]. To define the divergence indicator for a Choquet integral $Div (\mathscr{C}_\mu)$, previously the ascending quadratic weighted additive (AQWA) capacity has to be introduced as it was in Belles-Sampera et al [2].

Definition 1 (AQWA capacity). Let μ be a capacity on $N = \{1,...,n\}$. The ascending quadratic weighted additive (AQWA) capacity linked to μ is an additive capacity η on N defined by

(i) $\eta (\{j\}) := \dfrac{6(j-1)^2}{(n-1)n(2n-1)} \left[\mathbb{S}^\mu_{n-j+1} - \mathbb{S}^\mu_{n-j} \right]$, for all $j = 1,..,n$;

(ii) $\eta (A) := \displaystyle\sum_{k \in A} \eta (\{k\})$; and $\eta (\emptyset) := 0$.

Proof that η is a capacity on N is provided by Belles-Sampera et al [2].

The definition of the divergence indicator for a discrete Choquet integral is as follows:

$$Div (\mathscr{C}_\mu) := \frac{n(2n-1)}{3(n-1)} \boldsymbol{\omega}(\mathscr{C}_\eta) - [2 - \mu (N)] \boldsymbol{\omega}^2(\mathscr{C}_\mu). \tag{3}$$

The divergence indicator provides information related to the mean variability around the degree of orness of any input value to be aggregated. In other words, the value of the divergence indicator is associated to how the aggregation function returns scattered values around the one linked to the degree of orness. That means that, the larger the global divergence is, the weaker the effect of the degree of orness in the aggregation process is.

Proposition 1. *Definition (3) is equivalent to*

$$Div (\mathscr{C}_\mu) = \sum_{i=1}^{n} \left(\frac{i-1}{n-1} - \boldsymbol{\omega}(\mathscr{C}_\mu) \right)^2 \left[\mathbb{S}^\mu_{n-i+1} - \mathbb{S}^\mu_{n-i} \right]. \tag{4}$$

Proof. First of all, note that expression (4) can be written as

$$Div\left(\mathscr{C}_\mu\right) = \sum_{i=1}^n \left(\frac{i-1}{n-1}\right)^2 \left[\mathbb{S}^\mu_{n-i+1} - \mathbb{S}^\mu_{n-i}\right]$$
$$-2\boldsymbol{\omega}(\mathscr{C}_\mu)\sum_{i=1}^n \left(\frac{i-1}{n-1}\right)\left[\mathbb{S}^\mu_{n-i+1} - \mathbb{S}^\mu_{n-i}\right] + \boldsymbol{\omega}(\mathscr{C}_\mu)^2 \mu\,(N)\,.$$

Therefore, to prove the proposition it is enough to check the following:

(i) $\boldsymbol{\omega}(\mathscr{C}_\eta) = \dfrac{3(n-1)}{n(2n-1)}\sum_{i=1}^n \left(\dfrac{i-1}{n-1}\right)^2 \left[\mathbb{S}^\mu_{n-i+1} - \mathbb{S}^\mu_{n-i}\right]$; and

(ii) $\boldsymbol{\omega}(\mathscr{C}_\mu) = \sum_{i=1}^n \left(\dfrac{i-1}{n-1}\right)\left[\mathbb{S}^\mu_{n-i+1} - \mathbb{S}^\mu_{n-i}\right]$.

Let us see that item (i) is satisfied:

$$\boldsymbol{\omega}(\mathscr{C}_\eta) = \frac{1}{n-1}\sum_{i=1}^{n-1}[\mathbb{S}^\eta_i] = \frac{1}{n-1}\sum_{i=1}^{n-1}\binom{n}{i}^{-1}\sum_{\substack{A\subseteq N\\|A|=i}}\sum_{j\in A}\eta\left(\{j\}\right)$$

$$= \frac{1}{n-1}\sum_{i=1}^{n-1}\binom{n}{i}^{-1}\sum_{\substack{A\subseteq N\\|A|=i}}\sum_{j\in A}\frac{6(j-1)^2}{(n-1)n(2n-1)}\left[\frac{\displaystyle\sum_{\substack{B\subseteq N\\|B|=n-j+1}}\mu\,(B)}{\dbinom{n}{n-j+1}} - \frac{\displaystyle\sum_{\substack{B\subseteq N\\|B|=n-j}}\mu\,(B)}{\dbinom{n}{n-j}}\right]$$

$$= \frac{6}{n(2n-1)}\sum_{i=1}^{n-1}\sum_{j=1}^n\left(\frac{j-1}{n-1}\right)^2\left[\binom{n-1}{i-1}\binom{n}{i}^{-1}\mathbb{S}^\mu_{n-j+1} - \binom{n-1}{i-1}\binom{n}{i}^{-1}\mathbb{S}^\mu_{n-j}\right]$$

$$= \frac{6}{n(2n-1)}\sum_{i=1}^{n-1}\frac{i}{n}\sum_{j=1}^n\left(\frac{j-1}{n-1}\right)^2\left[\mathbb{S}^\mu_{n-j+1} - \mathbb{S}^\mu_{n-j}\right]$$

$$= \frac{3(n-1)}{n(2n-1)}\sum_{j=1}^n\left(\frac{j-1}{n-1}\right)^2\left[\mathbb{S}^\mu_{n-j+1} - \mathbb{S}^\mu_{n-j}\right].$$

Secondly, let us check item (ii):

$$\boldsymbol{\omega}(\mathscr{C}_\mu) = \frac{1}{n-1}\sum_{i=1}^{n-1}[\mathbb{S}^\mu_i] = \frac{1}{n-1}\sum_{j=2}^n\left[\mathbb{S}^\mu_{n-j+1}\right]$$

$$= \sum_{j=2}^n\left(\frac{j-1}{n-1} - \frac{j-2}{n-1}\right)\left[\mathbb{S}^\mu_{n-j+1}\right] = \sum_{j=1}^n\left(\frac{j-1}{n-1}\right)\left[\mathbb{S}^\mu_{n-j+1} - \mathbb{S}^\mu_{n-j}\right]\,.$$

Variance Indicator

Another indicator used to characterize the $\text{OWA}_{\vec{w}}$ operator is the variance indicator introduced by Fullér and Majlender [4]. This indicator computes the variance of the weighting vector $\vec{w}$ considering each component equally probable. It has been used, for instance, to determine the analytical expression of a minimum variability OWA operator. The variance indicator of a capacity linked to a Choquet integral may be defined as

$$D^2\left(\mathscr{C}_\mu\right) = \frac{1}{n}\sum_{i=1}^{n}\left[\mathbb{S}^\mu_{n-i+1} - \mathbb{S}^\mu_{n-i}\right]^2 - \frac{\mu(N)^2}{n^2} \quad. \tag{5}$$

Going a step forward, note that not only the uniform probability could be considered as a distribution of the components w_i of vector $\vec{w}$ when computing the variance indicator of an $OWA_{\vec{w}}$. Therefore, if instead of the uniform distribution another one is considered (denoted by p_i, for all $i = 1, ..., n$) then an alternative extension for the variance indicator to the Choquet integral level could be

$$\hat{D}^2\left(\mathscr{C}_\mu\right) = \left(\sum_{i=1}^{n}\left[\mathbb{S}^\mu_{n-i+1} - \mathbb{S}^\mu_{n-i}\right]^2 p_i\right)$$
$$- \left(\sum_{i=1}^{n}\left[\mathbb{S}^\mu_{n-i+1} - \mathbb{S}^\mu_{n-i}\right] p_i\right)^2 \quad, \tag{6}$$

a definition that includes as a particular case the $D^2\left(\mathscr{C}_\mu\right)$ defined before.

Rényi Entropies

Finally, alternative entropy measures than the dispersion may be used to characterize the weighting vector. Generalizations of the Shannon entropy that could be used are Rényi entropies [9, 13]. The Rényi entropy of degree $\alpha \in \mathbb{R}\backslash\{1\}$ for a Choquet integral with respect to the capacity μ may be defined as

$$H_\alpha\left(\mathscr{C}_\mu\right) = \frac{1}{1-\alpha}\log_2\left(\sum_{i=1}^{n}\left[\mathbb{S}^\mu_{n-i+1} - \mathbb{S}^\mu_{n-i}\right]^\alpha\right) \quad. \tag{7}$$

3 WOWA Operator and Inherited Indicators

The WOWA operator is the aggregation function introduced by Torra [15]. This operator unifies in the same formulation the weighted mean function and the OWA operator in the following way[3].

Definition 2 (WOWA operator). Let $\vec{v} = (v_1, v_2, ..., v_n) \in [0,1]^n$ and $\vec{q} = (q_1, q_2, ..., q_n) \in [0,1]^n$ such that $\sum_{i=1}^{n} v_i = 1$ and $\sum_{i=1}^{n} q_i = 1$. The weighted ordered weighted averaging (WOWA) operator with respect to $\vec{v}$ and $\vec{q}$ is a mapping from $\mathbb{R}^n$ to $\mathbb{R}$ defined by

---

[3] In the original definition $\vec{x}$ components are in descending order, while we use ascending order. More details on function h can be found in Torra and Lv [17].

$$\text{WOWA}_{\vec{v},\vec{q}}(x_1, x_2, \dots, x_n) := \sum_{i=1}^{n} x_{\sigma(i)} \left[h\left(\sum_{j \in A_{\sigma,i}} q_j \right) - h\left(\sum_{j \in A_{\sigma,i+1}} q_j \right) \right],$$

where σ is a permutation of $(1, 2, \dots, n)$ such that $x_{\sigma(1)} \leq x_{\sigma(2)} \leq \dots \leq x_{\sigma(n)}$, $A_{\sigma,i} = \{\sigma(i), \dots, \sigma(n)\}$ and $h : [0,1] \to [0,1]$ is a non-decreasing function such that $h(0) := 0$ and $h\left(\dfrac{i}{n} \right) := \sum_{j=n-i+1}^{n} v_j$; and if the points $\left(\dfrac{i}{n}, \sum_{j=n-i+1}^{n} v_j \right)$ lie on a straight line then h is linear.

Note that this definition implies that weights v_i can be expressed as $v_i = h\left(\dfrac{n-i+1}{n} \right) - h\left(\dfrac{n-i}{n} \right)$ and that $h(1) = 1$.

Remark 1. The WOWA operator generalizes the OWA operator. Given a WOWA$_{\vec{v},\vec{q}}$ operator on $\mathbb{R}^n$, if we define

$$w_i := h\left(\sum_{j \in A_{\sigma,i}} q_j \right) - h\left(\sum_{j \in A_{\sigma,i+1}} q_j \right),$$

and OWA$_{\vec{w}}$ where $\vec{w} = (w_1, \dots, w_n)$, then the following equality holds WOWA$_{\vec{v},\vec{q}} =$ OWA$_{\vec{w}}$.

Using Theorem 4 from page 3 in Torra [16], any WOWA operator may be understood as a Choquet integral. If particular weighting vectors $\vec{q}$ and $\vec{v}$ and non-decreasing function h are those of the WOWA$_{\vec{v},\vec{q}}$ operator under review, then the Choquet integral that replicates that WOWA operator is the one linked to the capacity μ such that $\mu(A) = h\left(\sum_{j \in A} q_j \right)$, for all $A \subseteq N$.

Therefore indicators for the WOWA$_{\vec{v},\vec{q}}$ operator may be defined as follows: $Bal_{[a,b]}(\text{WOWA}_{\vec{v},\vec{q}}) := Bal_{[a,b]}(\mathscr{C}_\mu)$, $Div(\text{WOWA}_{\vec{v},\vec{q}}) := Div(\mathscr{C}_\mu)$, $D^2(\text{WOWA}_{\vec{v},\vec{q}}) := D^2(\mathscr{C}_\mu)$ and $H_\alpha(\text{WOWA}_{\vec{v},\vec{q}}) := H_\alpha(\mathscr{C}_\mu)$.

4 Illustrative Example

In the following example two capacities are considered, κ and μ. The aim of this example is to compute some of the proposed indicators for the Choquet integral with respect to each one of these capacities. Both capacities are defined on $N = \{1, 2, 3\}$ as follows:

- $\kappa(\emptyset) = 0$, $\kappa(\{1\}) = \kappa(\{2\}) = \kappa(\{3\}) = 0.3$, $\kappa(\{1,2\}) = \kappa(\{1,3\}) = \kappa(\{2,3\}) = 0.85$ and $\kappa(N) = 1$;
- $\mu(\emptyset) = 0$, $\mu(\{1\}) = 0.3$, $\mu(\{2\}) = 0.883333$, $\mu(\{3\}) = 0.6$, $\mu(\{1,2\}) = 1$, $\mu(\{1,3\}) = 0.783333$, $\mu(\{2,3\}) = 1$, and $\mu(N) = 1$.

Let us compute the degree of orness of $\mathscr{C}_\kappa$ and $\mathscr{C}_\mu$. As it can be shown, $\omega(\mathscr{C}_\kappa) = 0 \times 0.15 + 0.5 \times 0.55 + 1 \times 1 = 0.575$ and $\omega(\mathscr{C}_\mu) = 0 \times 0.072222 + 0.5 \times 0.333333 + 1 \times 0.594444 = 0.761111$, so $\omega(\mathscr{C}_\kappa) \neq \omega(\mathscr{C}_\mu)$ and the $\mathscr{C}_\mu$ aggregation operator is going to generate aggregated values nearer the maximum of each input data set than the $\mathscr{C}_\kappa$ operator.

Let us calculate the divergence indicator for $\mathscr{C}_\kappa$ and $\mathscr{C}_\mu$, too. AQWA capacities linked to κ and μ are determined by the following values: $\text{AQWA}_\kappa(\{1\}) = 0$, $\text{AQWA}_\kappa(\{2\}) = 0.11$ and $\text{AQWA}_\kappa(\{3\}) = 0.24$; and $\text{AQWA}_\mu(\{1\}) = 0$, $\text{AQWA}_\mu(\{2\}) = 0.066667$ and $\text{AQWA}_\mu(\{3\}) = 0.475556$. The degrees of orness of these AQWA capacities are 0.175 and 0.271, respectively, and thus $Div(\mathscr{C}_\kappa) = 0.106875$ and $Div(\mathscr{C}_\mu) = 0.098488$. As long as a greater divergence produces a more scattered aggregation, the first operator $\mathscr{C}_\kappa$ is going to generate more disperse aggregated values around the value linked to the degree of orness.

Table 1 Values of several indicators regarding the example

Indicator	$\mathscr{C}_\kappa$	$\mathscr{C}_\mu$	AQWA - κ	AQWA - μ
Degree of orness	0.575	0.76111	0.175	0.27111
Degree of balance [-1,1]	0.15	0.15	0	0
Divergence	0.106875	0.098488	0.095302	0.118777
Variance indicator	0.027222	0.045453	0	0
Rényi entropy ($\alpha = 1.5$)	1.331768	1.154771	6.128682	4.234094

Some additional remarks can be made. For instance, $\mathscr{C}_\kappa$ is equivalent to an OWA operator because κ is a symmetric and normalized capacity. In fact, $\mathscr{C}_\kappa$ is equivalent to $\text{OWA}_{\vec{w}}$ with $\vec{w} = (w_1, w_2, w_3) = (0.15, 0.55, 0.3)$. It is easy to check that $\omega(\vec{w}) = \omega(\mathscr{C}_\kappa)$. Regarding $\mathscr{C}_\mu$, it is equivalent to a $\text{WOWA}_{\vec{v},\vec{q}}$ operator. The particular vectors $\vec{v}, \vec{q}$ and function h to obtain the equivalence are $\vec{v} = (0, 1/3, 2/3)$, $\vec{q} = \vec{w} = (0.15, 0.55, 0.3)$ and

$$h(t) = \begin{cases} 2t & if & 0 \leq t < 1/3 \\ t + 1/3 & if & 1/3 \leq t < 2/3 \\ 1 & if & 2/3 \leq t \leq 1 \end{cases}$$

Therefore, Table 1 shows the values of the indicators extended to the Choquet integral in two particular cases: when the Choquet integral corresponds to an OWA operator and when it corresponds to a WOWA one.

5 Discussion and Conclusions

New indicators for the discrete Choquet integral have been presented. The aim of these indicators is to complement the available ones in order to provide a more complete formulation that covers a wide range of situations. The need for this arises because sometimes the well-known degree of orness and the entropy of dispersion may not be enough. This paper has introduced the degree of balance, the divergence, the variance indicator and Rényi entropies in the Choquet integral framework. It has been shown that these indicators, commonly used for the OWA operator, can also be considered under the Choquet aggregation. We claim that these indicators may be easily computed for any aggregation operator that can be interpreted as a Choquet integral. As an application, the indicators for the WOWA operator have been implicitly deduced. We discuss the potential of these indicators to provide supplementary information to decision makers with the help of a numerical example.

We thought that the Choquet integral was an adequate starting point to investigate extensions of these indicators from their original OWA context to more general ones, but there is room for further research.

References

[1] Beliakov, G., Pradera, A., Calvo, T.: Aggregation Functions: A Guide to Practitioners. Springer, Berlin (2007)

[2] Belles-Sampera, J., Merigó, J.M., Santolino, M.: Indicators for the characterization of discrete Choquet integrals. IREA Working Papers, University of Barcelona, Research Institute of Applied Economics (2013)

[3] Choquet, G.: Theory of Capacities. Annales de l'Institute Fourier 5, 131–295 (1954)

[4] Fullér, R., Majlender, P.: On obtaining minimal variability OWA operator weights. Fuzzy Sets and Systems 136(2), 203–215 (2003)

[5] Grabisch, M.: On equivalence classes of fuzzy-connectives: The case of fuzzy integrals . IEEE Transactions on Fuzzy Systems 3(1), 96–109 (1995)

[6] Grabisch, M., Marichal, J.L., Mesiar, R., Pap, E.: Aggregation functions. Encyclopedia of Mathematics and its Applications, vol. 127. Cambridge University Press (2009)

[7] Grabisch, M., Marichal, J.L., Mesiar, R., Pap, E.: Aggregation functions: Means. Information Sciences 181(1), 1–22 (2011)

[8] Kojadinovic, I., Marichal, J.L., Roubens, M.: An axiomatic approach to the definition of the entropy of a discrete Choquet capacity. Information Sciences 172(1-2), 131–153 (2005)

[9] Majlender, P.: OWA operators with maximal Rényi entropy. Fuzzy Sets and Systems 155(3), 340–360 (2005)

[10] Marichal, J.L.: Entropy of discrete Choquet capacities. European Journal of Operational Research 137(3), 612–624 (2002)

[11] Marichal, J.L.: Tolerant or intolerant character of interacting criteria in aggregation by the Choquet integral. European Journal of Operational Research 155(3), 771–791 (2004)

[12] Mesiar, R., Mesiarová-Zemánková, A., Ahmad, K.: Discrete Choquet integral and some of its symmetric extensions. Fuzzy Sets and Systems 184(1), 148–155 (2011)

[13] Rényi, A.: On measures of entropy and information. In: Proceedings of the 4th Berkeley Symposium on Mathematical Statistics and Probability, pp. 547–561 (1961)

[14] Shannon, C.E.: A mathematical theory of communication. Bell System Technical Journal 27(3), 379–423 (1948)

[15] Torra, V.: The weighted OWA operator. International Journal of Intelligent Systems 12(2), 153–166 (1997)

[16] Torra, V.: On some relationships between the WOWA operator and the Choquet integral. In: Proceedings of the IPMU 1998 Conference, Paris, France, pp. 818–824 (1998)

[17] Torra, V., Lv, Z.: On the WOWA operator and its interpolation function. International Journal of Intelligent Systems 24(10), 1039–1056 (2009)

[18] Torra, V., Narukawa, Y.: Modeling Decisions: Information Fusion and Aggregation Operators. Springer, Berlin (2007)

[19] Yager, R.R.: On ordered weighted averaging operators in multicriteria decision-making. IEEE Transactions on Systems, Man and Cybernetics 18(1), 183–190 (1988)

[20] Yager, R.R.: Constrained OWA aggregation. Fuzzy Sets and Systems 81(1), 89–101 (1996)

[21] Yager, R.R.: On the entropy of fuzzy measures. IEEE Transactions on Fuzzy Systems 8(4), 453–461 (2000)

[22] Yager, R.R.: Heavy OWA operators. Fuzzy Optimization and Decision Making 1, 379–397 (2002)

[23] Yager, R.R., Kacprzyk, J., Beliakov, G. (eds.): Recent Developments in the Ordered Weighted Averaging Operators: Theory and Practice. STUDFUZZ, vol. 265. Springer, Heidelberg (2011)

Exponential Family of Level Dependent Choquet Integral Based Class-Conditional Probability Functions

Vicenç Torra and Yasuo Narukawa

Abstract. In a recent paper we introduced two new families of probability-density functions. We introduced first the exponential family of Choquet integral based class-conditional probability-density functions, and second the exponential family of Choquet-Mahalanobis integral based class-conditional probability-density functions. The latter being a generalization of the former, and also a generalization of the normal distribution. In this paper we study some properties of these distributions and define another generalization based on level-dependent Choquet integrals.

1 Introduction

Classification problems are often solved by means of the computation of the maximum-a-posteriori (MAP) classification decision rule. This rule classifies a data vector of a n-dimensional space (for example, $x \in \mathbb{R}^n$) in one of $\Omega = \{\omega_1, \ldots, \omega_k\}$ classes. When data in classes $\omega_i \in \Omega$ are generated from Gaussian distributions following mean $\bar{x}_i$ and covariance Σ_i, in such a way that $P(\omega_i) = P(\omega_j)$ and $\Sigma_i = \Sigma_j$, the decision rule is equivalent to the Mahalanobis distance between x and the means $\bar{x}_i$.

The Mahalanabis distance can be seen as a way to compute the distance between the point x and the mean of this class. It can be considered as a multivariate analog of the z-score [8]. Recall that the z-score of x is defined by the expression $(x - \bar{x})/\sigma$ where $\bar{x}$ is the mean of the population and σ its standard deviation. Therefore, the z-score is a dimensionless value that evaluates how far is x from the mean using

Vicenç Torra
IIIA-CSIC, Institut d'Investigació en Intel·ligència Artificial, Campus UAB s/n,
08193 Bellaterra, Catalonia, Spain
e-mail: vtorra@iiia.csic.es

Yasuo Narukawa
Toho Gakuen, 3-1-10, Naka, Kunitachi, Tokyo, 186-0004 Japan
e-mail: nrkwy@ybb.ne.jp

H. Bustince et al. (eds.), *Aggregation Functions in Theory and in Practise*,   477
Advances in Intelligent Systems and Computing 228,
DOI: 10.1007/978-3-642-39165-1_45, © Springer-Verlag Berlin Heidelberg 2013

as units the standard deviation. In other words, how many deviations is x from the mean.

In a recent paper [7] we have introduced two new families of probability distributions.

One was based on the Choquet integral [2]. In a way analogous to the normal distribution that is related to the Mahalanobis distance, we introduced a distribution that is related to a distance based on the Choquet integral. Both, the Mahalanobis distance and the Choquet integral distance, permit us to take into account some interactions between the variables. In the Mahalanobis distance the interactions are represented in terms of a matrix (the covariance matrix) while in the Choquet integral distance the interactions are represented by means of a fuzzy measure.

The other distribution is to take into account both types of interactions. That is, the one expressed in terms of the (covariance) matrix and the one expressed in terms of the fuzzy measure.

We named the former measure the exponential family of Choquet integral based class-conditional probability-density functions, and the second one the family of Choquet-Mahalanobis integral based class-conditional probability-density functions.

In this paper we study some properties of these probability density functions and we introduce another family of probability-density functions that generalize the previous ones. These families are based on level-dependent Choquet integrals [3].

The structure of the paper is as follows. In Section 2, we review some preliminaries needed in the rest of the paper. In Section 3, we introduce the new distributions, and in Section 4 we present some of their properties. The paper finishes with some conclusions.

2 Preliminaries

In this section we review the fuzzy integrals we need later on in this work. The section begins with the notation. From a formal point of view, fuzzy integrals integrate a function with respect to a fuzzy measure.

Let $X = \{x_1, \ldots, x_n\}$ be a set, and let $f : X \to [0, 1]$ be a function. Then, $f(x_i)$ is a set of values. For simplicity, we also use $a_i := f(x_i)$. Fuzzy integrals aggregate the values $f(x_i)$ with respect to a fuzzy measure μ.

Definition 1. Let $X = \{x_1, \ldots, x_n\}$ be a set; then, a set function $\mu : 2^X \to [0, \infty)$ is a fuzzy measure if it satisfies the following axioms:

(i) $\mu(\emptyset) = 0$ (boundary conditions)
(ii) $A \subseteq B$ implies $\mu(A) \leq \mu(B)$ (monotonicity)

Definition 2. Let μ be a fuzzy measure on X; then, the *Choquet integral* of a function $f : X \to \mathbb{R}^+$ with respect to the fuzzy measure μ is defined by

$$(C) \int f d\mu = \sum_{i=1}^{n} [f(x_{s(i)}) - f(x_{s(i-1)})] \mu(A_{s(i)}), \tag{1}$$

where $f(x_{s(i)})$ indicates that the indices have been permuted so that $0 \le f(x_{s(1)}) \le \cdots \le f(x_{s(n)}) \le 1$, and where $f(x_{s(0)}) = 0$ and $A_{s(i)} = \{x_{s(i)}, \dots, x_{s(n)}\}$.

We also use the notation $CI_\mu(a_1, \dots, a_n)$ to express the Choquet integral of $a_i := f(x_i)$.

The Choquet integral can be formulated alternatively as:

$$CI_\mu(f) = \int_{\min(\{a_1,\dots,a_n\})}^{+\infty} \mu(\{x_i \in X \,|\, f(x_i) \ge t\})dt + \min(\{a_1, \dots, a_n\}). \quad (2)$$

2.1 Level Dependent Choquet Integral

Level dependent Choquet integrals integrate a function with respect to *a set* of fuzzy measures. In short, we might have a different fuzzy measure for each level t in the Choquet integral of Equation 2. We begin reviewing the definition of generalized fuzzy measure as given by Greco, Matarazzo and Giove in [3].

Definition 3. [3] Let $X = \{x_1, \dots, x_n\}$ be a set. A generalized fuzzy measure is a function $\mu^G : 2^X \times (\alpha, \beta) \to [0,1]$, $(\alpha, \beta) \subseteq \mathbb{R}$, such that

1. for all $t \in (\alpha, \beta)$ and $A \subseteq B \subseteq X$, $\mu^G(A,t) \le \mu^G(B,t)$,
2. for all $t \in (\alpha, \beta)$, $\mu^G(\emptyset, t) = 0$ and $\mu^G(N, t) = 1$,
3. for all $A \subseteq X$, $\mu^G(A,t)$ considered as a function with respect to t is Lebesgue mesurable.

Definition 4. The generalized Choquet integral of $a = (a_1 = f(x_1), \dots, a_n = f(x_n))$ with $a \in (\alpha, \beta)^n$, $(\alpha, \beta) \subseteq \mathbb{R}$, with respect to the generalized capacity μ^G is defined as follows:

$$CI_{\mu^G}^G(a) = \int_{\min(a_1,\dots,a_n)}^{\max(a_1,\dots,a_n)} \mu^G(A(a,t),t)dt + \min(a_1, \dots, a_n), \quad (3)$$

where

$$A(a,t) = \{x_i \in X : a_i \ge t\}.$$

For simplicity in applications, [3] introduce interval level dependent fuzzy measures (capacities), and give a result that shows that the Choquet integral can be decomposed for this type of fuzzy measures into an addition of standard Choquet integrals. See Theorem 3 in [3].

2.2 Choquet Integral Based Distributions

We defined in a recent paper [7] two distributions based on the Choquet integral. We review below the first one. The other generalizes this one as well as the normal distribution.

Definition 5. Let $\{Y_1, \dots, Y_n\}$ be a set of random variables describing data on a $\mathbb{R}^n$ dimensional space. Let $(\alpha, \beta) \subseteq \mathbb{R}$, and $\mu : 2^Y \to [0,1]$ a fuzzy measure.

Let $\bar{x} \in \mathbb{R}^n$ be the mean of the random variables; then, the exponential family of Choquet integral based class-conditional probability-density functions is defined by:

$$P(x) = \frac{1}{K} e^{-\frac{1}{2} CI_\mu ((x-\bar{x}) \otimes (x-\bar{x}))}$$

where K is a constant that is defined so that the function is a probability, and where $v \otimes w$ denotes the elementwise product of vectors v and w (i.e., $(v \otimes w) = (v_1 w_1 \ldots v_n w_n)$).

Although the definition of the density function needs the constant K, the exact value of K is not rellevant in classification problems, or for studying the shape of the distribution function. In any case, the K is the value such that:

$$\int_{x \in X} P(x) = 1$$

So, K should be defined by

$$K = \int_{x \in X} e^{-\frac{1}{2} CI_\mu ((x-\bar{x}) \otimes (x-\bar{x}))}.$$

3 A Probability Distribution with a Level-Dependent Choquet Integral

In this section we introduce a new family of probability distributions using the level-dependent Choquet integral. This new family is an extension of the one in Definition 5 based on the Choquet integral.

Definition 6. Let $\{Y_1, \ldots, Y_n\}$ be a set of random variables describing data on a $\mathbb{R}^n$ dimensional space. Let $(\alpha, \beta) \subseteq \mathbb{R}$, and $\mu^G : 2^Y \times (\alpha, \beta) \to [0, 1]$ a generalized fuzzy measure.

Let $\bar{x} \in \mathbb{R}^n$ be the mean of the random variables; then, the exponential family of level dependent Choquet integral based class-conditional probability-density functions is defined by:

$$P(x) = \frac{1}{K} e^{-\frac{1}{2} CI^G_{\mu^G} ((x-\bar{x}) \otimes (x-\bar{x}))}$$

where K is a constant that should be defined so that the function is a probability, and where $v \otimes w$ denotes the elementwise product of vectors v and w (i.e., $(v \otimes w) = (v_1 w_1 \ldots v_n w_n)$).

Although the definition of the density function needs the constant K, the exact value of K is not rellevant in classification problems, or for studying the shape of the distribution function. In any case, the K is the value such that:

$$\int_{x \in X} P(x) = 1$$

So, K should be defined by

$$K = \int_{x \in X} e^{-\frac{1}{2} CI_{\mu G}^{G}((x-\bar{x}) \otimes (x-\bar{x}))}$$

Figure 1 displays two distributions based on the Choquet integral and another based on the level dependent Choquet integral. Each one of the ones based on the Choquet integral intensifies in one axis (on in the x Axis and the other in the y Axis). In contrast, the one based on the level dependent Choquet integral has both intensifications but on different levels of the z Axis. This is due to the definition of the fuzzy measure.

Lemma 1. *The probability distribution defined in Definition 6 generalizes the probability distribution introduced in [7].*

The same approach used in Definition 6 can be used to generalize the distribution based on the Choquet-Mahalanobis integral. This latter distribution was based on an operator that takes into account interactions expressed in terms of fuzzy measures as well as in terms of covariance matrices.

4 Some Basic Properties of the Choquet Integral Based Probability Distributions

In this section we study some properties of the probability distributions based on the Choquet integral.

Proposition 1. *Let $P(x)$ with $x \subset \mathbb{R}^n$ be an exponential Choquet integral probability-density function with mean $\bar{x} = 0$. Then, for any fuzzy measure μ, the mean vector $\bar{X} = [E[X_1], E[X_2], \ldots, E[X_n]]$ is zero (i.e., $\bar{X} = [0, 0, \ldots, 0]$) and $\Sigma = [Cov[X_i, X_j]]$ for $i = 1, \ldots, n$ and $j = 1, \ldots, n$ is zero for all $i \neq j$ and, thus, diagonal.*

Proposition 2. *Let $P(x)$ with $x \subset \mathbb{R}^n$ be an exponential Choquet-Mahalanobis integral probability-density function with mean $\bar{x} = 0$. Then, for any fuzzy measure μ and any diagonal matrix Σ, the mean vector $\mu = [E[X_1], E[X_2], \ldots, E[X_n]]$ is zero (i.e., $\mu = [0, 0, \ldots, 0]$) and $\Sigma = [Cov[X_i, X_j]]$ for $i = 1, \ldots, n$ and $j = 1, \ldots, n$ is zero for all $i \neq j$ and thus, diagonal.*

In the case that Σ is not diagonal and, thus, $\Sigma(X_i, X_j) \neq 0$ for $i \neq j$, we might have $Cov[X_i, X_j] \neq 0$. It is important to note that it is not at all required that $Cov[X_i, X_j] = \Sigma(X_i, X_j)$. The following example illustrates this fact.

Example 1. Let us consider the Choquet-Mahalanobis integral based distribution with a fuzzy measure $\mu(\emptyset) = 0$, $\mu(\{x\}) = 0.5$, $\mu(\{y\}) = 0.2$, $\mu(\{x, y\}) = 1$ and the matrix

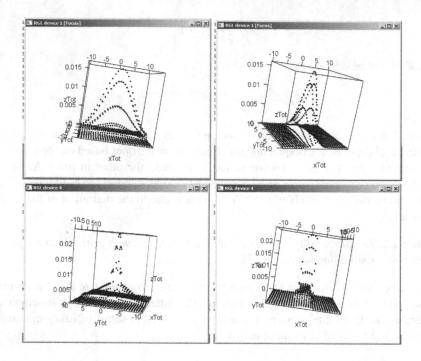

Fig. 1 On the top, two CI-based distributions. One with the measure $\mu^1(\{x\}) = 0.05$ and $\mu^1(\{y\}) = 0.95$, and the other with $\mu^2(\{x\}) = 0.95$ and $\mu^2(\{y\}) = 0.05$. On the bottom, two perspectives of the same level dependent Choquet integral-based distribution. This distribution is based on the same fuzzy measures μ^1 and μ^2 with intervals $(0,3)$, $(3,100)$. In the first interval μ^1 is used and in the second μ^2.

$$\Sigma = \begin{pmatrix} 1 & 0.9 \\ 0.9 & 1 \end{pmatrix}.$$

The covariance matrix of this distribution is:

$$\Sigma = \begin{pmatrix} 0.9548251 & 0.9262923 \\ 0.9262923 & 1.0293333 \end{pmatrix}.$$

The correlation coefficient between the two variables is 0.9343469.

4.1 Normality Tests

There are several approaches [6] to check whether a distribution follows a multivariate normal distribution. One of them is Mardia's test [5]. This test is based on multivariate extensions of skewness and kurtosis. In particular, for the multivariate skewness of a sample in a k dimensional space Mardia obtained the following expression:

$$b_{1,k} = \frac{1}{n^2} \sum_{i=1}^{n} \sum_{j=1}^{n} [(x_i - \bar{x})' \hat{\Sigma} (x_i - \bar{x})]^3. \tag{4}$$

In the case of the multivariate kurtosis, the expression obtained is the following one:

$$b_{2,k} = \frac{1}{n} \sum_{i=1}^{n} [(x_i - \bar{x})' \hat{\Sigma} (x_i - \bar{x})]^2. \tag{5}$$

Here, $\bar{x}$ is the sample mean vector and $\hat{\Sigma}$ is the covariance matrix. They correspond to:

$$\bar{x} = (1/n) \sum_{i=1}^{n} x_i \quad \hat{\Sigma} = \frac{1}{n} \sum_{i=1}^{n} (x_i - \bar{x})(x_i - \bar{x})^T \tag{6}$$

Then, when the distribution is a multivariate normal distribution (i.e., when the null hypothesis holds), the expression

$$A = n \cdot b_{1,k}/6$$

follows a chi-squared distribution with $k(k+1)(k+2)/6$ degrees of freedom, and the expression

$$B = \sqrt{\frac{n}{8k(k+2)}} \left(b_{2,k} - k(k+2) \right)$$

follows a standard normal random variable $N(0,1)$.

A preliminary analysis of the Mardia's test on the Choquet integral based distributions shows that at least some of these distributions pass the test for the skewness statistic (i.e., A). We need to further study this normality test, taking also into account that there are some results in the literature that seem to indicate that this test fails for some other distributions (see e.g. [1]). In addition, other normality tests might be applied in this setting.

5 Conclusions

In this paper we have introduced a new probability distribution based on the level dependent Choquet integral, and we have studied some properties for the distribution based on the Choquet integral.

As future work, we plan to further study the properties of this set of probability distribution functions, and consider probabilities distributions constructed from other transformations of the Choquet integral.

Acknowledgements. Partial support by the Spanish MEC projects ARES (CONSOLIDER INGENIO 2010 CSD2007-00004), eAEGIS (TSI2007-65406-C03-02), and COPRIVACY (TIN2011-27076-C03-03) is acknowledged.

References

[1] Baringhaus, L., Henze, N.: Limit distributions for measures of multivariate skewness and kurtosis based on projections. Journal of Multivariate Analysis 38(51) (1991), doi:10.1016/0047-259X(91)90031-V.edit

[2] Choquet, G.: Theory of capacities. Ann. Inst. Fourier 5, 131–295 (1953/1954)

[3] Greco, S., Matarazzo, B., Giove, S.: The Choquet integral with respect to a level dependent capacity. Fuzzy Sets and Systems 175, 1–35 (2011)

[4] Kankainen, A., Taskinen, S., Oja, H.: On Mardia's tests of multinormality. In: Hubert, M., Pison, G., Struyf, A., Van Aelst, S. (eds.) Theory and Applications of Recent Robust Methods, pp. 153–164. Birkhäuser, Basel (2004)

[5] Mardia, K.V.: Measures of multivariate skewness and kurtosis with applications. Biometrika 57(3), 519–530 (1970)

[6] Mecklin, C.J., Mundfrom, D.J.: An appraisal and bibliography of tests for multivariate normality. International Statistical Review 72(1), 123–138 (2004)

[7] Torra, V., Narukawa, Y.: On a comparison between Mahalanobis distance and Choquet integral: The Choquet-Mahalanobis operator. Information Sciences 190, 56–63 (2012)

[8] Wicklin, R.: What is Mahalanobis distance? Entry (2012), http://blogs.sas.com/content/iml/2012/02/15/what-is-mahalanobis-distance/ (February 15, 2012)

Axiomatic Foundations of the Universal Integral in Terms of Aggregation Functions and Preference Relations

Salvatore Greco, Radko Mesiar, and Fabio Rindone*

Abstract. The concept of universal integral has been recently proposed in order to generalize the Choquet, Shilkret and Sugeno integrals. We present two axiomatic foundations of the universal integral. The first axiomatization is expressed in terms of aggregation functions, while the second is expressed in terms of preference relations.

1 Basic Concepts

For the sake of simplicity in this note we present the result in a Multiple Criteria Decision Making (MCDM) setting (for a state of art on MCDM see [2]). Let $N = \{1, \ldots, n\}$ be the set of criteria and let us identify the set of possible alternatives with $[0,1]^n$. For all $E \subseteq N$, $\mathbf{1}_E$ is the vector of $[0,1]^n$ whose ith component equals 1 if $i \in E$ and equals 0 otherwise. For all $x = (x_1 \ldots, x_n) \in [0,1]^n$, the set $\{i \in N \mid x_i \geq t\}, t \in [0,1]$, is briefly indicated with $\{x \geq t\}$. For all $x, y \in [0,1]^n$ we say that x dominates y and we write $x \succeq y$ if $x_i \geq y_i$, $i = 1, \ldots, n$. An aggregation function $f : [0,1]^n \to \mathbb{R}$ is a function such that $f(x) \geq f(y)$ whenever $x \succeq y$ and $\inf_{x \in [0,1]^n} f = f(\mathbf{1_0}) = 0$,

Salvatore Greco · Fabio Rindone
Department of Economics and Business
95029 Catania, Italy
Department of Mathematics and Descriptive Geometry, Faculty of Civil Engineering
Slovak University of Technology
Bratislava, Slovakia
e-mail: {salgreco,frindone}@unict.it

Radko Mesiar
Institute of Theory of Information and Automation
Czech Academy of Sciences
Prague, Czech Republic
e-mail: radko.mesiar@stuba.sk

* Corresponding author.

H. Bustince et al. (eds.), *Aggregation Functions in Theory and in Practise*,
Advances in Intelligent Systems and Computing 228,
DOI: 10.1007/978-3-642-39165-1_46, © Springer-Verlag Berlin Heidelberg 2013

$\sup_{x \in [0,1]^n} f = f(\mathbf{1}_N) = 1$, [3]. Let $\mathscr{A}_{[0,1]^n}$ be the set of aggregation functions on $[0,1]^n$.

Let M denotes the set of all capacities m on N, i.e. for all $m \in M$ we have $m : 2^N \to [0,1]$ satisfying the following conditions:

- *boundary conditions*: $m(\emptyset) = 0, m(N) = 1$;
- *monotonicity*: $m(A) \leq m(B)$ for all $\emptyset \subseteq A \subseteq B \subseteq N$.

A *universal integral* on N [4] is a function $I : M \times [0,1]^n \to [0,1]$ satisfying the following properties:

(UI1) I is non-decreasing in each coordinate,

(UI2) there exists a pseudo-multiplication $\otimes$ (i.e. $\otimes : [0,1]^2 \to [0,1]$ is nondecreasing in its two coordinates and $\otimes(c,1) = \otimes(1,c) = c$) such that for all $m \in M$, $c \in [0,1]$ and $A \subseteq N$,

$$I(m, c\mathbf{1}_A) = \otimes(c, m(A)),$$

(UI3) for all $m_1, m_2 \in M$ and $x, y \in [0,1]^n$, if $m_1(\{x \geq t\}) = m_2(\{y \geq t\})$ for all $t \in]0,1]$, then $I(m_1, x) = I(m_2, y)$.

Given a universal integral I with respect to the pseudomultiplication $\otimes$, we shall write

$$I(m, x) = \int_{univ, \otimes} x \, dm$$

for all $m \in M, x \in [0,1]^n$.

Suppose $I(m, x)$ is a universal integral and consider $m^* \in M$ then the $I_{m^*} : [0,1]^n \to [0,1]$ defined by $I_{m^*}(\mathbf{x}) = I(m^*, \mathbf{x})$ for all $\mathbf{x} \in [0,1]^n$, is an aggregation function. Thus the universal integral $I(m, x)$ can be viewed as a family of aggregations functions, $I_m(\mathbf{x})$, one for each capacity $m \in M$.

2 Axiomatic Foundation in Terms of Aggregation Functions

Consider a family $\mathscr{F} \subseteq \mathscr{A}_{[0,1]^n}$ with $\mathscr{F} \neq \emptyset$ and consider the following axioms on $\mathscr{F}$:

(A$_1$) For all $f_1, f_2 \in \mathscr{F}$ and $x, y \in [0,1]^n$ such that for all $t \in [0,1]$

$$f_1\left(\mathbf{1}_{\{x \geq t\}}\right) \geq f_2\left(\mathbf{1}_{\{y \geq t\}}\right),$$

then $f_1(x) \geq f_2(y)$;

(A$_2$) Every $f \in \mathscr{F}$ is idempotent, i.e. for all $c \in [0,1]$ and $f \in \mathscr{F}$,

$$f(c \cdot \mathbf{1}_N) = c;$$

(A$_3$) For all $m \in M$ there exists $f \in \mathscr{F}$ such that $f(\mathbf{1}_A) = m(A)$ for all $A \subseteq N$.

Observe that axioms $(A_1) - (A_3)$ are independent as showed by the following examples, in which two of above axioms hold, but the remaining is not valid:

1) $\mathscr{F}$ is the set of all weighted averages on $[0,1]^n$, i.e. for any $f \in \mathscr{F}$ there are $w_1, \ldots, w_n \in [0,1], \sum_{i=1}^n w_i = 1$, such that $f(x) = \sum_{i=1}^n w_i x_i$ for all $x \in [0,1]^n$: in this case (A_1) and (A_2) are satisfied, but (A_3) does not hold;

2) $\mathscr{F}$ is the set of all aggregation functions f_m defined for all $m \in M$ and $x \in [0,1]^n$, as

$$f_m(x) = \int_0^1 m(\{x \geq \sqrt{t}\}) dt$$

for all $x \in [0,1]^n, m \in M$: in this case (A_1) and (A_3) hold, but (A_2) is not satisfied;

3) $\mathscr{F}$ is the set of all of all aggregation functions f_{m_1,m_2}, $m_1, m_2 \in M$ defined as follows:

$$f_{m_1,m_2} = \int_0^{0.5} m_1(\{x \geq t\}) dt + \int_{0.5}^1 m_2(\{x \geq t\}) dt$$

for all $x \in [0,1]^n, m_1, m_2 \in M$: in this case (A_2) and (A_3) hold, but (A_1) is not satisfied. Indeed in this case consider $N = \{1,2\}$, $x = (0.2, 0.4)$ and the capacities m_1, m_2, m_1', m_2' defined by $m_1(\{1\}) = m_1(\{2\}) = 0.5, m_2(\{1\}) = m_2(\{2\}) = 0,$ $m_1'(\{1\}) = 1, m_1'(\{2\}) = 0, m_2'(\{1\}) = 0, m_2'(\{2\}) = 0.5$. It results that $f_{m_1,m_2}(1_{\{x \geq t\}}) = f_{m_1',m_2'}(1_{\{x \geq t\}})$ for all $t \in [0,1]$ but $f_{m_1,m_2}(x) = 0.3 > f_{m_1',m_2'}(x) = 2$, which contraddicts axiom (A_1).

Proposition 1. *Axioms* (A_1), (A_2) *and* (A_3) *hold if and only if there exists a universal integral* $I : M \times [0,1]^n \to [0,1]$ *with a pseudo-multiplication* $\otimes_{\mathscr{F}}$ *such that, for all* $f \in \mathscr{F}$ *there exists an* $m_f \in M$ *for which*

$$f(x) = \int_{univ, \otimes_{\mathscr{F}}} x \, dm_f \qquad \text{for all} \quad x \in [0,1]^n .$$

More precisely, for all $f \in \mathscr{F}$ *and for all* $A \subseteq N$, $m_f(A) = f(1_A)$ *and for all* $a, b \in [0,1], \otimes_{\mathscr{F}}(a,b) = f(a1_B)$ *if* $f(1_B) = b$, *with* $B \subseteq N$.

Remark 1. One can weaken axiom (A_3) as follow.

$(\widetilde{A_3})$ For all $c \in [0,1]$ there exist $A \subseteq N$ and $f \in \mathscr{F}$ such that $f(1_A) = c$.

In this case the above Proposition 1 holds provided that the universal integral is no more defined as a function $I : M \times [0,1]^n \to [0,1]$, but as a function $I : M_{\mathscr{F}} \times [0,1]^n \to [0,1]$ with $M_{\mathscr{F}} \subseteq M$. More precisely, we have $M_{\mathscr{F}} = \{m_f | f \in \mathscr{F}\}$.

Remark 2. In order to prove the sufficient part of Proposition 1 it would be sufficient the following two axioms on the function $I : M \times [0,1]^n \to [0,1]$

(UI2) there exists a pseudo-multiplication $\otimes$ (i.e. $\otimes : [0,1]^2 \to [0,1]$ is nondecreasing in its two coordinates and $\otimes(c,1) = \otimes(1,c) = c$) such that for all $m \in M$, $c \in [0,1]$ and $A \subseteq N$,

$$I(m, c1_A) = \otimes(c, m(A)),$$

(UI3)' for all $m_1, m_2 \in M$ and $x, y \in [0,1]^n$, if $m_1(\{x \geq t\}) \geq m_2(\{y \geq t\})$ for all $t \in]0,1]$, then $I(m_1, x) \geq I(m_2, y)$.

On the converse, these two axioms are implied by condition $(A_1) - (A_3)$ and, then the definition of universal integral can be equivalently given substituting axioms $(UI1) - (UI3)$ with axioms $(UI2), (UI3)'$.

3   Axiomatic Foundation of Level Dependent Capacity-Based Universal Integral in Terms of Aggregation Functions

Definition 1. A level dependent capacity is a function $m_{LD} : 2^N \times [0,1] \to [0,1]$ such that its restriction $m_{LD}(\cdot, t) : 2^N \to [0,1]$ is a capacity for any $t \in [0,1]$.

Definition 2. (Mesiar) A level dependent capacity is a family $m_{LD} = (m_t)_{t \in]0,1]}$ of set functions $m_t : 2^N \to [0,1]$ where each m_t is a capacity.

We denote by M_{LD} the set of all level dependent capacities m_{LD}. Given two level dependent capacities $m_{LD}, m_{LD}^* \in M_{LD}$ we say that m_{LD} is smaller than m_{LD}^* and we write $m_{LD} \leq m_{LD}^*$ if $m_{LD}(E, t) \leq m_{LD}^*(E, t)$ for all $E \subseteq N$ and $t \in]0,1]$.
For each pair $(m_{LD}, x) \in M_{LD} \times [0,1]^n$ we can define the function

$$h^{(m_{LD}, x)} :]0,1] \to [0,1]$$

by

$$h^{(m_{LD}, x)}(t) = m_{LD}(\{x \geq t\}, t)$$

which, in general, is neither monotone nor Borel measurable. For a fixed $m_{LD} \in M_{LD}$ the vector $x \in [0,1]^n$ (or the function $x : N \to [0,1]$) is called m_{LD}−measurable if the function $h^{(m_{LD}, x)}$ is Borel measurable. The set of all m_{LD}−measurable vectors in $[0,1]^n$ will be denoted by $[0,1]^n_{m_{LD}}$. Moreover, we define

$$\mathscr{L}_{[0,1]} = \bigcup_{m_{LD} \in M_{LD}} m_{LD} \times [0,1]^n_{m_{LD}}.$$

Definition 3. A function $L : \mathscr{L}_{[0,1]} \to [0,1]$ is called a level dependent capacity-based universal integral [5] if the following axioms hold

(UIL1) L is non-decreasing in each component, i.e., for all $m_{LD}, m_{LD}^* \in M_{LD}$ satisfying $m_{LD} \leq m_{LD}^*$, and for all $x_1 \in [0,1]^n_{m_{LD}}$, $x_2 \in [0,1]^n_{m_{LD}^*}$ with $x_1 \preceq x_2$ we have

$$L(m_{LD}, x_1) \leq L(m_{LD}^*, x_2),$$

(UIL2) there is a universal integral $I : M \times [0,1]^n \to [0,1]^n$ such that for each capacity $m \in M$, for each level dependent capacity $m_{LD} \in M_{LD}$ satisfying $m_{LD}(\cdot, t) = m$ for all $t \in [0,1]$, and for each $x \in [0,1]^n_{m_{LD}}$ we have

$$L(m_{LD}, x) = I(m, x),$$

(UIL3) for all pairs (m_{LD}, x), $(m_{LD}^*, y) \in \mathscr{L}_{[0,1]}$ with $h^{(m_{LD}, x)} = h^{(m_{LD}^*, y)}$ we have

$$L(m_{LD}, x) = L(m_{LD}^*, y).$$

Observe that, because of axiom $[(UIL2)]$, each level dependent capacity-based universal integral L is an extension of some universal integral I.

Now, consider a family $\mathscr{F} \subseteq \mathscr{A}_{[0,1]}$ with $\mathscr{F} \neq \emptyset$ and consider the following axioms on $\mathscr{F}$:

$(AULD_1)$ There exists $\mathscr{F}_c \subseteq \mathscr{F}$ that satisfies above axioms $(A_1) - (A_3)$.

$(AULD_2)$ for each $f \in \mathscr{F}$ and for each $t \in [0,1]$ there exists $f^t \in \mathscr{F}_c$, such that for any $x, y \in [0,1]^n$

$$f_1^t(\mathbf{1}_{\{x \geq t\}}) \geq f_2^t(\mathbf{1}_{\{y \geq t\}}) \Rightarrow f_1(x) \geq f_2(y).$$

$(AULD_3)$ For any $m_{LD} \in M_{LD}$ there exists $f_{m_{LD}} \in \mathscr{F}$ such that $f_{m_{LD}}^t(\mathbf{1}_A) = m_{LD}(\mathbf{1}_A, t)$ for any $A \subseteq N$ and for any $t \in [0,1]$.

Proposition 2. *Axioms $(AULD_1)$, $(AULD_2)$ and $(AULD_3)$ hold if and only if there exists a level dependent capacity-based universal integral $L : \mathscr{L}_{[0,1]} \to [0,1]$ which is an extension of some universal integral I with a pseudo-multiplication $\otimes_{\mathscr{F}}$ such that, for all $f \in \mathscr{F}$ there exists an $m_{LD,f} \in M_{LD}$ for which*

$$f(x) = L\left(m_{LD,f}, x\right) \text{ for all } \quad x \in [0,1]^n_{m_{LD,f}}.$$

More precisely, for all $f \in \mathscr{F}$, for all $A \subseteq N$ and for all $t \in [0,1]$, $m_{LD,f}(A,t) = f^t(\mathbf{1}_A)$ and for all $a, b \in [0,1], \otimes_{\mathscr{F}}(a,b) = f(aI_B)$ if $f(\mathbf{1}_B) = b$, with $f \in \mathscr{F}_c$ and $B \subseteq N$.

Proof. see proofs of proposition 1 and of [5, Theorem 4.4]. □

4 Axiomatic Foundation in Terms of Preference Relations

We consider the following primitives:

- a set of outcomes X,
- a set of binary preference relations $\mathscr{R} = \{\succsim_t, t \in \mathbf{T}\}$ on $X^n, n \in \mathbb{N}$.

In the following

- we shall denote by α the constant vector $[\alpha, \alpha, \dots, \alpha] \in X^n$, with $\alpha \in X$;
- we shall denote by $\succ_t$ and $\sim_t$ the asymmetric and the symmetric part of $\succsim_t \in \mathscr{R}$, respectively;
- we shall denote by (α_A, β_{N-A}), $\alpha, \beta \in X, A \subset N, x \in X^n$ such that $x_i = \alpha$ if $i \in A$ and $x_i = \beta$ if $i \notin A$.

We consider the following axioms:

A1) $\succsim_t$ is a complete preorder on X^n for all $\succsim_t \in \mathscr{R}$.

A2) For all $\alpha, \beta \in X$ and for all $\succsim_t, \succsim_r \in \mathscr{R}$, $\alpha \succsim_t \beta \Rightarrow \alpha \succsim_r \beta$.

A3) X is infinite and there exists a countable subset $A \subseteq X$ such that for all $\succsim_t \in \mathscr{R}$, for all $\alpha, \beta \in X$ for which $\alpha \succ_t \beta$ there is $\gamma \in A$ such that $\alpha \succsim_t \gamma \succsim_t \beta$.

A4) There are $1, 0 \in X$ such that for all $\succsim_t \in \mathscr{R}$ $\mathbf{1} \succ_t \mathbf{0}$ and for all $\boldsymbol{x} \in X^n$,

$$\mathbf{1} \succsim_t \boldsymbol{x} \succsim_t \mathbf{0}.$$

A5) For each $\boldsymbol{x} \in X^n$ and for each $\succsim_t \in \mathscr{R}$, there exists $\alpha \in X$ such that $\boldsymbol{x} \sim_t \alpha$.

A6) For each $\boldsymbol{x} \in X^n$, for each $\alpha, \beta \in X$, for each $i \in N$, and for each $\succsim_r, \succsim_t \in \mathscr{R}$,

$$\alpha \succsim_r \beta \Rightarrow (\alpha_i, \boldsymbol{x}_{-i}) \succsim_t (\beta_i, \boldsymbol{x}_{-i}).$$

A7) For all $\boldsymbol{x}, \boldsymbol{y} \in X^n$, $\succsim_t, \succsim_r, \succsim_s \in \mathscr{R}$,

$$[(1_{\{i \in N : x_i \succsim_t \alpha\}}, 0_{N - \{i \in N : x_i \succsim_t \alpha\}}) \succsim_r \beta \Rightarrow (1_{\{i \in N : y_i \succsim_t \alpha\}}, 0_{N - \{i \in N : y_i \succsim_t \alpha\}}) \succsim_s \beta, \forall \alpha, \beta \in X]$$

$$\Rightarrow$$

$$[\boldsymbol{x} \succsim_r \gamma \Rightarrow \boldsymbol{y} \succsim_s \gamma, \forall \gamma \in X].$$

A8) For all $\mathscr{A} = \{\alpha_1, \ldots, \alpha_p\} \subseteq X$, $1 \leq p \leq 2^n - 2$, there exists $\succsim_t \in \mathscr{R}$ such that for all $\alpha \in \mathscr{A}$ there is A, $\emptyset \subset A \subset N$, for which $\alpha \sim_t 1_A$.

Theorem. Conditions $A1) - A7)$ hold if and only if there exist

- a function $u : X \to [0,1]$,
- a bijection between $\mathscr{R}$ and M for which each $\succsim_t \in \mathscr{R}$ corresponds to one capacity $\mu_t \in M$,
- a pseudo-multiplication $\otimes$,

such that, for all $\boldsymbol{x}, \boldsymbol{y} \in X^n$ and for all $\succsim_t \in \mathscr{R}$

$$\boldsymbol{x} \succsim_t \boldsymbol{y} \Leftrightarrow \int_{univ, \otimes} \mathbf{u}(\boldsymbol{x}) d\mu_t \geq \int_{univ, \otimes} \mathbf{u}(\boldsymbol{y}) d\mu_t,$$

where $\mathbf{u}(\boldsymbol{x}) = [u(x_1), \ldots, u(x_n)]$ and $\mathbf{u}(\boldsymbol{y}) = [u(y_1), \ldots, u(y_n)]$.

Proof. We leave the proof of the necessity to the reader and we give the proof of the sufficiency. By A1), each $\succsim_t \in \mathscr{R}$ induces a complete preorder $\succsim_t^c$ on the family of constant vectors $C = \{[\alpha, \ldots, \alpha] \in X^n | \alpha \in X\}$, i.e. for all $\succsim_t \in \mathscr{R}$ and for all $\alpha, \beta \in C, \alpha \succsim_t^c \beta$ iff $\alpha \succsim_t \beta$. By A2), we have $\succsim_{t_1}^c = \succsim_{t_2}^c$ for all $\succsim_{t_1}, \succsim_{t_2} \in \mathscr{R}$. Thus in the following we shall write $\alpha \succsim \beta$ if for some $\succsim_t \in \mathscr{R}$ $\alpha \succsim_t \beta$, $\alpha, \beta \in C$. Of course, $\succsim$ is a complete preorder on C. By A3), due to the Debreu open gap lemma [1] there is a function $u : C \to Y$ with Y an interval of real numbers $\mathfrak{R}$, such that, for all $\alpha, \beta \in C$

$$\alpha \succsim \beta \Leftrightarrow u(\alpha) \geq u(\beta).$$

By A4), there is a minimum and a maximum in Y, such that, without loss of the generality we can set $Y = [0,1]$, such that $u : C \to [0,1]$, $u(\mathbf{1}) = 1$ and $u(\mathbf{0}) = 0$. By A5), for each $\succsim_t \in \mathscr{R}$ one can define a function $h_t : X^n \to [0,1]$ as follows: for all $\boldsymbol{x} \in X^n$

$$h_t(x) = u(\alpha) \text{ with } \alpha \in X \text{ such that } x \sim_t \alpha.$$

By A6) we get that for all $x, y \in X^n$

$$[x_i \succsim y_i \text{ for all } i \in N] \Rightarrow [x \succsim_r y \text{ for all } \succsim_r \in \mathscr{R}].$$

For each $\succsim_t \in \mathscr{R}$ we can define a function $f_t : [0,1]^n \to [0,1]$ non decreasing in each argument such that $f(\mathbf{1}) = 1$ and $f(\mathbf{0}) = 0$, i.e. an aggregation function on $[0,1]^n$, as follows: for all $x \in X^n$,

$$f_t(u(x_1), \ldots, u(x_n)) = h_t(x).$$

Thus for all $x, y \in X^n$ and for all $\succsim_t \in \mathscr{R}$,

$$x \succsim_t y \Rightarrow f_t(u(x_1), \ldots, u(x_n)) \geq f_t(u(y_1), \ldots, u(y_n)).$$

Observe that axiom (A_2) is satisfied because for all $t \in T$ function f_t is idempotent because for all $\alpha \in X$,

$$f_t(u(\alpha), \ldots, u(\alpha)) = h_t(\alpha) = u(\alpha).$$

Also axiom (A_1) is satisfied because, using function $f_t, t \in T$, axiom A7) can be rewritten as follows: for all $x, y \in X^n$, $\succsim_r, \succsim_s \in \mathscr{R}$,

$$[f_r(\mathbf{1}_{\{i \in N : u(x_i) \geq u(\alpha)\}}) \geq u(\beta) \Rightarrow f_s(\mathbf{1}_{\{i \in N : u(y_i) \geq u(\alpha)\}}) \geq u(\beta), \forall \alpha, \beta \in X]$$

$$\Rightarrow$$

$$[f_r(x) \geq u(\gamma) \Rightarrow f_s(y) \geq u(\gamma), \forall \gamma \in X].$$

This is equivalent to

$$[f_r(\mathbf{1}_{\{i \in N : u(x_i) \geq t\}}) \geq f_s(\mathbf{1}_{\{i \in N : u(y_i) \geq t\}}) \forall t \in [0,1]]$$

$$\Rightarrow$$

$$[f_r(x) \geq f_s(y)].$$

Finally, observe that also axiom (A_3) is satisfied, because by A8) there is a function f_t for all capacity $m \in M$. Indeed, for all capacity $m \in M$ one can pick $\alpha_1, \ldots, \alpha_{2^n - 2}$, not necessarily all different each other and ordered from the smaller to the larger with possible ex aequo, such that $m(\mathbf{1}_{A_1}) = f_t(\alpha_1), \ldots, m(\mathbf{1}_{A_{2^n - 2}}) = f_t(\alpha_{2^n - 2})$, where $A_k \not\subseteq A_h$ if $h < k$. Since axioms $(A_1) - (A_3)$ are satisfied, by above Proposition 1, the set of aggregation functions $\mathscr{F} = \{f_t : \succsim_t \in \mathscr{R}\}$ defines a universal integral and we conclude our proof.

Acknowledgements. The work on this contribution was partially supported by the grants VEGA 1/0184/12 and GACRP 402/11/0378.

References

[1] Debreu, G.: Representation of a preference ordering by a numerical function. In: Decision Processes, pp. 159–165 (1954)

[2] Figueira, J., Greco, S., Ehrgott, M.: Multiple criteria decision analysis: state of the art surveys, vol. 78. Springer (2005)

[3] Grabisch, M., Marichal, J., Mesiar, R., Pap, E.: Aggregation functions: means. Information Sciences 181(1), 1–22 (2011)

[4] Klement, E., Mesiar, R., Pap, E.: A universal integral as common frame for Choquet and Sugeno integral. IEEE Transactions on Fuzzy Systems 18(1), 178–187 (2010)

[5] Klement, E.P., Kolesárová, A., Mesiar, R., Stupnanová, A.: A generalization of universal integrals by means of level dependent capacities. Knowledge-Based Systems 38, 14–18 (2013),
http://www.sciencedirect.com/science/article/pii/S0950705112002419,
doi:10.1016/j.knosys.2012.08.021

Part X
Incomplete Data

Aggregation of Incomplete Qualitative Information

Juan Vicente Riera and Joan Torrens

Abstract. In this article we propose a method to construct aggregation functions on the set of discrete fuzzy numbers whose support is any subset of natural numbers, from discrete aggregation functions defined on $L_n = \{0, 1, \cdots, n\}$. The interest on these discrete fuzzy numbers lies on the fact that, when their support is a closed interval, they can be interpreted as linguistic expert valuations that increase the flexibility of the elicitation of qualitative information based on linguistic terms. When the support is not an interval of L_n, the corresponding discrete fuzzy number can be interpreted as an incomplete linguistic expert valuation. From the results on this work we can manage this incomplete information in some different ways.

Keywords: aggregation function, linguistic variable, discrete fuzzy number, subjective evaluation, decision making.

1 Introduction

The process of merging some data into a representative output is usually carried out by the so-called aggregation functions that have been extensively investigated in the last decades [1, 2, 11]. Decision making, subjective evaluations, optimization and control are, among others, examples of concrete application fields where aggregation functions become an essential tool. In all these fields, it is well known that the data to be aggregated vary among many different kinds of information, from quantitative to qualitative information. Moreover, many times some uncertainty is inherent to such information.

Qualitative information is often interpreted to take values in a totally ordered finite scale like this:

$$\mathscr{L} = \{Extremely\ Bad,\ Very\ Bad,\ Bad,\ Fair,\ Good,\ Very\ Good,\ Extremely\ Good\}. \tag{1}$$

In these cases, the representative finite chain $L_n = \{0, 1, \ldots, n\}$ is usually considered to model these linguistic hedges and several researchers have developed an extensive

Juan Vicente Riera · Joan Torrens
University of the Balearic Islands, Palma de Mallorca 07122, Spain
e-mail: {jvicente.riera,jts224}@uib.es

H. Bustince et al. (eds.), *Aggregation Functions in Theory and in Practise*,
Advances in Intelligent Systems and Computing 228,
DOI: 10.1007/978-3-642-39165-1_47, © Springer-Verlag Berlin Heidelberg 2013

study of aggregation functions on L_n, usually called *discrete aggregation functions* (see [7, 8, 9]). Another approximation is based on assigning a fuzzy set to each linguistic term trying to capture its meaning. However, the modelling of linguistic information is limited because the information provided by experts for each variable must be expressed by a simple linguistic term. In most cases, this is a problem for experts because their opinion does not agree with a concrete term. On the contrary, experts' values are usually expressions like *"better than Good"*, *"between Fair and Very Good"* or even more complex expressions.

To avoid the limitation above (see [4, 5, 10]) the authors deal with the possibility of extending monotonic operations on L_n to operations on the set of discrete fuzzy numbers whose support is a set of consecutive natural numbers contained in L_n (i.e, an interval contained in L_n), usually denoted by $\mathscr{A}_1^{L_n}$. The idea lies on the fact that any discrete fuzzy number $A \in \mathscr{A}_1^{L_n}$ can be considered (identifying the scale $\mathscr{L}$ given in (1) with L_n with $n = 6$) as an assignment of a $[0, 1]$-value to each term in our linguistic scale. As an example, the above mentioned expression *"between Fair and Very Good"* can be performed, for instance, by a discrete fuzzy number $A \in \mathscr{A}_1^{L_6}$, with support given by the subinterval $[F, VG]$ (that corresponds to the interval $[3,5]$ in L_6). The values of A in its support should be described by experts, allowing in this way a complete flexibility of the qualitative valuation. A possible discrete fuzzy number A representing the expression mentioned above is given in Figure 1 (note that there are pictured only the values of A in its support).

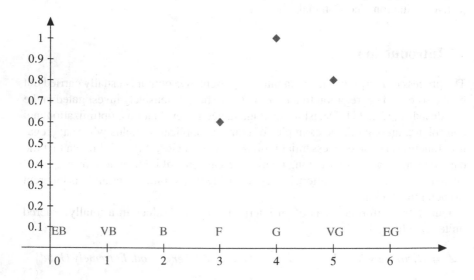

Fig. 1 Graphical representation of a discrete fuzzy number whose support is the interval $[3,5]$. In addition, note that this fuzzy set can be interpreted as expression "between Fair and Very Good", after identifying the linguistic scale $\mathscr{L}$ with the chain L_6.

Thus, aggregation functions on $\mathscr{A}_1^{L_n}$ will allow us to manage qualitative information in a more flexible way. In [4] t-norms and t-conorms on $\mathscr{A}_1^{L_n}$ are described and studied, as well as it is done for uninorms, nullnorms and general aggregation functions in [10]. In both cases, an example of application in decision making or subjective evaluation is included.

Thus, discrete fuzzy numbers in $\mathscr{A}_1^{L_n}$ can be interpreted as flexible qualitative information and they have been successfully used in decision making problems and subjective evaluation. On the other hand, note that when we take a discrete fuzzy number A whose support is not an interval of L_n (that is with A not in $\mathscr{A}_1^{L_n}$), say for instance $supp(A) = \{i_1, i_2, \ldots, i_k\}$, there are some items in the interval $[i_1, i_k]$ that have no assigned value. These gaps in the support can be interpreted as lacks of information and any discrete fuzzy number of this type as an incomplete qualitative information. This lack of information can be produced by many reasons. For instance, because some parts of the information have been lost during the process, because the expert was unable to perform a more detailed valuation, or many others.

In this paper we want to study how to aggregate this type of incomplete information and we will do it in two different ways. First in Section 3, we present a method to directly merging this incomplete information through aggregation functions on the set of all discrete fuzzy numbers, with the particularity that the obtained result is going to be again incomplete. Second, in Section 4, we will complete the compiled information in many different ways through discrete associations, and then we will aggregate the results, obtaining in this case a complete decision.

2 Preliminaries

In this section, we recall some definitions and the main results about discrete fuzzy numbers which will be used later. By a fuzzy subset of $\mathbb{R}$, we mean a function $A : \mathbb{R} \to [0,1]$. For each fuzzy subset A, let $A^\alpha = \{x \in \mathbb{R} : A(x) \geq \alpha\}$ for any $\alpha \in (0,1]$ be its α-level set (or α-cut). By $supp(A)$, we mean the support of A, i.e. the set $\{x \in \mathbb{R} : A(x) > 0\}$. By A^0, we mean the closure of $supp(A)$.

Definition 1. [13] A fuzzy subset A of $\mathbb{R}$ with membership mapping $A : \mathbb{R} \to [0,1]$ is called a *discrete fuzzy number* if its support is finite, i.e., there exist $x_1, \ldots, x_n \in \mathbb{R}$ with $x_1 < x_2 < \ldots < x_n$ such that $supp(A) = \{x_1, \ldots, x_n\}$, and there are natural numbers s, t with $1 \leq s \leq t \leq n$ such that:

1. $A(x_i)=1$ for any natural number i with $s \leq i \leq t$ (*core*)
2. $A(x_i) \leq A(x_j)$ for each natural number i, j with $1 \leq i \leq j \leq s$
3. $A(x_i) \geq A(x_j)$ for each natural number i, j with $t \leq i \leq j \leq n$

Remark 1. If the fuzzy subset A is a discrete fuzzy number then the support of A coincides with its closure, i.e. $supp(A) = A^0$.

From now on, we will denote the set of discrete fuzzy numbers by DFN and the abbreviation *dfn* will denote a discrete fuzzy number.

From now on, we will denote by $\mathscr{A}_1^{L_n}$ the set of discrete fuzzy numbers whose support is a subset of consecutive natural numbers contained in the finite chain L_n and by $\mathscr{D}_{L_n}$ the set of discrete fuzzy numbers whose support is a subset of natural numbers contained in L_n.

Remark 2. Note that $\mathscr{A}_1^{L_n}$ is a subset of $\mathscr{D}_{L_n}$.

Let $A, B \in \mathscr{A}_1^{L_n}$ be two discrete fuzzy numbers, and let $A^\alpha = [x_1^\alpha, x_p^\alpha]$, $B^\alpha = [y_1^\alpha, y_k^\alpha]$ be their α-level cuts for A and B respectively.

The following result holds for $\mathscr{A}_1^{L_n}$, but is not true for the set of discrete fuzzy numbers in general(see [3]).

Theorem 1. *[3] The triplet $(\mathscr{A}_1^{L_n}, MIN, MAX)$ is a bounded distributive lattice where $1_n \in \mathscr{A}_1^{L_n}$ (the unique discrete fuzzy number whose support is the singleton $\{n\}$) and $1_0 \in \mathscr{A}_1^{L_n}$ (the unique discrete fuzzy number whose support is the singleton $\{0\}$) are the maximum and the minimum, respectively, and where $MIN(A,B)$ and $MAX(A,B)$ are the discrete fuzzy numbers belonging to the set $\mathscr{A}_1^{L_n}$ such that they have the sets*

$$\begin{aligned} MIN(A,B)^\alpha &= \{z \in L_n \mid \min(x_1^\alpha, y_1^\alpha) \leq z \leq \min(x_p^\alpha, y_k^\alpha)\} \text{ and} \\ MAX(A,B)^\alpha &= \{z \in L_n \mid \max(x_1^\alpha, y_1^\alpha) \leq z \leq \max(x_p^\alpha, y_k^\alpha)\} \end{aligned} \quad (2)$$

as α-cuts respectively for each $\alpha \in [0,1]$ and $A, B \in \mathscr{A}_1^{L_n}$.

Remark 3. [3] Using these operations, we can define a partial order on $\mathscr{A}_1^{L_n}$ in the usual way:
$A \preceq B$ if and only if $MIN(A,B) = A$, or equivalently, $A \preceq B$ if and only if $MAX(A,B) = B$ for any $A, B \in \mathscr{A}_1^{L_n}$. Equivalently, we can also define the partial ordering in terms of α-cuts:
  $A \preceq B$ if and only if $\min(A^\alpha, B^\alpha) = A^\alpha$
  $A \preceq B$ if and only if $\max(A^\alpha, B^\alpha) = B^\alpha$

Similarly, using the same operations due by expression (2) and the corresponding relation $\preceq$ in the set $\mathscr{D}_{L_n}$, the following result holds.

Theorem 2. *The structure $\mathscr{D}_{L_n} = (\mathscr{D}_{L_n}, \preceq, 1_0, 1_n)$ is a bounded partially ordered set where $1_n \in \mathscr{D}_{L_n}$ and $1_0 \in \mathscr{D}_{L_n}$ represent the maximum and the minimum, respectively.*

Aggregation functions on L_n were extended to $\mathscr{A}_1^{L_n}$ (see [4, 10]) through the following theorem.

Theorem 3. *[4, 10] Let us consider a binary aggregation function F on the finite chain L_n. The binary operation on $\mathscr{A}_1^{L_n}$ defined as follows*

$$\begin{aligned} \mathscr{F} : \mathscr{A}_1^{L_n} \times \mathscr{A}_1^{L_n} &\longrightarrow \mathscr{A}_1^{L_n} \\ (A,B) &\longmapsto \mathscr{F}(A,B) \end{aligned}$$

being $\mathscr{F}(A,B)$ the discrete fuzzy number whose α-cuts are the sets

$$\{z \in L_n \mid \min F(A^\alpha, B^\alpha) \leq z \leq \max F(A^\alpha, B^\alpha)\}$$

for each $\alpha \in [0,1]$ *is an aggregation function on* $\mathscr{A}_1^{L_n}$. *This function will be called the extension of the discrete aggregation function F to* $\mathscr{A}_1^{L_n}$. *In particular, if F is a t-norm, a t-conorm, an uninorm or a nullnorm its extension* $\mathscr{F}$ *too.*

3 Aggregation Functions on $\mathscr{D}_{L_n}$

In [5, 10] the discrete fuzzy numbers on $\mathscr{A}_1^{L_n}$ were interpreted like subjective evaluations made by experts. In this work, the discrete fuzzy numbers defined on the set $\mathscr{D}_{L_n}$ which does not belong to $\mathscr{A}_1^{L_n}$ will be interpreted as an incomplete subjective valuation made by experts. In this section we will provide a procedure to aggregate this incomplete subjective evaluation.

Aggregation functions on bounded partially ordered sets have been deeply studied (see for instance [6, 12, 14]). Now, we will see that from an aggregation function F defined on the finite chain L_n it is possible to establish an aggregation function $\mathscr{F}$ on the bounded partially ordered set $\mathscr{D}_{L_n}$.

Proposition 1. *Let F be an aggregation function on* L_n *and* $A, B \in \mathscr{D}_{L_n}$. *For each* $\alpha \in [0,1]$ *let us consider the sets*

$$C_{F,A,B}^\alpha = \{z \in F(supp(A), supp(B)) \mid \min F(A^\alpha, B^\alpha) \leq z \leq \max F(A^\alpha, B^\alpha)\}.$$

There exists an unique discrete fuzzy number, denoted by $\mathfrak{F}(A,B)$, *such that its* α-*cut sets are the sets* $C_{F,A,B}^\alpha$ *for any* $\alpha \in [0,1]$.

Proposition 1 enables us to define a binary operation $\mathfrak{F}$ on $\mathscr{D}_{L_n}$ from the aggregation function F defined on L_n as follows,

Definition 2. Let F be an aggregation function on L_n. The binary operation on $\mathscr{D}_{L_n}$

$$\mathfrak{F} : \mathscr{D}_{L_n} \times \mathscr{D}_{L_n} \longrightarrow \mathscr{D}_{L_n}$$
$$(A,B) \longmapsto \mathfrak{F}(A,B)$$

will be called *the extension of the discrete aggregation function F to* $\mathscr{D}_{L_n}$, being $\mathfrak{F}(A,B)$ the discrete fuzzy number whose α-level sets are the sets

$$\{z \in F(supp(A), supp(B)) \mid \min F(A^\alpha, B^\alpha) \leq z \leq \max F(A^\alpha, B^\alpha)\}$$

for each $\alpha \in [0,1]$.

Now we wish to prove that $\mathfrak{F}$, as defined above, is an aggregation function on the bounded partially ordered set $\mathscr{D}_{L_n}$.

Proposition 2. *Let* $\mathfrak{F} : \mathscr{D}_{L_n} \times \mathscr{D}_{L_n} \to \mathscr{D}_{L_n}$ *be the extension of F to* $\mathscr{D}_{L_n}$. *Let* 1_0 *and* 1_n *the minimum and the maximum of* $\mathscr{D}_{L_n}$ *respectively. Then the following properties hold:*

1. $\mathfrak{F}$ *is increasing in each component,*
2. $\mathfrak{F}(1_0, 1_0) = 1_0,$
3. $\mathfrak{F}(1_n, 1_n) = 1_n.$

Example 1. Consider the discrete fuzzy numbers $A = \{0.5/0, 0.8/3, 1/5, 0.7/6\}$, $B = \{0.5/0, 0.8/4, 1/6, 0.7/8\} \in \mathscr{D}_{L_8}$. Let us consider the following t-norm

$$T(x,y) = \begin{cases} \max(0, x+y-3) & \text{if } x \in [0,3] \\ \max(3, x+y-8) & \text{if } x \in [3,8] \\ \min(x,y) & \text{otherwise} \end{cases} \tag{3}$$

on L_8.

A simple calculation shows that $\mathfrak{T}(A,B) = \{0.5/0, 1/3, 0.7/4, 0.7/5, 0.7/6\}$.

Remark 4. Consider the linguistic hedge $\mathfrak{L} = \{EB, VB, B, MB, F, MG, G, VG, EG\}$ where the letters refer to the linguistic terms Extremely Bad, Very Bad, Bad, More or Less Bad, Fair, More or Less Good, Good, Very Good and Extremely Good and they are listed in an increasing order:

$$EB \prec VB \prec B \prec MB \prec F \prec MG \prec G \prec VG \prec EG$$

It is obvious that we can consider a bijective application between this ordinal scale $\mathfrak{L}$ and the finite chain $L_8 = \{0, 1, 2, 3, 4, 5, 6, 7, 8\}$ of natural numbers which keep the order. Furthermore, each normal convex fuzzy subset defined on the ordinal scale $\mathfrak{L}$ can be considered like a discrete fuzzy number belonging to $\mathscr{A}_1^{L_8}$, and viceversa. Similarly, each element of $\mathscr{D}_{L_8}$ can be interpret as a convex fuzzy subset of the same ordinal scale $\mathfrak{L}$ too. Thus, if we interpret each one of the two discrete fuzzy numbers $A, B \in \mathscr{D}_{L_8}$ of the previous examples as incomplete subjective evaluations (formally understood as normal convex fuzzy subsets defined on the linguistic scale $\mathfrak{L}$ such that their support is not an interval of $\mathfrak{L}$), they can be expressed as

$$A = \{0.5/EB, 0.8/MB, 1/MG, 0.7/EG\} \tag{4}$$
$$B = \{0.5/EB, 0.8/F, 1/G, 0.7/EG\} \tag{5}$$

Observe that in the evaluation due by expression (4) above, there is a loss of information between the linguistic labels EB and MB, between MB and MG and between MG and EG too. Similarly, in the evaluation represented by the expression (5), there is a loss of information between the linguistic labels EB and F, F and G and finally G and EG.

Now the aggregations obtained in the previous example can be understood as an incomplete subjective evaluation too, and it can be written as

$$\mathfrak{T}(A,B) = \{0.5/EB, 1/MB, 0.7/F, 0.7/MG, 0.7/G\} \tag{6}$$

According to expression (6), it is worth noting that the process of aggregation of incomplete subjection evaluations provides incomplete information in general (see figure 2).

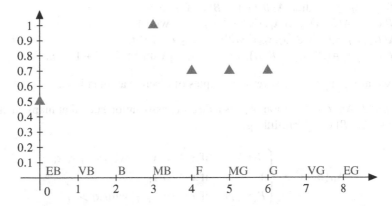

Fig. 2 Aggregation on the incomplete subjective evaluations A and B using the extension of the t-norm represented by formula (3) on L_8. The values of the labels VB and B are unknown or advisedly omitted.

In next section, we introduce a method that will allow us to construct complete subjective evaluations from incomplete subjective evaluations.

4 Aggregation of Incomplete Subjective Evaluations Based on Discrete Associations

In [4, 5, 10] the authors deal with the construction of aggregation functions defined on the set of all discrete fuzzy numbers whose support is a subset of consecutive natural numbers and the particular cases of t-norms, t-conorms, uninorms and nullnorms are studied in detail. These aggregation functions are constructed from discrete aggregation functions (defined on a finite chain) and they are applied to the aggregation of complete subjective evaluations. So, using theses results we will propose a procedure that will allow to aggregate incomplete subjective evaluations.

4.1 Discrete Association Functions

Definition 3. Let $B \in \mathscr{D}_{L_n}$ a discrete fuzzy number whose support is the set $supp(B) = \{x_1, ..., x_s, ..., x_t, ..., x_m\}$ with $x_1 < \cdots < x_s < \cdots < x_t < \cdots < x_m$ and $B(x_p) = 1$ for all p such that $s \leq p \leq t$. A *discrete association* is a mapping

$$\mathbf{A} : \mathscr{D}_{L_n} \to \mathscr{A}_1^{L_n}$$
$$B \mapsto \mathbf{A}(B)$$

such that maps each discrete fuzzy number $B \in \mathscr{D}_{L_n}$ into $\mathbf{A}(B) \in \mathscr{A}_1^{L_n}$ fulfilling the following properties:

1. If $x_i \in supp(B)$ then $\mathbf{A}(B)(x_i) = B(x_i)$ for each $i = 1, \ldots, m$.
2. $B(x_i) \leq \mathbf{A}(B)(x) \leq B(x_{i+1})$, $\forall x \in [x_i, x_{i+1}]$ with $1 \leq i \leq i+1 \leq s$.
3. $\mathbf{A}(B)(x_i) = 1$, $\forall x \in [x_i, x_{i+1}]$ with $s \leq i \leq i+1 \leq t$.
4. $B(x_{i+1}) \leq \mathbf{A}(B)(x_i) \leq B(x_i)$, $\forall x \in [x_i, x_{i+1}]$ with $t \leq i \leq i+1 \leq m$.

Now we are going to give several examples of discrete associations.

Example 2. An α-association, $\mathbf{A}_\alpha$, is a discrete association such that maps each $B \in \mathscr{D}_{L_n}$ into $\mathbf{A}_\alpha(B) \in \mathscr{A}_1^{L_n}$ fulfilling:

$$\mathbf{A}_\alpha(B)(x) = \begin{cases} B(x_i) & \text{if } x \in [x_i, x_{i+1}) \text{ with } x_{i+1} < x_s \\ 1 & \text{if } x \in [x_s, x_t] \\ B(x_{i+1}) & \text{if } x \in (x_i, x_{i+1}] \text{ with } x_i > x_t \end{cases}$$

For instance:
 Let $B = \{0.4/2, 1/5, 1/6, 0.8/8\} \in \mathscr{D}_{L_8}$, then

$$\mathbf{A}_\alpha(B)(x) = \begin{cases} 0.4 & \text{if } x \in \{2,3,4\} \\ 1 & \text{if } x \in \{5,6\} \\ 0.8 & \text{if } x \in \{7,8\} \end{cases}$$

Figure 3 shows the transformation process of the discrete fuzzy number B into $\mathbf{A}_\alpha(B)$ using the α-association defined above.

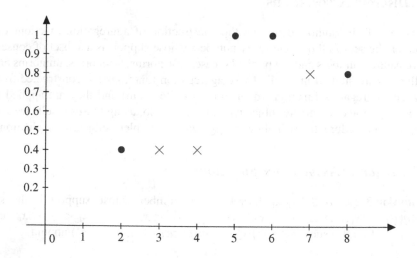

Fig. 3 The blue points represent the discrete fuzzy number B from Example 2 and the red cross corresponding to the points that are added to construct $\mathbf{A}_\alpha(B) \in \mathscr{A}_1^{L_8}$

Example 3. An ω-association, $\mathbf{A}_\omega$, is a discrete association such that maps each $C \in \mathscr{D}_{L_n}$ into $\mathbf{A}_\omega(C) \in \mathscr{A}_1^{L_n}$ defined as

$$\mathbf{A}_\omega(C)(x) = \begin{cases} C(x) & \text{if } x \in supp(C) \\ C(x_{i+1}) & \text{if } x \in (x_i, x_{i+1}) \text{ with } x_{i+1} \leq x_s \\ 1 & \text{if } x \in (x_s, x_t) \\ C(x_i) & \text{if } x \in (x_i, x_{i+1}) \text{ with } x_i \geq x_t \end{cases}$$

For instance:

Let $C = \{0.4/1, 0.6/4, 1/5, 0.8/6, 0.6/8\} \in \mathscr{D}_{L_8}$, then

$$\mathbf{A}_\omega(C)(x) = \begin{cases} 0.4 & \text{if } x \in \{1\} \\ 0.6 & \text{if } x \in \{2, 3, 4\} \\ 1 & \text{if } x \in \{5\} \\ 0.8 & \text{if } x \in \{6, 7\} \\ 0.6 & \text{if } x \in \{8\} \end{cases}$$

Figure 4 depicts the transformation process of $C \in \mathscr{D}_{L_8}$ into $\mathbf{A}_\omega(C)$ from the ω-association.

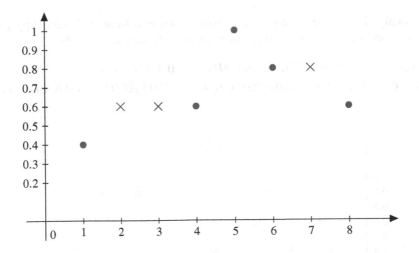

Fig. 4 Red points represent the discrete fuzzy number C from Example 3 and the blue points correspondint to the points that are added to construct $\mathbf{A}_\omega(C) \in \mathscr{A}_1^{L_8}$.

Remark 5. According to Remark 4 the discrete fuzzy numbers B and C (considered in Example 2 and Example 3 above) can be interpreted as the incomplete subjective evaluations (see figure 5)

$$B = \{0.4/B, 1/MG, 1/G, 0.8/EG\} \tag{7}$$
$$C = \{0.4/VB, 0.6/F, 1/MG, 0.8/G, 0.6/EG\} \tag{8}$$

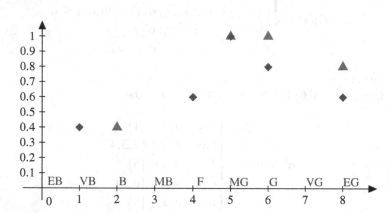

Fig. 5 The green triangles represent the incomplete subjective evaluation B and the red diamons show the incomplete subjective evaluation C

Finally, the complete subjective evaluation obtained from $B, C \in \mathcal{D}_{L_8}$ applying the α-association and the ω-association respectively are (see figure 6)

$$\mathbf{A}_\alpha(B) = \{0.4/B, 0.4/MB, 0.4/F, 1/MG, 1/G, 0.8/VG, 0.8/EG\} \tag{9}$$
$$\mathbf{A}_\omega(C) = \{0.4/VB, 0.6/B, 0.6/MB, 0.6/F, 1/MG, 0.8/G, 0.8/VG, 0.6/EG\} \tag{10}$$

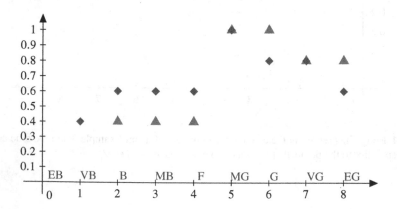

Fig. 6 The green triangles represent the complete subjective evaluation $\mathbf{A}_\alpha(B)$ and the red diamons show the complete subjective evaluation $\mathbf{A}_\omega(C)$.

4.2 The Aggregation Method

The idea is to complete any discrete fuzzy number $A_j \in \mathscr{D}_{L_n}$ (incomplete information) into a complete information $\mathbf{A}(A_j) \in \mathscr{A}_1^{L_n}$ using a fixed discrete association $\mathbf{A}$. Then we can aggregate these discrete fuzzy numbers in $\mathscr{A}_1^{L_n}$ through the extension of a discrete aggregation function F, obtaining the final complete aggregation $\mathscr{F}(\mathbf{A}(A_1), \cdots, \mathbf{A}(A_r)) \in \mathscr{A}_1^{L_n}$.

Specifically, the proposed method is as follows:

First step: We chose the linguistic scale $\mathscr{L}$ the experts will use in order to make the subjective evaluations. From this scale, let $L_n = \{0, 1, \cdots, n\}$ be the finite chain that is bijective with $\mathscr{L}$.

Second step: Suppose that $O_1, \cdots, O_j$ represent the incomplete subjective evaluations that we wish aggregate. We let $\tilde{O}_1, \cdots, \tilde{O}_j$ denote the discrete fuzzy numbers belonging to $\mathscr{D}_{L_n}$ which have built from the corresponding evaluations O_k, with $1 \leq k \leq j$.

Third step: Let us choose a discrete association $\mathbf{A}$ that will be used in order to transform each discrete fuzzy number $\tilde{O}_k \in \mathscr{D}_{L_n}$ with $1 \leq k \leq j$ into the discrete fuzzy number $\mathbf{A}(\tilde{O}_k) \in \mathscr{A}_1^{L_n}$ with $1 \leq k \leq j$.

Fourth step: We get an aggregation function F on L_n and according to Theorem 3 we consider its extension $\mathscr{F}$ to $\mathscr{A}_1^{L_n}$ that will use to compute the aggregation of all $\mathbf{A}(\tilde{O}_k) \in \mathscr{A}_1^{L_n}$ with $1 \leq k \leq j$, obtaining $\mathscr{F}(\mathbf{A}(\tilde{O}_1), \cdots, \mathbf{A}(\tilde{O}_j))$.

Fifth step: Finally, we can choose as final aggregation $\mathscr{F}(\mathbf{A}(\tilde{O}_1), \cdots, \mathbf{A}(\tilde{O}_j)) \in \mathscr{A}_1^{L_n}$ and then to interpret this discrete fuzzy number as a subjective evaluation explained in terms of the linguistic scale $\mathscr{L}$.

Remark 6. Note that the discrete fuzzy number $\mathscr{F}(\mathbf{A}(\tilde{O}_1), \cdots, \mathbf{A}(\tilde{O}_j))$ defined according to the fourth step above does not coincide in general with the discrete fuzzy number $\mathfrak{F}(\tilde{O}_1, \cdots, \tilde{O}_j)$, where $\mathfrak{F}$ denotes the extension of G to $\mathscr{D}_{L_n}$ following Definition 2. Of course we could complete the information $\mathfrak{F}(\tilde{O}_1, \cdots, \tilde{O}_j)$ using the same association $\mathbf{A}$ obtaining $\mathbf{A}(\mathfrak{F}(\tilde{O}_1, \cdots, \tilde{O}_j))$. In general this procedure does not coincide with our proposed method as we can see in the following example.

Example 4. Let us consider the discrete fuzzy numbers $A = \{0.5/0, 0.8/3, 1/5, 0.7/6\}$ and $B = \{0.5/0, 0.8/4, 1/6, 0.7/8\}$ of Example 1, and the t-norm T due by the expression (3). If we apply the ω-association $\mathbf{A}_\omega$ to $\mathfrak{T}(A, B)$ it results

$$\mathbf{A}_\omega(\mathfrak{T}(\mathfrak{A}, \mathfrak{B})) = \{0.5/0, 1/1, 1/2, 1/3, 0.7/4, 0.7/5, 0.7/6\} \qquad (11)$$

On the other hand, if we compute $\mathscr{T}(\mathbf{A}_\omega(A), \mathbf{A}_\omega(B))$ it results

$$\mathscr{T}(\mathbf{A}_\omega(A), \mathbf{A}_\omega(B)) = \{0.8/0, 0.8/1, 0.8/2, 1/3, 1/4, 0.7/5, 0.7/6\} \qquad (12)$$

Thus, we can see that in general $\mathbf{A}_\omega(\mathfrak{T}(\mathfrak{A}, \mathfrak{B})) \neq \mathscr{T}(\mathbf{A}_\omega(A), \mathbf{A}_\omega(B))$.

Note however that, when we use the α-association $\mathbf{A}_\alpha$ in the previous example, we obtain $\mathbf{A}_\alpha(\mathfrak{T}(\mathfrak{A}, \mathfrak{B})) = \mathscr{T}(\mathbf{A}_\alpha(A), \mathbf{A}_\alpha(B))$. We claim that this fact concerning $\mathbf{A}_\alpha$

is true for all $A, B \in \mathscr{D}_{L_n}$. Moreover, we claim that $\mathbf{A}_\alpha$ is the only association with this property and this is part of our future work on this topic.

Acknowledgements. This work has been partially supported by the MTM2009-10962 and MTM2009-10320 project grants, both with FEDER support.

References

[1] Beliakov, G., Pradera, A., Calvo, T.: Aggregation Functions: A Guide for Practicioners. STUDFUZZ, vol. 221. Springer, Heidelberg (2007)

[2] Calvo, T., Mayor, G., Mesiar, R.: Aggregation Operators: New Trends and Applications. STUDFUZZ, vol. 97. Springer, Heidelberg (2002)

[3] Casasnovas, J., Riera, J.V.: Lattice properties of discrete fuzzy numbers under extended min and max. In: Proceedings IFSA-EUSFLAT, Lisbon, Portugal, pp. 647–652 (July 2009)

[4] Casasnovas, J., Riera, J.V.: Extension of discrete t-norms and t-conorms to discrete fuzzy numbers. Fuzzy Sets and Systems 167(1), 65–81 (2011)

[5] Casasnovas, J., Riera, J.V.: Weighted means of subjective evaluations. In: Seising, R., Sanz González, V. (eds.) Soft Computing in Humanities and Social Sciences. STUDFUZZ, vol. 273, pp. 331–354. Springer, Heidelberg (2012)

[6] De Baets, B., Mesiar, R.: Triangular norms on product lattices. Fuzzy Sets and Systems 104, 61–75 (1999)

[7] Kolesarova, A., Mayor, G., Mesiar, R.: Weighted ordinal means. Information Sciences 177, 3822–3830 (2007)

[8] Mas, M., Mayor, G., Torrens, J.: t-operators and uninorms on a finite totally ordered set. International Journal of Intelligent Systems 14, 909–922 (1999)

[9] Mayor, G., Torrens, J.: Triangular norms on discrete settings. In: Klement, E.P., Mesiar, R. (eds.) Logical, Algebraic, Analytic, and Probabilistic Aspects of Triangular Norms, pp. 189–230. Elsevier, Netherlands (2005)

[10] Riera, J.V., Torrens, J.: Aggregation of subjective evaluations based on discrete fuzzy numbers. Fuzzy Sets and Systems 191, 21–40 (2012)

[11] Narukawa, Y., Torra, V.: Modeling Decisions: Information Fusion and Aggregation Operators. Springer (2007)

[12] Kormorníková, M., Mesiar, R.: Aggregation functions on bounded partially ordered sets and their classification. Fuzzy Sets and Systems 175(1), 48–56 (2011)

[13] Voxman, W.: Canonical representations of discrete fuzzy numbers. Fuzzy Sets and Systems 54, 457–466 (2001)

[14] Zhang, D.: Triangular norms on partially ordered sets. Fuzzy Sets and Systems 153, 195–209 (2005)

Consistency and Stability in Aggregation Operators: An Application to Missing Data Problems

Daniel Gomez, Karina Rojas, Javier Montero, and J. Tinguaro Rodríguez

1 Introduction

An *aggregation operator* [1, 5, 7, 8, 9, 12] is usually defined as a real function A_n such that, from n data items $x_1, \ldots, x_n$ in $[0, 1]$, produces an aggregated value $A_n(x_1, \ldots, x_n)$ in [0,1] [4]. This definition can be extended to consider the whole family of operators for any n instead of a single operator for an specific n. This has led to the current standard definition [4, 15] of a *family of aggregation operators (FAO)* as a set $\{A_n : [0,1]^n \to [0,1], n \in N\}$, providing instructions on how to aggregate collections of items of any dimension n. This sequence of aggregation functions $\{A_n\}_{n \in N}$ is also called *extended aggregation functions (EAF)* by other authors [15, 5].

In this work, we will deal with two different but related problems for *extended aggregation functions* or *family of aggregation operators*

On one hand, let us remark that in practice, it is frequent that some information can get lost, be deleted or added, and each time a cardinality change occurs a new aggregation operator A_m has to be used to aggregate the new collection of m elements. However, it is important to remark that a relation between $\{A_n\}$ and $\{A_m\}$ does not necessarily exist in a family of aggregation operators as defined in [4]. In this context, it seems natural to incorporate some properties to maintain the logical *consistency* between operators in a *FAO* when changes on the cardinality of the data occur, for which we need to be able to build up a definition of family of aggregation operators in terms of its logical consistency, and solve each problem of aggregation without knowing *apriori* the cardinality of the data. This is, the operators that compose a *FAO* have to be somehow related, so the aggregation process

Daniel Gómez
Faculty of Statistics, Av. Puerta de Hierro s/n 28040, Madrid, Spain
e-mail: dagomez@estad.ucm.es

Karina Rojas · Javier Montero · J. Tinguaro Rodríguez
Faculty of Mathematics, Plz. Ciencias s/n 28040, Madrid, Spain
e-mail: krpatuelli@yahoo.com, monty@mat.ucm.es, jtrodrig@mat.ucm.es

H. Bustince et al. (eds.), *Aggregation Functions in Theory and in Practise*,
Advances in Intelligent Systems and Computing 228,
DOI: 10.1007/978-3-642-39165-1_48, © Springer-Verlag Berlin Heidelberg 2013

remains *the same* throughout the possible changes in the dimension n of the data. Therefore, it seems logical to study properties giving sense to the sequences $A(2)$, $A(3), A(4), \ldots$, because otherwise we may have only a bunch of disconnected operators. With this aim, in [26, 27, 16] a notion of consistency based on the robustness of the aggregation process, i.e. *stability*, was studied. In this sense, the notion of *stability* for a family of aggregation operators is inspired in continuity, though our approach focuses in the cardinality of the data rather than in the data itself, so we can assure some robustness in the result of the aggregation process. Particularly, let $A_n(x_1, \ldots, x_n)$ be the aggregated value of the n-dimensional data $x_1, \ldots, x_n$. Now, let us suppose that a new element x_{n+1} has to be aggregated. If x_{n+1} is close to the aggregation result $A_n(x_1, \ldots, x_n)$ of the n-dimensional data $x_1, \ldots, x_n$, then the result of aggregating these $n+1$ elements should not differ too much with the result of aggregating such n items. Following the idea of stability for any mathematical tool, if $|x_{n+1} - A_n(x_1, \ldots, x_n)|$ is small, then $|A_{n+1}(x_1, \ldots, x_n, x_{n+1}) - A_n(x_1, \ldots, x_n)|$ should be also small. It is important to note that if the family $\{A_n\}$ is not symmetric (i.e. there exist a n for which the aggregation operator A_n is not symmetric), then the position of the new data is relevant to the final output of the aggregation process. From this observation, in [26, 27, 16] it some definitions of *stability* that extend the notion of self-identity defined in [29] were presented.

On the other hand, a problem that has not been received too much attention is how to obtain an aggregation when some of the variables to be aggregated are missing. If the aggregation operator function A_n present a clear definition for the case in which the dimension is lower, this problem is easily solved, but not always is a trivial task. Following the ideas of stability, in this paper we will deal with the problem of missing data for some well-known families of aggregation operators.

2   Consistency in Families of Aggregation Operators

As has been pointed out in the introduction, a *family of aggregation operators* (*FAO*) is a set of aggregation operators $\{A_n : [0,1]^n \to [0,1], n \in N\}$, providing instructions on how to aggregate collections of items of any dimension n. Few properties has been studied or defined to a *FAO* in general (see [27] for more details). In [15] it is shown that the aggregation functions of a family can be related by means of certain grouping properties. For example, continuity, symmetry or other well-known properties defined usually for aggregation functions can be defined in a general way for a family of aggregation operators imposing that these properties have to be satisfied for all n. Nevertheless, these kind of properties don't guarantee any consistency in the aggregation process since they don't establish any constraint among the different aggregation functions.

In the aggregation operators' literature it is possible to find some properties for aggregation operators that can be understood as properties for the whole family establishing some relations among the different aggregation operators. Here we recall some of them.

An important notion that establish some relations among members of different dimensions in a FAO is the notion of recursivity. Recursivity was introduced in [8] in the context of OWA operators aggregation functions. Further, and following [8], in [1, 9, 12, 18] recursivity of a FAO was also studied in a more general way to establish some consistency in the aggregation process. In order to understand this notion of recursivity, first it necessary to defined the concept of *ordering rule*.

Definition 1 (see [1, 9, 12] for more details). Let us denote $\pi_n(x_1,\ldots,x_n) = (x_{\pi_n(1)},\ldots,x_{\pi_n(n)})$. An *ordering rule* π is a consistent family of permutations $\{\pi_n\}_{n\geq 2}$ such that for any possible finite collections of numbers, each extra item x_{n+1} is allocated keeping previous relative positions of items, i.e.

$\pi_{n+1}(x_1,\ldots,x_n,x_{n+1})$ equal to $(x_{\pi_n(1)},\ldots,x_{\pi_n(j-1)},x_{n+1},x_{\pi_n(j)},\ldots x_{\pi_n(n)})$ for some $j \in \{1,2,\ldots,n+1\}$.

Definition 2 (see [1, 9, 12] for more details). A family of aggregation operators $\{A_n : [0,1]^n \longrightarrow [0,1]\}_{n>1}$ is left-recursive if there exist a family of binary operators $\{L_n : [0,1]^2 \longrightarrow [0,1]\}_{n>1}$ verifying $A_2(x_1,x_2) = L_2(x_{\pi(1)};x_{\pi(2)})$ and
$A_n(x_1,x_2,\ldots,x_n)$ equal to $L_n(A_{n-1}(x_{\pi(1)}),\ldots,x_{\pi(n-1)};x_{\pi(n)})$ for all $n \geq 2$,
where π is an ordering rule.

In a similar way, it is possible to define the right recursive rules.

Definition 3 (see [1, 9, 12] for more details). A family of aggregation operators $\{A_n : [0,1]^n \longrightarrow [0,1]\}_{n>1}$ is left-recursive if there exist a family of binary operators $\{R_n : [0,1]^2 \longrightarrow [0,1]\}_{n>1}$ verifying $A_2(x_1,x_2) = R_2(x_{\pi(1)};x_{\pi(2)})$ and
$A_n(x_1,x_2,\ldots,x_n)$ equal to $R_n(x_1,A_{n-1}(x_{\pi(2)},\ldots,x_{\pi(n-1)}))$ for all $n \geq 2$,
where π is an ordering rule.

From previous two definitions, it is possible to introduce the concept of LR recursivity in the following way.

Definition 4 (see [1, 9, 12] for more details). A family of aggregation operators $\{A_n : [0,1]^n \longrightarrow [0,1]\}_{n>1}$ is left-right recursive if there exist two families of binary operators $\{R_n : [0,1]^2 \longrightarrow [0,1]\}_{n>1}$ and $\{L_n : [0,1]^2 \longrightarrow [0,1]\}_{n>1}$ verifying the left and the right recursive conditions simultaneously.

A particular case of previous definitions (when the binary aggregation is the same and the recursivity is from the left) can be founded in [4, 5], in which it is said that a FAO $\{A_n\}_{n\in N}$ will be *recursive* if it verifies

$$A_n(x_1,x_2,..,x_n) = A_2(A_{n-1}(x_1,x_2,..,x_{n-1}),x_n).$$

Let us observe that previous definitions guarantee certain consistency in the family $\{A_n\}$ since the A_n function is build taking into account the previous function A_{n-1}. Taking this definition into account, the previously described situation, in which the different operators A_n have no relation among them, cannot hold.

Other properties that establish some conditions among the different members of the whole family are the following:

Definition 5. Decomposability. [see [4, 5] for more details]

A family of aggregation operators $\{A_n\}_{n\in N}$ satisfies the *decomposability* property if $\forall n, m = 1, 2, ...,\ \forall x \in [0,1]^m$ and $\forall y \in [0,1]^n$ the following holds:

$$A_{m+n}(x_1, x_2, .., x_m, y_1, y_2, .., y_n) =$$
$$A_{m+n}(\underbrace{A_m(x_1, x_2, .., x_m), ..., A_m(x_1, x_2, .., x_m)}_{m \text{ times}}, y_1, y_2, .., y_n)$$

Definition 6. Bisymmetry [see [4, 5] for more details]

A family of aggregation operators $\{A_n\}_{n\in N}$ satisfies the *bisymmetry* property if $\forall n, m = 1, 2, ...$ and $\forall x \in [0,1]^{mn}$ the following holds:

$$A_{mn}(x_1, x_2, .., x_{mn}) = A_m(A_n(x_{11}, x_{12}, .., x_{1n}), ..., A_n(x_{m1}, x_{m2}, .., x_{mn}))$$
$$= A_n(A_m(x_{11}, x_{21}, .., x_{m1}), ..., A_m(x_{1n}, x_{2m}, .., x_{mn}))$$

Although previous definitions impose some stability or consistency to a family of aggregation operators, these ones are more focused on the way in which it is possible to build the operator aggregation function of dimension n from aggregation operators of lower dimensions than on a general idea of stability or consistency. Moreover/However, pursuing the idea of consistency of a family of aggregation operators and based on the self-identity definition given by Yager in [29], in [26, 27, 16] the notion of *strict stability* of a FAO was defined in three different levels. The idea is simple: in a family of aggregation operators, A_n and A_{n+1} should be closely related, in the sense that if a new item has to be aggregated and such a new item is the result of the aggregation of the previous n items, then the result of the aggregation of these $n+1$ items should be close to the aggregation of the n previous ones. Otherwise, the aggregation operator function A_{n+1} would differ too much from the aggregation operator function A_n, producing and *unstable* family $\{A_n\}_{n\in N}$. Taking into account that in general FAOs are not necessarily symmetric, two possibilities (left and right stability) were analyzed in the definition of strict stability.

Definition 7. Let $\{A_n : [0,1]^n \rightarrow [0,1], n \in N\}$ be a family of aggregation operators. Then, it is said that:

1. $\{A_n\}_n$ is a R-strictly stable family if

$$A_n(x_1, x_2, ...x_{n-1}, A_{n-1}(x_1, x_2..., x_{n-1})) = A_{n-1}(x_1, x_2, ..., x_{n-1})$$

   holds $\forall n \geq 3$ and $\forall \{x_n\}_{n\in N}$ in $[0,1]$

2. $\{A_n\}_n$ is a L-strictly stable family if

$$A_n(A_{n-1}(x_1, x_2, ..., x_{n-1}), x_1, x_2, ...x_{n-1}) = A_{n-1}(x_1, x_2, ..., x_{n-1})$$

   holds $\forall n \geq 3$ and $\forall \{x_n\}_{n\in N}$ in $[0,1]$

Although previous definitions can be relaxed from an asymptotic and probabilistic point of view (see [27]), in this work we are going to focus on the strict stability conditions just exposed.

3 On j-L and i-R Stability

Previous definitions impose that the information that has to be aggregated appears in the last or in the first position. Obviously, this assumption could be relaxed. Taking into account that the stability concept presented in [27] could be relaxed, in [2], it is introduced the notion $j - L$ stability, imposing now that the new datum enters in the i-th position from the right. And similarly, we can define the $i - R$ strictly stability imposing that the new datum enters in the j-th position. Obviously, the relaxed versions of strict stability from an asymptotic and probabilistic point of view could be defined in a similar way.

Definition 8. Let $\{A_n : [0,1]^n \to [0,1], n \in N\}$ be a family of aggregation operators. Then, it is said that:

1. $\{A_n\}_n$ is a i-R-strictly stable family if

$$A_n(x_1, x_2, \dots x_{n-i}, A_{n-1}(x_1, x_2 \dots, x_{n-1}), \dots, x_{n-1}) = A_{n-1}(x_1, x_2, \dots, x_{n-1})$$

holds $\forall n \geq 3$ and $\forall \{x_n\}_{n \in N}$ in $[0,1]$.

2. $\{A_n\}_n$ is a j-L-strictly stable family if

$$A_n(x_1, \dots, x_{j-1}, A_{n-1}(x_1, x_2, \dots, x_{n-1}), x_j, \dots, x_{n-1}) = A_{n-1}(x_1, x_2, \dots, x_{n-1})$$

holds $\forall n \geq 3$ and $\forall \{x_n\}_{n \in N}$ in $[0,1]$.

Let us observe that the $i - L$ and $j - R$ strict stability conditions previously defined are equivalent (for any i and/or j) when the FAO is symmetric. But, in general, it is very difficult that a non-symmetric FAO satisfies simultaneously more than one condition (see [27] for more details). In our opinion, the conditions that a general FAO should satisfy to be strictly stable shouldtake into account the structure of the data that has to be aggregated (and of course also the way in which this family is defined).

In a similar way as symmetric FAOs impose indirectly that the structure of the data hasn't effect in the aggregation result (since the order in which the information is aggregated is not relevant), non-symmetric families of aggregation operators makes the assumption that the data has an inherent structure and thus the position of the data items in the aggregation process is relevant. Strict stability or consistency of an aggregation process (among other properties) should also take into account that the data may present some structure. In the following section, we will present some possible definitions of stability for non-symmetric FAO that will be dependent of the structure of the data that is aggregated.

4   Dealing with Weights and Missing Data: An Application of Stability

To illustrate an interesting application of the concept of stability let us introduce a very simple example. Suppose a multi-criteria decision problem having four criteria C_1, C_2, C_3, C_4. A jury, after some deliberations, evaluates the different alternatives on the four criteria and then uses a weighted mean operator as aggregation rule. Then, what happens if for one alternative some information related with the criteria C_4 has been lost or deleted? What should be the aggregation of the remaining information? As has been pointed out in the introduction, this dimensional problem has not received too much attention in the aggregation literature. This problem is related with the following question: what should be the relations between aggregation operators of different dimensions to be consistent? In this section, we will analyze the stability of some well-known families trying to deal with this problem.

Let us recall again that our aim is not to decide how the vector of weights $w^4 = (w_1^4, w_2^4, w_3^4, w_4^4)$ should be, but to guarantee some stability or consistence in the aggregation process. For example, it would seem rather inconsistent to choose $w^4 = (1/8, 4/8, 1/8, 2/8)$ if data is available regarding the four mentioned criteria, but also choosing $w^3 = (0.8, 0.2, 0)$ in case the criteria C_4 presents a missing value for one of the alternatives. From the point of view of consistency, this jury would not be stable.

We first focus our attention in the weighted mean aggregation family. This family, $\{W_n, n \in N\}$, is defined through a vector of weights $w^n = (w_1^n, ..., w_n^n) \in [0, 1]^n$ in such a way that $W_n(x_1, ..., x_n) = \sum_{i=1}^{n} w_i^n x_i$, where $\sum_{i=1}^{n} w_i^n = 1$ and $(x_1, ..., x_n) \in [0, 1]^n$ $\forall n$. The stability of this family was studied from a LR point of view in [27]. Nevertheless, as we will see below, this study can not be directly applicable to the missing value problem in aggregation problems. In the $\{W_n\}_n$ FAO, the weights associated to the elements being aggregated represent the *importance* of each one of these elements in the aggregation process. For this reason, the weighted mean surely is one of the most relevant and used aggregation operators in many different areas (e.g. statistics, knowledge representation problems, fuzzy logic, multiple criteria decision making, group decision making, etc.), and one of the most studied problems in all these areas is how to determine these *importance* weights.

A missing data problem appears when for a specific object $x = (x_1, ..., x_n)$ one of its values is missing. In the previous example, $n = 4$, the information regarding an alternative is aggregated through $W_4(x_1, ..., x_4) = \sum_{i=1,4} w_i^4 x_i$, and the importance of the four criteria has been established by means of the four dimensional vector $w^4 = (w_1^4, w_2^4, w_3^4, w_4^4) = (1/8, 2/4, 1/8, 1/4)$. Now, consider an alternative x that presents the values $x = (0.3, 1, 1, not\ evaluable)$. What should be the aggregation? What should be the aggregation operator A_3?

If we decide to use the weighted mean aggregation function for $n = 3$ (i.e. $A_3 = W_3$), the problem here is to determine the weights vector w^3. A possibility is to impose that W_3 and W_4 satisfy the strict stability conditions. Nevertheless, asstudied in [27], for non-symmetric FAOs, it is very difficult that more than one stability condition is satisfied simultaneously. Let us observe that the different strict stability conditions (L, R, $i - L$ or $j - R$ for different i and j positions) will give us different possibilities and solutions for the vector w^3. So, what stability condition should we choose? Taking into account that the 4-th value x_4 is the one missing, it seems reasonable to impose the R (or equivalently the 4-L or 1-R) strict stability condition, i.e.

$$W_4(x_1, x_2, x_3, W_3(x_1, x_2, x_3)) = W_3(x_1, x_2, x_3)$$

for any x_1, x_2 and x_3 in $[0,1]$ Concerning our example, this condition holds if and only if

$$\frac{1}{8}x_1 + \frac{4}{8}x_2 + \frac{1}{8}x_3 + 2/8(w_1^3 x_1 + w_2^3 x_2 + w_3^3 x_3) = w_1^3 x_1 + w_2^3 x_2 + w_3^3 x_3 \quad \forall x_1, x_2, x_3 \in [0,1],$$

which is equivalent to say that $w^3 = (\frac{1}{6}, \frac{4}{6}, \frac{1}{6})$. Let us observe that this vector maintains the relative proportions between the original weights for the non-missing values in the positions 1, 2 and 3.

In the previous example, the fourth value of the alternative $x = (0.3, 1, 1, not\ evaluable)$ is missing. But what should be the aggregation if the missing value is the second one? In general, and for non-symmetric FAOs where the position in which data appear is relevant, if there is some information $x = (x_1, \ldots, x_n)$ that has to be aggregated and we have a missing value x_j, we should impose strict $j - L$ stability or equivalently $(n - (j + 1)) - R$ strict stability to find the relations that should exist between the aggregation functions A_n and A_{n-1} in the whole family. In the following proposition it is established a condition that guarantees the strict j-L stability of the family $\{W_n\}_n$.

Proposition 1. *(see also [2]) Let $w^n = (w_1^n, \ldots, w_n^n) \in [0,1]^n, n \in N$, be a sequence of weights of a weighted mean family $\{W_n\}_{n \in N}$ such that $\sum_{i=1}^{n} w_i^n = 1$ holds $\forall n \geq 2$. Then, the family $\{W_n\}_{n \in N}$ is a j-L-strict stable family if and only if the sequence of weights satisfies*

$$\begin{cases} w_k^n = (1 - w_j^n) \cdot (w_k^{n-1}) & for\ k = 1, \ldots, j-1 \\ w_{k+1}^n = (1 - w_j^n) \cdot (w_k^{n-1}) & for\ k = j, \ldots, n-1 \end{cases} \quad \forall n \in N.$$

Proof

Note that for a generic weighted mean FAO $\{W_n\}_{n \in N}$ with weights w^n, $n \in N$, the j-L-strict stability property can be restated as

$$0 = |A_n(x_1, \ldots, x_{j-1}, A_{n-1}(x_1, x_2, \ldots, x_{n-1}), x_j, \ldots, x_{n-1}) - A_{n-1}(x_1, x_2, \ldots, x_{n-1})|$$

which is equivalent to

$$\sum_{i=1}^{j-1}(w_i^n-(1-w_j^n)w_i^{n-1})x_i+\sum_{i=j}^{n-1}(w_{i+}^n-(1-w_j^n)w_i^{n-1})x_i=0 \forall x_1,\ldots x_{n-1}\in[0,1].$$

From previous equation it is straightforward to conclude that the proposition holds.

In order to extend the previous properties to a more general class of FAO, we will analyze the j-L strict stability for transformations of the original FAO. But, let us first introduce the following notations and definitions.

Definition 9. Let $f:[0,1]\to A$ be a continuous and injective function, and let $\{\phi_n: A\to A, n\in N\}$ be a family of aggregation operators defined in the domain A. Then, the transformed aggregation operator family $\{M_f^{\phi_n}\}_{n\in N}$ is defined as:

$$M_f^{\phi_n}(x_1,\ldots,x_n)=f^{-1}(\phi_n(f(x_1),\ldots,f(x_n)))$$

Let us observe that if f is the identity function, then the transformation family coincides with the original family. If $\{\phi_n\}_{n\in N}$ is the mean or the weighted mean then $M_f^{\phi_n}$ is called quasi-arithmetic mean or weighted quasi-arithmetic mean. The quasi-arithmetic mean functions are very important in many aggregation analysis. Some well-known quasi-arithmetic aggregation families are: the geometric mean (when $f(x)=log(x)$), the harmonic mean (when $f(x)=1/x$) and the power mean (when $f(x)=x^p$), among others. It is important to remark that some of the most usual aggregation operators families (as for example the productory $\{P_n\}_{n\in N}$), can not be transformed or extended directly. For example if $f(x)=5x$, then $A=[0,5]$, but we can not guarantee that for all $n\in N$, $P_n(f(x_1),\ldots,f(x_n))=\prod_{i=1}^n f(x_i)$ belongs to the interval $[0,5]$.

In the following proposition, we show that j-L-strict stability remains after transformation.

Proposition 2. *Let $\{\phi_n\}_{n\in N}$ and $\{M_f^{\phi_n}\}_{n\in N}$ be a family of aggregation operators and its extension or transformed aggregation. Then:*

$\{M_f^{\phi_n}\}_{n\in N}$ *is a j-L-strictly stable family if and only if $\{\phi_n\}_{n\in N}$ is a j-L -strictly stable family in the domain A.*

Proof:
Taking into account that $M_f^{\phi_n}\left(x_1,\ldots,x_{j-1},M_f^{\phi_n}(x_1,\ldots,x_{n-1}),\ldots,x_n\right)$ can be rewritten as

$$f^{-1}\left(\phi_n\left(f(x_1),\ldots,f(x_{j-1}),\phi_{n-1}(f(x_1),\ldots,f(x_{n-1})),\ldots,f(x_n)\right)\right),$$

the j-L strict stability condition for $\{M_f^{\phi_n}\}_{n\in N}$ can be formulated as

$$f^{-1}\left(\phi_n\left(f(x_1),\ldots,f(x_{j-1}),\phi_{n-1}(f(x_1),\ldots,f(x_{n-1})),\ldots,f(x_n)\right)\right) =$$

$$f^{-1}\left(\phi_{n-1}(f(x_1),\ldots,f(x_{n-1}))\right).$$

Hence, since f is a continuous and injective function, such a condition holds if and only if $\{\phi_n\}_n$ is an strictly stable family in A. And thus, the proposition holds.

Corollary. The weighted quasi-arithmetic aggregation operators family is a j-L-strict stable family if and only if the sequence of weights satisfies

$$\begin{cases} w_k^n = (1 - w_j^n)\cdot(w_k^{n-1}) & for\ k = 1,\ldots,j-1 \\ w_{k+1}^n = (1 - w_j^n)\cdot(w_k^{n-1}) & for\ k = j,\ldots,n-1 \end{cases} \quad \forall n \in N.$$

To conclude the study of the missing values in aggregation operators from a stability point of view, we will try to extend the previous analysis to a situation in which more than one value could be missing. Let us suppose that we have two missing values in the positions $r < s$. So we have $x = (x_1,\ldots,x_{r-1},missing,\ldots,mising,x_{s+1},\ldots,x_n)$. Let us observe that the j-L strict stability condition can be stated in a more general way by imposing conditions between the aggregation functions A_n and A_{n-2} for this purpose.

Definition 10. Let $\{A_n : [0,1]^n \to [0,1], n \in N\}$ be a family of aggregation operators. Then, it is said that $\{A_n\}_n$ is a $r - s$-L-strictly stable family if

$$A_n(x_1,\ldots,x_{r-1},A_{n-2}(x_1,x_2,\ldots,x_{n-2}),x_r,\ldots,A_{n-2}(x_1,x_2,\ldots,x_{n-2}),x_{s-1},\ldots,x_{n-2})$$

coincides with $A_{n-2}(x_1,x_2,\ldots,x_{n-2})\ \forall n \geq 3$ and $\forall\{x_n\}_{n\in N}$ in $[0,1]$

Following this equation of the $r - s$-L strict stability, it is possible to build the aggregation operator A_{n-2} from A_n for a given n. Let us continue with the example of the four criteria. If now for one alternative the values associated with the criteria 2 and 3 are missing, and we decide to use the weighted mean aggregation function for $n = 2$ (i.e. $A_2 = W_2$), the problem is to determine the weights vector w^2 from w^4 (which is the available information). Then, it seems reasonable to impose the $2 - 3$-L stability condition to find the weights associated with the aggregation operator W_2 i.e:

$$W_4(x_1,W_2(x_1,x_2),W_2(x_1,x_2),x_2) = W_2(x_1,x_2)$$

for any x_1, x_2 in $[0,1]$.

For notational convenience, we have denoted by x_1 the value for the first variable and by x_2 the value for the fourth variable. So it is $x = (x_1,missing,missing,x_2)$. Then, the condition above holds if and only if

$$\frac{1}{8}x_1 + \frac{4}{8}(w_1^2x_1 + w_2^2x_2) + \frac{1}{8}(w_1^2x_1 + w_2^2x_2) + 2/8(x_2) = w_1^2x_1 + w_2^2x_2\ \forall x_1,x_2,x_3 \in [0,1],$$

which is equivalent to say that $w^2 = (\frac{1}{3}, \frac{2}{3})$. Let us observe that this vector maintains the relative proportions between the original weights for the non-missing values in the positions 1 and 4.

We would like to conclude this section pointing out that it is possible to define strict stability for a sequence of positions $r_1, \ldots r_k$ in a similar way as done above for two positions, allowing us to establish consistency conditions between the aggregation functions A_n and A_{n-k}.

5 Final Comments

In this work, we have continued with the key issue of the relationship that should hold between the operators in a family $\{A\}_n$ in order to understand they properly define a *consistent*. The basic concepts of consistency addressed as stability of a family of aggregation operators was presented in [27, 16, 26] in which it is defined the L and R strict stability in different levels. In this work we have extended some of these previous definitions into a more general framework defining the $i - L$ and $j - R$ strict stability for a family of aggregation operators and some of its analysis to the weighted quasi-arithmetic means families. In addition, we present an interesting application of the strict stability conditions to deal with missing data problems in an aggregation operator framework.

References

[1] Amo, A., Montero, J., Molina, E.: Representation of consistent recursive rules. European Journal of Operational Research 130, 29–53 (2001)

[2] Beliakov, G., James, S.: Stability of weighted penalty-based aggregation functions. Fuzzy Sets and Systems, doi:10.1016/j.fss.2013.01.007

[3] Beliakov, G., Pradera, A., Calvo, T.: Aggregation Functions, a Guide to Practitioners. STUDFUZZ, vol. 221. Springer, Heidelberg (2007)

[4] Calvo, T., Kolesarova, A., Komornikova, M., Mesiar, R.: Aggregation operators, properties, classes and construction methods. In: Calvo, T., et al. (eds.) Aggregation Operators New Trends and Aplications, pp. 3–104. Physica-Verlag, Heidelberg (2002)

[5] Calvo, T., Mayor, G., Torrens, J., Suñer, J., Mas, M., Carbonell, M.: Generation of weighting triangles associated with aggregation fuctions. International Journal of Uncertainty, Fuzziness and Knowledge-Based Systems 8(4), 417–451 (2000)

[6] Carlsson, C.H., Fuller, R.: Fuzzy reaasoning in decision making and optimization. Springfield-Verlag, Heidelberg (2002)

[7] Cutello, V., Montero, J.: Hierarchical aggregation of OWA operators: basic measures and related computational problems. Uncertainty, Fuzzinesss and Knowledge-Based Systems 3, 17–26 (1995)

[8] Cutello, V., Montero, J.: Recursive families of OWA operators. In: Proceedings FUZZ-IEEE Conference, pp. 1137–1141. IEEE Press, Piscataway (1994)

[9] Cutello, V., Montero, J.: Recursive connective rules. International Journal of Intelligent Systems 14, 3–20 (1999)

[10] Dubois, D., Gottwald, S., Hajek, P., Kacprzyk, J., Prade, H.: Terminological difficulties in fuzzy set theory - the case of intuitionistic fuzzy sets. Fuzzy Sets and Systems 156, 485–491 (2005)

[11] Fung, L.W., Fu, K.S.: An axiomatic approach to rational decision making in a fuzzy environment. In: Zadeh, L.A., et al. (eds.) Fuzzy Sets and their Applications to Cognitive and Decision Processes, pp. 227–256. Academic Press, New York (1975)

[12] Gómez, D., Montero, J.: A discussion of aggregation functions. Kybernetika 40, 107–120 (2004)

[13] Gómez, D., Montero, J., Yáñez, J., Poidomani, C.: A graph coloring algorithm approach for image segmentation. Omega 35, 173–183 (2007)

[14] Gómez, D., Montero, J., Yánez, J.: A coloring algorithm for image classification. Information Sciences 176, 3645–3657 (2006)

[15] Grabisch, M., Marichal, J., Mesiar, R., Pap, E.: Aggregation Functions. Encyclopedia of Mathematics and its Applications (2009)

[16] Gómez, D., Montero, J., Tinguaro Rodríguez, J., Rojas, K.: Stability in aggregation operators. In: Greco, S., Bouchon-Meunier, B., Coletti, G., Fedrizzi, M., Matarazzo, B., Yager, R.R. (eds.) IPMU 2012, Part III. CCIS, vol. 299, pp. 317–325. Springer, Heidelberg (2012)

[17] Bustince, H., Barrenechea, E., Fernández, J., Pagola, M., Montero, J., Guerra, C.: Aggregation of neighbourhood information by means of interval type 2 fuzzy relations

[18] Bustince, H., de Baets, B., Fernández, J., Mesiar, R., Montero, J.: A generalization of the migrativity property of aggregation functions. Information Sciences 191, 76–85 (2012)

[19] Kolesárová, A.: Sequential aggregation. In: González, M., et al. (eds.) Proceedings of the Fifth International Summer School on Aggregation Operators, AGOP, Universitat de les Illes Balears, Palma de Mallorca, pp. 183–187 (2009)

[20] López, V., Garmendia, L., Montero, J., Resconi, G.: Specification and computing states in fuzzy algorithms. Uncertainty, Fuzziness and Knowledge-Based Systems 16, 301–336 (2008)

[21] López, V., Montero, J.: Software engineering specification under fuzziness. Multiple-Valued Logic and Soft Computing 15, 209–228 (2009)

[22] López, V., Montero, J., Rodriguez, J.T.: Formal Specification and implementation of computational aggregation Functions. In: Computational Intelligence, Foundations and Applications Proceedings of the 9th International FLINS Conference, pp. 523–528 (2010)

[23] Montero, J.: A note on Fung-Fu's theorem. Fuzzy Sets and Systems 13, 259–269 (1985)

[24] Montero, J., Gómez, D., Bustince, H.: On the relevance of some families of fuzzy sets. Fuzzy Sets and Systems 158, 2429–2442 (2007)

[25] Montero, J., López, V., Gómez, D.: The role of fuzziness in decision making. In: Wang, P.P., Ruan, D., Kerre, E.E. (eds.) Fuzzy Logic. STUDFUZZ, vol. 215, pp. 337–349. Springer, Heidelberg (2007)

[26] Rojas, K., Gómez, D., Rodríguez, J.T., Montero, J.: Some properties of consistency in the families of aggregation operators. In: Melo-Pinto, P., Couto, P., Serôdio, C., Fodor, J., De Baets, B. (eds.) Eurofuse 2011. AISC, vol. 107, pp. 169–176. Springer, Heidelberg (2011)

[27] Rojas, K., Gómez, D., Rodríguez, J.T., Montero, J.: Strict Stability in Aggregation Operators. Fuzzy sets and Systems (2013), doi:10.1016/j.fss.2012.12.010
[28] Yager, R.R.: On ordered weighted averaging aggregation operators in multi-criteria decision making. IEEE Transactions on Systems, Man and Cybernetics 18, 183–190 (1988)
[29] Yager, R.R., Rybalov, A.: Nonconmutative self-identity aggregation. Fuzzy Sets and Systems 85, 73–82 (1997)
[30] Yager, R.R.: Prioritized aggregation operators. International Journal of Approximate Reasoning 48, 263–274 (2008)

Part XI
Aggregation on Lattices

Quasi-OWA Operators on Complete Lattices

I. Lizasoain[*]

Abstract. In this paper the concept of an ordered weighted quasi-average (Quasi-OWA or QOWA) operator is extended from $[0,1]$ to any complete lattice endowed with a t-norm and a t-conorm. In the case of a complete distributive lattice it is shown to agree with some OWA operator and consequently with a particular case of the discrete Sugeno integral. As an application, we show several ways of aggregating either restricted equivalence functions or closed intervals by using QOWA operators.

1 Introduction

The problem of selecting among a set of alternatives based on their satisfaction to several criteria is reaching a growing interest due to its many applications in fuzzy set theory, data fusion, multicriteria decision making, etc. ([5], [12], [13]).

The satisfaction degree of alternative A to criterion C_i is commonly represented by a value a_i belonging to a complete lattice $(L, \leq_L)$. If we consider n criteria $\{C_1, \ldots, C_n\}$, each alternative is represented by a vector $(a_1, \ldots, a_n) \in L^n$.

Aggregation functions allow us to combine the data $(a_1, \ldots, a_n)$ in order to obtain a single value that represents the *global* satisfaction degree of alternative A to the collection of criteria $\{C_1, \ldots, C_n\}$. One of the most common aggregation functions is the weighted average $F : L^n \to L$ given by

$$F(a_1, \ldots, a_n) = S[T(\alpha_1, a_1), \cdots, T(\alpha_n, a_n)], \quad (a_1, \ldots, a_n) \in L^n,$$

I. Lizasoain
Universidad Pública de Navarra, 31006 Pamplona, Spain
e-mail: ilizasoain@unavarra.es

[*] The author is partially supported by MTM2010-19938-C03-03.

where T and S are respectively a t-norm and a t-conorm defined on L and the weights $(\alpha_1, \ldots, \alpha_n) \in L^n$ satisfying $S(\alpha_1, \cdots, \alpha_n) = 1_L$ are chosen so that the greatest weights correspond to the prioritized criteria (see [8]).

In [10] Yager introduces an ordered weighted average operator (OWA) defined on $L = [0,1]$ in which the weight associated to each value a_i does not depend on the priority given to criterion C_i, but on the place the value a_i occupies in the ordered sequence of the values $(a_1, \ldots, a_n)$:

If $b_1 \geq \cdots \geq b_n$ is the rearrangement of $(a_1, \ldots, a_n)$, Yager's OWA operator associated to a weighting vector $(\alpha_1, \ldots, \alpha_n) \in [0,1]^n$ with $\alpha_1 + \cdots + \alpha_n = 1$, is given by

$$F_\alpha(a_1, \ldots, a_n) = \sum_{i=1}^{n} \alpha_i b_i, \ (a_1, \ldots, a_n) \in [0,1]^n \tag{1}$$

Unlike the usual weighted average, Yager's OWA operator is symmetric, which means that the decision function obtained does not depend on the order in which the different criteria are considered. OWA operators have been seen as a particular case of a more general kind of operators, the so-called ordered weighted quasi-average operators (or QOWA operators) defined on $[0,1]$ in [6] as

$$F_\alpha(a_1, \ldots, a_n) = g^{-1} \left(\sum_{i=1}^{n} \alpha_i g(b_i) \right), \ (a_1, \ldots, a_n) \in [0,1]^n$$

for any strictly monotonic bijection $g : [0,1] \to [0,1]$.

As a particular case, generalizad OWA operators (GOWA) are defined by Yager in [11] as

$$F_\alpha(a_1, \ldots, a_n) = \left(\sum_{i=1}^{n} \alpha_i b_i^\lambda \right)^{\frac{1}{\lambda}}, \ (a_1, \ldots, a_n) \in [0,1]^n \ (\lambda > 0).$$

In [9] Yager's OWA operators are extended to the case in which the satisfaction degrees belong to any complete lattice L endowed with a t-norm T and a t-conorm S and applied to aggregate closed intervals contained in $[0,1]$.

This paper is devoted to that generalization for QOWA operators. In the case of a distributive lattice, any QOWA operator is shown to agree with some OWA operator and consequently with a discrete Sugeno integral (see [9]).

We also study how to use QOWA operators to aggregate either closed intervals contained in $[0,1]$ or restricted equivalence functions defined on any complete lattice.

The paper is organized as follows. Section 2 is devoted to revisit well-known results concerning to aggregation functions defined on complete lattices. Section 3 introduces the concept of a QOWA operator on any complete lattice and analyzes the case of a distributive lattice. In Section 4, QOWA operators are used to aggregate restricted equivalence functions. Finally, Section 5 shows how to use QOWA operators to aggregate intervals contained in $[0,1]$.

2 Preliminaries

Throughout this paper $(L, \leq_L)$ will denote a complete lattice, i.e., a partially ordered set in which all subsets have both a supremum and an infimum. 0_L and 1_L will respectively stand for the least and the greatest elements of L.

Definition 1 (see [3])

1. A map $T : L \times L \to L$ is said to be a t-norm on $(L, \leq_L)$ if it is commutative, associative, increasing in each component and has a neutral element 1_L.
2. A map $S : L \times L \to L$ is said to be a t-conorm on $(L, \leq_L)$ if it is commutative, associative, increasing in each component and has a neutral element 0_L.

Notation: Throughout this paper $(L, \leq_L, T, S)$ will denote a complete lattice endowed with a t-norm T and a t-conorm S. If the t-norm and the t-conorm considered on a complete lattice $(L, \leq_L)$ are respectively the meet (greatest lower bound) and the join (least upper bound), we will write $(L, \leq_L, \wedge, \vee)$.

Remark 1. The t-norm given by the meet is not necessarily distributive with respect to the t-conorm given by the join (see [7]). Indeed, the following distributive properties are equivalent for any complete lattice $(L, \leq_L)$:

1. $a \wedge (b \vee c) = (a \wedge b) \vee (a \wedge c)$ for all $a, b, c \in L$.
2. $a \vee (b \wedge c) = (a \vee b) \wedge (a \vee c)$ for all $a, b, c \in L$.

A complete lattice $(L, \leq_L, \wedge, \vee)$ in which one (and then both) of the previous distributive properties holds is called *a complete distributive lattice*.

Definition 2. Let $(L, \leq_L)$ be a bounded lattice. Call $(L^n, \leq_{L^n})$ the bounded lattice given by

$$(a_1, \ldots, a_n) \leq_{L^n} (c_1, \ldots, c_n) \text{ if and only if } a_i \leq_L c_i \text{ for every } 1 \leq i \leq n.$$

Notice that $1_{L^n} = (1_L, \ldots, 1_L)$ and $0_{L^n} = (0_L, \ldots, 0_L)$. An n-ary aggregation function is a function $M : L^n \to L$ such that:

1. $M(a_1, \ldots, a_n) \leq_L M(c_1, \ldots, c_n)$ whenever $(a_1, \ldots, a_n) \leq_{L^n} (c_1, \ldots, c_n)$.
2. $M(0_{L^n}) = 0_L$ and $M(1_{L^n}) = 1_L$.

An n-ary aggregation function M is said to be *idempotent* if $M(a, \ldots, a) = a$ for every $a \in L$. It is said to be *symmetric* if, for every permutation σ of the set $\{1, \ldots, n\}$, $M(a_1, \ldots, a_n) = M(a_{\sigma(1)}, \ldots, a_{\sigma(n)})$.

Definition 3 ([9]). Let $(L, \leq_L, T, S)$ be a complete lattice. A vector $(\alpha_1, \ldots, \alpha_n) \in L^n$ is said to be a

1. *weighting vector in* $(L, \leq_L, T, S)$ if $S(\alpha_1, \ldots, \alpha_n) = 1_L$ and it is said to be a
2. *distributive weighting vector in* $(L, \leq_L, T, S)$ if it also satisfies that

$$a = T(a, S(\alpha_1, \ldots, \alpha_n)) = S(T(a, \alpha_1), \ldots, T(a, \alpha_n)) \text{ for any } a \in L.$$

Remark 2.   1. If $(\alpha_1,\ldots,\alpha_n) \in L^n$ satisfies that $\alpha_k = 1_L$ for some $1 \leq k \leq n$ and
$\alpha_i = 0_L$ for any $i \neq k$, then $(\alpha_1,\ldots,\alpha_n)$ is a distributive weighting vector in any
complete lattice $(L,\leq_L,T,S)$.
2. If $(L,\leq_L,\wedge,\vee)$ is a complete lattice and $(\alpha_1,\ldots,\alpha_n) \in L^n$ satisfies that $\alpha_i = 1_L$
for some $1 \leq i \leq n$, then $(\alpha_1,\ldots,\alpha_n)$ is a distributive weighting vector in L.
3. If $(L,\leq_L,\wedge,\vee)$ is a complete distributive lattice, then any weighting vector
$(\alpha_1,\ldots,\alpha_n) \in L^n$, with $\alpha_1 \vee \cdots \vee \alpha_n = 1_L$, is distributive.
4. If L is the real interval $[0,1]$ with the usual order $\leq$, $T(a,b) = ab$ for every
$a,b \in [0,1]$ and $S(a,b) = \min\{a+b,1\}$ for every $a,b \in [0,1]$, then $(\alpha_1,\ldots,\alpha_n) \in$
$[0,1]^n$ with $S(\alpha_1,\ldots,\alpha_n) = 1$ is not necessarily a distributive weighting vector
in $[0,1]^n$. Indeed, $(\alpha_1,\ldots,\alpha_n) \in [0,1]^n$ is a distributive weighting vector if and
only if $\alpha_1 + \cdots + \alpha_n = 1$. (See [9]).

3   QOWA Operators on Any Complete Lattice

In this section QOWA (ordered weighted quasi-average) operators are extended
from $[0,1]$ to the more general case of any complete lattice $(L,\leq_L,T,S)$. The essential properties of QOWA operators on $[0,1]$ are shown to hold on this new setting.

Definition 4. Let $(L,\leq_L,T,S)$ be a complete lattice. For any vector $(a_1,\ldots,a_n) \in$
L^n,

1. a totally ordered vector $\tau_L(a_1 \ldots a_n) = (b_1,\ldots,b_n) \in L^n$ is built by means of

   - $b_1 = a_1 \vee \cdots \vee a_n \in L$.
   - $b_2 = [(a_1 \wedge a_2) \vee \cdots \vee (a_1 \wedge a_n)] \vee [(a_2 \wedge a_3) \vee \cdots \vee (a_2 \wedge a_n)] \vee$
     $\cdots \vee [a_{n-1} \wedge a_n] \in L$.
     $\vdots$
   - $b_k = \bigvee\{a_{j_1} \wedge \cdots \wedge a_{j_k} \mid \{j_1,\ldots,j_k\} \subseteq \{1,\ldots,n\}\} \in L$.
     $\vdots$
   - $b_n = a_1 \wedge \cdots \wedge a_n \in L$.

2. Let $(\alpha_1,\ldots,\alpha_n) \in L^n$ be a distributive weighting vector in $(L,\leq_L,T,S)$ and $g :$
$L \to L$ an strictly monotonic bijection. The function $F_{(\alpha,g)} : L^n \to L$ given by

$$F_{(\alpha,g)}(a_1,\ldots,a_n) = g^{-1}[S(T(\alpha_1,g(b_1)),\cdots,T(\alpha_n,g(b_n)))], \quad (a_1,\ldots,a_n) \in L^n$$

is called an *n-ary QOWA operator.*

Lemma 1. *In   the   conditions   of   Def.   4,   let $(a_1,\ldots,a_n) \in L^n$ and*
$(b_1,\ldots,b_n) = \tau_L(a_1,\ldots,a_n)$. *Then*

1. *For any $i \neq j$, $g(a_i \wedge a_j) = g(a_i) \wedge g(a_j)$ and $g(a_i \vee a_j) = g(a_i) \vee g(a_j)$.*
2. $(g(b_1),\ldots,g(b_n)) = \tau_L(g(a_1),\ldots,g(a_n)) \in L^n$.

Proof. 1. Let $i \neq j$. If $a_i = a_j$, the identities are trivial. If $a_i < a_j$, the strict monotonicity of g gives $g(a_i) < g(a_j)$. Hence,

$$g(a_i \wedge a_j) = g(a_i) = g(a_i) \wedge g(a_j).$$

The case of $a_j < a_i$ is analogous. Otherwise, if $a_i \wedge a_j < a_i$ and $a_i \wedge a_j < a_j$, then $g(a_i \wedge a_j) < g(a_i)$ and $g(a_i \wedge a_j) < g(a_j)$. In addition, there cannot be any $c \in L$ with $g(a_i \wedge a_j) < c < g(a_i)$ and $g(a_i \wedge a_j) < c < g(a_j)$ because g is a bijection on L. Therefore $g(a_i \wedge a_j) = g(a_i) \wedge g(a_j)$.

The $\vee$-identity can be proven in a similar way.

2. For any $1 \leq k \leq n$, consider $b_k = \bigvee \{a_{j_1} \wedge \cdots \wedge a_{j_k} \mid \{j_1, \ldots, j_k\} \subseteq \{1, \ldots, n\}\} \in L$. Then

$$g(b_k) = g(\bigvee \{a_{j_1} \wedge \cdots \wedge a_{j_k}) \mid \{j_1, \ldots, j_k\}) \subseteq \{1, \ldots, n\}\}$$
$$= \bigvee \{g(a_{j_1} \wedge \cdots \wedge a_{j_k}) \mid \{j_1, \ldots, j_k\} \subseteq \{1, \ldots, n\}\}$$
$$= \bigvee \{g(a_{j_1}) \wedge \cdots \wedge g(a_{j_k}) \mid \{j_1, \ldots, j_k\} \subseteq \{1, \ldots, n\}\}$$

and the result follows.

The following results show that any n-ary QOWA operator defined on $(L, \leq_L, T, S)$ satisfies the main properties of Yager's.

Proposition 1. *Let $(\alpha_1, \ldots, \alpha_n) \in L^n$ be a distributive weighting vector in $(L, \leq_L, T, S)$, $g : L \to L$ an strictly monotonic bijection and $F_{(\alpha, g)}$ the corresponding QOWA operator. Then*

1. *$F_{(\alpha, g)}$ is a symmetric n-ary aggregation function.*
2. *$F_{(\alpha, g)}$ is idempotent.*
3. *$a_1 \wedge \cdots \wedge a_n \leq_L F_{(\alpha, g)}(a_1, \ldots, a_n) \leq_L a_1 \vee \cdots \vee a_n$ for every $(a_1, \ldots, a_n) \in L^n$.*

Proof. 1. If $a_i \leq_L a_i'$ for every $1 \leq i \leq n$, it is easy to prove that $b_k \leq_L b_k'$ for every $1 \leq k \leq n$ and then $g(b_k) \leq_L g(b_k')$. Hence, $T(\alpha_k, g(b_k)) \leq_L T(\alpha_k, g(b_k'))$ for any $1 \leq k \leq n$ and consequently

$$S(T(\alpha_1, g(b_1)), \cdots, T(\alpha_n, g(b_n))) \leq_L S(T(\alpha_1, g(b_1')), \cdots, T(\alpha_n, g(b_n'))).$$

Finally,

$$g^{-1}[S(T(\alpha_1, g(b_1)), \cdots, T(\alpha_n, g(b_n)))] \leq_L$$
$$g^{-1}[S(T(\alpha_1, g(b_1')), \cdots, T(\alpha_n, g(b_n')))].$$

In addition, since $g(0_L) = 0_L$ and $g(1_L) = 1_L$,

$$F_{(\alpha, g)}(0_L, \ldots, 0_L) = g^{-1}(F_\alpha(0_L, \ldots, 0_L)) = g^{-1}(0_L) = 0_L$$

and $F_{(\alpha, g)}(1_L, \ldots, 1_L) = g^{-1}(F_\alpha(1_L, \ldots, 1_L)) = g^{-1}(1_L) = 1_L$.

It is immediate to check that $F_{(\alpha, g)}$ is symmetric.

2. If $a_1 = \cdots = a_n = a \in L$, then $b_k = a$ for every $1 \leq k \leq n$ and hence

$$F_{(\alpha,g)}(a,\ldots,a) = g^{-1}[S(T(\alpha_1,g(a)),\cdots,T(\alpha_n,g(a)))]$$
$$= g^{-1}[T(S(\alpha_1,\ldots,\alpha_n),g(a))] = g^{-1}[T(1_L,g(a))] = g^{-1}(g(a)) = a.$$

3. Let $(a_1,\ldots,a_n) \in L^n$. For any $1 \leq k \leq n$, $b_n \leq_L b_k \leq_L b_1$ and then $g(b_n) \leq_L g(b_k) \leq_L g(b_1)$. Therefore,

$$g(a_1 \wedge \cdots \wedge a_n) = g(b_n) = T(g(b_n),1_L) = T(g(b_n),S(\alpha_1,\cdots,\alpha_n)) =$$
$$S(T(\alpha_1,g(b_n)),\cdots,T(\alpha_n,g(b_n))) \leq_L S(T(\alpha_1,g(b_1)),\cdots,T(\alpha_n,g(b_n)))$$
$$\leq_L S(T(\alpha_1,g(b_1)),\cdots,T(\alpha_n,g(b_1)))$$
$$= T(S(\alpha_1,\cdots,\alpha_n),g(b_1)) = T(1_L,g(b_1)) = g(b_1) = g(a_1 \vee \cdots \vee a_n).$$

Hence,

$$a_1 \wedge \cdots \wedge a_n \leq_L g^{-1}[S(T(\alpha_1,g(b_1)),\cdots,T(\alpha_n,g(b_n)))] \leq_L a_1 \vee \cdots \vee a_n.$$

Remark 3. If $g = \mathrm{id}_L$ and $(\alpha_1,\cdots,\alpha_n)$ is any weighting distributive vector, then the QOWA operator $F_{(\alpha,g)}$ agrees with the OWA operator F_α defined in [9] by

$$F_\alpha(a_1,\ldots,a_n) = S(T(\alpha_1,b_1),\cdots,T(\alpha_n,b_n))] \quad (a_1,\ldots,a_n) \in L^n \qquad (2)$$

Yager's OWA operator given by (1.1) is a particular case of (1.2) when $(L,\leq_L,T,S)$ is the lattice described in item 4 of Remark 2.

The next result shows that the minimum, maximum and, if $(L,\leq_L)$ is a chain, the k-th order statistic with $1 \leq k \leq n$, are particular cases of QOWA operators $F_{(\alpha,g)}$ even when g is a monotonic bijection different from the identity map.

Proposition 2. *Let $(L,\leq_L,T,S)$ be any complete lattice endowed with a t-norm T and a t-conorm S and $g : L \to L$ an strictly monotonic bijection.*

1. *If $(\alpha_1,\ldots,\alpha_n) \in L^n$ satisfies that $\alpha_k = 1_L$ for some $1 \leq k \leq n$ and $\alpha_i = 0_L$ for any $i \neq k$, then $F_{(\alpha,g)}(a_1,\ldots,a_n) = b_k$, the k-th component of $\tau_L(a_1,\ldots,a_n)$ for any $(a_1,\ldots,a_n) \in L^n$.*
2. *In particular, if $(\alpha_1,\ldots,\alpha_n) = (1_L,0_L,\ldots,0_L) \in L^n$, then*

$$F_{(\alpha,g)}(a_1,\ldots,a_n) = a_1 \vee \cdots \vee a_n \text{ for any } (a_1,\ldots,a_n) \in L^n.$$

3. *If $(\alpha_1,\ldots,\alpha_n) = (0_L,\ldots,0_L,1_L) \in L^n$, then*

$$F_{(\alpha,g)}(a_1,\ldots,a_n) = a_1 \wedge \cdots \wedge a_n \text{ for any } (a_1,\ldots,a_n) \in L^n.$$

Proof. For any $(a_1,\ldots,a_n) \in L^n$, call $(b_1,\ldots,b_n) = \tau_L(a_1,\ldots,a_n)$, the totally ordered vector defined in Def. 4

1. Let $\alpha_k = 1_L$ and $\alpha_i = 0_L$ for any $i \neq k$.

$$F_{(\alpha,g)}(a_1,\ldots,a_n) = g^{-1}[S(T(0_L,g(b_1)),\cdots,T(1_L,g(b_k)),\cdots,T(0_L,g(b_n)))]$$
$$= g^{-1}[S(g(b_k))] = g^{-1}(g(b_k)) = b_k.$$

2. If $k = 1$, $b_k = b_1 = a_1 \vee \cdots \vee a_n$.
3. If $k = n$, $b_k = b_n = a_1 \wedge \cdots \wedge a_n$.

Remark 4. If $(L, \leq_L)$ is a chain, then the k-th component of parameter τ_L agrees with the k-th order statistic.

The next result shows that, if $(L, \leq_L, \wedge, \vee)$ is a complete distributive lattice, any QOWA operator is indeed an OWA operator.

Proposition 3. *Let* $(L, \leq_L, \wedge, \vee)$ *be a complete distributive lattice. Consider a weighting vector* $(\alpha_1,\ldots,\alpha_n) \in L^n$, *which is always distributive, and an strictly increasing bijection* $g : L \to L$.

1. $(g^{-1}(\alpha_1),\ldots,g^{-1}(\alpha_n))$ *is a weighting vector in* L^n *denoted by* $g^{-1}(\alpha)$.
2. $F_{(\alpha,g)}(a_1,\ldots,a_n) = F_{g^{-1}(\alpha)}(a_1,\ldots,a_n)$ *for any* $(a_1,\ldots,a_n) \in L^n$.
3. *The QOWA operator* $F_{(\alpha,g)}$ *is indeed an OWA operator on* L.

Proof. 1. $g^{-1}(\alpha_1) \vee \cdots \vee g^{-1}(\alpha_n) = g^{-1}(\alpha_1 \vee \cdots \vee \alpha_n) = g^{-1}(1_L) = 1_L$.
2. Let $(a_1,\ldots,a_n) \in L^n$ and $(b_1,\ldots,b_n) = \tau_L(a_1,\ldots,a_n)$. Then

$$F_{(\alpha,g)}(a_1,\ldots,a_n) = g^{-1}((\alpha_1 \wedge g(b_1)) \vee \cdots \vee (\alpha_n \wedge g(b_n))) =$$
$$= g^{-1}((g(g^{-1}(\alpha_1)) \wedge g(b_1)) \vee \cdots \vee ((g(g^{-1}(\alpha_n)) \wedge g(b_n)))$$
$$= ((g^{-1}(\alpha_1) \wedge b_1) \vee \cdots \vee (g^{-1}(\alpha_n) \wedge b_n)) = F_{g^{-1}(\alpha)}(a_1,\ldots,a_n)$$

as defined in (1.2).
3. It is a direct consequence of (ii).

Corollary 1. *Let* $(L, \leq_L, \wedge, \vee)$ *be a complete distributive lattice. Any QOWA operator is a particular case of the discrete Sugeno integral.*

Proof. It is shown in [9] that any OWA operator defined on a complete distributive lattice is a particular case of the discrete Sugeno integral. Therefore, the assertion is a direct consequence of the previous proposition.

4 QOWA Operators to Aggregate Restricted Equivalence Relations

Bustince et al. define in [1] a restricted equivalence function as a map $REF : [0,1] \times [0,1] \to [0,1]$ satisfying certain conditions. If we extend $([0,1], \leq)$ to any complete lattice $(L, \leq_L)$, we get the following

Definition 5. A *restricted equivalence function* defined on a complete lattice $(L, \leq_L)$ is a map $\rho : L \times L \to L$ satisfying

1. $\rho(a,b) = \rho(b,a)$ for all $a, b \in L$.
2. $\rho(a,b) = 1_L$ if and only if $a = b$.
3. $\rho(a,b) = 0_L$ if and only if $a = 1_L$ and $b = 0_L$ or $a = 0_L$ and $b = 1_L$.
4. If $n : L \to L$ is a strong negation (an strictly decreasing bijection with $n(n(a)) = a$ for any $a \in L$,

$$\rho(a,b) = \rho(n(a),n(b)) \text{ for all } a, b \in L.$$

5. Whenever $a \leq_L b \leq_L c$, then $\rho(a,c) \leq_L \rho(a,b) \wedge \rho(b,c)$.

Theorem 1. *Consider any complete lattice* $(L, \leq_L, T, S)$. *Let* $(\alpha_1, \ldots, \alpha_n) \in L^n$ *be a distributive weighting vector,* $g : L \to L$ *an strictly monotonic bijection and* $\rho : L \times L \to L$ *a restricted equivalence function on* L.

Then the function $\tilde{\rho} : L^n \times L^n \to L$ *given for each* $(a_1, \ldots, a_n), (c_1, \ldots, c_n) \in L^n$ *by*

$$\tilde{\rho}((a_1, \ldots, a_n), (c_1, \ldots, c_n)) = g^{-1}[S(T(\alpha_1, g(b_1)), \cdots, T(\alpha_n, g(b_n)))],$$

where $(b_1, \ldots, b_n) = \tau_L(\rho(a_1, c_1), \ldots, \rho(a_n, c_n))$, *satisfies*

1. $\tilde{\rho}((a_1, \ldots, a_n), (c_1, \ldots, c_n)) = \tilde{\rho}((c_1, \ldots, c_n), (a_1, \ldots, a_n))$.
2. $\tilde{\rho}((a_1, \ldots, a_n), (a_1, \ldots, a_n)) = 1_L$.
3. $\tilde{\rho}((1_L, \ldots, 1_L), (0_L, \ldots, 0_L)) = 0_L$.
4. *If* $n : L \to L$ *is a strong negation, then the map* $\tilde{n} : L^n \to L^n$ *given for any* $(a_1, \ldots, a_n) \in L^n$ *by* $\tilde{n}(a_1, \ldots, a_n) = (n(a_1), \ldots, n(a_n))$, *is a strong negation on* L^n *and*

$$\tilde{\rho}((a_1, \ldots, a_n), (c_1, \ldots, c_n)) = \tilde{\rho}(\tilde{n}(a_1, \ldots, a_n), \tilde{n}(c_1, \ldots, c_n)).$$

5. *Whenever* $(a_1, \ldots, a_n) \leq_{L^n} (c_1, \ldots, c_n) \leq_{L^n} (d_1, \ldots, d_n)$, *then*

$$\tilde{\rho}((a_1, \ldots, a_n), (d_1, \ldots, d_n)) \leq_L$$
$$\tilde{\rho}((a_1, \ldots, a_n), (c_1, \ldots, c_n)) \wedge \tilde{\rho}((c_1, \ldots, c_n), (d_1, \ldots, d_n))$$

Proof. Let $(a_1, \ldots, a_n), (c_1, \ldots, c_n) \in L^n$.

1. Since $\rho(a_i, c_i) = \rho(c_i, a_i)$ for any $1 \leq i \leq n$, it is clear that

$$\tilde{\rho}((a_1, \ldots, a_n), (c_1, \ldots, c_n)) = \tilde{\rho}((c_1, \ldots, c_n), (a_1, \ldots, a_n)).$$

2. As $\rho(a_i, a_i) = 1_L$ for any $1 \leq i \leq n$,

$$\tilde{\rho}((a_1, \ldots, a_n), (a_1, \ldots, a_n)) = g^{-1}[S(T(\alpha_1, g(1_L)), \cdots, T(\alpha_n, g(1_L)))]$$
$$= g^{-1}[S(T(\alpha_1, 1_L), \cdots, T(\alpha_n, 1_L))] = g^{-1}(S(\alpha_1, \cdots, \alpha_n))$$
$$= g^{-1}(1_L) = 1_L.$$

3. Since $\rho(1_L, 0_L) = 0_L$, then

$$\tilde{\rho}((1_L,\ldots,1_L),(0_L,\ldots,0_L)) = g^{-1}[S(T(\alpha_1,g(0_L)),\cdots,T(\alpha_n,g(0_L)))]$$
$$= g^{-1}[S(T(\alpha_1,0_L),\cdots,T(\alpha_n,0_L))] = g^{-1}(S(0_L,\cdots,0_L)) = g^{-1}(0_L) = 0_L.$$

4. It is clear that the map $\tilde{n} : L^n \to L^n$ is bijective, strictly decreasing and satisfying $\tilde{n}(\tilde{n}(a_1,\ldots,a_n)) = (a_1,\ldots,a_n)$ for any $(a_1,\ldots,a_n) \in L^n$. In addition,

$$\tilde{\rho}((a_1,\ldots,a_n),(c_1,\ldots,c_n)) = \tilde{\rho}(\tilde{n}(a_1,\ldots,a_n),\tilde{n}(c_1,\ldots,c_n))$$

because $\rho(a_i,c_i) = \rho(n(a_i),n(c_i))$ for any $1 \le i \le n$.

5. Let $(a_1,\ldots,a_n) \le_{L^n} (c_1,\ldots,c_n) \le_{L^n} (d_1,\ldots,d_n)$. Then, for any $1 \le i \le n$, $a_i \le_L c_i \le_L d_i$ and so, $\rho(a_i,d_i) \le_L \rho(a_i,c_i)$ and $\rho(a_i,d_i) \le_L \rho(c_i,d_i)$. Hence,

$$\tau_L(g(\rho(a_1,d_1)),\ldots,g(\rho(a_n,d_n))) \le_{L^n} \tau_L(g(\rho(a_1,c_1)),\ldots,g(\rho(a_n,c_n)))$$

and then, by Lemma 1,

$$\tilde{\rho}((a_1,\ldots,a_n),(d_1,\ldots,d_n)) \le_L \tilde{\rho}((a_1,\ldots,a_n),(c_1,\ldots,c_n)).$$

Analogously,

$$\tau_L(g(\rho(a_1,d_1)),\ldots,g(\rho(a_n,d_n))) \le_{L^n} \tau_L(g(\rho(c_1,d_1)),\ldots,g(\rho(c_n,d_n)))$$

and then

$$\tilde{\rho}((a_1,\ldots,a_n),(d_1,\ldots,d_n)) \le_L \tilde{\rho}((c_1,\ldots,c_n),(d_1,\ldots,d_n)).$$

5 QOWA Operators to Aggregate Intervals

Throughout this section $\mathbb{I}$ will denote the set of all the closed intervals contained in the real interval $[0,1]$. The order relation considered will be given by

$$[a_1, c_1] \le [a_2, c_2] \iff a_1 \le a_2 \text{ and } c_1 \le c_2,$$

where the $\le$ on the right of the arrow denotes the usual order in $[0,1]$. Notice that $(\mathbb{I},\le)$ is a complete lattice which is not totally ordered.

Proposition 4 ([9]). *Let* $([a_1, c_1],\ldots,[a_n, c_n]) \in \mathbb{I}^n$. *The totally ordered vector of* $\mathbb{I}^n$ *given by Def. 4,* $\tau_{\mathbb{I}}([a_1, c_1],\ldots,[a_n, c_n])$ *is equal to* $([b_1, d_1],\ldots,[b_n, d_n])$, *where* $(b_1,\ldots,b_n) = \tau_{[0,1]}(a_1,\ldots,a_n)$ *and* $(d_1,\ldots,d_n) = \tau_{[0,1]}(c_1,\ldots,c_n)$.

Example 1. Consider the complete lattice $(\mathbb{I},\le,T_t,S)$ of all the closed intervals contained in $[0,1]$ with the t-conorm S given by the join and the t-norm T_t (with $t \in [0,1]$) given for any $[a_1,c_1], [a_2,c_2] \in \mathbb{I}$ by

$$T_t([a_1,c_1],[a_2,c_2]) = [a_1 \wedge a_2, (t \wedge c_1 \wedge c_2) \vee (a_1 \wedge c_2) \vee (c_1 \wedge a_2)].$$

It is shown in [4] that T_t is distributive with respect to the join on $(\mathbb{I},\le)$. In addition, call $g : \mathbb{I} \to \mathbb{I}$ the monotonic bijection given by

$$g([a,b]) = [\frac{a+b}{2},b] \text{ for any } [a,b] \in \mathbb{I}.$$

Consider the following set of four alternatives considered under three criteria, appearing in [2]

	C_1	C_2	C_3
A_1	$[0'45,0'65]$	$[0'50,0'70]$	$[0'20,0'45]$
A_2	$[0'65,0'75]$	$[0'65,0'75]$	$[0'45,0'85]$
A_3	$[0'45,0'65]$	$[0'45,0'65]$	$[0'45,0'80]$
A_4	$[0'75,0'85]$	$[0'35,0'80]$	$[0'65,0'85]$

Fix the weighting vector $([\alpha_1,\beta_1], [\alpha_2,\beta_2], [\alpha_3,\beta_3])$ given by

$$\alpha_1 = \beta_1 = \frac{1}{3}; \ \alpha_2 = \beta_2 = \frac{2}{3}; \ \alpha_3 = \beta_3 = 1.$$

For alternative A_1, $[a_1,c_1] = [0'45,0'65]$, $[a_2,c_2] = [0'50,0'70]$ and $[a_3,c_3] = [0'20,0'45]$. So, $[b_1,d_1] = [0'50,0'70]$, $[b_2,d_2] = [0'45,0'65]$ and $[b_3,d_3] = [0'20,0'45]$. Hence, $g([b_1,d_1]) = [0'60,0'70]$, $g([b_2,d_2]) = [0'55,0'65]$, $g([b_3,d_3]) = [0'325,0'45]$ and

$$F_{(\alpha,g)}([a_1,c_1],[a_2,c_2],[a_3,c_3]) =$$
$$g^{-1}(T_t([\alpha_1,\beta_1],g([b_1,d_1])) \vee T_t([\alpha_2,\beta_2],g([b_2,d_2])) \vee T_t([\alpha_3,\beta_3],g([b_3,d_3])))$$
$$= g^{-1}\left([\frac{1}{3},\frac{1}{3}] \vee [0'55,0'65] \vee [0'325,0'45]\right) = g^{-1}([0'55,0'65]) = [0'45,0'65].$$

For alternative A_2, $[a_1,c_1] = [0'65,0'75]$, $[a_2,c_2] = [0'65,0'75]$ and $[a_3,c_3] = [0'45,0'85]$. So, $[b_1,d_1] = [0'65,0'85]$, $[b_2,d_2] = [0'65,0'75]$ and $[b_3,d_3] = [0'45,0'75]$. Hence, $g([b_1,d_1]) = [0'75,0'85]$, $g([b_2,d_2]) = [0'70,0'75]$, $g([b_3,d_3]) = [0'60,0'75]$ and

$$F_{(\alpha,g)}([a_1,c_1],[a_2,c_2],[a_3,c_3]) =$$
$$g^{-1}(T_t([\alpha_1,\beta_1],g([b_1,d_1])) \vee T_t([\alpha_2,\beta_2],g([b_2,d_2])) \vee T_t([\alpha_3,\beta_3],g([b_3,d_3])))$$
$$= g^{-1}\left([\frac{1}{3},\frac{1}{3}] \vee [\frac{2}{3},\frac{2}{3}] \vee [0'60,0'75]\right) = g^{-1}([\frac{2}{3},0'75]) = [\frac{7}{12},0'75].$$

For alternative A_3, $[a_1,c_1] = [0'45,0'65]$, $[a_2,c_2] = [0'45,0'65]$ and $[a_3,c_3] = [0'45,0'80]$. So, $[b_1,d_1] = [0'45,0'80]$, $[b_2,d_2] = [0'45,0'65]$ and $[b_3,d_3] = [0'45,0'65]$. Hence, $g([b_1,d_1]) = [0'625,0'80]$, $g([b_2,d_2]) = [0'55,0'65]$, $g([b_3,d_3]) = [0'55,0'65]$ and

$$F_{(\alpha,g)}([a_1,c_1],[a_2,c_2],[a_3,c_3]) =$$
$$g^{-1}\left(T_t([\alpha_1,\beta_1],g([b_1,d_1])) \vee T_t([\alpha_2,\beta_2],g([b_2,d_2])) \vee T_t([\alpha_3,\beta_3],g([b_3,d_3]))\right)$$
$$= g^{-1}\left([\tfrac{1}{3},\tfrac{1}{3}] \vee [0'55,0'65] \vee [0'55,0'65]\right) = g^{-1}([0'55,0'65]) = [0'45,0'65].$$

Finally, for alternative A_4, $[a_1,c_1] = [0'75,0'85]$, $[a_2,c_2] = [0'35,0'80]$ and $[a_3,c_3] = [0'65,0'85]$. So, $[b_1,d_1] = [0'75,0'85]$, $[b_2,d_2] = [0'65,0'85]$ and $[b_3,d_3] = [0'35,0'80]$. Hence, $g([b_1,d_1]) = [0'80,0'85]$, $g([b_2,d_2]) = [0'75,0'85]$, $g([b_3,d_3]) = [0'575,0'80]$ and

$$F_{(\alpha,g)}([a_1,c_1],[a_2,c_2],[a_3,c_3]) =$$
$$g^{-1}\left(T_t([\alpha_1,\beta_1],g([b_1,d_1])) \vee T_t([\alpha_2,\beta_2],g([b_2,d_2])) \vee T_t([\alpha_3,\beta_3],g([b_3,d_3]))\right)$$
$$= g^{-1}\left([\tfrac{1}{3},\tfrac{1}{3}] \vee [\tfrac{2}{3},\tfrac{2}{3}] \vee [0'575,0'80]\right) = g^{-1}([\tfrac{2}{3},0'80]) = [\tfrac{8}{15},0'80].$$

This approach allows us to assign a unique interval to each alternative. In this example, the assigned intervals are not totally ordered. Indeed,

$$F_{(\alpha,g)}(A_1) = F_{(\alpha,g)}(A_3) \leq F_{(\alpha,g)}(A_2) \text{ and}$$
$$F_{(\alpha,g)}(A_1) = F_{(\alpha,g)}(A_3) \leq F_{(\alpha,g)}(A_4)$$

but it is not true either $F_{(\alpha,g)}(A_2) \leq F_{(\alpha,g)}(A_4)$ or $F_{(\alpha,g)}(A_4) \leq F_{(\alpha,g)}(A_2)$. This arrangement differs from that obtained in [9] for the OWA F_α:

$$F_\alpha(A_3) \leq F_\alpha(A_1) \leq F_\alpha(A_2) \leq F_\alpha(A_4).$$

References

[1] Bustince, H., Barrenechea, E., Pagola, M.: Image thresholding using restricted equivalence functions and maximizing the measures of similarity. Fuzzy Sets and Systems 158, 496–516 (2007)

[2] Bustince, H., Fernandez, J., Sanz, J., Galar, M., Mesiar, R., Kolesárová, A.: Multicriteria Decision Making by Means of Interval-Valued Choquet Integrals. In: Melo-Pinto, P., Couto, P., Serôdio, C., Fodor, J., De Baets, B. (eds.) Eurofuse 2011. AISC, vol. 107, pp. 269–278. Springer, Heidelberg (2011)

[3] De Baets, B., Mesiar, R.: Triangular norms on product lattices. Fuzzy Sets and Systems 104, 61–75 (1999)

[4] Deschrijver, G., Kerre, E.E.: Classes of intuitionistic fuzzy t-norms satisfying the residuation principle. International Journal of Uncertainty, Fuzziness and Knowledge-Based Systems 11(6), 691–709 (2003)

[5] Dubois, D., Prade, H.: On the use of aggregation operations in information fusion processes. Fuzzy Sets and Systems 142, 143–161 (2004)

[6] Fodor, J., Marichal, J.L., Roubens, M.: Characterizations of the ordered weighted averaging operator. IEEE Transactions on Fuzzy Systems 3, 236–240 (1995)

[7] Grätzer, G.: General lattice theory. Birkhäuser Verlag, Basel (1978)
[8] Komorníková, M., Mesiar, R.: Aggregation functions on bounded partially ordered sets and their classification. Fuzzy Sets and Systems 175, 48–56 (2011)
[9] Lizasoain, I., Moreno, C.: OWA operators defined on complete lattices. Fuzzy Sets and Systems (in press)
[10] Yager, R.R.: On ordered weighting averaging aggregation operators in multicriteria decision-making. IEEE Transaction on Systems, Man and Cybernetics 18, 183–190 (1988)
[11] Yager, R.R.: Generalized OWA aggregation operators. Fuzzy Optimization and Decision Making 2, 93–107 (2004)
[12] Yager, R.R., Gumrah, G., Reformat, M.: Using a web Personal Evaluation Tool-PET for lexicographic multi-criteria service selection. Knowledge-Based Systems 24, 929–942 (2011)
[13] Zhou, S.-M., Chiclana, F., John, R.I., Garibaldi, J.M.: Type-1 OWA operators for aggregating uncertain information with uncertain weights induced by type-2 linguistic quantifiers. Fuzzy Sets and Systems 159, 3281–3296 (2008)

Obtaining Multi-argument Fuzzy Measures on Lattices

S. Cubillo, T. Calvo, and E.E. Castiñeira

Abstract. Measures have been used to establish the degree in which a property holds. When this property is applied on a undetermined number of elements, multi-argument measures should be considered. More, if the values are in $[0,1]$, we finally have to handle multi-argument fuzzy measures. This paper is devoted to obtain new multi-argument fuzzy measures on bounded lattices, using aggregation functions in two different ways. Developing the proposed methods, several functions measuring some properties with specified conditions are obtained.

1 Introduction

Measures have been studied throughout the literature in many backgrounds. They have been used to establish the degree in which a property holds. This property could refer to a single or more elements in a set. In previous papers, the authors have studied some functions measuring the contradiction, the incompatibility and other properties between two fuzzy sets ([7, 9, 10]), and later, they have extended their study to multi-argument fuzzy measures ([8]). It is necessary to quote that the previous research could be applied not only to the bounded lattice of the fuzzy sets on a universe X, but also to any bounded lattice. This is the main purpose of the present work.

The first section is devoted to give some definitions necessary to properly understand the rest of the paper. Section two focuses on some results that allow to obtain new multi-argument measures. Next, these results are used to construct some functions measuring properties with different conditions, and the relation with previous papers is showed. Finally some conclusions are set out.

S. Cubillo · E.E. Castiñeira
Polytechnical University of Madrid
e-mail: {scubillo,ecastineira}@fi.upm.es

T. Calvo
University of Alcalá de Henares
e-mail: tomasa.calvo@uah.es

H. Bustince et al. (eds.), *Aggregation Functions in Theory and in Practise*,   533
Advances in Intelligent Systems and Computing 228,
DOI: 10.1007/978-3-642-39165-1_50, © Springer-Verlag Berlin Heidelberg 2013

2 Basic Notions

Let $\mathscr{L} = (L, \leq_L, 0_L, 1_L)$ be a bounded lattice [4, 5] whose minimum and maximum elements are denoted by 0_L and 1_L, respectively. For each $n \in \mathbb{N}$, let us consider the set

$$L^n = \{(a_1, \ldots, a_n) \mid a_i \in L, \forall i \in \{1, \ldots, n\}\}$$

and the order relation $\leq_{L^n}$ induced by $\leq_L$, that is, given $\bar{a} = (a_1, \ldots, a_n), \bar{b} = (b_1, \ldots b_n) \in L^n$,

$$\bar{a} \leq_{L^n} \bar{b} \iff a_i \leq_L b_i, \forall i \in \{1, \ldots, n\}.$$

We have that L^n with the order relation $\leq_{L^n}$ is also a bounded lattice, whose minimum and maximum elements are $0_{L^n} = (0_L, \overset{n)}{\ldots}, 0_L)$ and $1_{L^n} = (1_L, \overset{n)}{\ldots}, 1_L)$, respectively. We say that $\mathscr{L}^n = (L^n, \leq_{L^n}, 0_{L^n}, 1_{L^n})$ is induced by $\mathscr{L}$.

Moreover, if $\mathscr{L}$ is complete, then $\mathscr{L}^n$ is also complete.

Throughout the paper the bounded and complete lattice of real numbers $\mathscr{I} = ([0, 1], \leq, 0, 1)$ will be considered.

Definition 1. Let $\mathscr{L} = (L, \leq_L, 0_L, 1_L)$ be a bounded lattice and, for each $n \in \mathbb{N}$, $\mathscr{L}^n = (L^n, \leq_{L^n}, 0_{L^n}, 1_{L^n})$ the lattice induced by $\mathscr{L}$. A multi-argument function $\mathscr{F} : \bigcup_{n \in \mathbb{N}} L^n \to L$ is said to be n-increasing if

$$\mathscr{F}(a_1, \ldots, a_n) \leq_L \mathscr{F}(a_1, \ldots, a_n, a_{n+1})$$

holds for all $n \in \mathbb{N}$ and for all $a_1, \ldots, a_n, a_{n+1} \in L$. Similarly, it is n-decreasing if

$$\mathscr{F}(a_1, \ldots, a_n, a_{n+1}) \leq_L \mathscr{F}(a_1, \ldots, a_n)$$

holds for all $n \in \mathbb{N}$ and for all $a_1, \ldots, a_n, a_{n+1} \in L$

The following definition captures the axioms of fuzzy measures given in [12], considering lattices instead of measurable spaces.

Definition 2. Let us consider a bounded lattice $\mathscr{L} = (L, \leq_L, 0_L, 1_L)$, the lattice $\mathscr{L}^n$ induced by $\mathscr{L}$, for a fixed $n \in \mathbb{N}$, and the lattice of real numbers $\mathscr{I}$. A map $M : L^n \to [0, 1]$ is said to be a *fuzzy measure* on $\mathscr{L}^n$ or a *fuzzy n-measure* on $\mathscr{L}$ (or on L), if it satisfies:

  i) $M(0_{L^n}) = 0$ and $M(1_{L^n}) = 1$ (*boundary conditions*).
  ii) M is increasing with respect to the orders of the lattices $\mathscr{L}^n$ and $\mathscr{I}$, that is, for all $\bar{a}, \bar{b} \in L^n$ such that $\bar{a} \leq_{L^n} \bar{b}$, the inequality $M(\bar{a}) \leq M(\bar{b})$ holds (*monotony condition*).

In particular, if $n = 1$ then M is said to be a *fuzzy measure* on $\mathscr{L}$ (or on L).

Nevertheless, in most of the applications the properties are applied to an unsettled number of elements. So it is necessary to extend the definition of fuzzy measure by considering any number of arguments as follows.

Definition 3. ([7]) Let $\mathscr{L} = (L, \leq_L, 0_L, 1_L)$ be a bounded lattice and, for each $n \in \mathbb{N}$, let $\mathscr{L}^n$ be the lattice induced by $\mathscr{L}$. A map $M : \bigcup_{n \in \mathbb{N}} L^n \to [0,1]$ is said to be a *multi-argument fuzzy measure on* $\mathscr{L}$ (or on L) if, for each $n \in \mathbb{N}$, the function M restricted to L^n, $M|_{L^n}$, is a fuzzy n-measure.

Moreover,

iii) M is *increasing with respect to the argument* n or n-*increasing* if $M(a_1, \ldots, a_n) \leq M(a_1, \ldots, a_n, a_{n+1})$ holds for all $n \in \mathbb{N}$ and for all $a_1, \ldots, a_n, a_{n+1} \in L$.

iv) M is *decreasing with respect to the argument* n or n-*decreasing* if $M(a_1, \ldots, a_n) \geq M(a_1, \ldots, a_n, a_{n+1})$ holds for all $n \in \mathbb{N}$ and for all $a_1, \ldots, a_n, a_{n+1} \in L$.

Example 1. Any aggregation function is a fuzzy multi-measure on $\mathscr{I}$. Indeed, recall that an aggregation function [3, 6, 13, 14] is a map $\mathscr{A} : \bigcup_{n \in \mathbb{N}} [0,1]^n \to [0,1]$ such that

1. $\mathscr{A}(0, \ldots, 0) = 0$ and $\mathscr{A}(1, \ldots, 1) = 1$.
2. $\mathscr{A}(a) = a$ for all $a \in [0,1]$.
3. For each $n \in \mathbb{N}$, $\mathscr{A}(a_1, \ldots, a_n) \leq \mathscr{A}(b_1, \ldots, b_n)$ holds provided $a_i \leq b_i$ for all $i \in \{1, \ldots, n\}$.

We denote $S_n = \{\pi : \{1, \ldots, n\} \to \{1, \ldots, n\} \mid \pi \text{ is a bijection}\}$, that is, S_n is the set of *permutations* of $\{1, \ldots, n\}$. Then

Similarly to the case of symmetric operators, the following definition is given.

Definition 4. A multi-argument fuzzy measure M on a bounded lattice $\mathscr{L} = (L, \leq_L, 0_L, 1_L)$ is *symmetric* if, for each $n \in \mathbb{N}$, the function $M|_{L^n}$ is symmetric, that is, $M(a_1, \ldots, a_n) = M(a_{\pi(1)}, \ldots, a_{\pi(n)})$ holds for any $\pi \in S_n$ and for any $(a_1, \ldots, a_n) \in L^n$.

Example 2. Recall that a t-conorm ([1, 15]) is a function $S : [0,1]^2 \to [0,1]$, conmutative, associative, increasing and with neutral element 0. Because of the associativity, it can be extended to $S : \bigcup_{n \in \mathbb{N}} [0,1]^n \to [0,1]$, where $S(a_1, \ldots, a_n) = S(a_1, S(a_2, \ldots, a_n)), \forall n > 1$. If, moreover, we fix $S(a) = a$ for all $a \in [0,1]$, we obtain that any t-conorm is an aggregation function that is also a symmetric n-*increasing* multi-argument fuzzy measure.

Similarly, the extension of any t-norm T ($T : [0,1]^2 \to [0,1]$, conmutative, associative, increasing and with neutral element 1) to $\bigcup_{n \in \mathbb{N}} [0,1]^n$ is an aggregation function that is also a symmetric n-*decreasing* multi-argument fuzzy measure.

Example 3. The function $\mathscr{M} : \bigcup_{n \in \mathbb{N}} [0,1]^n \to [0,1]$, given by

$$\mathscr{M}(a) = a$$

$$\mathscr{M}(a_1, \ldots, a_n) = a_1^2 \prod_{i=2}^{n} a_i, \ \forall n > 1$$

is an aggregation function, that is a non symmetric n-*decreasing* multi-argument fuzzy measure on $([0,1], \leq)$.

3   Some Ways to Obtain Multi-argument Fuzzy Measures

The two following results whose proof is straightforward, provide two ways to construct fuzzy multi-measures from a given fuzzy measure, using aggregation functions.

One way is to aggregate measures of lattice elements, whereas the other is to measure the aggregation of lattice elements.

Proposition 1. ([7]) Let $\mathscr{L} = (L, \leq_L, 0_L, 1_L)$ be a bounded lattice and $m : L \to [0,1]$ be a fuzzy measure on $\mathscr{L}$. If $\mathscr{A}$ is a fuzzy multi-measure on $\mathscr{I}$, then $M_{\mathscr{A}} : \bigcup_{n \in \mathbb{N}} L^n \to [0,1]$, defined for each $(a_1, \ldots, a_n) \in L^n$ by

$$M_{\mathscr{A}}(a_1, \ldots, a_n) = \mathscr{A}(m(a_1), \ldots, m(a_n)),$$

is a fuzzy multi-measure on $\mathscr{L}$.

Moreover, if $\mathscr{A}$ is n-increasing (n-decreasing), then $M_{\mathscr{A}}$ is n-increasing (n-decreasing).

Example 4. Let X be a non-empty and finite set, and let $\mathscr{P}(X)$ denotes the set of all subsets of X, that is, the power set of X. Let us consider the bounded lattice $(\mathscr{P}(X), \subseteq, \emptyset, X)$, which is, in fact, a Boolean algebra. Let $m : \mathscr{P}(X) \to [0,1]$ be the fuzzy measure defined for each $A \in \mathscr{P}(X)$ as $m(A) = \frac{card(A)}{card(X)}$, and $\mathscr{A}$ the n-increasing fuzzy multi-measure on $\mathscr{I}$, defined as $\mathscr{A}(a_1, \ldots a_n) = Max(a_1, \ldots a_n)$. Then $M_{\mathscr{A}} : \bigcup_{n \in \mathbb{N}} \mathscr{P}(X)^n \to [0,1]$, defined for each $(A_1, \ldots, A_n) \in \mathscr{P}(X)^n$ by

$$M_{\mathscr{A}}(A_1, \ldots, A_n) = Max(m(A_1), \ldots, m(A_n)) = Max_{i=1\ldots n} \left\{ \frac{card(A_i)}{card(X)} \right\},$$

is an n-increasing fuzzy multi-measure on $\mathscr{P}(X)$.

In a similar way, considering the n-decreasing fuzzy multi-measure Min, we obtain the n-decreasing fuzzy multi-measure on $\mathscr{P}(X)$

$$M_{\mathscr{A}}(A_1, \ldots, A_n) = Min(m(A_1), \ldots, m(A_n)) = Min_{i=1\ldots n} \left\{ \frac{card(A_i)}{card(X)} \right\},$$

Let us observe that in the first case we could have changed the Max by any t-conorm to obtain n-increasing fuzzy multi-measures, and in the second one, the Min by any t-norm to obtain n-decreasing fuzzy multi-measures.

Proposition 2. ([7]) Let $\mathscr{L} = (L, \leq_L, 0_L, 1_L)$ be a bounded lattice and $m : L \to [0,1]$ be a fuzzy measure on $\mathscr{L}$.
If $\mathscr{F} : \bigcup_{n \in \mathbb{N}} L^n \to L$ is a multi-argument function such that $\mathscr{F}(0_L, \ldots, 0_L) = 0_L$, $\mathscr{F}(1_L, \ldots, 1_L) = 1_L$ and $\mathscr{F}(a_1, \ldots, a_n) \leq_L \mathscr{F}(b_1, \ldots, b_n)$, whenever $a_i \leq_L b_i$ for each $i \in \{1, \ldots, n\}$ and for all $n \in \mathbb{N}$, then the multi-argument function $M_{\mathscr{F}} : \bigcup_{n \in \mathbb{N}} L^n \to [0,1]$, defined for each $(a_1, \ldots, a_n) \in L^n$ by

$$M_{\mathscr{F}}(a_1, \ldots, a_n) = m(\mathscr{F}(a_1, \ldots, a_n)),$$

is a fuzzy multi-measure on $\mathscr{L}$.

Moreover, if $\mathscr{F}$ is *n-increasing* (*n-decreasing*), then $M_{\mathscr{F}}$ is *n-increasing* (*n-decreasing*).

Example 5. 1. Consider the *n-decreasing* multi-argument function on the bounded lattice $(\mathscr{P}(X), \subseteq, \emptyset, X)$, $\mathscr{F} : \bigcup_{n \in \mathbb{N}} \mathscr{P}(X)^n \to \mathscr{P}(X)$, defined for each $(A_1, \ldots, A_n) \in \mathscr{P}(X)^n$ by $\mathscr{F}(A_1, \ldots, A_n) = A_1 \cap \ldots \cap A_n$, and the fuzzy measure m of the example 4. Then,

$$M_{\mathscr{F}}(A_1, \ldots, A_n) = m(\mathscr{F}(A_1, \ldots, A_n)) = \frac{card(A_1 \cap \ldots \cap A_n)}{card(X)}$$

is an *n-decreasing* multi-argument fuzzy measure on $\mathscr{P}(X)$.

2. Consider the *n-increasing* multi-argument function $\mathscr{G} : \bigcup_{n \in \mathbb{N}} \mathscr{P}(X)^n \to \mathscr{P}(X)$ defined for each $(A_1, \ldots, A_n) \in \mathscr{P}(X)^n$ by $\mathscr{G}(A_1, \ldots, A_n) = A_1 \cup \ldots \cup A_n$. Then,

$$M_{\mathscr{G}}(A_1, \ldots, A_n) = m(\mathscr{G}(A_1, \ldots, A_n)) = \frac{card(A_1 \cup \ldots \cup A_n)}{card(X)}$$

is an *n-increasing* multi-argument fuzzy measure on $\mathscr{P}(X)$.

Next results provide multi-argument fuzzy measures on L with the order reverse throughout strong negations. To show them, we first need some definitions.

Given a non-empty set L and an order relation $\leq_L$ defined on L, the reverse order induced by $\leq_L$ is defined as follows: for all $a, b \in L$,

$$a \geq_L b \quad \text{if and only if} \quad b \leq_L a$$

Note that with this order, the minimum element is 1_L, and the maximum is 0_L. Then, we have the bounded lattice $\mathscr{L}^* = (L, \geq_L, 1_L, 0_L)$.

Definition 5. ([2]) Let $\mathscr{L} = (L, \leq_L, 0_L, 1_L)$ be a bounded lattice. The function $N : L \to L$ is a negation if

1. $N(0_L) = 1_L$
2. $N(1_L) = 0_L$
3. If $a \leq_L b$, then $N(b) \leq_L N(a)$, for any $a, b \in L$

More, if $N(N(a)) = a$ holds for any $a \in L$, it is said that N is a strong negation.

Proposition 3. *Let $M : \bigcup_{n \in \mathbb{N}} L^n \to [0, 1]$ be a multi-argument fuzzy measure on $\mathscr{L} = (L, \leq_L, 0_L, 1_L)$, and $N : L \to L$ a strong negation on L. Then the function $M_N : \bigcup_{n \in \mathbb{N}} L^n \to [0, 1]$ defined as $M_N(a_1, \ldots, a_n) = M(N(a_1), \ldots, N(a_n))$ for any $(a_1, \ldots, a_n) \in L^n$ is a multi-argument fuzzy measure on $\mathscr{L}^* = (L, \geq_L, 1_L, 0_L)$. Furthermore, if M is n-increasing (n-decreasing), M_N is n-increasing (n-decreasing).*

Proof. 1. $M_N(1_L, \ldots, 1_L) = M(N(1_L), \ldots, N(1_L)) = M(0_L, \ldots, 0_L) = 0$.
2. Similarly, $M_N(0_L, \ldots, 0_L) = M(1_L, \ldots, 1_L) = 1$.

3. $(a_1,\ldots,a_n) \geq_L (b_1,\ldots,b_n) \Leftrightarrow (b_1,\ldots,b_n) \leq_L (a_1,\ldots,a_n)$
   $\Leftrightarrow b_i \leq_L a_i \ \forall i = 1,\ldots,n \Leftrightarrow N(a_i) \leq_L N(b_i) \ \forall i = 1,\ldots,n$
   $\Leftrightarrow M_N(a_1,\ldots,a_n) = M(N(a_1),\ldots,N(a_n)) \leq M(N(b_1),\ldots,N(b_n)) = M_N(b_1,\ldots,b_n)$.

4. $M_N(a_1,\ldots,a_{n+1}) = M(N(a_1),\ldots,N(a_{n+1})) \geq M(N(a_1),\ldots,N(a_n)) = M_N(a_1,\ldots,a_n)$. Analogously the case of M being n-decreasing.

Example 6. Let $(\mathscr{P}(X),\subseteq,\emptyset,X)$ be the lattice of the power set of X, with X finite. Let $N : \mathscr{P}(X) \to \mathscr{P}(X)$ the function complement, that is, for any $A \subseteq X$ it is $N(A) = A^c = X - A$. In fact, this function is a strong negation on $\mathscr{P}(X)$. And let $M_{\mathscr{F}}$ the n-decreasing multi-argument fuzzy measure on $\mathscr{P}(X)$ obtained in the example 5, $M_{\mathscr{F}}(A_1,\ldots,A_n) = \frac{card(A_1 \cap \ldots \cap A_n)}{card(X)}$.

Then the function $M_{\mathscr{F}N} : \mathscr{P}(X) \to [0,1]$, defined as

$$M_{\mathscr{F}N}(A_1,\ldots,A_n) = \frac{card(N(A_1) \cap \ldots \cap N(A_n))}{card(X)} = \frac{card(A_1^c \cap \ldots \cap A_n^c)}{card(X)}$$

is an n-decreasing multi-argument fuzzy measure on $(\mathscr{P}(X),\supseteq,X,\emptyset)$.

In a similar way, considering in the same example the n-increasing measure $M_{\mathscr{G}}(A_1,\ldots,A_n) = \frac{card(A_1 \cup \ldots \cup A_n)}{card(X)}$ we obtain the n-increasing multi-argument fuzzy measure on $(\mathscr{P}(X),\supseteq,X,\emptyset)$

$$M_{\mathscr{G}N}(A_1,\ldots,A_n) = \frac{card(N(A_1) \cup \ldots \cup N(A_n))}{card(X)} = \frac{card(A_1^c \cup \ldots \cup A_n^c)}{card(X)}.$$

Proposition 4. *Let $M : \bigcup_{n \in \mathbb{N}} L^n \to [0,1]$ be a multi-argument fuzzy measure on $\mathscr{L} = (L,\leq_L,0_L,1_L)$, and $N : [0,1] \to [0,1]$ a strong negation. Then the function $M^N : \bigcup_{n \in \mathbb{N}} L^n \to [0,1]$ defined as $M^N(a_1,\ldots,a_n) = N(M(a_1,\ldots,a_n))$ for any $(a_1,\ldots,a_n) \in L^n$ is a multi-argument fuzzy measure on $\mathscr{L}^* = (L,\geq_L,1_L,0_L)$.*
Furthermore, if M is n-increasing (n-decreasing), M^N is n-decreasing (n-increasing).

Example 7. Let $(\mathscr{P}(X),\subseteq,\emptyset,X)$ be the lattice of the power set of X. Let $N : [0,1] \to [0,1]$ be the strong negation given by $N(x) = 1 - x$ for all $x \in [0,1]$. If $M_{\mathscr{F}}$ is the n-decreasing multi-argument fuzzy measure on $\mathscr{P}(X)$ obtained in the example 5, $M_{\mathscr{F}}(A_1,\ldots,A_n) = \frac{card(A_1 \cap \ldots \cap A_n)}{card(X)}$, then the function $M_{\mathscr{F}}{}^N : \mathscr{P}(X) \to [0,1]$, defined as

$$M_{\mathscr{F}}{}^N(A_1,\ldots,A_n) = 1 - \frac{card(A_1 \cap \ldots \cap A_n)}{card(X)}$$

is an n-increasing multi-argument fuzzy measure on $(\mathscr{P}(X),\supseteq,X,\emptyset)$.

In a similar way, considering in the same example the n-increasing measure $M_{\mathscr{G}}(A_1,\ldots,A_n) = \frac{card(A_1 \cup \ldots \cup A_n)}{card(X)}$ we obtain the n-decreasing multi-argument fuzzy measure on $(\mathscr{P}(X),\supseteq,X,\emptyset)$

$$M_{\mathscr{G}N}(A_1,\ldots,A_n) = 1 - \frac{card(A_1 \cup \ldots \cup A_n)}{card(X)}.$$

4 Searching Functions to Measure Any Property

Let P be a property that a set of elements of any bounded lattice $\mathscr{L} = (L, \leq_L, 0_L, 1_L)$ could gradually satisfy. Let us suppose that we have to establish a function on L measuring this graduation. Then we should define a function $M_P : \bigcup_{n \in \mathbb{N}} L^n \to [0,1]$, satisfying some conditions depending on the characteristics of this property.

Firstly, if we have that a set of elements $\{a_1, \ldots, a_n\}$ do not satisfy at all the property, we should establish that $M_P(a_1, \ldots, a_n) = 0$. For the rest of the sets of elements, we should fix the requirements M_P has to satisfy.

If the degree of property increases as the elements are greater (in the order $\leq_L$) we can define $M_P(a_1, \ldots, a_n) = \mathscr{A}(a_1, \ldots, a_n)$, provided $\mathscr{A}$ is any multi-argument fuzzy function; but if property P decreases as the elements are greater, we could define $M_P(a_1, \ldots, a_n) = \mathscr{A}(N(a_1), \ldots, N(a_n))$ (Proposition 3), provided $\mathscr{A}$ is any multi-argument fuzzy function, and N any strong negation on L.

More, if the degree of the property increases (or decreases) depending on a characteristic of the elements, and it is possible to establish a function m measuring this characteristic, we can establish $M_P(a_1, \ldots, a_n) = \mathscr{A}(m(a_1), \ldots, m(a_n))$, provided $\mathscr{A}$ is an aggregation function. But if the degree increases depending of a characteristic (measurable by m) of a relation between the elements (represented by a multi-argument function $\mathscr{A}$), then we could establish $M_P(a_1, \ldots, a_n) = m(\mathscr{F}(a_1, \ldots, a_n))$.

Finally, if the degree of the property P increases as the number n of elements increases, we can choose $\mathscr{A}$ as any n-increasing multi-argument function, and if the degree decreases as the number n of elements increases, $\mathscr{A}$ could be any n-decreasing multi-argument function.

In previous papers, the authors have considered the lattice of the fuzzy sets $\mathscr{L} = ([0,1]^X, \leq, \mu_\emptyset, \mu_X)$, introducing some functions in order to measure different properties. Furthermore, if $T : \bigcup_{n \in \mathbb{N}} [0,1]^n \to [0,1]$ is a t-norm, we can establish $T : \bigcup_{n \in \mathbb{N}} ([0,1]^X)^n \to [0,1]^X$ as $(T(\mu_1, \cdots, \mu_n))(x) = T(\mu(x), \cdots, \mu_n(x))$. Also if $S : \bigcup_{n \in \mathbb{N}} [0,1]^n \to [0,1]$ is a t-conorm, we can establish $S : \bigcup_{n \in \mathbb{N}} ([0,1]^X)^n \to [0,1]^X$ as $(S(\mu_1, \cdots, \mu_n))(x) = S(\mu(x), \cdots, \mu_n(x))$.

Example 8. In [8] for any t-norm T, the T-compatibility is studied. This property is increasing on $\mathscr{L}$ and n-decreasing. Then, according to Proposition 1, a way to measure the T-compatibility of a number of fuzzy sets could be

$$C(\mu_1, \ldots, \mu_n) = \begin{cases} 0, & \text{if } T(\mu_1, \ldots, \mu_n) = \mu_\emptyset \\ T^0(m(\mu_1), \ldots, m(\mu_n)), & \text{otherwise.} \end{cases}$$

provided T^0 is a t-norm, and m a measure on $[0,1]^X$. For example, the supremum $(m(\mu) = Sup_{x \in X} \mu(x))$, or the minimum $(m(\mu) = Min_{x \in X} \mu(x))$.

Note that $T(\mu_1,\ldots,\mu_n) = \emptyset$ means that $\mu_1,\ldots,\mu_n$ are not T-compatible.
Moreover, according to Proposition 2, a new measure of T-compatibility will be:

$$C(\mu_1,\ldots,\mu_n) = \begin{cases} 0, & \text{if } T(\mu_1,\ldots,\mu_n) = \mu_\emptyset \\ m(T^0(\mu_1,\ldots,\mu_n)), & \text{otherwise.} \end{cases}$$

provided T^0 is a t-norm, and m a measure on $[0,1]^X$

Example 9. Measures of supplementarity were introduced in [11]. Supplementarity is an increasing property on $\mathscr{L}$ and n-increasing. Then according to Proposition 1, given a t-conorm S, a measure of S-supplementarity could be

$$M(\mu_1,\ldots,\mu_n) = \begin{cases} 0, & \text{if } S(\mu_1,\ldots,\mu_n) \neq \mu_X \\ S^0(m(\mu_1),\ldots,m(\mu_n)), & \text{otherwise.} \end{cases}$$

provided S^0 is a t-conorm, and m a measure on $[0,1]^X$.

And according to Proposition 2, a new measure of S-supplementarity will be:

$$M(\mu_1,\ldots,\mu_n) = \begin{cases} 0, & \text{if } S(\mu_1,\ldots,\mu_n) \neq \mu_X \\ m(S^0(\mu_1,\ldots,\mu_n)), & \text{otherwise.} \end{cases}$$

provided S^0 is a t-conorm, and m a measure on $[0,1]^X$.

Before giving the following examples, note that in Definition 5, when considering the particular lattice $([0,1],\leq,0,1)$, the usual definition of strong negation on $[0,1]$ ([16]) is captured. Besides, a strong negation n on $[0,1]$ determines a strong negation on the lattice $[0,1]^X$, $N : [0,1]^X \to [0,1]^X$, with $(N(\mu))(x) = N(\mu(x))$ for all $x \in X$.

Example 10. Measures of incompatibility were introduced in [9]. This is a decreasing property on and n-increasing. Then, according to Proposition 3 it is possible to establish for any t- norm T, the T-incompatibility measure:

$$I(\mu_1,\ldots,\mu_n) = \begin{cases} 0, & \text{if } T(\mu_1,\ldots,\mu_n) \neq \mu_\emptyset \\ S(N(\mu_1),\ldots,N(\mu_n)), & \text{otherwise.} \end{cases}$$

provided S is a t-conorm, and N a strong negation on $[0,1]^X$.

Note that $T(\mu_1,\ldots,\mu_n) \neq \mu_\emptyset$ means that $\mu_1,\ldots,\mu_n$ are not T-incompatible.
More, taking into account Proposition 4 we also could establish as a T-incompatibility measure:

$$I(\mu_1,\ldots,\mu_n) = \begin{cases} 0, & \text{if } T(\mu_1,\ldots,\mu_n) \neq \mu_\emptyset \\ N(T^0(\mu_1,\ldots,\mu_n)), & \text{otherwise.} \end{cases}$$

provided T^0 is a t-norm, and N a strong negation on $[0,1]^X$.

Example 11. Measures of unsupplementarity were introduced in [8]. This is a decreasing property on $\mathscr{L}$ and n-decreasing. Then, according to Proposition 3 it is possible to establish as S-unsupplementarity measure:

$$U(\mu_1,\ldots,\mu_n) = \begin{cases} 0, & \text{if } S(\mu_1,\ldots,\mu_n) = \mu_X \\ T(N(\mu_1),\ldots,N(\mu_n)), & \text{otherwise.} \end{cases}$$

provided T is a t-norm, and N a strong negation on $[0,1]^X$.

More, taking into account Proposition 4 we also could establish as a S-unsupplementarity measure:

$$U(\mu_1,\ldots,\mu_n) = \begin{cases} 0, & \text{if } S(\mu_1,\ldots,\mu_n) = \mu_X \\ N(S(\mu_1,\ldots,\mu_n)), & \text{otherwise.} \end{cases}$$

provided S is a t-conorm, and N a strong negation on $[0,1]^X$.

Conclusions. In this paper, multi-argument fuzzy functions on a bounded lattice have been faced. In particular, aggregation functions constitute a special class of them. Four propositions have provided some ways to obtain multi-measures satisfying different porperties. Examples on the lattice of power set of any finite set, have been showed. When focusing on measures on fuzzy sets the methods proposed allow to construct functions measuring properties applying on a subset of them. Depending on the characteristics of each property, the suitable method will be chosen. In this a way, some measures introduced by the authors in previous papers, are captured.

Acknowledgements. This paper has been partially supported by the Spanish Ministry of Science and Innovation, under projects TIN2011-29827-C02-01, TIN2009-07901, TIN2012-32482, MTM 2009-10962, and by UPM-CAM.

References

[1] Alsina, C., Frank, M., Schweizer, B.: Associative Functions: Triangular Norms and Copulas. World Scientific, Singapore (2006)

[2] Bedregal, B., Beliakov, G., Bustince, H., Fernandez, J., Pradera, A., Reiser, R.H.S.: Negations Generated by Bounded Lattices t-Norms. In: Greco, S., Bouchon-Meunier, B., Coletti, G., Fedrizzi, M., Matarazzo, B., Yager, R.R. (eds.) IPMU 2012, Part III. CCIS, vol. 299, pp. 326–335. Springer, Heidelberg (2012)

[3] Beliakov, G., Pradera, A., Calvo, T.: Aggregation Functions: A Guide for Practitioners. STUDFUZZ, vol. 221. Springer, Berlin (2007)

[4] Birkhoff, G.: Lattice Theory. American Mathematical Society, Providence (1940)

[5] Blyth, T.S.: Lattices and Ordered Algebraic Structures. Springer, London (2005)

[6] Calvo, T., Kolesarová, A., Komorníková, M., Mesiar, R.: Aggregation operators: properties, Classes and Construction Methods. In: Calvo, T., Mesiar, R., Mayor, G. (eds.) Aggregation Operators: New Trends and Applications. STUDFUZZ, vol. 97, pp. 3–104. Physica-Verlag, Heilderberg (2002)

[7] Castiñeira, E.E., Calvo, T., Cubillo, S.: Multi-arguments fuzzy measures on some special lattices. In: Proc. of 11th International Conference on Computational and Mathematical Methos in Science and Engineering, CMMS 2011, Benidorm, Spain, pp. 319–330 (2011)

[8] Castiñeira, E.E., Calvo, T., E.Cubillo, S.: Multi-argument fuzzy measures on lattices of fuzzy sets. Sent to Knowledge Based Systems

[9] Castiñeira, E.E., Cubillo, S., Montilla, W.: Measuring incompatibility between Atanassov's intuitionistic fuzzy sets. Information Sciences 180, 820–833 (2011)

[10] Cubillo, S., Castiñeira, E.: Measuring Contradiction in Fuzzy Logic. International Journal of General Systems 34(1), 39–59 (2005)

[11] Cubillo, S., Castiñeira, E.E., Montilla, W.: Supplementarity measures on fuzzy sets. In: Proc. of 7th Conference of the European Society for Fuzzy Logic and Technology (EUSFLAT-LFA 2011), Aix-Les-Bains, France, pp. 897–903 (2011)

[12] Grabisch, M.: Fuzzy integral in multicriteria decision making. Fuzzy Sets and Systems 69, 279–298 (1995)

[13] Grabisch, M., Marichal, J.L., Mesiar, R., Pap, E.: Aggregation Functions. Cambridge University Press, Cambridge (2009)

[14] Klement, E.P., Mesiar, R., Pap, E.: Triangular Norms. Kluwer Academic Publisher, Dordrecht (2000)

[15] Schweizer, B., Sklar, A.: Associative functions and statistical triangle inequalities. Public. Math. Debrecen 8, 169–186 (1961)

[16] Trillas, E.: Sobre funciones de negación en los conjuntos difusos. Stochastica 3(1), 47–60 (1979) (in Spanish); Reprinted (English version). In: Barro, S., et al. (ed.) Advances of Fuzzy Logic. Universidad de Santiago de Compostela, pp. 31–43 (1998)

[17] Wang, Z., Klir, G.J.: Fuzzy Measure Theory. Plenum Press, New York (1992)

Aggregation of Convex Intuitionistic Fuzzy Sets

Vladimír Janiš and Susana Montes

Abstract. Aggregation of intuitionistic fuzzy sets is studied from the point of view of preserving convexity. We focus on those aggregation functions for IF-sets, that are results of separate aggregation of the membership and of nonmembership functions, that is, the representable aggregation functions. A sufficient and necessary condition for an aggregation function is given in order to fulfil that the aggregation of two IF-sets preserves the convexity of cuts.

1 Preliminaries

The object of our study are the intuitionistic fuzzy sets (IF-sets) introduced by Atanassov in [1] and [2] and in more details studied in [3]. Although they were seen at the beginning justa as an extension of the ordinary fuzzy set, quickly they were revealed with essentially different properties. An adequate example to show that was proposed in [3]:

Example 1. Let X be the set of all countries with elective governments. Assume that we know for every country $x \in X$ the percentage of the electorate that have voted for the corresponding government. Denote it by $M(x)$ and let $\mu(x) = M(x)/100$ (degree of membership, validity, etc.). Let $v(x) = 1 - \mu(x)$. This number corresponds to the part of electorate who have not voted for the government. By fuzzy set theory alone we cannot consider this value in more detail. However, if we define $v(x)$ (degree of non-membership, non-validity, etc.) as the number of votes given to parties or

Vladimír Janiš
Department of Mathematics, Faculty of Natural Sciences,
Matej Bel University, Slovak Republic
e-mail: vladimir.janis@umb.sk

Susana Montes
Department of Statistics and O.R., University of Oviedo, Spain
e-mail: montes@uniovi.es

H. Bustince et al. (eds.), *Aggregation Functions in Theory and in Practise*,
Advances in Intelligent Systems and Computing 228,
DOI: 10.1007/978-3-642-39165-1_51, © Springer-Verlag Berlin Heidelberg 2013

persons outside the government, then we can show the part of electorate who have not voted at all or who have given bad voting-paper and the corresponding number will be $\pi(x) = 1 - \mu(x) - v(x)$ (degree of indeterminacy, uncertainty, etc.). Thus we can construct the set

$$\{(x, \mu(x), v(x)) | x \in X\}$$

and it is trivial that

$$0 < \mu(x) + v(x) < 1.$$

Obviously, for every ordinary fuzzy set $\pi_A(x) = 0$ for each $x \in X$ and these sets have the form $\{(x, \mu(x), 1 - \mu(x)) | x \in X\}$. However, it is clear that IFS can be different from ordinary fuzzy sets.

We will use the following definition of an IF-set:

Definition 1. Let X be a universe. A pair $A = (\mu_A, v_A)$ where μ_A and v_A are two functions from X to $[0,1]$ fulfilling that $\mu_A(x) + v_A(x) \leq 1$ for all $x \in X$ is called an intuitionistic fuzzy set (IF-set). The functions μ_A, v_A are its membership and nonmembership functions, respectively.

Suppose that T is an arbitrary triangular norm (see, for instance, [8]). The definitions of T-based union, T-based intersection, complementation and inclusion given by Atanassov in [2] are the following:

- Intersection: $A \cap_T B = \langle \mu_{A \cap_T B}, v_{A \cap_T B} \rangle$ where $\mu_{A \cap_T B}(x) = T(\mu_A(x), \mu_B(x))$ and $v_{A \cap_T B}(x) = S(v_A(x), v_B(x))$, for all $x \in X$;
- Union: $A \cup_S B = \langle \mu_{A \cup_S B}, v_{A \cup_S B} \rangle$ where $\mu_{A \cup_S B}(x) = S(\mu_A(x), \mu_B(x))$ and $v_{A \cup_S B}(x) = T(v_A(x), v_B(x))$, for all $x \in X$;
- Complement: if $A = \langle \mu_A, v_A \rangle$, then $\overline{A} = \langle v_A, \mu_A \rangle$;
- Inclusion: $A \subseteq B$ iff $\mu_A(x) \leq \mu_B(x)$ and $v_A(x) \geq v_B(x)$, for all $x \in X$.

But we are not working in general with IF-sets, but a particular case of them: the convex IF-sets. In order to study the properties connected to convexity, throughout the paper we assume that X is a linear space. From the above it follows that the IF-set $(0, 1)$ is the zero element and the IF-set $(1, 0)$ is the unit element in the partially ordered set of all IF-sets on X.

We devoted our study to this class of sets, since convexity is one of the most important aspects in the study of geometric properties of not only crisp (usual) sets, but also fuzzy and IF-sets, mainly in applications connected to optimization and control (see [4, 13]). In [15], Zadeh introduced the concept of convex fuzzy set, which is an important kind of extension of classical convex sets from the point of view of cut sets.

Thus, following directly the concept of convexity for crisp sets we obtain that a fuzzy set μ from X to $[0,1]$ is convex, if for all $x, y \in \text{supp}\,\mu$ and $\lambda \in [0,1]$ there is

$$\mu(\lambda x + (1 - \lambda)y) \geq \lambda\mu(x) + (1 - \lambda)\mu(y).$$

Although this definition of convexity is sometimes used, it has at least two weak points. The first one is that it is not suitable for the case of a lattice valued fuzzy set, as the addition in the lattice is not defined. And the second one is that under this definition a fuzzy set for which all its level sets are convex, need not be convex. Recall that by a level set or α-cut of a fuzzy set μ, with $\alpha \in (0, 1]$, we understand the crisp set $\mu_\alpha = \{x \in X; \mu(x) \geq \alpha\}$.

Therefore we will consider the following definition of convexity for fuzzy sets:

Definition 2. A fuzzy set $\mu : R^n \to R$ is convex if for all $x, y \in R^n, \lambda \in [0, 1]$ there is $\mu(\lambda x + (1 - \lambda)y) \geq \min\{\mu(x), \mu(y)\}$.

This property has been (under the name of quasiconvexity) introduced in [4]. It can be easily shown that it overcomes both mentioned problems, it can be used also for mappings into partially ordered spaces and the class of all convex fuzzy sets is exactly the class of those fuzzy sets, for which all their cuts are convex.

The notion of quasiconvexity has been widely studied and applied. However, it could be too restrictive in several situations, especially in a framework of a fuzzy logic model in which a t-norm other than minimum is used. By this reason, the notion of T-convexity was proposed in [5]. There, T assumed the role of the minimum in the previous definition, but it was any t-norm. It is clear that quasiconvexity implies T-convexity for any t-norm T. Moreover, it can be proven that T-convexity of a normal fuzzy set (there exists an x in X such that the membership function of the fuzzy set assumes the value one in x) implies its quasiconvexity. Thus, both concepts are very related and even they are just the same in the case of normal fuzzy sets.

Although the theory and application of convex fuzzy sets have been studied intensively, the corresponding research for convex intuitionistic fuzzy sets is rather scarce, which restricts its application greatly. Thus, in [16], Zhang et al. define 16 kinds of intuitionistic convex fuzzy sets from the point view of cut sets and neighbourhood relations between a fuzzy point and an IF-set. In [14], Xu et al. introduced the concept of quasi-convex IF-set based on convex fuzzy sets and concave fuzzy sets and discussed the relations among them. In a similar way interval-valued convex fuzzy sets were introduced in [17] and even some generalization to graded convex intuitionistic fuzzy sets were considered in [11]. We will try to consider these approaches in order to define and study some properties for convex IF-sets. In particular we will interested in the study of the intersection of two convex IF-sets.

The paper is organized as follows. In Section 2, we introduce the convex IF-sets and we establish some equivalent definitions. In Section 3, we analyse the behaviour of a generalization of the intersection of two convex IF-sets. Finally, in the last section, we present some concluding remarks and we comment some open problems.

2 Convex Intuitionistic Fuzzy Sets

Since we would like the concept of convex intuitionistic fuzzy sets fulfils the cutworthy property (for the notion of cutworthiness see, for instance, the book by Klir and Yuan [9]), we will start by considering the notion of alpha-cut of an IF-set.

In [3] the α-cut of an IF-set $A = (\mu_A, \nu_A)$ was defined as the set of those $x \in X$ for which $\mu_A(x) \geq \alpha$ and $\nu_A(x) \leq 1 - \alpha$. We will denote it by $A_\alpha = (\mu_A, \nu_A)_\alpha$.

As an starting point, we will consider this definition along this paper, but some generalization where proposed in [6] and [10]. In the second case the conjunction was replaced by a triangular norm and the negation for the non-membership degree by a fuzzy negation. In the fist case only the standard negation was used.

Using cuts we can define convexity for IF-sets, just from the cutworthy property.

Definition 3. An IF-set A is convex if its cuts A_α are convex subsets of X for all $\alpha \in (0,1]$.

In fact, this concept can be characterized only by means of the convexity of the first component of the IF-set, if we consider it as the membership function of a fuzzy set. Thus,

Proposition 1. *Let A be an intuitionistic fuzzy set on X. The following statements are equivalent:*

1. *A is a convex IF-set.*
2. *μ_A is the membership function of quasi-convex fuzzy set.*

Proof. Hence the α-cut of $A = (\mu_A, \nu_A)$ is the intersection of the α-cuts of μ_A and $1 - \nu_A$.

However, from the condition for IF-sets we see that $\mu_A(x) \leq 1 - \nu_A(x)$ for all $x \in X$, hence $(\mu_A)_\alpha \subseteq (1 - \nu_A)_\alpha$ and so the intersection of these cuts is $(\mu_A(x))_\alpha$. Since the quasi-convexity of a fuzzy set is equivalent to require that all its alpha-cuts are convex (see [7]), this implies that the convexity of A is equivalent to the convexity of μ_A.   □

On the other hand, in [14] the concept of quasi-convexity of IF-sets was introduced as follows:

Definition 4. An IF-set A is called quasi-convex if

$$\mu_A(\lambda(x-y)+y) \geq \min(\mu_A(x), \mu_A(y))$$

$$\nu_A(\lambda(x-y)+y) \leq \max(\nu_A(x), \nu_A(y))$$

for all $x, y \in X, \lambda \in [0,1]$.

It is immediate to prove that

Proposition 2. *Let A be an intuitionistic fuzzy set on X. The following statements are equivalent:*

1. *A is a quasi-convex IF-set.*
2. *The α-cuts of the fuzzy sets μ_A and $1 - \nu_A$ are convex crisp sets, for any $\alpha \in (0,1]$.*

Although we could think that both concepts, convexity and quasi-convexity, are equivalent, this is not right and we only have one implication.

Proposition 3. *Let A be an intuitionistic fuzzy set on X. If A is quasi-convex, then it is also convex, but the convex is not true in general.*

Proof. If A is a quasi-convex IF-set, then the α-cuts of the fuzzy sets μ_A and $1 - \nu_A$ are convex crisp sets, for any $\alpha \in (0,1]$. This implies the quasi-convexity of the fuzzy set defined by means of μ_A and therefore, by Proposition 1, the convexity of A.

To prove that the converse is not true in general, let us consider X is the real line and A is the IF-set defined by

$$\mu_A(x) = \nu_A(x) = \begin{cases} 0.5 & \text{if } x \in [1,2], \\ 0 & \text{otherwise.} \end{cases}$$

As

$$A_\alpha = \{x | \mu_A(x) \geq \alpha \text{ and } \nu_A(x) \leq 1 - \alpha\} = \begin{cases} [1,2] & \text{if } \alpha \in (0,0.5] \\ \emptyset & \text{if } \alpha \in (0.5,1] \end{cases}$$

and it is trivial that is is convex for any $\alpha \in (0,1]$, that is, A is a convex IF-set.

However, the 0.7-cut of $1 - \nu_A$ is not a convex crisp set, since $(1 - \nu_A)_{0.7} = (-\infty,1) \cup (2,\infty)$, and therefore, by Proposition 2, A is not a quasi-convex IF-set. $\square$

Since we would like to consider the most general case, we will work along this paper with the notion of convex IF-set. One of the properties that make convexity so important is its preserving under intersections. We will study this question for the case of IF-sets. However, instead of the intersection we will consider a wider class of aggregation functions. As the intersection of IF-sets is defined by a triangular norm, which is a particular case of an aggregation operator, our result will be valid also for intersections.

3 Aggregation Functions and Convexity

We restrict our considerations to binary aggregation functions on a unit interval, that is, the maps $W : [0,1] \times [0,1] \to [0,1]$ such that $W(0,0) = 0, W(1,1) = 1$ and W is increasing in each component. In particular we will restrict our study to representable aggregation functions.

Definition 5. An aggregation function W on $[0,1]$ is representable if there exists a pair of aggregation functions (W_1, W_2) on a unit interval such that for each $\alpha, \beta, \gamma, \delta \in [0,1], \alpha + \beta \leq 1, \gamma + \delta \leq 1$ there is

$$W((\alpha,\beta),(\gamma,\delta)) = (W_1(\alpha,\gamma), W_2(\beta,\delta)).$$

In the following proposition we clarify the connection between a representable aggregation function and its representing functions. By a dual W^d of an aggregation function W we denote the mapping $W^d : [0,1] \to [0,1]$ such that

$$W^d(\alpha,\beta) = 1 - W(1 - \alpha, 1 - \beta).$$

Proposition 4. *The aggregation function* W *on IF-sets is representable by a pair* (W_1, W_2) *if and only if* $W_2 \leq W_1^d$.

Proof. Let W be represented by (W_1, W_2). Suppose that the inequality $W_2 \leq W_1^d$ does not hold. Then there is a pair $(\alpha, \beta) \in [0,1]^2$ such that

$$W_2(\alpha, \beta) > W_1^d(\alpha, \beta) = 1 - W_1(1 - \alpha, 1 - \beta).$$

Then

$$W((1 - \alpha, \alpha), (1 - \beta, \beta)) = (W_1(1 - \alpha, 1 - \beta), W_2(\alpha, \beta)),$$

but using the above inequality we have

$$W_1(1 - \alpha, 1 - \beta) + W_2(\alpha, \beta) > 1,$$

what is a contradiction.

Conversely let $W_2 \leq W_1^d$, let $W((\alpha, \beta), (\gamma, \delta)) = (W_1(\alpha, \gamma), W_2(\beta, \delta))$. Our aim is to show that the sum of both these components does not exceed 1.

$$W_1(\alpha, \gamma) + W_2(\beta, \delta) \leq W_1(\alpha, \gamma) + 1 - W_1(1 - \beta, 1 - \delta).$$

It is sufficient to show that the right-hand side of this inequality does not exceed 1. To get this we use the inequalities $\alpha \leq 1 - \beta, \gamma \leq 1 - \delta$ and the monotony of W_1. Then

$$W_1(\alpha, \gamma) \leq W_1(1 - \beta, 1 - \delta)$$

and we have

$$W_1(\alpha, \gamma) + 1 - W_1(1 - \beta, 1 - \delta) \leq 1.$$

$\square$

Now we are able to use these functions to defined the composition of two IF-sets as follows:

Definition 6. Let $A = (\mu_A, \nu_A)$ and $B = (\mu_B, \nu_B)$ be two IF-sets on X and let W be a representable aggregation function. The aggregation of A and B by W is the IF-set $W(A, B)$ whose membership degree is $W_1(\mu_A, \mu_B)$ and whose non-membership degree is $W_2(\nu_A, \nu_B))$.

A crucial role for the preserving convexity in aggregation of IF-sets is played by the proposition proved in [7] in more general context for fuzzy sets, here we reformulate it for our purposes:

Proposition 5. *Let* $W : [0,1]^2 \to [0,1]$ *be an aggregation function and let* $\mu, \nu : X \to [0,1]$ *be convex fuzzy sets. Then the following are equivalent:*

1. *The fuzzy set* $W(\mu, \nu)$ *is convex,*
2. $W(\min\{\alpha, \gamma\}, \min\{\beta, \delta\}) = \min\{W(\alpha, \beta), W(\gamma, \delta)\}$ *for each* $\alpha, \beta, \gamma, \delta \in [0,1]$.

Note that the condition 2 from this proposition means that W commutes (see [12]) with the minimum. In particular, since any t-norm commutes with itself, the above condition is fulfilled when $W = \min$.

Now we formulate the result for preserving convexity for (representable) aggregation of IF-sets.

Proposition 6. *Let W be a representable aggregation function represented by a pair (W_1, W_2). Let A, B are convex IF-sets. Then the following are equivalent:*

1. *The IF-set $W(A, B)$ is convex,*
2. $W_1(\min\{\alpha, \gamma\}, \min\{\beta, \delta\}) = \min\{W_1(\alpha, \beta), W_1(\gamma, \delta)\}$ *for each $\alpha, \beta, \gamma, \delta \in [0, 1]$.*

Proof. If $A = (\mu_A, \nu_A), B = (\mu_B, \nu_B)$ are IF-sets, then their aggregation under W is the IF-set

$$W(A, B) = (W_1(\mu_A, \mu_B), W_2(\nu_A, \nu_B)).$$

By Proposition 1, its convexity is equivalent to the convexity of $W_1(\mu_A, \mu_B)$.

But as W_1 is an aggregation function on fuzzy sets, using Proposition 5 we see that the necessary and sufficient condition for $W(A, B)$ to be convex is commuting of W_1 with the minimum. $\qquad\square$

As we already commented, the minimum commutes with itself. Thus, the previous result is true if $W_1 = \min$. As it dual is the maximum t-conorm, by Proposition 4, any aggregation function W_2 can be considered to define W and, in this case, when the minimum is the first component, the aggregation of two convex IF-sets is always convex.

4 Further Topics

Clearly the obvious question following from this research is to characterize the class of all (not necessarily representable) aggregation functions on IF-sets preserving convexity. However, also the notion of an α-cut of an IF-set can be understood from the point of view of fuzzy logic as it was done in [6] and [10]. It might also be interesting to find conditions for an aggregation function preserving convexity of the above defined cuts.

Acknowledgements. The first author acknowledges the support of the grant 1/0297/11 provided by Slovak grant agency VEGA. Also the second author would express that the research reported on in this contribution has been partially supported by Project MTM2010-17844 of the Spanish Ministry of Economy and Competitiveness.

References

[1] Atanassov, K.: Intuitionistic Fuzzy Sets, VII ITKR Session, Sofia (1983) (in Bulgarian)
[2] Atanassov, K.: Intuitionistic fuzzy sets. Fuzzy Sets and Systems 20, 87–96 (1986)

[3] Atanassov, K.: Intuitionistic Fuzzy Sets, Theory and Applications. Physica-Verlag, Heidelberg (1999)

[4] Ammar, E., Metz, J.: On convexity and parametric optimization. Fuzzy Sets and Systems 49, 135–141 (1992)

[5] Iglesias, T., Montes, I., Janiš, V., Montes, S.: T-convexity for lattice-valued fuzzy sets. In: Proceedings of the ESTYLF Conference (2012)

[6] Janiš, V.: T-norm based cuts of intuitionistic fuzzy sets. Information Sciences 180(7), 1134–1137 (2010)

[7] Janiš, V., Král', P., Renčová, M.: Aggregation operators preserving quasiconvexity. Information Sciences 228, 37–44 (2013)

[8] Klement, E.P., Mesiar, R., Pap, E.: Triangular norms. Kluwer Academic Publishers, Dordrecht (2000)

[9] Klir, G., Yuan, B.: Fuzzy Sets and Fuzzy Logic. Prentice-Hall, New Jersey (1995)

[10] Martinetti, D., Janiš, V., Montes, S.: Cuts of intuitionistic fuzzy sets respecting fuzzy connectives. Information Sciences (2013), doi:http://dx.doi.org/10.1016/j.ins.2012.12.026

[11] Pan, X.: Graded Intuitionistic Fuzzy Convexity with Application to Fuzzy Decision Making. In: Zeng, D. (ed.) Advances in Information Technology and Industry Applications. LNEE, vol. 136, pp. 709–716. Springer, Heidelberg (2012)

[12] Saminger-Platz, S., Mesiar, R., Dubois, D.: Aggregation operators and Commuting. IEEE Transactions on Fuzzy Systems 15(6), 1032–1045 (2007)

[13] Syau, Y.-R., Lee, E.S.: Fuzzy Convexity with Application to Fuzzy Decision Making. In: Proceedings of the 42nd IEEE Conference on Decision and Control, pp. 5221–5226 (2003)

[14] Xu, W., Liu, Y., Sun, W.: On Starshaped Intuitionistic Fuzzy Sets. Applied Mathematics 2, 1051–1058 (2011)

[15] Zadeh, L.A.: Fuzzy sets. Inform. and Control 8, 338–353 (1965)

[16] Zhang, C., Xiao, P., Wang, S., Liu, X. (s,t]-intuitionistic convex fuzzy sets. In: Cao, B.-Y., Wang, G.-J., Guo, S.-Z., Chen, S.-L. (eds.) Fuzzy Information and Engineering 2010. AISC, vol. 78, pp. 75–84. Springer, Heidelberg (2010)

[17] Zhang, C., Su, Q., Zhao, Z., Xiao, P.: From three-valued nested sets to interval-valued (or intuitionistic) fuzzy sets. International Journal of Information and Systems Sciences 7(1), 11–21 (2011)

Index